Allele Mining for Genomic Designing of Vegetable Crops

This book deliberates on the concept, strategies, tools, and techniques of allele mining in vegetable crops and its application potential in genome elucidation and improvement, including studying allele evolution, discovery of superior alleles, discerning new haplotypes, assessment of intra- and interspecific similarity, and studies of gene expression and gene prediction. Available gene pools in global germplasm collections specifically consisting of wild allied species and local landraces for almost all major crops have facilitated allele mining. Development of advanced genomic techniques including PCR-based allele priming and Eco-TILLING-based allele mining are being widely used now for mining superior alleles. Allele's discovery has become more relevant now for employing molecular breeding to develop designed crop varieties matching consumer needs and with genome plasticity to adapt the climate change scenarios. All these concepts and strategies along with precise success stories are presented in the chapters dedicated to the major vegetable crops.

The features of this book are as follows:

1. The first book on the novel strategy of allele mining in vegetable crops for precise breeding.
2. Presents genomic strategies for mining superior alleles underlying agronomic traits from genomic resources.
3. Depicts case studies of PCR-based allele priming and Eco-TILLING-based allele mining.
4. Elaborates on gene discovery and gene prediction in major vegetable crops

This book will be useful to the students and faculties in various plant science disciplines including genetics, genomics, molecular breeding, agronomy, and bioinformatics; the scientists in seed industries; and, the policymakers and funding agencies interested in crop improvement.

Preface to the Series

Agriculture is going through a phase of transformation now. It has both challenges and opportunities. One of the major challenges it is facing is obviously food security. A population of about 9.3 billion projected by 2050 necessitates an increase of crop production by 60–70%. Another 10–15% increase would be needed to counter the reduction of crop production due to global warming and climate change. Nutrition security is another challenge with more than 800 million people suffering from malnutrition. Depletion of fossil fuel and increasing dependence on biofuel from crop residues is another challenge. Loss of biodiversity and land degradation also add to the list of challenges. The cultivable area is also decreasing fast due to housing, transport, and industrialization. Above all, climate change and global warming have a huge impact on agriculture, health, and environment.

Several novel concepts and strategies are emerging to address the above challenges. For example, practice of precision farming will lead to precise and timely use of inputs resulting in higher yield and better quality of produce on the one hand and reduction of loss of inputs on the other. Similarly, urban agriculture and vertical agriculture will add a lot of cultivable areas in future. Protected farming will facilitate growing crops under controlled conditions leading to reduction in biotic and abiotic stresses and precise utilization of inputs.

Agriculture requires maintenance of the philosophy farmers have been following since inception over thousands of years which are being sold today in new packages with sophisticatedly coined labels such as traditional agriculture, natural agriculture, sustainable agriculture, conservation agriculture, integrated agriculture, regenerative agriculture, organic agriculture, etc.

In the changing scenarios, next generation agriculture will need designed crop varieties and breeds produced through designed practices. This book series aims at deliberations on several novel concepts and innovative strategies developed for next generation agriculture and will hopefully benefit agricultural education, research, and outreach.

Chittaranjan Kole
Kolkata
January 10, 2024

NextGen Agriculture: Novel Concepts and Innovative Strategies
Chittaranjan Kole and Phullara Kole
Series Editor

Allele Mining for Genomic Designing of Fruit Crops
Chittaranjan Kole, Kenta Shirasawa, and Anil Kumar Singh

Allele Mining for Genomic Designing of Cereal Crops
Chittaranjan Kole, Sharat Kumar Pradhan, and Vijay K Tiwari

Allele Mining for Genomic Designing of Vegetable Crops
Chittaranjan Kole, Tusar Kanti Behera, and Prashant Kaushik

Allele Mining for Genomic Designing of Oilseed Crops
Chittaranjan Kole, Manish Kumar Pandey, and Naveen Puppala

Sustainable Urban Agriculture: New Frontiers
Kheir Al-Kodmany, Madhav Govind, Sharmin Khan, and Chittaranjan Kole

Allele Mining for Genomic Designing of Grain Legume Crops
Chittaranjan Kole, C. Bharadwaj, and Abhimanyu Sarkar

The Vertical Farm: Scientific Advances and Technological Developments
Kheir Al-Kodmany, Andrew Keong Ng, Abel Tablada, and Chittaranjan Kole

For more information, please visit our website: https://www.routledge.com/Nextgen-Agriculture/book-series/NGA

Allele Mining for Genomic Designing of Vegetable Crops

Edited by
Chittaranjan Kole, Tusar Kanti Behera,
and Prashant Kaushik

CRC Press
Taylor & Francis Group
Boca Raton London New York

CRC Press is an imprint of the
Taylor & Francis Group, an **informa** business

First edition published 2024
by CRC Press
2385 NW Executive Center Drive, Suite 320, Boca Raton FL 33431

and by CRC Press
4 Park Square, Milton Park, Abingdon, Oxon, OX14 4RN

CRC Press is an imprint of Taylor & Francis Group, LLC

Library of Congress Cataloging-in-Publication Data
Names: Kole, Chittaranjan, editor. | Behera, Tusar Kanti, editor. | Kaushik, Prashant, editor.
Title: Allele mining for genomic designing of vegetable crops / edited by
Chittaranjan Kole, Tusar Kanti Behera, and Prashant Kaushik
Other titles: Nextgen agriculture: novel concepts and innovative strategies.
Description: First edition | Boca Raton, FL : CRC Press, 2024 |
Series: Nextgen agriculture: novel concepts and innovative strategies |
Includes bibliographical references and index
Identifiers: LCCN 2023057735 (print) | LCCN 2023057736 (ebook) |
ISBN 9781032453187 (hardback) | ISBN 9781032453194 (paperback) |
ISBN 9781003376439 (ebook)
Subjects: LCSH: Vegetables–Genetic engineering. |
Plant molecular genetics. | Plant genomes.
Classification: LCC SB324.7 .A54 2024 (print) |
LCC SB324.7 (ebook) | DDC 635/.0423–dc23/eng/20240109
LC record available at https://lccn.loc.gov/2023057735
LC ebook record available at https://lccn.loc.gov/2023057736

ISBN: 9781032453187 (hbk)
ISBN: 9781032453194 (pbk)
ISBN: 9781003376439 (ebk)

DOI: 10.1201/ 9781003376439

Typeset in Times
by Newgen Publishing UK

Dedicated to

Dr. Norman Ernest Borlaug

The Father of Green Revolution, Nobel Peace Prize winner and the Founder of the World Food Prize

Who inspired my research and academic works with his generous appreciations and advice.

Contents

Abbreviations

2b-RAD	Type IIB endonucleases restriction-site associated DNA
AA	Antioxidant activity
ABA	Abscisic acid
ABA8H4	ABA 8'-hydroxylase 4
ACO	1-Aminocyclopropane-1-carboxylic acid oxidase
ACS2	1-Aminocyclopropane-1-carboxylic acid synthase 2
AFLP	Amplified fragment length polymorphism
ALDH	Aldehyde dehydrogenase
AM	Allele mining
AMT	Ammonium transporter
ANN1	Annexins 1
AP1	*Apetala 1*
AP2	*Apetala 2*
AP2A	Activating enhancer binding protein 2 alpha
AP3	*APETALA 3*
APX	Ascorbate peroxidase
APY	Apyrase
ARF6	Auxin response factor 6
ATP	Adenosine triphosphate
Avr/AVR	Avirulence
AVRDC	Asian Vegetable Research and Development Center
B	B gene for flesh color
BAC	Bacterial artificial chromosome
BADH	Betaine aldehyde dehydrogenase
BAM	β-amylase
BC	Backcross
BDGEH	β-D-glucan exohydrolase
BED	BEAF and DREF
bHLH	Basic helix-loop-helix
BL	Bayesian lasso
BLAST	Basic local alignment search tool
bp	Base pair
BPYV	Beet pseudo-yellows virus
BR	Brassinosteroid/ Brassinolide
BRAD	Brassicaceae Database
BRR	Bayesian ridge regression
BSA	Bulked segregant analysis
Bu	Bush growth habit
BWA	Burrows-wheeler aligner
C. hystrix	Cucumis hystrix
C. melo	*Cucumis melo*
C. sativus	*Cucumis sativus*
CABYV	Cucurbit aphid-borne yellows virus
CAPS	Cleaved amplified polymorphic sequences
CAPS	Cleaved amplified polymorphisms
CAS	CRISPR-associated protein
CAS	Cycloartenol synthase

Cas9	CRISPR-associated protein 9
CASP	Cleaved amplified sequence polymorphism
CBF	C–repeat binding factor
CCYV	Cucurbit chlorotic yellows virus
cDNA	Complementary DNA
CGMMV	Cucumber green mottle mosaic virus
CGN	Centre for Genetic Resources
Chr	Chromosome
Chy2	Beta-carotene hydroxylase 2
CIP	International Potato Center
CIP	International Potato Center, Peru
Cl	Core Length
cM	CentiMorgan
CmaCh15G005270	MYB transcription factor in *C. maxima*
CMO-1	*C. moschata* inbred line 1
CMO-97	*C. moschata* inbred line 97
CmoYABBY	YABBY gene in *C. moschata*
CMS	Cytoplasmic male-sterility
CMV	Cucumber mosaic virus
CmV1	Gene for vine elongation in *C. moschata*
CNR	Colorless Non-Ripening
CNV	Copy number variation
CODDLE	Codons programmed to identify deleterious lesions
Cp4.1LG12g04350	NAC domain-containing protein
CPRI	Central Potato Research Institute
CREs	Cis-regulatory elements
CRISPR	Clustered regularly interspaced short palindromic repeats
CuGenDB	Cucurbit Genomics Database
CVYV	Cucumber vein yellowing virus
CYS1	Cystatin 1
CYSDV	Cucurbit yellow stunting disorder virus
DAS-ELISA	Double-antibody sandwich enzyme-linked immunosorbent assay
DBP4	DNA binding protein 4
dCAPs	Derived cleaved amplified polymorphic sequences
DcPAR1	*PHYTOCHROME RAPIDLY REGULATED1* gene
ddRAD	Double digest restriction-site associated DNA
DEG	Differentially expressed gene
DER	Dihydroflavonol Reductase
DH	Doubled haploid
DL	Deep learning
DM	Downy mildew
DNA-A	Circular single-stranded DNA of ToLCNDV
DNA-B	Circular single-stranded DNA of ToLCNDV
DOF	DNA binding with one finger
DPPH	2,2-diphenyl-1-picrylhydrazyl
DREB1	Dehydration-responsive element-binding protein 1
DREB2	Dehydration-responsive element-binding protein 2
DWF4	DWARF4
ECO-TILLING	Ecotype-TILLING
EIL2	Ethylene-Insensitive3-Like 2

EIN2	Ethylene insensitive 2
EIX	Ethylene-inducing xylanase
EMS	Ethylmethane sulfonate
ENU	N-ethyl-N-nitrosoure
eQTL	Expression QTL
ER	Extreme resistance
EREBP1	Ethylene-responsive element-binding protein 1
ERECTA	Family of receptor kinase genes
ERF	Ethylene response factor
ERF186	Ethylene response factor 186
ERF3	Ethylene response transcription factor 3
ERS1	Ethylene response sensor 1
EST	Expressed sequence tag
ETO1	Ethylene-overproduction protein 1
ETP1	EIN2 targeting protein 1
ETR1	Ethylene receptor 1
EXA1	Potexvirus accumulation 1
F2	Second filial generation
FAO	Food and Agriculture Organization
FAOSTAT	Food and Agriculture Organization Corporate Statistical Database
FIGS	Focused identification of germplasm strategy
FISH	Fluorescence in situ hybridization
FLC	*Flowering Locus C*
FLS2	*FLAGELLIN-SENSING 2*
FLT1	Flowering locus T-like protein 1
FLT2	Flowering locus T-like protein 2
FLX	Federal Learning Exchange
FPKM	Transcript Sequence Per Millions of Base Pairs of Sequencing Data
FUL	Fruitful
FW	Fruit Weight
GA	Gibberellic acid/ Gibberellin
GA20ox	Gibberellin synthesis gene
GA2ox1	Gibberellin 2-beta-dioxygenase 1
GA3ox	Gibberellin 3-oxidase
GBLUP	Genomic best linear unbiased prediction
GBS	Genotyping-by-sequencing
GCA	General combining ability
GP	Gene pool
GPAT2	Glycerol-3-Phosphate Acyltransferase)
GP-I	Primary gene pool
GP-II	Secondary gene pool
GP-III	Tertiary gene pool
GR	Glutathione reductase
Gr	Rind color of mature fruit
Gro1	*Globodera rostochiensis* race Ro
GS	Genomic selection
GS	Green stripe
GSL	Glucosinolate
GST	Glutathione-S-transferase
GWA	Genome wide association

GWAS	Genome wide association study/studies
GWD	Glucan/water dikinase
HM-S2	Variety of *C. pepo* subsp. *pepo* var. Zucchini
HR	Hypersensitive resistance
HRM	High resolution melting
HSc70	Heat-shock cognate protein 70
HSP	Heat shock protein
HSP20	Heat-shock protein 20
HT	High temperatures
HTS	High-throughput sequencing
Hvd	Head vertical diameter
ICA	Independent cost analysis
ICARDA	International Center for Agricultural Research in Dry Areas
ICGR	Institute of Crop Germplasm Resources
ICRISAT	International Crops Research Institute for the Semi-Arid Tropics
InDel	Insertion/ deletion
INIAP	Instituto Nacional de Investigaciones Agropecuarias
INRA	French National Institute for Agricultural Research
INRAE	Institut National de la Recherche Agronomique
INTA-Balcarce	Argentine National Institute of Agricultural Technology
IPB-UPLB	Institute of Plant Breeding University of The Philippines Los BaNOs
IPK	Institute of Plant Genetics and Crop Plant Research
ISSR	Inter-simple sequence repeats
ITC	Isocyanates
JA	Jasmonic acid
JA2	Jasmonic acid 2
JERF1	Jasmonate and ethylene response factor 1
KAO	Ent-kaurenoic acid oxidase
KASP	Kompetitive allele-specific PCR
KIN	Kinase
LC	Locule number
Lcye	Lycopene epsilon cyclase
LD	Linkage disequilibrium
LEA	Late embryogenesis abundant
LEA5	Late embryogenesis abundant 5
LFY	LEAFY
LG	Linkage group
LMI1	Late Meristem Identity 1
LOX	Lipoxygenase
LR PCR	Long range polymerase chain reaction
LRR	Leucine rich repeat
MAB	Marker-assisted breeding
MADS-box	Minichromosome maintenance factor1 Agamous Deficiens Serum response factor–box
MAF	Missing allele frequency
MAGIC	Multi-parent advanced generation intercross
MAM	Methylthioalkylmalate
MAS	Marker-assisted selection
MBS	Mapping-by-sequencing
MetRS	Methionyl-tRNA synthetase

MHCREC	Mountain Horticultural Crops Research and Extension Center
miRNA	MicroRNA
MMD1	Male Meiocyte Death 1-like
MNs	Maganucleases
MOCOS	Molybdenum cofactor sulfurase
mQTL	Metabolite QTL
MRP-ABC	Multidrug resistance protein ATP binding cassette
MS2	Male sterility 2
MSA	Multiple sequence alignment
MSG	Multiplexed shotgun genotyping
MU16	Variety of *Cucurbita pepo* ssp. *pepo* var. Zucchini
MU-CU-16	Variety of *C. pepo* subsp. *pepo*
MWMV	Moroccan watermelon mosaic virus
MYB/Myb	Myeloblastosis
MYSV	Melon yellow spot virus
NAC	N-acetyl-cysteine
NAC2	NAC transcription factor 2
NAM	Nested association mapping
NAM	No apical meristem
NBPGR	National Bureau of Plant Genetic Resources
NBS	Nucleotide-binding site
NBS–LRR	Nucleotide binding site–Leucine-rich repeat
NCBI	National Center for Biotechnology Information
NCED	9-Cis-epoxycarotenoid dioxygenase
NDPK2	Nucleoside diphosphate protein kinase 2
NDVI	Normalized difference vegetation index
NE9	Northeast
NGS	Next-generation sequencing
NI	Neutral invertase
NIAS	National Institute of Advanced Studies
NIAS	National Institute of Agrobiological Sciences
NILs	Near isogenic lines
NMU	N-nitroso-N-methylurea
NR	Never ripe
NSGT1	Non-Smoky Glycosyltransefrase1
P5CR	Pyrroline-5-carboxylate reductase
P5CS	Pyrroline-5-carboxylate synthase
PAL	Phenylalanine
PARG	Poly (ADP-ribose) glycohydrolase)
PARSESNP	Project compatible linked sequences and SNPs evaluation
PC	Principal Components
PCR	Polymerase chain reaction
PHP	Polymerase and histidinol phosphatase
PHS	Pre-harvest sprouting
PI	Plant introduction
PKMT	Protein-lysine methyltransferase
PlantCARE	Plant cis-acting regulatory elements, enhancers and repressors
P-Loop NTase	P-Loop containing nucleoside triphosphate hydrolase superfamily protein
PLRV	Potato leaf roll virus
PM	Powdery mildew

Pm-0	Powdery mildew resistance gene
PMR	Powdery mildew resistant
PP2C	Protein phosphatase-2C
PPR	Pentatricopeptide repeat-containing
PPVFR	Protection of Plant Varieties and Farmar's Right
pQTL	Protein QTL
PRODH	Proline dehydrogenase
ProTs	Proline transporters
PRSV	Papaya ring spot virus
PRSV-W	Papaya ringspot virus-watermelon strain
PSY1	Phytoene synthase 1
PVA	Potato virus A
PVE	Phenotypic variation explained
PVM	Potato virus M
PVP	Plant variety protection
PVS	Potato virus S
PVX	Potato virus X
PVY	Potato virus Y
PYL4	Pyrabactin resistance-like 4
qCmB2	QTL associated with dwarf phenotype in *C. maxima*
qRT-PCR	Quantitative real-time PCR
QTL	Quantitative trait locus
QtlPC-C04	QTL for crown rot resistance in *C. moschata* on chromosome 4
QtlPC-C11	QTL for crown rot resistance in *C. moschata* on chromosome 11
QtlPC-C13	QTL for crown rot resistance in *C. pepo* on chromosome 13
QtlPC-C14	QTL for crown rot resistance in *C. moschata* on chromosome 14
QtlPRSV-C09	QTL for PRSV resistance in *C. moschata* accession, Nigerian Local
QTLs	Quantitative trait loci
QtlZYMV-C02	QTL for ZYMV resistance in Nigerian Local on chromosome 2
QtlZYMV-C04	QTL for ZYMV resistance in Nigerian Local on chromosome 4
QtlZYMV-C08	QTL for ZYMV resistance in Nigerian Local on chromosome 8
QtlZYMV-C20	QTL for ZYMV resistance in Nigerian Local on chromosome 20
RAD	Restriction site-associated DNA
RAPD	Random/ly amplified polymorphic DNA
Rc	Rind color
Rf	Fertility restoration gene
RF	Root forks
RFLP	Restriction fragment length polymorphism
RGA	R gene analogue
RGA	Rapid generation advance
R-gene	Resistance gene
RIL	Recombinant inbred line
RIN	Ripening Inhibitor
RING	Really interesting new gene
RNAi	RNA interference
RNA-seq	RNA sequencing
ROS	Reaction oxygen species
RPK	Receptor protein kinase
RPRV-HN	Hemagglutinin neuraminidase gene of peste des petits ruminants' virus
RPVH	Hemagglutinin gene of render pest virus

rrBLUP	Tool for genomic prediction
S1225-548	RAPD marker
SA	Salicylic acid
SAMDC	S-adenosylmethionine decarboxylase
SBAP	Sequence-based amplified polymorphism
SBT4B	Subtilase 4B
SCAR	Sequence characterized amplified region
SCR	SCARECROW
SEAN	Strategic environment analysis
SGN	Sol genomic networks
sgRNA	Single guide RNA
SIFT	Sorting intolerant from tolerant
SIPP	Soluble inorganic pyrophosphatase
SiR	Sulfite reductase
SLAF	Specific locus amplified fragment
SLAF-seq	SLAF-sequencing
SLC	Semidomesticated *S. lycopersicum*
SLDML2	*Solanum lycopersicum* DNA demethylases 2
SLEF1	*Solanum lycopersicum* elongated fruit1
SLSG6	S-locus specific glycoprotein S6
SMS	Single molecule sequencing
SN-1	Snaking-1
SNP	Single nucleotide polymorphism
SOD	Superoxide dismutase
SolCAP	Solanaceae Coordinated Agricultural Project
SOLiD	Supported Oligonucleotide Ligation and Detection System
SP	*Solanum Pimpinellifolium*
SPAC	Soil–plant–atmosphere continuum
SPP	Single position polymorphism
Sq026	Pumpkin inbred line with dwarf vine
SRA	Sequence read archive
SRAP	Sequence-related amplified polymorphism
SRF	Strubbelig receptor
SSAP	Sequence-specific amplification polymorphism
SSR	Simple sequence repeat
St	*Solanum tuberosum*
STMS	Sequence tagged microsatellite site
StnsLTP1	Non-specific lipid transfer protein-1
STPK	Serine/threonine protein kinase
STS	Sequence tagged site
SUPER	Simply unified pair-end read
SV	Structural variant
TALEN	Transcription activator-like effector nuclease
TE	Transposable element
TED	Tomato expression database
TF	Transcription factor
TFBM	Transcription factor binding motif
TFGD	Tomato functional genomics database
TG10	*C. moschata* heirloom
TGMV	Tomato golden mosaic virus

TGM-Zym-1	Marker associated with resistance to ZYMV in *C. pepo* accession 381e
TGRD	Tomato genomic resource database
TGS-Zym-2	Marker associated with resistance to ZYMV in *C. pepo* accession 381e
TILLING	Targeting induced local lesions in genomes
TIR-NBS-LRR	Toll interleukin-1 receptor–NBS-LRR
TMV	Tobacco mosaic virus
TNL-DR	TIR-NBS-LRR disease resistance
ToLC9NDV	Tomato leaf curl New Delhi virus
ToLCNDV	Tomato leaf curl New Delhi virus
TPS1	Trehalose-6-phosphate synthase 1
TRANSFAC	TRANScription FACtor database
TRV	Tobacco rattle virus
UGGT	UDP-glucose glycosyltransferase
UGPase	UDP-glucose-1-phosphate uridylyltransferase
UGT	UDP-glucosyltranferase
USDA	United States Department of Agriculture
VIGS	Virus induced gene silencing
VIR	Vavilov Research Institute of Plant Industry
VIR	Virtually integrated repository
VPSP	Vacuolar protein sorting protein
WB	Waltham Butternut
WGS	Whole genome sequencing
WMV	Watermelon mosaic virus
WOX	WUSCHEL-related homeobox
WRKY	WRKY transcription factor
YABBY1	Transcription factor associated with Bush locus in *C. moschata*
Zep/ZEP	Zeaxanthin epoxidase
ZFN	Zinc finger nuclease
ZFP1	Zinc finger protein 1
ZGMMV	Zucchini green mottle mosaic virus
ZYFV	Zucchini yellow fleck virus
Zym-0	Resistance gene to ZYMV in Nigerian Local
Zym-4	Resistance gene to ZYMV in Nigerian Local
ZYMV	Zucchini yellow mosaic virus

Preface

The affordability, precision, and high-throughput nature of sequencing technologies in the last two decades have provided an unprecedented understanding of the genomes and target plant traits of economic importance. Further, the advances in the third-generation sequencing technologies including Hi-C especially in the last decade have made it possible for developing high-quality reference genomes, pangenomes, and gene annotations, including genome-wide structural variations. These developments have also increased the precision in gene and superior haplotype discovery followed by mining allelic diversity among breeding lines and genebank collections. The high-throughput and cost-effective genotyping platforms have further boosted the adoption rate of genomic tools in the ongoing breeding programs across the world. Deployment of these advances in crops and making them well equipped to scale up the integration of genomic interventions in the breeding programs will facilitate development of future-ready crops.

The above advances have made gene/allele mining more precise and cost-effective for the target traits preferred by farmers, consumers, and industry as well as efforts towards mitigation of climate change impacts and achieving a nutritional security prospective. It is important to accelerate the process of genomic designing to develop future-ready crops by bringing new allelic combination(s) at one or more loci underlying desired genes using modern breeding approaches, such as marker-assisted breeding, haplotype-based breeding, and gene/allele pyramiding; and also now through gene editing.

These new alleles originate through natural or induced mutations also at one or more genomic regions at a time. The traditional plant breeding methods practiced in the twentieth century depended solely on phenotypes for the development of new genotypes primarily through selection and hybridization. The advent of DNA-based molecular markers in the 1980s facilitated indirect identification of the alleles by using linked molecular markers using genetic linkage mapping. Molecular breeding as a supplement to traditional breeding has successfully been employed in almost all major crops for their genetic improvement in a short time. It is worthwhile to mention that transgenic breeding has also contributed significantly to crop improvement at least for some major input target traits, such as herbicide tolerance, insect resistance, etc., however, this technology is still facing challenges in societal adoption in different parts of the world, especially for food crops. Nevertheless, gene editing has been receiving positive vibes from the policymakers of different countries making a conducive environment for accelerating the use of this technology on large scale.

Despite the huge loss of diversity during domestication, several national and international gene banks have millions of diverse germplasms to explore variability for the traits that are needed to breed climate-resilient and future-ready crops. Genetic diversity played a pivotal role during the entire period of conventional plant breeding, traditional and molecular included. However, the plant breeders have been using mainly the primary gene pool constituting the indigenous germplasm lines and local landraces as a genetic resource of new alleles. Meantime, a drastic dearth of new alleles in the primary gene pool happened due to genetic erosion caused naturally or due to domestication over time but aggravated because of the injudicious and irrational implementation of the so-called green revolution. Precisely, this highlighted the importance of allied wild crop relatives (WCRs) and thereby a mission of collection, conservation, characterization, and utilization of biodiversity. The emergence of advanced bioinformatics and the availability of cheaper and innovative genome sequencing technologies have empowered us to explore germplasm and identify a large number of donor alleles and allele mining has become a reality!

The strategy of association mapping, the most popular trait mapping approach in animal and human sciences, has received huge adoption in plant genomics studies. The availability of large sequenced diverse panels with multilocation phenotyping data has provided much-needed insights into trait genomics through genome-wide association studies (GWAS) leading to the discovery of genes and superior haplotypes in many crops. This has led to the possibility of mining a large number of genes in the available accessions in crops. Mutations occur in the genome as single nucleotide polymorphisms (SNPs) or insertion/deletions (In/Dels) resulting in the generation of new alleles or changes in existing alleles and allele combinations. Using exactly this principle, the approach of Eco-TILLING has facilitated large-scale allele mining in several major crops. On the other hand, sequence-based allele mining has led to the discovery of allelic diversity both at the target gene and at the whole genome. Recent advances in modern genomic tools and technologies such as genome engineering, speed breeding, etc., facilitate the deployment of these new alleles in breeding programs in a relatively short time to accelerate the rate of genetic gains. Truly speaking, allele mining has revolutionized the strategy of utilizing genetic as well as genomic resources and become an important component of next-generation precision breeding and biotechnology!

Allele discovery has become more relevant now for employing molecular breeding to develop designed crop varieties matching consumer needs and with genome plasticity to adapt the climate change scenarios using targeted and precision breeding tools validated by both genomics and phenomics studies. All these concepts and strategies along with precise success stories are presented in over 50 chapters allocated under five volumes dedicated to the five crop groups, including cereals, oilseeds, grain legumes, fruits, and vegetables under this book series.

This volume entitled, "Allele Mining for Genomic Designing of Vegetable Crops" includes 12 chapters focused on Potato, Tomato, Cucumber, Brassica oleracea Crops, Carrot, Capsicums, Cucurbita, and Radish contributed by 49 scientists from nine countries including Benin, Ghana, India, Italy, Mexico, Pakistan, South Korea, Turkey, and the USA. We express our thanks for their excellent contributions and timely cooperation.

This book will hopefully be useful for students, research scholars, and teaching faculties in academia to learn about fundamentals and applications of allele mining; the scientists practicing plant breeding in public and private institutes in improvising their programs; and also, the policymakers and funding agencies in prioritizing research fields.

Chittaranjan Kole, Kolkata, India
Tusar Kanti Behera, Varanasi, India
Prashant Kaushik, Shizuoka, Japan

Contributors

Shailesh Acharya
Horticultural Sciences Department and the Tropical Research & Education Center, University of Florida, Homestead, FL, USA

Joseph Adjebeng-Danquah
CSIR-Savanna Agricultural Research Institute, Nyankpala, Ghana

Richard Yaw Agyare
CSIR-Savanna Agricultural Research Institute, Nyankpala, Ghana

Victor Acheampong Amankwaah
CSIR-Crops Research Institute, Kumasi, Ghana

Marwen Amari
Department of Agri culture, Food and Envitonment (Di3A), University of Catania, Italy

Hajer Ben Ammar
Department of Agri culture, Food and Envitonment (Di3A), University of Catania, Italy

Mahdi Badri Anarjan
Plant Genomics Laboratory, Department of Bio-resource Engineering, College of Life Sciences, Sejong University, 209 Neungdong-ro, Gwanjing-gu, Seoul, Republic of Korea

Fredrick Justice Awuku
CSIR-Savanna Agricultural Research Institute, Nyankpala, Ghana

Gayatri B. K.
Division of Vegetable Science, ICAR-Indian Agricultural Research Institute, New Delhi, India

Sumit Kumar Bag
Molecular Biology and Biotechnology Division, CSIR-National Botanical Research Institute, Rana Pratap Marg, Lucknow, India

Archana Bhardwaj
Molecular Biology and Biotechnology Division, CSIR-National Botanical Research Institute, Rana Pratap Marg, Lucknow, India

Ferdinando Branca
Department of Agri culture, Food and Envitonment (Di3A), University of Catania, Italy

Tanuja Buckseth
ICAR-Central Potato Research Institute, Shimla, Himachal Pradesh, India

Mehmet Emin Çalışkan
Department of Agricultural Genetic Engineering, Faculty of Agricultural Sciences and Technologies Niğde Ömer Halisdemir University, Niğde, Türkiye

Clarissa Challam
ICAR-Central Potato Research Institute, Regional Station, Shillong, Meghalaya, India

Shyam Sundar Dey
Division of Vegetable Science, ICAR-Indian Agricultural Research Institute, New Delhi, India

Jingwei Fu
Horticultural Sciences Department, University of Florida, Gainesville, FL, USA

Corina Hayano-Kanashiro
Departamento de Investigaciones Científicas y Tecnológicas de la Universidad de Sonora, Universidad de Sonora, Hermosillo, Sonora, Mexico

Pritam Kalia
Former Head, Division of Vegetable Science, ICAR-Indian Agricultural Research Institute, New Delhi, India

Yogesh P. Khade
ICAR-Directorate of Onion and Garlic Research, Pune, Maharashtra, India

Anil Khar
ICAR-Indian Agricultural Research Institiute, Pusa Campus, New Delhi, Delhi, India.

Sanghyeob Lee
Plant Genomics Laboratory, Department of Bio-resource Engineering, College of Life Sciences, Sejong University, 209 Neungdong-ro, Gwanjing-gu, Seoul, Republic of Korea &
Plant Engineering Research Institute, Sejong University, 209 Neungdong-ro, Gwanjing-gu, Seoul, Republic of Korea

Haisu Li
Plant Genomics Laboratory, Department of Bio-resource Engineering, College of Life Sciences, Sejong University, 209 Neungdong-ro, Gwanjing-gu, Seoul, Republic of Korea

Octavio Martinez
Centro de Investigación y de Estudios Avanzados del Instituto Politécnico Nacional (Cinvestav), Unidad de Genómica Avanzada (Langebio), Irapuato, Guanajuato, Mexico

Leander Dede Melomey
University of Ghana, West Africa Centre for Crop Improvement, Legon, Ghana

Isaac Newton Boakye Mensah
CSIR-Crops Research Institute, Kumasi, Ghana

Geoffrey Meru
Horticultural Sciences Department and the Tropical Research & Education Center, University of Florida, Homestead, FL, USA

Muhammad Naeem
Agriculture Extension and Adaptive Research Wing, Punjab Agriculture Department, Pakistan

Nandhakumar Natarajan
ICAR-Central Potato Research Institute, Regional Station, Shillong, Meghalaya, India

Michael Kwabena Osei
CSIR-Crops Research Institute Horticulture division, Kumasi, Ghana

Isaac Osei-Bonsu
CSIR–Crops Research Institute, Kumasi, Ghana

Gangadhara Rao P.
College of Horticulture, Venkataramannagudem, Dr. YSR Horticultural University Andhra Pradesh, India

Dominick Padilla
Horticultural Sciences Department, University of Florida, Gainesville, FL, USA

Priti Prasad
Redcliffe Lifesciences Pvt. Ltd., Noida, Uttar Pradesh, India

Doris Kanvenaa Puozaa
CSIR-Savanna Agricultural Research Institute, Nyankpala, Ghana

Bala Rathinasabapathi
Horticultural Sciences Department, University of Florida, Gainesville, FL, USA

Prerna Sabharwal
Horticultural Sciences Department and the Tropical Research & Education Center, University of Florida, Homestead, FL, USA

Raman Selvakumar
Center for Protected Cultivation Technology, ICAR-Indian Agricultural Research Institute, New Delhi, India

Swati Shrestha
Horticultural Sciences Department and the Tropical Research & Education Center, University of Florida, Homestead, FL, USA

Jagesh Kumar Tiwari
ICAR-Indian Institute of Vegetable Research, Varanasi, Uttar Pradesh, India

Simone Treccarichi
Department of Agri culture, Food and Envitonment (Di3A), University of Catania, Italy

Caner Yavuz
Department of Agricultural Genetic Engineering, Faculty of Agricultural Sciences and Technologies Niğde Ömer Halisdemir University, Niğde, Türkiye

Herbaud Phanuel Fortunat Zohoungbogbo
World Vegetable Centre, Cotonou, Benin

Editor Biography

Chittaranjan Kole is an internationally reputed scientist with an illustrious professional career spanning over 40 years and globally recognized for his original contributions in the fields of plant genomics, biotechnology, and molecular breeding, leading to the publication of more than 150 quality research articles. He has edited over 180 books for the leading publishers of the world including Springer, Wiley-Blackwell, and Taylor and Francis Group. His scientific contributions and editing acumen have been appreciated by seven Nobel Laureates including Profs. Norman Borlaug, Arthur Kornberg, Werner Arber, Phillip Sharp, Günter Blobel, Lee Hartwell and Roger Kornberg. He has been honored with a number of fellowships, honorary fellowships, and national and international awards including the "Outstanding Crop Scientist Award" conferred by the International Crop Science Society. Prof. Kole has served in all prestigious positions in academia including as Vice-Chancellor of the BC Agricultural University and Project Coordinator of Indo-Russian Center of Biotechnology in India, and Director of Research of the Institute of Nutraceutical Research of Clemson University, in the USA. He worked also in the Pennsylvania State University and Clemson University as a Visiting Professor in the USA. He was awarded with the Raja Ramanna Fellow by the Department of Energy, Government of India. He is also heading the International Climate-Resilient Crop Genomics Consortium, International Phytomedomics and Nutriomics Consortium and Genome India International as their Founder President. Prof. Kole has recently established the Prof. Chittaranjan Kole Foundation for Science and Society and is working as its Founding Chairman.

Tusar Kanti Behera has 25 years' experience in research on vegetable crops and is presently working as Director, ICAR-Indian Institute of Vegetable Research, Varanasi, Uttar Pradesh, India. He bred 27 varieties and hybrids in cucurbitaceous and solanaceous vegetable crops and identified six trait specific breeding lines in bitter gourd, *Luffa* and cucumber. He is involved in genome sequencing of *Momordica balsamina* L. and comparative study with *M. charantia* and identified QTLs for gynoecy, fruit length and earliness in bitter gourd using GBS, QTL-seq and GWAS. He also identified candidate gene *CsCLV3* for carpel number in cucumber and *SMARCA3*-like Chromatin remodelling factor for regulating HYPOCOTYL elongation in cucumber. He introgressed *F* locus and derived 11 Indian tropical gynoecious cucumber lines using MAS and *gy-1* (gynoeciuos) gene into two commercial bitter gourd varieties.

Dr. Behera is recipient of Fulbright-Nehru Academic and Professional Excellence Fellow and fellow of National Academy of Agricultural Sciences (NAAS), India; Indian Academy of Horticultural Sciences; Indian Society of Genetics and Plant Breeding, Indian Society of Vegetable Science. He received several awards including Dr. B.R. Barwale Award for Excellence in Plant Genetic Resources; Hari Krishna Shastri Memorial Award and Best Teacher Award in Higher Education.

Dr. Behera published 125 research papers and 12 review papers of international repute and edited four books. He guided 11 Ph.D. and two M.Sc. students as major advisor at the Post Graduate School of ICAR-Indian Agricultural Research Institute, New Delhi.

Dr. Prashant Kaushik, an expert in genomics, molecular biology, and bioinformatics, boasts over a decade of academic and research excellence. He is currently serving as a District Extension Specialist in Vegetable Science. Dr. Kaushik has an extensive publication record with more than 150 peer-reviewed articles, 11 scholarly books, and numerous contributions to scientific advancements in crop resilience, nutritional quality, and stress tolerance. His work has significantly impacted the field of agricultural innovation, demonstrated by securing competitive research grants, membership on several editorial and advisory boards, and a strong track record in teaching and mentoring. Dr. Kaushik's education includes a Ph.D. in Biotechnology, an M.Sc. in Vegetable Science, an MBA in Production Management, and degrees in Agriculture. His commitment to science communication and collaboration across academia and industry marks him as a pivotal figure in contemporary agricultural research and education.

1 Allele Mining in Potato Genomic Design

Nandhakumar Natarajan,[1] *Jagesh Kumar Tiwari,*[*,2]
Clarissa Challam,[1] *and Tanuja Buckseth*[3]
[1]ICAR-Central Potato Research Institute, Regional Station, Shillong, Meghalaya, India
[2]ICAR-Indian Institute of Vegetable Research, Varanasi, Uttar Pradesh, India
[3]ICAR-Central Potato Research Institute, Shimla, Himachal Pradesh, India
[*]Correspondence author: jageshtiwari@gmail.com

1.1 INTRODUCTION

The cultivated potato, *Solanum tuberosum* L., was domesticated nearly 8000 years ago, from its origin and domestication in the southern Andes of Peru, where it was essential for feeding a growing population (Tiwari 2023). It is now widely recognized as the most important non-cereal food crop and is central to global food security (Spooner et al. 2005; Pearsall 2008). It has been declared as the "Food for the Future" (FAO 2019), and it was estimated that more than two billion people worldwide will depend on potatoes for food, feed, or income by 2020 (Ortiz and Mares 2017). Over the last decade, potato production has increased at an annual rate of 4.5%, outpacing the rate of new varieties' generation. The crop can produce high yields with limited inputs, and supplies at the same time a good source of energy and health-promoting nutrients (Birch et al. 2012). However, increasing biotic and abiotic stresses poses a serious and constant risk for food security, jeopardizing stable production (Bradshaw 2007). Domestication and selection have significantly reduced the genetic diversity of cultivated potatoes, which may provide allelic resources for controlling such stresses. Only a few clones of tetraploid cultivated *S. tuberosum* from the Andes were introduced to Europe, and despite the fact that they must have contained a significant amount of genetic variation, the available biodiversity was only partially captured (Ramsay and Bryan 2011; Birch et al. 2012; Kloosterman et al. 2013). This limited genetic diversity was exacerbated by genetic bottlenecks during photoperiod adaptation, as well as losses caused by viruses and the late blight caused by *Phytophthora infestans*, which was responsible for the infamous Irish potato famine in 1845–1846 (Bethke et al. 2017). However, cultivated potato and their wild relatives provide an extraordinary resource of genes and allelic diversity that modern cultivars lack (Spooner and Hijmans 2001; Machida-Hirano and Niino 2017; Tiwari et al. 2022). Their value as breeding material stems from their wide geographical distribution and great range of ecological adaptation (Hawkes 1994), together with their availability through the Inter-gene bank Potato Database (IPD) (http: //germplasmdb.cip.cgiar.org) established by the CIP (International Potato Centre) and the Association for Potato Intergene bank Collaboration. Introgressions of novel alleles from wild relatives of crop plants into cultivated varieties (McCouch et al. 2007) have clearly demonstrated that certain alleles and their combinations potentially make dramatic changes in trait expression when moved to a suitable genetic background by overcoming the genetic bottlenecks, which restricted their introgression to cultivars. Hence, the vast germplasm resources are potentially valuable sources of variation for identifying genes and allelic variants of agronomic, nutritional, and industrial quality for potato improvement (Andre et al. 2007).

DOI: 10.1201/9781003376439-1

Furthermore, the publicly available genome sequence of potato (Potato Genome Sequencing Consortium 2011) will facilitate mining for novel alleles or genes. Based on gene, genome sequences, and target trait, several mining strategies are devised to characterize and identify useful alleles of genes from a wide range of species. Identification of allelic variants from germplasm collections not only provides new germplasm for delivering novel alleles to targeted trait for crop improvement but also categorizes germplasm entries for their conservation (Tiwari 2023). Realizing the importance of allele mining in genomics-driven plant breeding era, this chapter highlights allele mining as a research field aimed at identification of allelic variation of genes and loci of relevant traits in potato and their applications in the genetic enhancement and potato improvement.

1.2 ALLELE MINING AND TRUE ALLELE MINING

Allele is an alternative form of a gene found at a given locus on a chromosome. Mining is nothing but searching the new alleles in the wild germplasm. So, allele mining refers to the search for useful alleles of genes from a diverse set of cultivars, related species, and even across species. In other words, it is a research field that seeks to identify allelic variation in relevant traits within genetic resource collections (Kumari et al. 2018).

1.2.1 "True" Allele Mining

Early studies of allele mining have focused only on the identification of SNPs/InDels at coding sequences or exons of the gene, since these variations were expected to affect the encoded protein structure and/or function. There are numerous examples available to demonstrate the effect of such sequence variations in genic regions in altering the phenotypes. However, many reports indicate that the nucleotide changes in non-coding regions (5′ UTR including promoter, introns and 3′ UTR) also have significant effect on transcript synthesis and accumulation, altering trait expression. Vleeshouwers et al. (2008) demonstrated true allele mining in potato by isolating the functional alleles of *Rpi-blb1, Rpi-sto1* and *Rpi-pta1* present in *S. stoloniferum*. Long-range PCR was used to isolate the entire genes, with primers up and downstream of the coding regions. The cloned genes' specificity was demonstrated using different *P. infestans* isolates as well as effector IpiO −1 and 2, which are recognized by *Rpi-blb1*, *Rpi-sto1*, and *Rpi-pta1*. An allelic relationship between the three genes was also shown using marker (CT88) segregation studies (Wang et al. 2008). Sequence analyses revealed that the putative functional homologs *Rpi-sto1* and *Rpi-pta1* are nearly identical to *Rpi-blb1*, with only three and five non-synonymous nucleotide substitutions inside the coding sequence, respectively. Champouret (2010) provided a slightly different example of allele mining for *R3a* and *R2* alleles using a start-stop codon approach. The screening of *R3a* revealed alleles with identical activity in distantly related species using a start-stop codon approach. This is considered as true allele mining. Similarly, *R2* also revealed many genes with identical activity, however, a few genes were also identified which had slightly different recognition specificities (Champouret 2010).

1.2.2 Promoter Mining

Mining for sequence variation, particularly in regulatory regions of genes, has grown in importance in the context of gene expression in recent years. Since it is well understood that promoter elements play an important role in gene regulation, any changes in their sequences will have a significant impact on gene expression, resulting in variable trait expression. Such differential gene expression is usually due to sequence polymorphism in upstream cis-acting/trans-acting regulatory element binding sites. Thus, "true" allele mining should include analysis of non-coding and regulatory regions of candidate genes, in addition to sequence variations in the coding regions, in order to cover the majority of functional variations of relevance in the genes (Latha et al. 2004). Realizing the importance of "mining" for variation in regulatory regions, promoters of genes have been chosen as the potential target for allele mining, which is termed as "promoter mining".

Promoter mining coupled with *in silico* analysis aimed to identify nucleotide variation in transcription factor binding motif (TFBM), number/frequency, and location of regulatory elements binding sites in promoter regions and their corresponding transcription factors, which are associated with a specific expression pattern in response to constitutive, developmental, tissue-specific, hormonal, and environmental regulation (Veerla and Hoglund 2006). Such an analysis of promoter sequences also provides the knowledge on conservation of regulatory motifs specific to particular differentiated cell/tissue types. Ultimately, promoter mining would offer insights into the molecular mechanism of gene expression and isolation of novel and efficient promoters for use in genetic engineering applications. As the global gene expression experiments to identify sets of genes with similar expression patterns are rapidly increasing, it is highly essential to incorporate approaches like *in vitro* and *in silico* promoter mining in regular allele mining programs to identify and categorize cis-regulatory elements and to understand their role in gene regulation.

1.3 TECHNIQUES FOR ALLELE MINING

Dependent on the research question but also dependent on genetic, genomic, and financial resources available, several techniques can be used for allele mining, ranging from a rapid and inexpensive polymerase chain reaction (PCR) to next gen sequencing and everything in between. For some applications, only partial sequence information or only molecular polymorphism of the alleles is sufficient. For other applications, the entire allele must be cloned. In general, all DNA-based tools require careful selection of target genes. The target gene model may need to be validated, and successively, a careful design of selective primers will allow the identification of novel alleles at candidate loci in the entire or core germplasm collection. In Figure 1.1 a pipeline for novel allele discovery from germplasm collections is presented.

1.3.1 Molecular Tools for Allele Mining

All molecular marker techniques include a PCR amplification of one or multiple alleles. Single-strand specific nucleases could be used to identify polymorphisms between amplified alleles. Using this technique, that is often used in TILLING approaches, nicking of heteroduplexes of PCR products can be easily detected. The results may be visible on an agarose gel as a DNA fragment of a specific size. However, when gene-specific markers are used in different germplasm material, often a specific annealing of the primers can occur and therefore it will always be necessary to sequence the resulting PCR fragments to confirm their identity and homogeneity.

1.3.1.1 Modified TILLING or Eco-TILLING

TILLING (Targeting Induced Local Lesions IN Genome) is a technique that uses heteroduplex analysis to identify novel alleles (more specifically artificially induced point mutations) in a target gene (Colbert et al. 2001). TILLING was originally designed to detect mutations in *Arabidopsis thaliana* treated with the chemical mutagen EMS (Till et al. 2003). Eco-TILLING, a modified version of TILLING entails the detection of naturally occurring allelic variation in populations, such as wild relatives, landraces, cultivars etc. It allows the rapid detection of variation in many individuals and is cost effective because only one individual for each haplotype needs to be sequenced. Like TILLING, Eco-Tilling also relies on the enzymatic cleavage of heteroduplexed DNA (formed due to single nucleotide mismatch in sequence between reference and test genotype) with a single-strand specific nuclease (i.e., Cel-1, mung bean nuclease, S1 nuclease, etc.) under specific conditions followed by detection through Li-Cor genotypers (Li-Cor, USA). At point mutations, the nuclease will cleave the

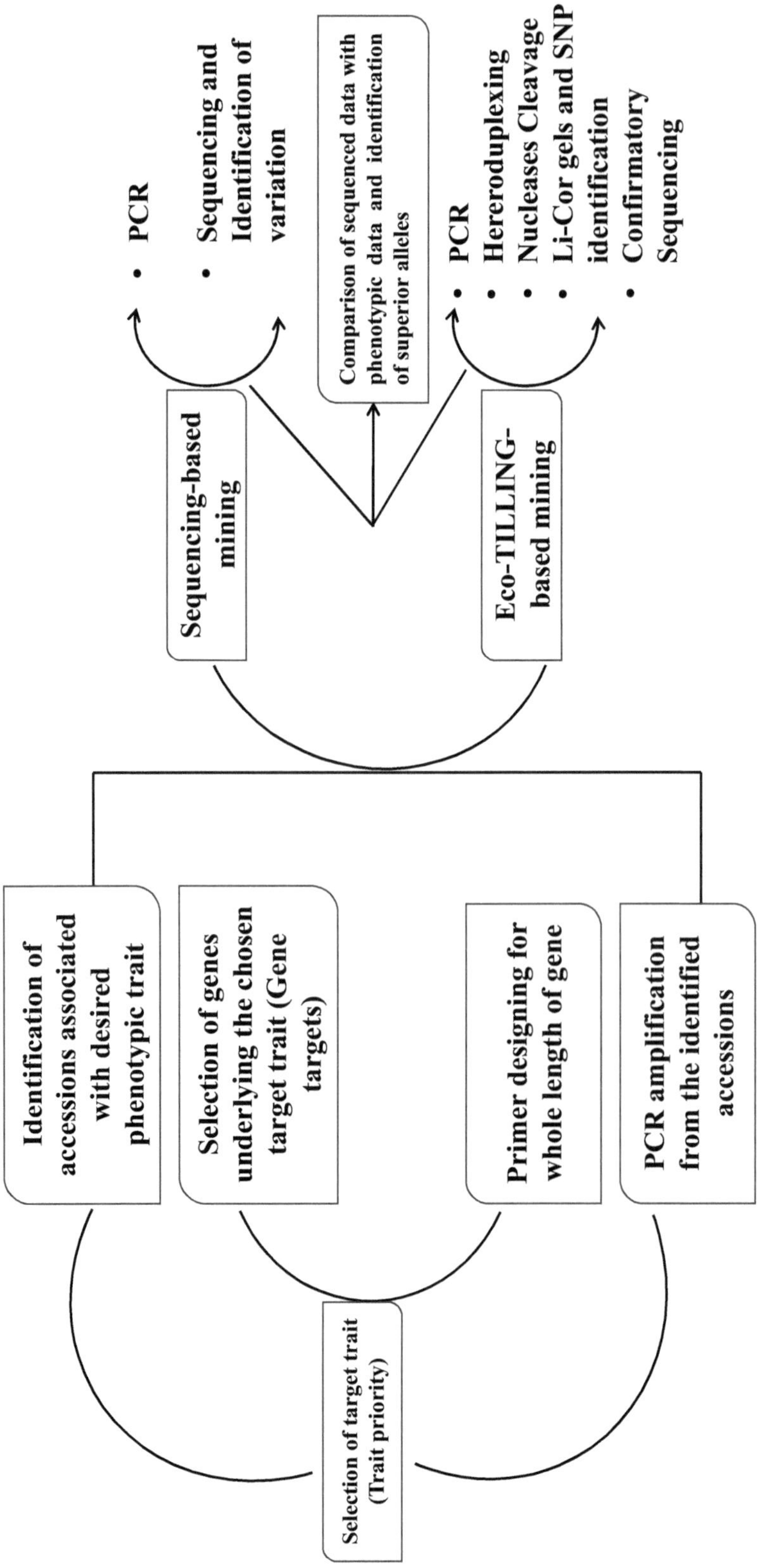

FIGURE 1.1 Steps involved in allele mining.

product into two cleaved products of equal size to the full-length product. Sequencing the amplicon from the test genotype that carries the mutation will confirm the presence, type, and location of the point mutation or SNP.

TILLING has been used successfully to identify both artificially induced and natural mutations (Eco-TILLING): for example, in *Arabidopsis thaliana*, barley, wheat, and rice (McCallum et al. 2000; Slade et al. 2005; Mejlhede et al. 2006; Till et al. 2007). However, the use of TILLING to identify mutations is more challenging in potato, because the genome is autotetraploid and shows considerable allelic diversity. Up to eight alleles for a given locus may exist within a crossing population, represented by an approximate 1: 10 mixture of insertion/deletion (InDels) polymorphisms and single nucleotide polymorphisms (SNPs). In one collection of 17 tetraploid and 11 diploid potato clones, an InDels polymorphism occurred, on average, once every 273 bp and an SNP occurred, on average, once every 21 bp (Rickert et al. 2003). Although TILLING and Eco-Tilling were proposed as cost-effective approaches for haplotyping and SNP discovery, these techniques require more sophistication and involve several steps starting from making DNA pools of reference and test genotypes, specific conditions for efficient cleavage by nuclease, detection in polyacrylamide gels using Li-Cor genotype and confirming through sequencing.

1.3.1.2 PCR-Based or Sequencing-Based Allele Mining

This technique involves PCR amplification of alleles in different genotypes, followed by DNA sequencing to identify nucleotide variation. In general, sequences in coding regions will be more conserved than primers in flanking sequences. Depending on the downstream application (sequence comparison, in plant expression), primers are chosen to amplify the entire gene, a portion of the gene, or just the open reading frame. Because even single nucleotide polymorphisms can have relevant differences between alleles, the DNA polymerase should ideally have proofreading activity. A long-range polymerase chain reaction (LR-PCR) polymerase is also preferred because long stretches of the target gene are frequently amplified. *Pfu*-Turbo from Invitrogen and Phusion from Fermentas are two enzymes that exhibit both characteristics. One approach was to use primers annealing to start and stop codon regions to amplify the entire coding sequence of the gene of interest. Following that, the amplicon is sequenced, and it can be cloned in a vector containing heterologous regulatory sequences for expression studies. Tiwari et al. (2015) selected 14 potato *R* genes and successfully PCR amplified using gene-specific primers, cloned and sequenced analyzed for homology in wild species. The *R* genes includes (1) *Rpi-pta1* (in *S. cardiophyllum* and *S. microdontum*); (2) *Rpi-edn1.1* (in *S. berthaultii*); (3) EDNR2GH7 (in *S. trifidum* and *S. verrucosum*); (4) SNKR2GH5 (in *S. iopetalum* and *S. lesteri*); and (5) *Rpi-snk1*.1 (in *S. jamesii*). Such allele mining in potato germplasm will accelerate genomics-assisted potato improvement (Tiwari et al. 2015). A possible lack of amplification can be expected for some accessions due to the lack of a coding gene or low sequence homology at the primer annealing sites. A drawback of this approach is that the promotor and terminator regions of the novel alleles are missing, so variation in these regulatory regions are neglected. In "true"allele mining, the use of primers matching the promotor and terminator regions is feasible when sequence conservation is sufficient.

1.3.1.3 Next-Generation Sequencing for Allele Mining

Most genome and transcriptome sequencing projects that previously used Sanger sequencing methodology are now being replaced by next-generation sequencing (NGS) technologies. These NGS technologies can generate data at a lower cost and at a rate that is several orders of magnitude faster than traditional technologies. There are several next-generation sequencers on the market right now (Voelkerding et al. 2009). Although the underlying biochemistry of most of these systems differs, all of these technologies sequence populations of PCR-amplified DNA molecules. The exceptions are the Oxford Nanopore and Pacific Biosciences, which sequence single molecules. Given the availability of low-cost, longer-read-length and high-throughput sequencing platforms, sequencing-based allele mining would result in faster generation of allelic data at a lower cost in the future. Furthermore, sequencing-based allele mining of specific genes in identified accessions and their association with phenotypic variations will provide a significant boost to precision breeding programs in crop plant.

Sequencing-based genotyping is rapidly becoming a more important tool for characterizing genomic variation and identifying SNPs. Several studies show that high-throughput short-read sequencing with Illumina can yield usable genotypic information in potatoes (Uitdewilligen et al. 2013; Slater et al. 2014). The estimation of dosage classes with this read information is a critical requirement prior to downstream application. The dosage of a SNP in an individual is estimated using relative counts of reads belonging to either the reference allele or alternative allele of that SNP. To determine these dosages, a polyploid variant caller such as FreeBayes is commonly used. In general, these tools require a high read depth to reliably assign dosage classes. Previously, it was determined that a read depth of > 80 is required in practise to achieve 95% accuracy (Uitdewilligen et al. 2013), primarily to differentiate between simplex (ABBB), duplex (AABB), and triplex (AABB) genotype calls.

Solis et al. (2022) determined the allelic diversity at the D, P, and R loci, which control anthocyanin biosynthesis and consequently potato tuber pigmentation in the 96 accessions with high and low levels of anthocyanins. The different allelic variants did not cluster for anthocyanin content, according to NGS analysis. SNP584(G) in the D locus, on the other hand, was a base mutation found in all high-anthocyanin accessions as well as 12 accessions with low levels of these metabolites. Greater allelic diversity was found at the P locus, where alleles positively correlated with purple pigmentation in the majority of the accessions. Because there were fewer red-pigmented accessions, the R locus had lower allelic diversity.

Resequencing of several genomes (Cao et al. 2011) followed by the comparison of all candidate genes is now feasible in *Arabidopsis*. In heterozygous species like potato, resequencing parts of the genome with duplicated sequences still remains a challenge (Potato Genome Sequencing Consortium 2011). However, the availability of third-generation sequencing technologies in recent years, allows the sequencing of very long DNA fragments. These long-read, single molecule technologies, such as those provided by Oxford Nanopore and Pacific Biosciences, have the potential to sequence DNA molecules with lengths of several 100 kb. Application of these new technologies will provide numerous opportunities for heterozygous species, such as potato to characterize highly complex regions, which are characterized by complex repeats and/or inversions.

1.3.1.4 Functional Analysis of Newly Identified Alleles

If candidate genes have been identified through allele mining studies, their functionality must be confirmed. For this purpose, both transient and stable transformations are valid. Agroinfiltration, an *Agrobacterium tumefaciens*-based method, is currently the best developed and most reliable method for transient expression in plants (Vleeshouwers et al. 2011). Using this method *R* gene alleles and the *Avr* genes can be coexpressed in *N. benthamiana* (Bos et al. 2006) or other plant species (Rietman et al. 2010). As a result, the infiltrated leaf area experiences a hypersensitive response (HR). This method is only applicable when the cognate *Avr* gene is available. Agro infiltration can also be accomplished by expressing only the *R* gene in a host plant, followed by pathogen challenge inoculations (Lokossou et al. 2009). Agro infection is another type of transient expression that allows for high-throughput screenings. A gene of interest is cloned into a viral genome. Successively, the viral genome is introduced into plant cells through *A. tumefaciens*. Only a few cells need to be infected after which viral particles are formed that spread through the plant. The gene of interest is expressed in the plant alongside the virus. However, there is a restriction on the size of the gene that can be expressed. Fragments larger than 500 bp in size will not express sufficiently.

Song et al. (2003) conferred resistance to late bight in both transient and stable expression systems when RGA2 (*Rpi-blb1/RB*) was cloned from *S. bulbocastanum*. The stable transformation into plant is still considered the most clear and definitive evidence for functional analysis. For example, potato transformation with RB homologues isolated from *S. bulbocastanum* (*Rpi-bt1*), *S. stoloniferum* (sensu Spooner et al. 2005; *Rpi-sto1*, *Rpi-pta1*) and *S. verrucosum* (*RB ver*) confirmed specificity of these genes against broad spectrum of *P. infestans* races (Vleeshouwers et al. 2008).

1.3.2 Bioinformatic Tools for Allele Mining

Regardless of the methods used, identifying polymorphism through comparative analysis of sequences from different genotypes is a critical step in allele mining. Several software tools are available viz., PLACE, plantCARE, TRANSFAC, JASPAR, MEME, Plantprom DB, DCPD, SCPD, BioEdit, ClustalW, etc. These tools are useful for handling the complex nucleotide data, prediction of putative function or structural components of complex macromolecules, prediction of TFBM, identification of sequence polymorphisms, and to predict the amino acid changes that are responsible for changes in encoded protein structure and/or function. These tools make it easier to access and analyze DNA sequences, aid in the detection of sequence polymorphism, and probe if the variance corresponds to the variation in TFBM. These tools are also effective for identifying overrepresented motifs in upstream regions while delimiting the main promoter region, which narrows the focus to uncover new variation in regulatory elements. A typical output of many such computational methods will be a list of motifs, which are most likely to correspond to the actual binding sites, along with a significant number of random variations. It should be noted that, the computational-based predictions depend on the pre-designed algorithms, which are largely developed based on the pre-characterized sequences available in the databases. Hence, such predictions should always be validated through systematic wet-lab experimental approaches.

1.4 CONSIDERATIONS FOR ALLELE MINING

Successful and efficient allele mining activity would depend mainly on the existing genetic base and the availability of genome and gene sequences information of a particular crop species. Several other important factors include availability of efficient and reliable phenotyping techniques; genomic resources; availability of high-throughput techniques for quick and easy generation of allelic data points; cost-effective sequencing platforms; efficient bioinformatic tools for identification of nucleotide variation and molecular marker construction technologies for marker-assisted selection (MAS).

1.4.1 Potato Genetic Resources

Genetic resources offer a vast reservoir of important novel traits and allelic variation for traits. Potato germplasm consists of diverse types of collections, such as primitive cultivars, natural hybrids, wild and weedy relatives, wild species, landraces, obsolete varieties, elite lines, breeding lines, mutants and polyploids, interspecific and intergeneric hybrids developed systematically (Haussmann et al. 2004). The value of germplasm is determined by its genetic diversity, availability, and utility. In this sense, potato stands out among all other crops. The high utility of wild and landrace potatoes has led to a series of collection expeditions in the centers of origin, which are currently available for breeders and researchers through gene banks. All over the globe, there are 98,285 potato accessions, which includes 15% wild species, 20% landraces, 16% breeding lines, 14% advanced cultivars, and 35% others (unknown or mixture of two or more types; Machida-Hirano 2015). Within these, 22,888 potato accessions are registered in GENESYS (www.genesys-pgr.org/ accessed Jan 10, 2023). A list of various potato gene banks carrying a diverse number of wild and cultivated accessions is shown in Table 1.1. These repositories assist in phenotypic screening of diverse potato accessions and will

TABLE 1.1
International Research Centres and Gene Banks Related to Wild, Landraces, and Cultivated Accessions of Potato

Research Centers/Gene Banks	Potato Collections
Centre for Genetic Resources (CGN), Wageningen University and Research Centre, Netherlands (www.wur.nl/en/Research-Results/Statutory-research-tasks/Centre-for-Genetic-Resources-the-Netherlands-1/Plant-Genetic-Resources/Genebank/CGN-crop-collections/CGN-potat o-collection.htm); (https: //potato.cgn.wur.nl/species.htm)	Working on 1,240 potato accessions, including 112 wild species and 220 Andean cultivated accessions.
The French National Institute for Agricultural Research (INRA), France (www6.rennes.inrae.fr/igepp_eng/AboutIGEPP/Platforms/BrACySol)	13,400 wild and cultivated accessions
The James Hutton Institute, Invergowrie, Dundee, UK. (www.hutton.ac.uk/about/facilities/commonwealth-potato-collection)	1,531 accessions including landraces and 47% wild potato species
Institute of Plant Genetics and Crop Plant Research (IPK), Germany. (www.ipk-gatersleben.de/en/genebank/satellite-collections-north/gross-luesewitz-potato-collections/glks/)	6,200 accessions; 46% (2,850) wild and 600 South American Landraces
Bonin Research Center, Laboratory of Potato Gene Resources and Tissue Culture, Plant Breeding and Acclimatization Institute, Radzików, Blonie, Poland. (http: //egiset.ihar.edu.pl/)	About 170 wild potato accessions
Vavilov Research Institute of Plant Industry (VIR), Department of Potato Genetic Resources, Bolshaya Morskaya St. Petersburg, Russia. (www.vir.nw.ru/en/)	4,600 accessions; 36% wild
US Potato Gene bank (NRSP-6), Sturgeon Bay, USA (https: //npgsweb.ars-grin.gov/gringlobal/site?id=29)	5,700 accessions; 70% wild
Potato Collection of the International Potato Center (CIP), Peru (https: //cipotato.org/genebankcip/potato-cultivated/); (https: //cipotato.org/genebankcip/potato-wild/)	Cultivated non-redundant 4,727 accessions, including 4,354 accessions of traditional landraces from 17 Andean countries; 4,052 accessions of wild potatoes from 16 countries of South America, Mexico and USA
INIAP (Instituto Nacional de Investigaciones Agropecuarias), Ecuador	350 varieties of *S andigenum, Phureja, chaucha* and *stenotomum*
BNGTRA-PROINPA, Bolivia (www.proinpa.org/web/)	2,393 accessions including 74% landraces and 26% wild species
Central Potato Research Institute (CPRI), India (https: //cpri.icar.gov.in/)	Total 2,628 accessions including 395 wild species and 924 native landraces
National Institute of Agrobiological Sciences (NIAS), Japan (www.gene.affrc.go.jp/databases-plant_search_en.php)	Total 1,843 accessions including 127 wild species and 25 native landraces

provide valuable alleles to breeders. The materials are conserved either as botanical seeds or in a vegetative form (tubers and *in vitro* plantlets).

The ability to exploit genetic diversity is critical to modern crop breeding, as it enables the introduction of new, useful genes and allelic variants into existing germplasm. Germplasm diversity offers a source of novel traits of agronomic value and can increase the variety of alleles within breeding populations (Hardigan et al. 2015). A large extent of genetic diversity can be found in wild and landrace relatives of cultivated crop species (Bradshaw et al. 2006; Pavek and Corsini 2001) and evaluation of the diversity in these wild species and landraces at the sequence level facilitates its use in developing new cultivars. The value of wild relatives and landraces for germplasm enhancement has been well established in potato and they have been exploited for the introduction or improvement of numerous traits, including resistance to bacterial pathogens bacterial pathogens, viral pathogens, and insect pests; tolerance of nonbiological stresses, such as salinity and frost; and traits related to the market quality of harvested tubers (Pavek and Corsini 2001). Andean landraces are also thought to have potential for broadening the genetic base of North American and European breeding populations, which are thought to house a small portion of the available diversity (Bradshaw et al. 2006; Jansky et al. 2013; Pavek and Corsini 2001). Hence, the vast germplasm resources are potentially valuable sources of variation for identifying genes and allelic variants of agronomic, nutritional, and industrial quality for potato improvement (Andre et al. 2007).

One might argue that resorting to potato wild relatives as donors of desirable traits seems no longer necessary with modern technologies, such as cisgenesis or the CRISPR-Cas9 genome editing, as it allows researchers to transfer directly the gene of interest or to change the native sequence into a tailor-made version, respectively. Nevertheless, it is first necessary to identify the original genes conferring the trait of interest and to mine their allele diversity in order to isolate them, clone them, and accurately modify them or transfer them into targeted cultivars.

1.4.2 Potato Genome and Sequence Information

The completed potato genome sequencing has expanded the scope of potato research. The cultivated potatoes are autotetraploid with a basic chromosome number of 12 and an estimated genome size of 840 million base pairs and 39,031 predicted protein-coding genes regulating various growth and development in the potato crop (Potato Genome Sequencing Consortium 2011). The availability of potato genome sequence and several transcriptomes from a diverse set of potato genotypes, organs and developmental conditions, have produced genomic tools for the analysis of wild and semi-cultivated potato species.

To date, genome assemblies of potato include a reference genome sequence of the doubled monoploid *S. tuberosum* Gp Phureja (DM 1-3 R44) (Potato Genome Sequencing Consortium 2011), an inbred diploid *S. tuberosum* Gp Tuberosum (van Lieshout et al. 2020), multiple diploid and dihaploid *S. tuberosum* Gp. Tuberosum lines (Freire et al. 2021; Zhang et al. 2021), and several wild diploid species (Aversano et al. 2015; Leisner et al. 2018). Available tetraploid genome sequences are limited to whole-genome shotgun sequencing data (Hardigan et al. 2017; Pham et al. 2017) or highly fragmented assemblies (Kyriakidou et al. 2020).

Whole genome resequencing can reveal important differences between cultivated potato varieties and related wild species, especially at a large scale. The genomes of 12 (Kyriakidou et al. 2019) and six (Kyriakidou et al. 2020) potatoes with different levels of ploidy genomes have recently been sequenced. Using the two publicly available reference genomes *S. tuberosum* Group *Phureja* (DM1-3) and *S. chacoense* M6 clone, copy number variation (CNV) and SNP analyses were performed. The study showed the great diversity of potato genomes across this panel. Efforts by the PGSC resulted in a genome sequence of high quality that can be used as a reference for resequencing of additional wild, cultivated, semi-cultivated species, lines, and cultivars facilitating mining for novel alleles or genes in potato germplasm.

Spud DB (http: // potato.plantbiology.msu.edu/), a potato database developed for potato community to access the potato genome sequence and associated annotation datasets, along with phenotypic and genotypic data from a diversity panel of 250 potato clones. The SNP finder tool in the Breeder's Assistant allows one to search for SNP based on genome location, annotated locus, or transcript identifier, and it supports both the PGSC and ITAG transcript identifiers. Returned data consist of a sortable table with SNP names, the chromosome and position of the SNP, the alleles, and the origin of the SNP. To provide the genome context of the SNP, all SNP names and any transcript identifiers are linked to their genome browser location. Collectively, this resource links potato genomic data with phenotypic and genotypic data from a large collection of potato lines for use by the potato community, especially breeders and geneticists.

1.5 APPLICATIONS OF ALLELE MINING IN POTATO

Allele mining can be used effectively to discover superior alleles by "mining" the gene of interest from diverse genetic resources. It can also shed light on the molecular basis of novel trait variations and identify nucleotide sequence changes associated with superior alleles. Furthermore, the rate of evolution of alleles, allelic similarity/dissimilarity at a candidate gene, and allelic synteny with other members of the family can be investigated. Allele mining may also pave the way for molecular discrimination among related species, the development of allele-specific molecular markers, the facilitation of novel allele introgression via MAS, or deployment via genetic engineering. While general applications of allele mining are described in Figure 1.2, the applications of allele mining in potato is detailed in Table 1.2.

1.5.1 Characterization of Allelic Diversity in Gene Banks

Identification and access to allelic variation that affects plant phenotype is critical for crop improvement using genetic resources. The efficient utilization of gene banks is dependent on knowledge of genetic diversity in general and allelic diversity at candidate gene(s) of interest, in particular. As a result, characterization of genetic diversity or allelic/genic diversity among the collection's accessions is critical for establishing genetic relationships and assessing genetic worthiness in terms of utility for improving a target trait. The availability of the sequences of candidate genes underlying agronomic traits will enable a pragmatic gene-based use of genetic resources, which, in theory, can be extended to any given trait. Realizing the potential of allele mining in potato genetic resources management applications, many research institutes with germplasm collections have initiated studies to characterize the crop's allelic diversity (Table 1.1). Although many challenges remain, the use of allele mining strategies is expected to result in a quantum leap in the use of genetic resources.

The utilization of germplasm for allele mining should contain maximum allelic variation at the loci of interest, in the smallest possible number of samples (Reeves et al. 2012). For over a century, wild potatoes have been used in disease resistance breeding programs, particularly late blight (Hawkes 1994). The resistance genes *R1–R11* from the wild Mexican species *Solanum demissum, which was* previously deployed for breeding late blight-resistant potato cultivars were allegedly defeated in the field by rapidly evolving *P. infestans* races (Fry 2008); nevertheless, many potato

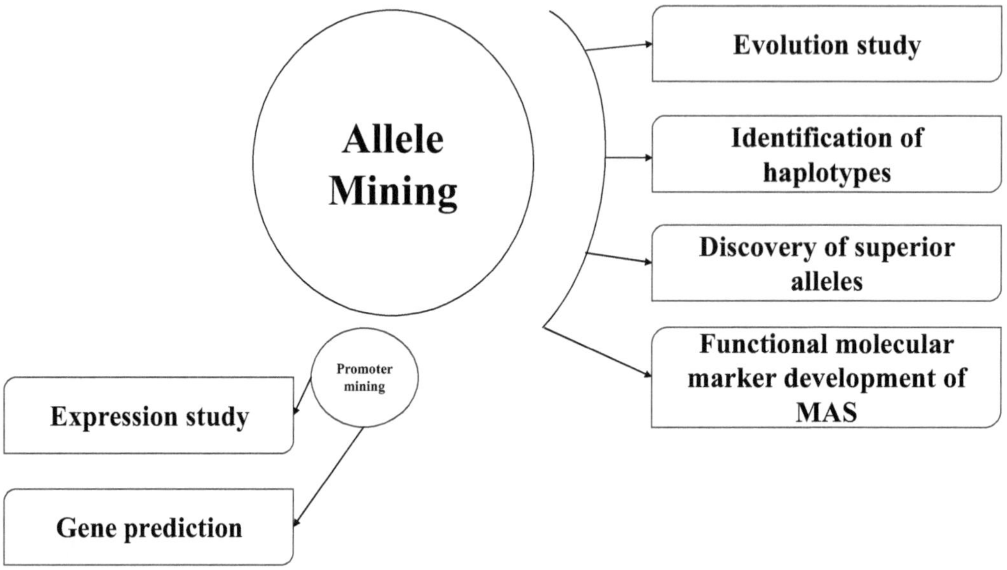

FIGURE 1.2 Major applications of allele mining.

TABLE 1.2
Allele Mining in Potato

S. No	Genes /Alleles	Species	Chr.	References
		Late blight		
1	*Rpi-amr3 and Rpi-amr1*	*S. americanum*	11	Witek et al. (2016); (2021)
2	*Rpi-Blb1/RB, Rpi-blb2, Rpi-blb3, Rpi-apbt*	*S. bulbocastanum*	8, 4	van der Vossen et al. 2005, Park et al. 2005a and 2005b
3	*Rpi-edn1.1 and Rpi-edn1.2*	*S. × edinense*	4	De Vetten et al. (2014)
4	*Rpi-chc1.1; Rpi-chc1.2; Rpi-chc2*	*S. chacoense*	10	Monino-Lopez et al. (2021)
5	*Rpi-cap1 and Rpi-qum1*	*S. capsicibaccatum*	11	Verzaux et al. (2012)
6	*Rpi-tar1 and Rpi-tar1.3*	*S. tarijense*	10	Monino-Lopez et al. (2021)
7	*Rpi-ver1*	*S. verrucosum*	9	Chen et al. (2018)
8	*Rpi-hcb1.1; Rpi-hcb1.2*	*S. huancabambense*	9	Aguilera-Galvez et al. (2020)
9	*Rpi-Smira1, Rpi-Smira2 (R8)a*	*S. tuberosum cv. Sárpo Mira*	11,9	Tomczyńska et al. (2014); Vossen et al. (2016)
		Potato Cyst Nematode		
1	*Gro1–4 (G. rostochiensis)*	*S. spegazzinii*	7	Paal et al. (2004)
2	*Ro2_A and Ro2_B (G. rostochiensis)* *Pa2/3_A and Pa2/3_B (G. pallida)*	*S. tuberosum ssp. andigena and S. vernei*	5 5	Park et al. (2019)
3	*Gro 1.2 (G. rostochiensis)*	*S. spegazzinii*	10	Park et al. (2019)
4	*GpaV (G. pallida)*	*S. vernei*	5	van Eck et al. (2017); Varypatakis et al. (2020)
5	*Gpa^{XII} tar (G. pallida)*	*S. tarijense*	11	Tan et al. (2009)
6	*Gpa^{V} spl* *Gpa^{XI} spl*	*S. sparsipilum*	5, 11	Caromel et al. (2005)
7	*H1(G. pallida)*	*S. multidissectum*	5	Strachan et al. (2019)
		Tuber quality and quantity traits		
1	*StOFP20*	*S. tuberosum*	10	van Eck et al. (2022)
2	*Solcap-snp-c2-39889 and solcap-snp-c2-5385*	*S. tuberosum*	11,12	Zia et al. (2020)
3	*Chy2* and *Zep*	*S. tuberosum*	3, 12	Wolters et al. (2010)
4	*D*, *P* and *R*	*S. tuberosum*	2, 10, 11	Solis et al. (2022)
5	*R2R3-MYB*	*S. tuberosum*	10	Endelman and Jansky (2016)
6	*Anthocyanin 1*	*S. tuberosum*	10, 11	

cultivars containing these *R* genes have higher late bight resistance than the genotypes lacking *R* genes (Stewart et al. 2003). Allele mining in *Solanum* species identified conserved homologous types of *Rpi-blb1* in *S. stoloniferum* (Wang et al. 2008) and many other wild *Solanum* species (Tiwari et al. 2015). Using an allele mining approach in potato, Pankin et al. (2011) analyzed wild *Solanum* species for homologs of *RB/Rpi-blb1* gene conferring durable late blight resistance. In addition, several already characterized *R* genes for late bight resistance from diverse wild *Solanum* species encode coiled coil-nucleotide binding site-leucine rich repeat kinases. Once such *R* genes are characterized, their orthologs found in other species using an allele-mining approach would help broaden the range of candidate genes for breeding for late bight resistance (Hein et al. 2009).

Flesh color in most tetraploid potato cultivars ranges from white via cream and yellow to dark yellow. The landraces grown by indigenous farmers in the Andean region belong to the diploid species *S. stenotomum, S. goniocalyx*, and *S. phureja* (Brown et al. 2007; Burgos et al. 2009). Brown et al. (2007) observed a relationship between ploidy level and total carotenoid content in 38 native South American cultivars. Significantly higher mean levels of total carotenoids were observed in diploid cultivars compared with tetraploid cultivars. Morris et al. (2004) describe a diploid high carotenoid-accumulating *S. phureja* accession (DB375\1, or "Inca Dawn") that predominantly contains zeaxanthin, but has a lower yield than tetraploid *S. tuberosum* cultivars (Bradshaw and Ramsay 2005).

1.5.2 Genetic Mapping Using Allele Mining

Understandably, most attention has been devoted to mapping resistance to late blight, the most important potato disease. The most remarkable source for resistance to *P. infestans* is *S. bulbocastanum* (Lokossou et al. 2010). In potato, mapping of *R* genes is strongly facilitated by allele mining through nucleotide binding site (NBS) profiling. Pel et al. (2009) and Jo et al. (2011) provided typical examples of *R* gene mapping approach. The first step in the approach consists of producing small (n = 20–100) populations segregating for *P. infestans* resistance, followed by phenotyping the populations for resistance, and composing bulks of resistant and susceptible individuals. Then, the bulks are genotyped using NBS profiling to identify markers that co-segregate with resistance. Following that, co-segregating NBS fragments are sequenced and BLAST analysis is used to identify the fragments. Combining this information with the literature and genome sequence data on mapping of resistance genes will suggest a putative map position. Finally, known flanking markers are used to confirm the map positions.

Jo et al. (2011) identified the genetic position of the late blight resistance gene *R8* using NBS profiling and successive marker sequence comparison to potato and tomato draft genome sequences. According to this work, *R8* was located on the long arm of chromosome IX. Through comparison of known markers in the tomato genetic map to the draft sequence, scaffolds were anchored to the tomato genetic map (anchored scaffold approach). Using the *R* gene cluster directed profiling approaches, the *R9* mediated-late blight resistance was also mapped near the *R8* locus on chromosome IX (Jo et al. 2015).

1.5.3 Uncovering Allelic Variation for Specific Genes

Allele mining can also be used to uncover genetic variation for a particular gene of interest and identify germplasm containing functional alleles from the same or different species. Nunziata et al. (2007) studied the variability of one cluster of genes at the *Gro1* locus responsible for resistance to Potato cyst nematodes, *Globodera rostochiensis* race *Ro1* in several potato species. The cluster is known to comprise ten different paralogs, among which only the *Gro1–4* gene has been demonstrated to confer resistance against *G. rostochiensis* race *Ro1*. Using available sequence information, three primer pairs were designed that target different regions of the *Gro1* sequence. The first was designed in a highly conserved region and allowed the presence of at least one member of the gene cluster to be identified in 16 wild species analyzed. The second primer pair was designed on a *Gro1–4* specific region and its use demonstrated that no gene identical to *Gro1–4* was present in any wild potato species analyzed. Finally, the major part of the LRR coding sequence of the *Gro1* gene was amplified and sequenced in 16 wild species. In total, 409 SNPs were identified, varying between species from 12 SNPs in *S. demissum* to 35 in *S. stoloniferum*. These data could be used to identify evolutionary selection pressure since the non-synonymous/synonymous ratio (Ka/Ks) in most species was different from 1.

A similar type of screen was performed by Wang et al. (2008) and Lokossou et al. (2010). They analyzed the presence and allelic diversity of the late blight *R* genes *Rpi-blb1*, *Rpi-blb2*, and *Rpi-blb3* in 196 different taxa of tuber-bearing *Solanum* species. The *Rpi-blb1* gene is part of a resistance gene analog (RGA) cluster of four members on chromosome VIII, *Rpi-blb2* resides in a locus harboring at least 15 tomato *Mi* gene homologs on chromosome VI and *Rpi-blb3* originates from a cluster on chromosome IV. For all genes, primers were designed that would allow amplification of a specific fragment of the gene. The genes were only present in some Mexican diploid as well as polyploid species closely related to *S. bulbocastanum*, although differences in the distribution existed among the three genes. The *Rpi-blb1* gene was only found in *S. bulbocastanum*, *S. cardiophyllum* subsp. *cardiophyllum*, and *S. stoloniferum*, the *Rpi-blb2* only in *S. bulbocastanum*, and the *Rpi-blb3* gene in *S. pinnatisectum*, *S. bulbocastanum* (including some subspecies), *S. hjertingii*, *S. nayaritense*, *S. brachistotrichum*, and *S. stoloniferum*. Sequence analysis of part of the *Rpi-blb1* and *Rpi-blb3* gene suggests an evolution through recombination and point mutations. For *Rpi-blb2* only sequences identical to the cloned gene were found, suggesting that it has emerged recently. The three *R* genes occurred in different combinations and frequencies in *S. bulbocastanum* accessions and their spread is confined to Central America (Lokossou et al. 2010). A practical outcome of the allele mining study by Wang et al. (2008) was the discovery of conserved homologs of *Rpi-blb1* in an EBN 2 tetraploid potato species, e.g., *S. stoloniferum*. The *Rpi-blb1* is present in the diploid tuber-bearing *S. bulbocastanum*, which is not directly crossable with the tetraploid *S. tuberosum*. *S. stoloniferum* can be directly crossed to cultivated potato, thus facilitating an easy transfer of a gene with exactly the same specificity and functionality as *Rpi-blb1*. Resistance to late blight and potato cyst nematodes have been identified in Indian native collection as well (Dalamu et al. 2023) and characterized using simple sequence repeat markers in wild species (Tiwari et al. 2019) and cultivated potato varieties (Tiwari et al. 2018).

The *StCDF1* locus, a master regulator of potato maturity, affecting the timing of tuberization and leaf senescence, which in turn affects yield, tuber quality traits, and field resistance to some diseases (Kloosterman et al. 2013). Hoopes et al. (2022) identified 12 unique alleles in 24 sequenced *StCDF1* haplotypes. Nine are classified as wild-type *StCDF1.1* alleles because they encode full-length StCDF1 proteins and the remaining three alleles (*StCDF1.2*, *StCDF1.3*, and *StCDF1.4*) encode truncated proteins due to structural variation in the second exon. Two cultivars, Altus and Avenger, contain only *StCDF1.1* alleles and show very late maturity at temperate latitudes. On the other extreme, Atlantic and Colombia have truncated alleles in three of the four homologs and are well known for their early maturity. The two other cultivars have intermediate maturity and intermediate dosage of truncated alleles (Spunta, 2 copies; Castle Russet, 1 copy). From a segregating population of 967 F1 progeny of Altus and Colomba, the effect of StCDF1.3 was estimated on early maturity was stronger than StCDF1.2 (Hoopes et al. 2022).

Flesh color in most tetraploid potato cultivars ranges from white via cream and yellow to dark yellow. This yellow color is caused by the presence of specific carotenoids. Wolters et al. (2010) observed that among 11 beta-carotene hydroxylase 2 (*Chy2*) alleles analyzed, only one dominant allele has a major effect, changing white into yellow flesh color. In contrast, none of the lycopene epsilon cyclase (*Lcye*) alleles seemed to have a large effect on flesh color. Analysis of zeaxanthin epoxidase (*Zep*) alleles showed that all (diploid) genotypes with orange tuber flesh were homozygous for one specific *Zep* allele. This *Zep* allele showed a reduced level of expression. The complete genomic sequence of the recessive *Zep* allele, including the promoter, was determined, and compared with the sequence of other *Zep* alleles. The most striking difference was the presence of a non-LTR retrotransposon sequence in the first intron of the recessive *Zep* allele, which was absent in all other *Zep* alleles investigated (Wolters et al. 2010).

1.5.4 Development of Haplotype in Tetraploid Potato

Haplotypes are more suitable as markers than bi-allelic SNP markers nevertheless, most of these molecular marker technologies result in bi-allelic markers, which means that a single marker distinguishes only between two haplotypes. Haplotype assembly makes use of SNP-alleles that are jointly present in a single sequencing fragment, to allow phase estimation based on physical linkage in sequencing data. In short, before haplotype assembly, sequencing fragments are aligned to the reference genome, after which genotyping is performed. Subsequently each alignment is interrogated for the presence of each SNP-allele, resulting in knowledge of physical linkage between SNPs. This information is exploited for the reconstruction of haplotypes. The advantage of sequencing-based haplotype reconstruction is the use of data of a single variety, avoiding the need for a large cohort of samples.

For diploids, many tools have been developed that allow characterization of allelic variation by haplotype assembly, however efforts for haplotype assembly in polyploids have been limited, and studies that performed large-scale haplotype reconstruction have so far not been reported. As already mentioned above, the cultivated potato shows not only tremendous phenotypic diversity, but also has a highly heterozygous genome as revealed by the potato genome sequencing efforts in 2011. Indeed, a single tetraploid variety has a SNP density of one variant in every 42 bp (Uitdewilligen et al. 2013), and exhibits extensive copy-number variations (Hardigan et al. 2016). Despite targeted resequencing efforts to uncover allelic variation associated with agronomical traits (Uitdewilligen et al. 2013; Schreiber et al. 2014), knowledge about the number of alleles per locus and haplotype diversity in the tetraploid potato gene pool is limited. Based on the estimation of linkage disequilibrium between markers, and therefore linkage decay, potato would contain anywhere between 6–12 founder haplotypes (Vos et al. 2017). This suggests that the potato haplotype landscape is characterized by a mosaic of large haplotype blocks. Efforts to start building a haplotype map of potato started with the study of Uitdewilligen et al. (2013), where 129,156 genomic variants were identified in approximately 800 genes. The application of these haplotype markers in QTL detection will likely result in a stronger association signal and therefore improve QTL detection power.

1.5.5 Establishment of Evolutionary Relationship

Knowledge on the evolution and distribution of agronomical important genes is imperative for a better understanding of the dynamics of genes in nature. Several factors can contribute to the extensive genome diversity among phylogenetically related species, among which, the pressure of artificial selection on DNA rearrangement is key. *Solanum phureja* exhibits a relatively strong impact of artificial selection on traits, including rapid maturation and lack of tuber dormancy, during a quick domestication process, resulting in abundant genomic structural variation, with a high level of genome diversity and gene sets with accelerated rates of protein sequence evolution (Yan et al. 2021). Nevertheless, *S. phureja* had relatively greater differences in speciation compared with *S. stenotomum* and the wild relative *S. chacoense*. The domestication process of the modern tetraploid potato *S. tuberosum* Andigenum Group is believed to have started as a single domestication event of wild progenitors from the *S. brevicaule* complex in southern Peru. The result of domestication was diploid *S. stenotomum*, from which other cultivated species were derived. Indeed, cultivated species were genetically enriched by incorporating wild species germplasm, resulting in gene flow from wild species. Recent comparative analyses with closely related *Solanaceae* species identified 358 *S. stenotomum* specific gene families, 885 gene families with expansion along the *S. stenotomum* lineage, and 149 genes experiencing accelerated rates of protein sequence evolution in *S. stenotomum*, the functions of which were mainly associated with defense responses, particularly against bacterial and fungal infection. Insights into the *S. stenotomum* genome and the genomic variation of cultivated potato taxa are valuable in elaborating the impact of potato evolution in early landrace diploid and facilitate modern potato breeding. As much within-species genetic variation is case-dependent, it is essential to sequence additional original cultivated potato species to capture a representative proportion of the potato's gene pool and resolve the allelic structure and gene variation within the species, thereby facilitating investigation of domestication and environmental adaptation (Yan et al. 2021).

1.6 CHALLENGES IN ALLELE MINING

Genetic resources collections are deemed to harbor a wealth of undisclosed allelic variants given the huge number of accessions that are held collectively in different potato gene banks. The challenge is to efficiently identify and exploit the useful variation for potato improvement. Here, we outline the challenges associated with allele mining and ways to overcome them in order to increase the efficiency.

1.6.1 Selection of Genotypes

The foremost challenge in unlocking the existing variation is the selection of germplasm to be "mined". Given a reliable protocol for characterizing a gene, screening the entire collection would certainly be helpful to find rare alleles, but this is an enormous and inefficient way of screening. Hence, we need to find a way forward from genotyping the whole composite collections to efficiently screening the accessions to discover new alleles and allele combinations while minimizing the number of accessions to be screened. Several approaches through which the prioritization of genotypes for allele mining can be done are described hereunder.

1.6.1.1 Development of Core/Mini Core Collections

Handling the entire germplasm is a Herculean task, be it for conventional plant breeding or for allele mining and hence requires sampling strategies to narrow down to a manageable size while maintaining the variability. Development of core and mini collections out of the entire collection has been proposed to simplify the conservation of germplasm resources and effective utilization of the existing variation in gene banks. A core collection is a subset of accessions from the entire collection that capture most of the available genetic diversity of the species (Brown 1989). The establishment of core and mini collections that enable an effective use of the existing variation in gene banks while maintaining the variability, as has been proposed before (Hoekstra 2009). Allele mining requires the assembly of a reasonably sized core germplasm collection usually comprising approximately 1,000 accessions representative of genetic diversity existing in the global population (Hofinger et al. 2009). Such collections can effectively be constructed using the Focussed Identification of Germplasm Strategy (FIGS) approach (Bhullar et al. 2009).

1.6.1.2 Accurate Phenotyping Methods

Phenotyping would be the key for success in short listing the genotypes for allele mining as phenotypes are considered as the best clues for genotypes. Increase in precision in phenotyping techniques maximizes the chances of "mining" the prospective genotypes. Also, with the development of trait-specific subsets obtained from phenotyping experiments the number of genotypes for "mining" is significantly reduced. Hence, there is a need to refine phenotyping protocols to increase the efficiency of allele mining.

Several methods are available to carry out phenotypic screens for target traits in gene bank collections. For example, screening of potato germplasm for late blight resistance different methods are being used. Inoculation of entire *in vitro* plantlets or inoculation of detached leaves can be used as high-throughput screening methods (Vleeshouwers 2011b). In the case of race-specific resistance, the selection of the pathogen isolates is an important issue in the identification of major *R* genes. Complementary to working with the entire pathogen, effector responsiveness can also be used to identify and classify R gene alleles in a germplasm core collection (Rietman et al. 2010). Effectors that displayed HR response in germplasm screens are potential *Avr* gene(s) recognized by the cognate *R* gene.

1.6.1.3 Flexible Computational Tools

Ever increasing number of plant genetic resources in gene banks created the problem of repetitiveness and genetic redundancy within the existing crop gene pools, which poses a challenge to effective management and potential utilization of these resources. Therefore, efficient computational tools are necessary to ensure the access to useful alleles. A major challenge that exists is to develop flexible computational tools for the selection of desirable alleles, which are seen as an avenue for the rapid expansion and extension of GenBank DNA data. Park (2007) has developed a computational tool called " 'Power Core" to support the development of core and allele mining sets by enhancing the richness and reducing the redundancy of useful alleles. Many such tools are essential to accelerate the future development of core subsets with all the observed characteristics or classes during the deployment of useful alleles from gene banks.

1.6.2 Handling Genomic Resources

To keep pace with rapid accumulation of nucleotide and gene expression data, computational tools need to be developed for analyzing the functional nucleotide diversity and to predict specific nucleotide changes responsible for altered function. Exploiting the developments in allele mining, association genetics, and comparative genomics by combining expertise from several disciplines, including molecular genetics, statistics, and bioinformatics is the suggested way.

1.6.3 Demarcation of Promoter Region

Demarcation of putative promoter regions poses one of the challenges in promoter mining analysis unlike defining the coding region for allele mining. Promoters and regulatory elements are spread across the upstream region and their locations are variable from one gene to another. Available literature clearly suggests the lack of consensus in fixing the size of the region for mining promoter and regulatory elements. This window size varies from as low as −100 bp to −3000 bp (Brazma et al. 1998; Park et al. 2002; Veerla and Hoglund 2006). However, high densities of regulatory elements have been reported only between −100 and −600 bp in the upstream region. Still, there is a need to develop specific software tools, which can accurately predict the core promoter region based on the representation/over-representation of consensus regulatory motifs.

1.6.4 Characterization of Regulatory Region

In contrast to identifying variation in coding regions of the genome, characterizing the extent of cis-acting regulatory variation presents a much greater challenge, since it is not possible to discern this even in the fully sequenced genomes. Screening for regulatory variants based on differences in transcript levels between individuals is confounded by potential trans-acting factors or environmental differences (Dai et al. 2007). As the promoter regions can function in bi-directional pathway, prediction of the orientation will be difficult even by the software. Moreover, complexity of analyzing these promoters is increased due to the fact that the transcription factors usually control more than one gene (Hehl and Wingender 2001). We suggest using a suit of different computational methods to get the snapshot of the regulatory elements which can be further examined through suitable experiments.

1.6.5 Higher Sequencing Costs

One of the important challenges is to minimize the time and efforts required while reducing the cost per data point. These challenges may partly be overcome by resorting to cheaper and faster sequencing platforms for high-throughput detection of allelic variations.

1.7 CONCLUSION

A large number of allelic variants of known genes have been discovered in potato, and in several cases, functionality of the variants has been demonstrated. As more genes are discovered and cloned, the likelihood that novel genes will reside at well-known and well-characterized loci increases, allowing for the use of comparative genomics and, as a result, the development of efficient allele mining strategies. The availability of potato and tomato genome sequences, combined with a steady decrease in sequencing costs, will boost allele mining even further. The rapid (r)evolution of high-throughput sequencing technologies, particularly the expected increase in read lengths from the third generation of single molecule sequencing platforms, will provide a comprehensive survey of the distribution of genes and gene clusters in potato, allowing for a dramatic acceleration in the process of identifying agronomically important genes.

REFERENCES

Aguilera-Galvez C, Chu Z, Omy SH, Wouters D, Gilroy EM, Vossen JH, Visser RGF, Birch P, Jones JDG, Vleeshouwers VGAA (2020) The *Rpi-mcq1* resistance gene family recognizes *Avr2* of *Phytophthora infestans* but is distinct from R2. *Stud Mycol* 89: 105–115.

Andre CM, Ghislain M, Bertin P, Oufir M, Del Rosario Herrera M, Hoffmann L, Hausman JF, Larondelle Y, Evers D (2007) Andean potato cultivars (*Solanum tuberosum* L.) as a source of antioxidant and mineral micronutrients. *J Agric Food Chem* 55: 366–378.

Aversano R, Contaldi F, Ercolano MR, Grosso V, Iorizzo M (2015) The *Solanum commersonii* genome sequence provides insights into adaptation to stress conditions and genome evolution of wild potato relatives. *Plant Cell* 27: 954–968.

Bethke PC, Halterman DA, Jansky SH (2017) Are we getting better at using wild potato species in light of new tools? *Crop Sci* 57: 1241–1258.

Bhullar NK, Street K, Mackay M, Yahiaoui N, Keller B (2009) Unlocking wheat genetic resources for the molecular identification of previously undescribed functional alleles at the *Pm3* resistance locus. *Proc Natl Acad Sci* 106: 9519–9524.

Birch PRJ, Bryan G, Fenton B (2012) Crops that feed the world 8: Potato: are the trends of increased global production sustainable. *Food Sec* 4: 477–508.

Bos JI, Kanneganti TD, Young C, Cakir C, Huitema E, Win J, Armstrong MR, Birch PR, Kamoun S (2006) The C-terminal half of *Phytophthora infestans* RXLR effector AVR3a is sufficient to trigger R3a-mediated hypersensitivity and suppress INF1-induced cell death in *Nicotiana benthamiana*. *Plant J* 48(2): 165–76.

Bradshaw J, Bryan G, Ramsay G (2006) Genetic resources (including wild and cultivated *Solanum* species) and progress in their utilisation in potato breeding. *Potato Res.* 49: 49–65.

Bradshaw JE (2007) Breeding potato as a major staple crop. In: MS Kang, PM Priyadarshan (eds) *Breeding major food staples*. Wiley-Blackwell, Oxford, pp 277–332.

Bradshaw JE, Ramsay G (2005) Utilisation of the commonwealth potato collection in potato breeding. *Euphytica* 146: 9–19.

Brazma A, Jonassen I, Vilo J, Ukkonen E (1998) Predicting gene regulatory elements in silico on a genomic scale. *Genomic Res* 8: 1202–1215.

Brown AHD (1989) Core collections: A practical approach to genetic resources management. *Genome* 31(2): 818–824.

Brown CR, Culley D, Bonierbale M, Amorós W (2007) Anthocyanin, carotenoid content, and antioxidant values in native South American potato cultivars. *Hort Sci* 42: 1733–1736.

Burgos G, Salas E, Amoros W, Auqui M, Muñoa L, Kimura M, Bonierbale M (2009) Total and individual carotenoid profiles in *Solanum phureja* of cultivated potatoes: I. Concentrations and relationships as determined by spectrophotometry and HPLC. *J Food Compos* 22: 503–508.

Cao J, Schneeberger K, Ossowski S, Günther T, Bender S (2011) Whole-genome sequencing of multiple *Arabidopsis thaliana* populations. *Nat Genet* 43: 956–963.

Caromel B, Mugniery D, Kerlan MC, Andrzejewski S, Palloix A, Ellisseche D (2005) Resistance quantitative trait loci originating from *Solanum sparsipilum* act independently on the sex ratio of *Globodera pallida* and together for developing a necrotic reaction. *Mol Plant Microbe Interact* 18: 1186–1194.

Champouret (2010) Functional genomics of Phytophthora infestans effectors and Solanum resistance genes. PhD thesis, Wageningen University.

Chen X, Lewandowska D, Armstrong MR, Baker K, Lim TY, Bayer M, Harrower B, McLean K, Jupe F, Kamil Witek K, Lees AK, Jones JD, Bryan GJ, Hein I (2018) Identification and rapid mapping of a gene conferring broad-spectrum late blight resistance in the diploid potato species *Solanum verrucosum* through DNA capture technologies. *Theor Appl Genet* 131: 1287–1297.

Colbert T, Till BJ, Tompa R, Reynolds S, Steine MN, Yeung AT, McCallum CM, Comai L, Henikoff S (2001) High-throughput screening for induced point mutations. *Plant Physiol* 126: 480–484.

Dai LY, Liu XL, Xiao HY, Wang GL (2007) Recent advances in cloning and characterization of disease resistance genes in rice. *J Integ Plant Biol* 49: 112–9.

Dalamu Tiwari JK, Bairwa A, Bhatia N, Zinta R, Kaushal N, Kumar V, Sharma AK Sharma S, Choudhary B, Luthra SK, Buckseth T, Singh RK, Thakur AK, Kumar M, Kumar D (2023) Resistance evaluation for native potato accessions against late blight disease and potato cyst nematodes by molecular markers and phenotypic screening in India. *Life* 13: 33.

De Vetten NCMH, Verzaux EC, Vossen JH, Rietman H, Jacobsen E (2014) *Cloning and Exploitation of a Functional R-Gene from Solanum x edinense*. US Patent WO 2011152722 A3.

Endelman JB, Jansky SH (2016) Genetic mapping with an inbred line-derived F2 population in potato. *Theor Appl Genet* 129: 935–943.

FAO (2019) http: //faostat.fao.org/.

Freire R, Weisweiler M, Guerreiro R, Baig N, Hüttel B, Obeng-Hinneh E, Renner J, Hartje S, Muders K, Truberg B, Rosen A, Prigge V, Bruckmüller J, Lübeck J, Stich B (2021) Chromosome-scale reference genome assembly of a diploid potato clone derived from an elite variety. *G3 (Bethesda)* 8: 11(12).

Fry W. (2008) Phytophthora infestans: the plant (and R gene) destroyer. *Molecular Plant Pathology*, 9: 385–402.

Hardigan MA, Bamberg J, Buell CR, Douches DS (2015) Taxonomy and genetic differentiation among wild and cultivated germplasm of *Solanum* sect. *The Plant Genome* 8(1): 1–16.

Hardigan MA, Crisovan E, Hamiltion JP, Kim J, Laimbeer P, Leisner CP, Manrique-Carpintero NC, Newton L, Pham GM, Vaillancourt B, Yang X, Zeng Z, Douches, DS, Jiang J, Veilleux RE, Buell CR (2016) Genome reduction uncovers a large dispensable genome and adaptive role for copy number variation in asexually propagated *Solanum tuberosum. The Plant Cell* 28(2): 388–405.

Hardigan MA, Laimbeer FPE, Newton L (2017) *Genome diversity of tuber-bearing Solanum uncovers complex evolutionary history and targets of domestication in the cultivated potato*. Proc Natl Acad Sci USA (46): E9999–E10008.

Haussmann BIG, Parzies HK, Presterl T, Susic Z, Miedaner T (2004) Plant genetic resources in crop improvement. *Plant Genet Resour. Characterisation Util.* 2(1): 3–21.

Hawkes JG (1994) Origins of cultivated potatoes and species relationships. In: Bradshaw JE, Mackay GR (eds) *Potato genetics*. CAB International, Wallingford, pp 3–42.

Hehl R, Wingender E (2001) Database-assisted promoter analysis. *Trends Plant Sci* 6: 251–5.

Hein I, Birch PRJ, Danan S (2009) Progress in mapping and cloning qualitative and quantitative resistance against *Phytophthora infestans* in potato and its wild relatives. *Potato Res* 52: 215–227.

Hoekstra R (2009) Exploring the natural biodiversity of potato for late blight resistance. *Potato Res* 52(3): 237–244.

Hofinger BJ, Jing HC, Hammond-Kosack KE, Kanyuka K (2009) High-resolution melting analysis of cDNA-derived PCR amplicons for rapid and cost-effective identification of novel alleles in barley. *Theor Appl Genet* 119: 851–865.

Hoopes G, Meng X, Hamilton JP, Achakkagari SR, Guesdes FAF, Bolger ME, Coombs JJ, Danny Esselink, Kaiser NR, Kodde L (2022) Phased, chromosome-scale genome assemblies of tetraploid potato reveal a complex genome, transcriptome, and predicted proteome landscape underpinning genetic diversity. *Mol Plant* 15(3): 520–536.

Jansky S, Dempewolf H, Camadro E, Simon R, Zimnoch-Guzowska E, Bisognin D (2013) A case for crop wild relative preservation and use in potato. *Crop Sci* 53: 746–754.

Jo KR, Arens M, Kim TY et al (2011) Mapping of the *S. demissum* late blight resistance gene *R8* to a new locus on chromosome IX. *Theor Appl Genet* 123: 1331–1340

Jo KR, Visser RG, Jacobsen E, Vossen JH (2015) Characterisation of the late blight resistance in potato differential *MaR9* reveals a qualitative resistance gene, *R9*a, residing in a cluster of Tm-2 (2) homologs on chromosome IX. *Theor Appl Genet* 128(5): 931–941.

Kloosterman B, Abelenda JA, Gomez MDMC, Oortwijn M, de Boer JM, K. Kowitwanich, K, Horvath BM, van Eck HJ, Smaczniak CS, Prat RGF, Bachem CWB (2013) Naturally occurring allele diversity allows potato cultivation in northern latitudes. *Nature* 495: 246–250.

Kumari R, Kumar P, Sharma, V, Kumar H (2018). Allele mining for crop improvement. *Int J Pure App Biosci* 6 (1): 1456–1465.

Kyriakidou M, Anglin N, Tai H, Ellis D, Strömvik MV (2019) Current strategies of polyploid plant genome sequence assembly. *Front Plant Sci* 9: 1660.

Kyriakidou M, Anglin NL, Ellis D, Tai HH, StrÃmvik, MV (2020) Genome assembly of six polyploid potato genomes. *Sci Data* 7(1): 88.

Latha R, Rubia L, Bennett J, Swaminathan MS (2004) Allele mining for stress tolerance genes in *Oryza* species and related germplasm. *Mol Biotech* 27: 101–108.

Leisner CP, Hamilton JP, Crisovan E, Manrique-Carpintero NC, Marand AP, Newton L, Pham GM (2018) Genome sequence of M6, a diploid inbred clone of the high-glycoalkaloid-producing tuber-bearing potato species *Solanum chacoense*, reveals residual heterozygosity. *Plant J* 94: 562–570.

Lokossou AA, Park T-H, van Arkel G, Arens M, Ruyter-Spira C, Morales J, Whisson SC, Birch PRJ, Visser RGF, Jacobsen E, van der Vossen EAG (2009) Exploiting knowledge of *R/Avr* genes to rapidly clone a new LZ-NBS-LRR family of late blight resistance genes from potato linkage group IV. *Mol Plant Microbe Interact* 22: 630–641.

Lokossou AA, Rietman H, Wang M (2010) Diversity, distribution and evolution of *Solanum bulbocastanum* late blight resistance genes. *Mol Plant Microbe Interact* 23: 1206–1216.

Machida-Hirano R (2015) Diversity of potato genetic resources. *Breed Sci* 65(1): 26–40.

Machida-Hirano R, Niino T (2017) Potato genetic resources. In: S Kumar Chakrabarti, C Xie, TJ Kumar (eds) *The potato genome. Compendium of plant genomes*. Springer, pp 11–29.

McCallum CM, Comai L, Greene EA, Henikoff S (2000) Targeted screening for induced mutations. *Nat Biotechnol* 18(4): 455–457.

McCouch SR, Sweeney M, Li J, Jiang H, Thomson M, Septinginsih E (2007) Through the genetic bottleneck: *O. rufipogon* as a source of trait-enhancing alleles for *O. sativa. Euphytica* 154: 317–339.

Mejlhede N, Kyjovska Z, Backes G, Burhenne K, Rasmussen SK, Jahoor A (2006) EcoTILLING for the identification of allelic variation in the powdery mildew resistance genes *Mlo* and *Mla* of barley. *Plant Breed* 125: 1–8.

Monino-Lopez D, Nijenhuis M, Kodde L, Kamoun S, Salehian H, Schentsnyi K, Stam R, Lokossou A, Abd-El-Haliem A, Visser RGF, Vossen JH (2021) Allelic variants of the NLR protein *Rpi-chc1* differentially recognize members of the *Phytophthora infestans* PexRD12/31 efector superfamily through the leucinerich repeat domain. *Plant J* 107: 182–197.

Morris WL, Ducreux L, Griffiths DW, Stewart D, Davies HV, Taylor MA (2004) Carotenogenesis during tuber development and storage in potato. *J Exp Bot* 55: 975–982.

Nunziata A, Ruggieri V, Frusciante L, Barone A (2007) Allele mining at the locus *Gro 1* in *Solanum* wild species. *Acta Hortic.* 745: 449–456.

Ortiz O, Mares V (2017) The historical, social, and economic importance of the potato crop. In: S Kumar Chakrabarti, C Xie, J Kumar Tiwari (eds) *The potato genome.* Cham (Switzerland), Springer, pp 1–10.

Paal J, Henselewski H, Muth J, Meksem K, Menendez CM, Salamini F (2004) Molecular cloning of the potato *Gro1–4* gene conferring resistance to pathotype *Ro1* of the root cyst nematode *Globodera rostochiensis*, based on a candidate gene approach. *Plant J* 38: 285–297.

Pankin A, Sokolova E, Rogozina E, Kuznetsova M, Deahl K, Jones R, Khavkin E (2011) Allele mining in the gene pool of wild *Solanum* species for homologues of late blight resistance gene *RB/Rpi-blb1. Plant Genet Res Charact Utiliz* 9: 305–308.

Park J, Hackett C, Dandurand L, Wang X, Jong WDD (2019) QTL for resistance to *Globodera rostochiensis* pathotype *Ro2* and *G. pallida* pathotype Pa2/3 in autotetraploid potato. *Am J Potato Res* 96: 552–563.

Park PJ, Butte AJ, Kohane IS (2002) Comparing expression profiles of genes with similar promoter regions. *Bioinform* 12: 1576–1584.

Park TH, Gros J, Sikkema A, Vleeshouwers VGAA, Muskens M, Allefs S (2005a) The late blight resistance locus *Rpi-blb3* from *Solanum bulbocastanum* belongs to a major late blight *R* gene cluster on chromosome 4 of potato. *Mol. Plant Microbe Interact* 18: 722–729.

Park TH, Vleeshouwers VGAA, Hutten RCB, van Eck HJ, van der Vossen E, Jacobsen E (2005b) High-resolution mapping and analysis of the resistance locus *Rpi-abpt* against *Phytophthora infestans* in potato. *Mol Breed* 16: 33–43.

Park YJ (2007) *Allele mining experience on rice germplasm through advanced M strategy using modified heuristic search, W9.* San Diego, CA: Plant and Animal Genomes XV Conference, pp 13–17.
Pavek J, Corsini D (2001) Utilization of potato genetic resources in variety development. *Am. J Potato Res* 78: 433–441.
Pearsall DM (2008) Plant Domestication and the Shift to Agriculture in the Andes. In: H Silverman, WH Isbell (eds) *The Handbook of South American Archaeology (New York).* Springer, pp 105–119.
Pel MA, Foster SJ, Park TH (2009) Mapping and cloning of late blight resistance genes from *Solanum venturii* using an interspecific candidate gene approach. *Mol Plant Microbe Interact* 22: 601–615.
Pham GM, Newton L, Wiegert-Rininger K, Vaillancourt B, Douches DS, Buell CR (2017) Extensive genome heterogeneity leads to preferential allele expression and copy number-dependent expression in cultivated potato. *Plant J* 92: 624–637.
Potato Genome Sequencing Consortium (2011) Genome sequence and analysis of the tuber crop potato. *Nature* 475: 189–195.
Ramsay G, Bryan G (2011) Solanum. In C Kole (ed) *Wild crop relatives: Genomic and breeding resources, vegetables.* Berlin, Heidelberg, Springer, pp 259–271.
Reeves P, Panella L, Richards C (2012) Retention of agronomically important variation in germplasm core collections: Implications for allele mining. *TAG. Theor Appl Genet* 124: 1155–71.
Rickert AM, Kim JH, Meyer S, Nagel A, Ballvora A (2003) First-generation SNP/InDel markers tagging loci for pathogen resistance in the potato genome. *Plant Biotechnol J* 1: 399–410.
Rietman H, Champouret N, Hein I et al (2010) *Plants and oomycetes, an intimate relationship: coevolutionary principles and impact on agricultural practice.* In D Hemming (ed), *CAB reviews.* CABI, UK, pp 5: 1–17.
Schreiber L, Nader-Nieto AC, Schönhals EM, Walkemeier B, Gebhardt C (2014) SNPs in Genes functional in starch-sugar interconversion associate with natural variation of tuber starch and sugar content of potato (*Solanum Tuberosum* L.). *G3* 4 (10): 1797–1811.
Slade AJ, Fuerstenberg SI, Loeffler D, Steine MN, Facciotti D (2005) A reverse genetic, nontransgenic approach to wheat crop improvement by TILLING. *Nat Biotechnol* 23: 75–81
Slater MJ, Coffee P, Barker JB, Evans AL (2014) Promoting shared meanings in 21 group memberships: A social identity approach to leadership in sport. *Reflective Practice: Int J Multidiscip Res Dev* 15(5): 672–685.
Solís J, Muth J, Canales Javier, Lizana C, Pruefer D, Riegel R, Behn A (2022) Allelic diversity of three anthocyanin synthesis genes in accessions of native *Solanum tuberosum* L. ssp. tuberosum at the Potato Genebank of the Universidad Austral de Chile. *Genet Resour Crop Evol* 69: 297–314.
Song J, Bradeen JM, Naess SK, Raasch JA, Wielgus SM, Haberlach GT, Liu J, Kuang H, Austin-Phillips S, Buell CR, Helgeson JP, Jiang J (2003) Gene RB cloned from *Solanum bulbocastanum* confers broad spectrum resistance to potato late blight. *Proc Natl Acad Sci* 100(16): 9128–9133.
Spooner DM, Hijmans RJ (2001) Potato systematics and germplasm collecting 1989–2000. *Am J Potato Res* 78: 237–268.
Spooner DM, Peralta IE, Knapp S (2005) Comparison ofAFLPs with other markers for phylogenetic inference in wild tomatoes (*Solanum* L. section *Lycopersicon* (Mill.) Wettst.) *Taxon* 54: 43–61
Stewart H, Bradshaw, JE, Pande, B (2003) The effect of the presence of *R*-genes for resistance to late blight (*Phytophthora infestans*) of potato (*Solanum tuberosum*) on the underlying level of field resistance. *Plant Pathol* 52: 193–198.
Strachan SM, Armstrong MR, Kaur A, Wright KM, Lim TY, Baker K (2019) Mapping the H2 resistance effective against *Globodera pallida* pathotype *Pa1* in tetraploid potato. *Theor Appl Genet* 132: 1283–1294.
Tan AMY, Park TH, Alles R, Hutten RC, Visser RG, van Eck HJ (2009) GpaXIl tar originating from *Solanum tarijense* is a major resistance locus to *Globodera pallida* and is localised on chromosome 11 of potato. *Theor Appl Genet* 119: 1477–1487.
Till BJ, Cooper J, Tai TH, Colowit P, Greene EA, Henikoff S (2007) Discovery of chemically induced mutations in rice by TILLING. *BMC Plant Biol* 7: 19.
Till BJ, Reynolds SH, Greene EA, Codomo CA, Enns LC, Johnson JE, Burtner C, Odden AR, Young K, Taylor NE, Henikoff JG, Comai L, Henikoff S (2003) Large-scale discovery of induced point mutations with high-throughput TILLING. *Genome Res* 13(3): 524–30.
Tiwari JK (2023) *Potato Improvement in the Post-Genomics Era.* CRC Press, Florida, USA, p 336.
Tiwari JK, Ali N, Devi S, Kumar V, Zinta R, Chakrabarti SK (2018) Development of microsatellite markers set for identification of Indian potato varieties. *Sci Hort* 231: 22–30.

Tiwari JK, Ali S, Devi S, Zinta R, Kumar V, Chakrabarti SK (2019) Analysis of allelic variation in wild potato (Solanum) species by simple sequence repeat (SSR) markers. *3 Biotech* 9: 262.

Tiwari JK, Buckseth T, Zinta R, Bhatia N, Dalamu D, Naik S, Poonia AK, Kardile HB, Challam C, Singh RK, Luthra SK, Kumar V and Kumar M (2022) germplasm, breeding, and genomics in potato improvement of biotic and abiotic stresses tolerance. *Front Plant Sci* 13: 805671.

Tiwari JK, Devi S, Sharma S, Chandel P, Rawat S Singh BP (2015) Allele mining in *Solanum* germplasm: cloning and characterization of RB homologous gene fragments from late blight resistant wild potato species. *Plant Mol Biol Rep* 33: 1584–1598.

Tomczyńska I, Stefańczyk E, Chmielarz M, Karasiewicz B, Kamiński P, Jones JDG, Lees AK, Śliwka J (2014) A locus conferring efective late blight resistance in potato cultivar Sárpo Mira maps to chromosome XI. *Theor Appl Genet* 127: 647–657.

Uitdewilligen JGAML, Wolters AMA, D'hoop BB, Borm TJA, Visser RGF, van Eck, HJ, Lukens L (2013) A next-generation sequencing method for genotyping-by-sequencing of highly heterozygous autotetraploid potato. *PLoS ONE* 8(5): e62355.

van der Vossen EAG, Gros J, Sikkema A, Muskens M, Wouters D, Wolters P (2005) The *Rpi-blb2* gene from *Solanum bulbocastanum* is an *Mi-1* gene homolog conferring broad-spectrum late blight resistance in potato. *Plant J* 44: 208–222.

van Eck HJ, Oortwijn ME, Terpstra IR, van Lieshout NH, van der Knaap E, Willemsen JH, Bachem CW (2022) *Engineering of tuber shape in potato (Solanum tuberosum) with marker assisted breeding or genetic modification using StOFP20*. Res Squ 1–19.

van Eck HJ, Vos PG, Valkonen JPT, Uitdewilligen JGAML, Lensing H, de Vetten N (2017) Graphical genotyping as a method to map Ny(o,n)sto and Gpa5 using a reference panel of tetraploid potato cultivars. *Theor Appl Genet* 130: 515–528.

van Lieshout N, van der Burgt A, de Vries ME, Ter Maat M, Eickholt D, Esselink D, van Kaauwen MPW, Kodde LP, Visser RGF, Lindhout P, Finkers R. Solyntus (2020) The new highly contiguous reference genome for potato (*Solanum tuberosum*). *G3 (Bethesda)* 10(10): 3489–3495.

Varypatakis K, Véronneau PY, Thorpe P, Cock PJA, Lim JTY, Armstrong MR, Janakowski S, Sobczak M, Hein I, Mimee B, Jones JT, Blok VC (2020) The Genomic Impact of Selection for Virulence against Resistance in the Potato Cyst Nematode, *Globodera pallida*. *Genes 11* 1429.

Veerla S, Hoglund M (2006) Analysis of promoter regions of co-expressed genes identified by microarray analysis. *BMC Bioinform* 7: 384.

Verzaux E, Van Arkel G, Vleeshouwers VGAA, Van der Vossen EAG, Niks RE, Jacobsen E, Vossen J, Visser RGF (2012) High-resolution mapping of two broad-spectrum late blight resistance genes from two wild species of the *Solanum circaeifolium* group. *Potato Res* 55: 109–123.

Vleeshouwers VG, Raffaele S, Vossen JH, Champouret N, Oliva R, Segretin ME, Rietman H, Cano LM, Lokossou A, Kessel G, Pel MA, Kamoun S (2011) Understanding and exploiting late blight resistance in the age of effectors. *Annu Rev Phytopathol* 49: 507–531.

Vleeshouwers VG, Rietman H, Krenek P, Champouret N, Young C, Oh SK, Wang M, Bouwmeester K, Vosman B, Visser RG, Jacobsen E, Govers F, Kamoun S, Van der Vossen EA (2008) Effector genomics accelerates discovery and functional profiling of potato disease resistance and *phytophthora infestans* avirulence genes. *PLoS One* 6(8): e2875.

Vleeshouwers VGAA, Raffaele S, Vossen J, Champouret N, Oliva R, Segretin ME, Rietman H, Cano LM, Lokossou AA, Kessel GJT, Pel M, Kamoun S (2011b) Understanding and exploiting late blight resistance in the age of effectors. *Annu Rev Phytopathol* 49: 507–531.

Voelkerding K, Dames S, Durtschi J (2009) Next-Generation Sequencing: From basic research to diagnostics. *Clin Chem* 55: 641–658.

Vos PG, Paulo MJ, Voorrips RE, Visser RG, van Eck HJ, van Eeuwijk FA (2017) Evaluation of LD decay and various LD-decay estimators in simulated and SNP-array data of tetraploid potato. *Theor Appl Genet* 130(1): 123–135.

Vossen JH, Van Arkel G, Bergervoet M, Jo KR, Jacobsen E, Visser RG (2016) The *Solanum demissum* R8 late blight resistance gene is an Sw-5 homologue that has been deployed worldwide in late blight resistant varieties. *Theor Appl Genet* 129: 1785–1796.

Wang M, Allefs S, van den Berg RG, Vleeshouwers VGAA, van der Vossen EAG (2008) Allele mining in *Solanum*: conserved homologues of *Rpi-blb1* are identified in *Solanum stoloniferum*. *Theor Appl Genet* 116(7): 933–943.

Witek K, Jupe F, Witek AI, Baker D, Clark MD, Jones JD (2016) Accelerated cloning of a potato late blight-resistance gene using RenSeq and SMRT sequencing. *Nat Biotechnol* 34: 656–660.

Witek K, Lin X, Karki HS (2021) A complex resistance locus in *Solanum americanum* recognizes a conserved *Phytophthora* effector. *Nat Plants* 7: 198–208.

Wolters AMA, Uitdewilligen JGAML, Kloosterman BA (2010) Identification of alleles of carotenoid pathway genes important for zeaxanthin accumulation in potato tubers. *Plant Mol Biol* 73: 659–671.

Yan L, Zhang Y, Cai G, Qing Y, Song J, Wang H, Tan Xi, Liu C, Yang M, Fang Z, Lai X (2021) *Genome assembly of primitive cultivated potato Solanum stenotomum provides insights into potato evolution G3*: 11(10): jkab262.

Zhang C, Yang Z, Tang D, Zhu Y, Wang P (2021) Genome design of hybrid potato. *Cell* 184: 3873–3883.

Zia MAB, Demirel U, Nadeem MA, Çaliskan ME (2020) Genome-wide association study identifies various loci underlying agronomic and morphological traits in diversified potato panel. *Physiol Mol Biol Plants* 26: 1003–1020.

2 Allele Mining for Molecular Breeding in Potato

Muhammad Naeem,[1] Caner Yavuz,[2] and Mehmet Emin Çalışkan[2]
[1]Agriculture Extension and Adaptive Research Wing, Punjab Agriculture Department, Pakistan
[2]Department of Agricultural Genetic Engineering, Faculty of Agricultural Sciences and Technologies Niğde Ömer Halisdemir University, Niğde, Türkiye

2.1 INTRODUCTION

The distribution and frequency of crop plants on planet earth has been altered by a process known as domestication. Domestication of crops is driven by anthropogenic force (humans) involving complex and dynamic processes of crop evolution. Crop domestication leads to genetic bottleneck (i.e., reduction of genetic diversity and decrease in the robustness of population leading to a drop in the fitter genes within the gene pool) and several traits of adaptive nature from wild plant relatives have been lost (Smýkal et al. 2018; Naeem et al. 2021). Plant breeders rely upon genetic variation in various crop species and the absence of variation means no possibility of selection for their steady improvement and poses a serious hurdle in the development of new improved cultivars. Over the past few decades, modern technologies, such as marker-assisted selection (MAS), genome sequencing technologies, genome wide association mapping (GWAS), genomic selection (GS), and allele mining (AM), have offered novel dynamics to assess the genetic diversity and unlock variation possessed by wild crop species. Allele mining refers to the identification and validation of a specific allele pertaining to the related phenotype connected with a trait of interest using next-generation sequencing (NGS) and polymerase chain reaction (PCR) technologies. This process discovers markers (for instance, single nucleotide polymorphisms – SNPs and/or insertion/deletions – InDels) in available genetic resources (viz., landraces, germplasm, obsolete varieties, mutant, and wild species) associated with various abiotic tolerance, biotic resistance, and quality characteristics (Premkumar et al. 2020). The hunt for allelic variation left behind during domestication was examined in germplasm collections to identify genes with known functions along with their DNA sequences for their use in breeding programs (Tiwari et al. 2017)

Potato is the third most important crop after rice and wheat in terms of human consumption (FAOSTAT 2022). Potato germplasm consists of wild and weedy relatives, natural hybrids, primitive cultivars, landraces, elite lines, obsolete varieties, mutants, breeding lines, polyploids, and systematically developed interspecific and intergeneric hybrids (Haussmann et al. 2004). The analysis of genetic diversity of potato germplasm gave insights into the temporal and geographic details of domestication. For instance, only a part of variation is possessed by modern-day cultivars as compared to those found in the Andean center (Brush et al. 1995). Most of the modern-day potato cultivars are separated from mid-20th century cultivars by only a few meiotic events (Naeem et al. 2021). Allele mining in potato involves transferring desirable alleles from wild potato relatives and landraces to high yielding genotypes without linkage drag. It entails assessment of genetic variation of desirable traits like biotic and abiotic tolerance/resistance and tuber quality characteristics,

DOI: 10.1201/9781003376439-2

FIGURE 2.1 Applications of allele mining in potato.

Source: Modified from Kumar et al. 2010.

investigation of gene evolution, synteny analysis, identification of new haplotypes and finally development of an allele specific marker for crop improvement and potato molecular breeding program (Figure 2.1). The present chapter focused on allele mining and its impact on molecular breeding of potato for traits of economic value.

2.2 ABIOTIC STRESS TOLERANCE

The introgression of stress tolerance genes in the cultivated germplasm is the basic principle of crop breeding to achieve abiotic stress tolerance in plants. A comparatively high productivity can be achieved with the augmentation of abiotic tolerance in plants. Potato is among the most vulnerable crops to abiotic stresses, however the huge genetic diversity of potato (reviewed by Aksoy et al. 2021) showed that potato landraces and wild relatives consist of potentially untapped and adapted genes under natural conditions to cope with climate adversities, without compromising growth and yield. Here, we emphasize on the allele mining in potato and highlight numerous genomic regions conferring tolerance against several abiotic factors, such as drought, salinity, heat, and cold stress.

2.2.1 Allele Mining for Drought Tolerance

Drought stress seriously affects physiological and biochemical processes in potato and results in a significant decrease in tuber yield and quality depending on severity and duration of the stress. Breeding drought tolerant cultivars is an effective strategy to overcome this constraint. To devise an efficient breeding program against a water deficit situation, it is advisable to consider the huge genetic diversity of potato that is greater in comparison to other crop species, viz., cotton, soybean, tomato. There are 98,285 potato accessions, which includes 20% landraces, 15% wild species, 14% advanced cultivars, 16% breeding lines, and 35% others (unknown or a mixture of two or more types) (Machida-Hirano 2015). Potato landraces such as *S. ajanhuiri*, *S. curtilobum*, *S. phureja*, *S. stenotomum*, and *S. tuberosum* Group *Andigenum*, along with wild potato relatives, including *S. acaule*, *S. brachycarpum*, *S. bulbocastanum*, *S. candolleanum*, *S. chacoense*, *S. demissum*, *S. fendleri*, *S. kurtzianum*, *S. microdontum*, *S. multidissectum*, *S. megistacrolobum*, *S. pinnatisectum*, *S. polytrichon*, *S. papita*, *S. polyadenium*, *S. sparsipilum*,

S. raphanifolium, S. spegazzinii, S. sanctae-rosae, S. stoloniferum, S. tarijense, S. verrucosum, and S. vernei possess genes/alleles conferring drought tolerance (Spooner and Hijmans 2001; Bradshaw and Ramsey 2005; Jansky et al. 2009; Machida-Hirano 2015; Aksoy et al. 2021). Allele mining in potato utilizing these sources of germplasm will result in the identification of novel alleles conferring drought tolerance. Unfortunately, allele mining for drought tolerance in potato is still at the nascent stage due to the polygenic effect of this trait. TILLING and Eco-TILLING techniques for drought tolerance in potato are rarely studied, however sequence-based studies utilizing a NGS platform can hunt alleles against drought stress.

A detailed investigation utilizing a sequence-based 12K SolCAP array was used for genotyping of drought tolerant wild potatoes *S. tarijense* and *S. chacoense* coupled with clones from CIP (International Potato Center, Peru) and INTA-Balcarce (Argentine National Institute of Agricultural Technology) for mining of genes/SNPs associated with drought tolerance in potato (Tagliotti et al. 2021). Five groups were revealed by Q+K (population structure and kinship) analysis, which confirmed the presence of genetic diversity in a panel, satisfactory for allele mining against a water deficit situation. Four SNPs namely solcap_snp_c1_13430 (T/C), solcap_snp_c1_10750 (C/T), solcap_snp_c2_12106 (C/T), and solcap_snp_c1_14992 (G/A) were identified positioned on chromosomes 1, 4, and 10. These SNPs were present within the candidate genes involved in starch and poly/oligosaccharide hydrolysis (*β-amylase*), triterpenoid and steroid biosynthesis (*cycloartenol synthase*), homeostatic balance in plants [*major facilitator superfamily with SPX (SYG1/Pho81/XPR1) domain-containing protein*], DNA repair and transport of proteins (*P-Loop containing nucleoside triphosphate hydrolase superfamily protein*). Two SNPs located on chromosome 1 [solcap_snp_c1_13430 (T/C)] and 4 [solcap_snp_c1_10750 (C/T)] were regarded as proline markers owing to their association with proline concentration under a water deficit environment (Tagliotti et al. 2021). The inclusion of wild potato relatives and clones from the North and South America (diversity centers) in a mapping population enhanced the effectiveness of SNPs discovered by Tagliotti et al. (2021) and holds the future for marker-assisted breeding of drought tolerance in potato. Similarly, *S. phureja* has been studied under drought conditions for identification of genomic loci/markers linked to drought tolerance, and it is a diploid landrace of potato commonly used for potato genome sequencing studies (PGSC 2011). A genotyping by sequencing (GBS) approach proves to be beneficial for mining of quantitative trait loci (QTLs) in a pool of *S. phureja* accessions derived from a breeding program at Universidad Nacional de Colombia (Díaz et al. 2021). Using a matrix of 47K SNPs in a mapping population of 104 diploid *S. phureja* accessions, 38 QTLs governing drought tolerance were identified following GBS and 2b-RAD (type IIB endonucleases restriction-site associated DNA) approaches (Juyo Rojas et al. 2019). In total, 22 candidate genes relevant to these SNPs/alleles have been revealed to be involved in several biological processes, viz., sugar transporter, proline transporter, transcription factor (*WRKY*), oxidative stress and heat shock protein. Two QTLs, i.e., *WDp9.15* and *WDp9.21* (60.57 Mbp) (C/T) located on chromosome 9 were associated with glucose and tuber weight under water deficit conditions, respectively (Díaz et al. 2021). Both these genomic regions colocalizes with *Glucan/water dikinase (GWD)* gene (Q9SAC6; Proteome: UP000006548). This gene control starch degradation, which provides necessary energy and carbon during the times of restricted photosynthetic activity under water stress conditions, thus enhancing drought tolerance (Szalonek et al. 2015). Allele mining for drought tolerant indices in mapping population of *S. phureja* accessions showed that QTLs *DTIp3.5*, *DTIp3.6*, and *DTIp5.7* were significantly associated with glucose content, with a gene annotation of *UDP-glucosyltranferase* (*UGT*), *Proline transporter* 2, and *MYB transcription factor* (*StMYB1R-1*) (Díaz et al. 2021). *UGT* is known to play a crucial role in abscisic acid (ABA) homeostasis under various stresses such as cold, drought, and salinity (Dong et al. 2014). Transcription factors such as *myb* is directly involved in the regulation of proline metabolism. Proline transporters (ProTs) interact with *myb* through cis-regulatory elements (CREs) binding sites and regulate cell signaling of proline and glycine betaine under drought stress (Akbudak and Filiz 2020).

Several transcription factors (TFs) have been mined from wild relatives and breeding populations in a search for superior alleles responding against drought stress. *StDREB* showed enhanced drought tolerance in potato due to its roles in enhancing hydrogen peroxide, total soluble sugars and free proline (Bouaziz et al. 2015). Similarly, *NAC* TFs performed pivotal role in improving osmotic tolerance in maize and Arabidopsis (Hu et al. 2006; Lu et al. 2012). Several MYB transcription factors are involved in plant secondary metabolism-driven responses against environmental stresses. The expression of *StMYB1R-1* has been up-regulated under drought stress resulting in enhanced drought tolerance in potato (*cv.* superior) (Chen et al. 2019). Mining of genomic regions pertaining to drought stress was done in a population of 103 European potato cultivars using association mapping approach. 14K Infinium array highlighted a significant region on chromosome 3 (PotVar0030768; scaffold PGSC0003DMB000000062) positioned at 55.67 Mbp. This SNP corresponds to marketable fraction of tuber fresh weight and tuber number under drought stress. This SNP can be integrated into a breeding program for enhanced tolerance against water deficit stress, since it contains pentatricopeptide repeat-containing protein (PPR) protein. A homolog of this gene family in Arabidopsis is involved in tolerance to abiotic stresses especially drought (Jiang et al. 2015; Aliche et al. 2019).

A candidate gene approach was developed by Schumacher et al. (2021), which result in the development and validation of functional simple sequence repeats (SSRs) markers for drought tolerance in potato. Six SSR markers (viz., HRO_ACS3_D, HRO_PP2C_1_B, HRO_ALDH_H, HRO_ETRTF_5a_D, and HRO_PARG_1A_B) showed allelic differences with respect to drought sensitivity. Thus, they can be integrated into the breeding programs to confer drought tolerance (Schumacher et al. 2021). The candidate gene approach can be validated in a pool of wild germplasm for marker-assisted enhancement of potato against drought stress. Several researchers believe that candidate genes, alleles, and casual polymorphic molecular markers would be potentially helpful to deal with abiotic stress factors especially drought (Szalonek et al. 2015; Krannich et al. 2015; Sprenger et al. 2016; Schumacher et al. 2021). An overview of the potential candidate genes significantly important under water deficit conditions is outlined in Table 2.1.

2.2.2 Allele Mining for Salinity Tolerance

Salinity in the vicinity of potato roots enhances osmotic pressure of soil water, reduces water infiltration, impairs soil structure, and resulted in restricted uptake of water by the roots. Moreover, uptake of other elements, such as chloride, sodium and boron, may cause toxicity (Levy and Veilleux 2007). Osmoprotectants (viz., glycine betaine and proline) and antioxidants (superoxide dismutase, glutathione, and catalase) are crucial in inducing salinity tolerance in potato (Blokhina et al. 2003; Koyro et al. 2012). Another strategy for limiting the uptake of toxic ions/elements is sequestering toxic elements in the vacuole and upregulating ion specific transporters at the tonoplast (Munns and Tester 2008; Munns et al. 2012).

Breeding approaches for improving salt tolerance in potato involves mining of tolerant alleles in a pool of wild relatives and primitive species. These offer tremendous potential in addressing the problem of salinity by providing a gene pool to be exploited in potato breeding for salinity tolerance. Andean potatoes possess good variability for salt tolerance. The trials carried out at CIP revealed seven accessions (10908–05, 10909–18, F20-1D, 9506–04, 11374–01, 10602–02, 9788–03) that were tolerant against salt stress including one wild species *Solanum juzepczukii*. These clones were derived through composite of crosses followed by selection among diploid *S. tuberosum* L., *S. chacoense*, *S. phureja* L., and *S. microdontum*, which depicts *in vitro* salt tolerance (Shaterian et al. 2008). CIP, Peru has 21 accessions, which are tolerant to saline conditions and can be requested for breeding purposes at https://genebank.cipotato.org/gringlobal/search.aspx. Furthermore, *in vitro* and greenhouse screening tests (under 80 mM NaCl and 120 mM NaCl) held at the Centre for Genetic Resources, The Netherlands (CGN) revealed four wild species to be tolerant against salinity stress.

TABLE 2.1
An Insight into Several Metabolic Processes Highlighting Important Candidate Genes for Molecular Breeding of Drought Tolerant Cultivars

Processes	Candidate Genes Involved (*Abbreviations followed by Expansions of Gene Abbreviations*)
Plant growth, development, and reproduction	*SLSG6* (S-locus specific glycoprotein S6), *LOX* (Lipoxygenase), *ARF6* (Auxin response factor 6), *SN-1* (Snaking-1), *NAM* (No apical meristem), *APY* (Apyrase), *APY-2, APY-3*
Plant hormones	*JA2* (Jasmonic acid 2), *EIX* (Ethylene-inducing xylanase), *StERF1, StERF3, StERF4, StERF5, StERF7, StERF8, StERF84, StERF130, StERF141* (Ethylene response factors), *JERF1* (Jasmonate and Ethylene Response Factor 1), *EIN2* (Ethylene insensitive 2), *ERS1* (Ethylene response sensor 1), *EREBP1* (Ethylene-responsive element-binding protein 1), *ETR1* (Ethylene receptor 1), *ETO1* (Ethylene-overproduction protein 1), *ETP1* (EIN2 targeting protein 1), *EIL2* (Ethylene-Insensitive3-Like 2), *ACS2, ACS3* (1-Aminocyclopropane-1-carboxylic acid synthase 2), *ACO* (1-Aminocyclopropane-1-carboxylic acid oxidase), *PYL4* (Pyrabactin resistance-like 4), *PP2C* (Protein phosphatase-2C), *ABA8H, ABA8H4* (ABA 8'-hydroxylase 4), *KAO* (Ent-kaurenoic acid oxidase), *GA3ox* (Gibberellin 3-oxidase), *MOCOS* (Molybdenum Cofactor sulfurase), *ZEP* (Zeaxanthin epoxidase), *NCED* (9-Cis-epoxycarotenoid dioxygenase)
Protection of cellular structures	*LEA5* (Late embryogenesis abundant 5)
Osmolytes	*BADH* (Betaine aldehyde dehydrogenase), *PRODH* (Proline dehydrogenase), *P5CS* (Pyrroline-5-carboxylate synthase), *P5CR* (Pyrroline-5-carboxylate reductase), *SCR* (SCARECROW), *TPS1* (Trehalose-6-phosphate synthase 1)
Detoxification/ DNA repair	*APX* (Ascorbate peroxidase), *SOD* (Superoxide dismutase), *GR* (Glutathione reductase), *ALDH* (Aldehyde dehydrogenase), *PARG* (Poly (ADP-ribose) glycohydrolase), P-Loop NTase (P-Loop containing nucleoside triphosphate hydrolase superfamily protein)
Defense/transport	*FLS2* (Flagellin-sensing 2), *TNL-DR* (TIR-NBS-LRR disease resistance), *VPSP* (Vacuolar protein sorting protein), *GST* (Glutathione-S-transferase), *BED* (BEAF and DREF), *NBS–LRR protein* (Nucleotide binding site–Leucine-rich repeat,), *MRP-ABC transporter* (Multidrug resistance protein ATP binding cassette)
Kinases/phosphatases	*LRR* (Leucine rich repeats), *STPK, STPK-plant type* (Serine/threonine protein kinase), *SIPP* (Soluble inorganic pyrophosphatase), *RPK* (Receptor protein kinase), *NDPK2* (Nucleoside diphosphate protein kinase 2), *KIN* (Kinase)
Transcription factors	*DBP4* (DNA binding protein 4), *WRKY* (WRKY transcription factor), *StDREB1* (Dehydration-Responsive Element-Binding 1), *STERF186* (Ethylene response factor 186)
CHO metabolism/ cell wall component/ Hydrolysis of poly/ oligosaccharides	*Extensin, UGPase* (UDP-glucose-1-phosphate uridylyltransferase), *BA, GT, UGGT* (UDP-glucose glycosyltransferase), *BDGEH* (β-D-glucan exohydrolase), *NI* (Neutral invertase), *BAM* (β-amylase)
Protein biosynthesis/ modification	*SBT4B* (subtilase 4B), *MetRS* (Methionyl-tRNA synthetase)
Calcium and phospholipid binding proteins	*STANN1* (Annexins 1)
Triterpenoid and steroid biosynthesis	*CAS* (cycloartenol synthase)

Source: Szalonek et al. 2015; Krannich et al. 2015; Sprenger et al. 2016; Schumacher et al. 2021.

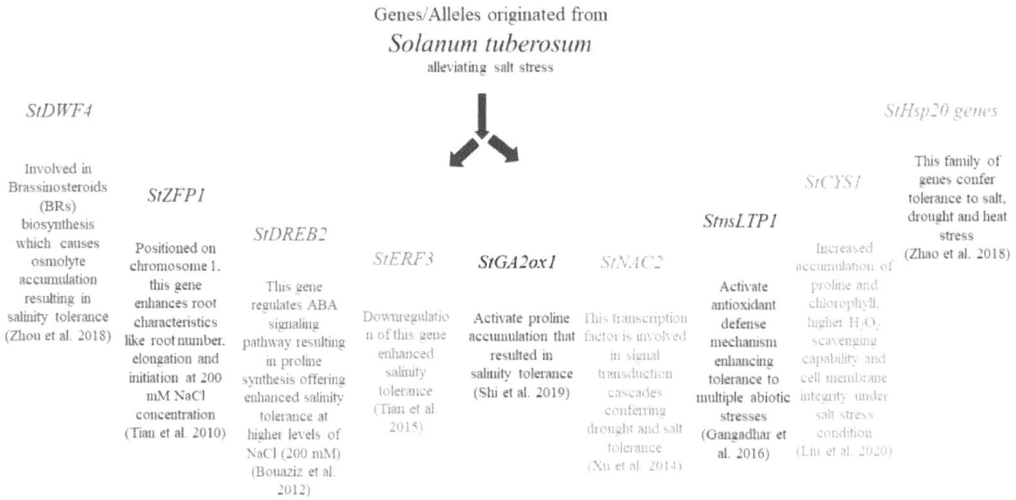

FIGURE 2.2 Significant genes in *S. tuberosum* pertaining to their roles in the alleviation of salt stress. *St: Solanum tuberosum*; *DWF4* (*DWARF4*); *ZFP1* (*Zinc finger protein 1*); *DREB2* (*dehydration-responsive element-binding protein 2*); *ERF3* (*ethylene response transcription factor 3*); *GA2ox1* (*Gibberellin 2-beta-dioxygenase 1*); *NAC2* (*NAC transcription factor*); *StnsLTP1* (*non-specific Lipid Transfer Protein-1*); *CYS1* (*cystatin 1*); *HSP20* (*heat-shock protein 20*).

It includes *S. sparsipilum* (CGN#18099), *S. chacoense* (CGN#17898), *S. vernei* (CGN#17836), and *S. gourlayi* (CGN#17853) (Shaterian et al. 2008). The tests also revealed moderate tolerance of *S. tarijense* to saline conditions. A list of gene banks holding wild potato relatives were outlined by Aksoy et al. (2021). The halophytic potential of wild potato relatives and various accessions from Bolivia, Mexico, Argentina, and Peru were evaluated under *in vitro* and greenhouse conditions (Huaman et al. 2000), but large-scale field testing is still at preliminary stages.

Allele mining studies for salt tolerance in potato are limited, however, several salt tolerance genes have been identified that can be successfully introgressed into cultivated species. As we know that QTLs contain several hundred genes that could be of high interest for breeders, therefore mining of candidate genes responsible for salinity tolerance is crucial. Many of these genes/alleles have been harbored by wild potato accessions. A snapshot overview of important genes originating from potato along with their function under salt stress conditions has been outlined in Figure 2.2. These candidate genes code for functional proteins such as glyceraldehyde-3-phosphate dehydrogenase (GPD), proline and trehalose synthesis protein and osmotin-like protein that perform vital regulatory functions (Byun et al. 2007; Chourasia et al. 2021; Bouaziz et al. 2012; Gangadhar et al. 2016; Liu et al. 2020; Shi et al. 2019; Tian et al. 2015; Tian et al. 2010).

2.2.3 Allele Mining for Heat Tolerance

Elevated temperature suppresses tuberization and minimizes the partitioning of assimilated carbon to tuber starch (Hancock et al. 2014). While, potato is a cool season crop, there exists considerable variation among potato clones and cultivars in response to heat stress. Wild potato relatives and landraces offers tolerance to elevated temperatures. For instance, *S. sogarandinum*, *S. kurtzianum*, *S. berthaultii*, *S. jamesii*, *S. pinnatisectum*, *S. stoloniferum*, *S. chacoense*, *S. phureja*, *S. demissum*. *S. tarijense*, *S. verrucosum*, and *S. vernei* can be effectively utilized in potato breeding programs for making heat-tolerant cultivars (Muthoni and Kabira 2015; Guedes et al. 2019). A diploid 06H1 population derived from a cross between *S. tuberosum* and *S. phureja* revealed four allelic variants (A1, A2, A3, A4) following *HSc70* CAPS (Cleaved Amplified Polymorphic Sequences) assay. A2 *HSc70* allele confers tolerance to high temperature stress in potato. It is noted that an upstream region of 495 bp from putative ATG start codon, in allele A2 consists of unique TA repeats, implying that the phenotype A2 results from variations in expression patterns under high temperature stress (Trapero-Mozos et al. 2018). The gene *HSc70* (heat-shock cognate protein 70) belongs to the HSP70 (heat-shock protein 70) family. *HSc70* is categorized with highly conserved SBDs (substrate binding domains) and ATPase, thus connected with the acquisition of thermotolerance in several crop species viz., Arabidopsis, cabbage, and potato (Sung and Guy 2003; Mayer 2013; Park et al. 2013; Trapero-Mozos et al. 2018). A recent investigation showed that the *HSc70* gene is correlated with tolerance to elevated temperature in diverse wild potato relatives. Interestingly, this gene is only 0.69 Mbp way from a SNP marker (Solcap_snp_c2_11571), which is located at 7.1 Mbp on chromosome 4. This region is quite significant in terms of tuber yield in addition to providing heat tolerance. *HSc70* gene provides protection against photooxidative injury to photosystem II (PS II) induced by abiotic stress. It is one of the adaptive features possessed by wild potato relatives (like *S. phureja*, *S. jamesii*, *S. pinnatisectum*) to survive in agroecological zones with elevated temperatures (Campbell et al. 2022). It may be concluded here that potato genotypes carrying *HSc70* A2 allele were protected against heat stress and maintains high tuber yield despite the stress. Furthermore, the promoter region of the *HSc70* gene carry TA repeat elements responsible for elevated *HSc70* expression levels.

GWAS analysis coupled with the candidate gene approach resulted in the identification of an important genomic region in a tetraploid population of PIN184 breeding clones. Chromosome 5 (solcap_snp_c2_11924) possess *FRUCTOSE-BISPHOSPHATE ALDOLASE* (*FBA*) genes that perform functional roles in various metabolic pathways, viz., gluconeogenesis, glycolysis, and Calvin cycle. FBA gene in Arabidopsis showed differential expression patterns under abiotic stress. So, this gene may play a crucial role in heat tolerance (Mosquera et al. 2016). A 4x-2x hybrid (*S. tuberosum* and *phureja-stenotomum*) was utilized to find out the QTLs responsible for internal heat necrosis and highlighted their presence on chromosomes 4, 5 and linkage groups 7 and 10 (McCord et al. 2011). While another study suggested SNPs positioned on chromosomes 1, 5, 9, and 12 to be significantly linked with internal heat necrosis in a mapping population derived from USDA-ARS Beltsville potato breeding program (Schumann et al. 2017). In potato, genes like *HEXOKINASE*, and *SP6A* provide heat tolerance through enhancing tuberization by assimilating photo assimilates to the developing tubers under heat and drought stress (Lehretz et al. 2021). *ERTF RAP2-7-like* gene, acts as a negative regulator of both drought and heat stress in potato. This gene helps in the regulation of DNA-binding transcription factor activity during single and combined abiotic stresses and involved in transitions between developmental stages of the shoot apical meristem. *ERTF RAP2-7-like* gene is suppressed by an overexpression of miR172b, enhancing the activity of immune receptor gene *FLAGELLIN-SENSING2* (*FLS2*) conferring tolerance against heat and drought (Asim et al. 2021). Genes which negatively regulate tuberization includes *StPHYB*, *StSUT4*, *StCO*, *StSP5G*, *StTOC1*, *StRAP1*, whilst *StTFL1*, *StSP6A*, *StBEL5*, *StPOTH1*, *StPA2Ac*, *POTM1*, *miR172*, *StCDF*, *StTUB7*, *StTUB19*, *AtABF4*, *AtABF2* are known to positively regulate tuberization. Additionally, these genes respond indirectly or directly to heat stress (Singh et al. 2015; Dutt et al. 2017; Morris et al. 2019; Singh et al. 2020). Mining of beneficial alleles offering resilience to heat stress required continuous efforts from both the potato breeders and geneticists at the universities and research institutes to develop heat tolerant cultivars during the era of global warming.

2.2.4 Allele Mining for Cold/Frost Tolerance

Potato is sensitive to damage by freezing temperatures and frost. No variety of *S. tuberosum* grown under temperate conditions can withstand freezing temperatures of –5°C to –2.8°C. However, wild species of potato possesses considerable tolerance to frost and freezing temperatures as low as –5°C such as *S. acaule* (Chang et al. 2014). A detail investigation regarding the assessment of frost tolerance of wild potato species pertaining to the geographical area and ecological factors revealed that species found in south and central Peruvian Andes, central Chilean Andes and lowlands of Argentina have high level of frost tolerance. It was also hypothesized that wild potatoes exhibited higher frost tolerance in ecological zones with average minimum temperatures of 3°C as compared to warmer regions. The wild potatoes commonly surviving in these geographical and ecological niches includes *S. acaule*, *S. albicans*, *S. commersonii*, *S. demissum*, *S. etuberosum*, *S. boliviense*, *S. sogarandinum*, *S. acroscopicum*, *S. raphanifolium*, *S. bukasovii*, *S. tarnii*, *S. palustre*, *S. megistacrolonum*, *S. malmeanum*, *S. sanctae-rosae*, *S. paucissectum*, *S. vidaurrei*, and *S. vernei* (Hijmans et al. 2003; Luthra et al. 2007). It is believed that all these wild potato relatives contain valuable allelic sources for breeding of frost tolerance potato cultivars.

Though allele mining for frost/cold tolerance is still lagging behind, but a recent research attempted for the introgression of freezing tolerance capacity of *S. malmeanum* into dihaploid *S. tuberosum* through protoplast fusion. Somatic hybrids were then backcrossed with the tetraploid cultivars to obtain BC1 progeny (i.e., FT069-16, FT069-32, FT070-22, FT070-1, FT071-69, and FT071-57) that offer strong freezing tolerance without affecting the tuberization capacity. Hence, both freezing tolerance and tuberization were controlled by independent genetic loci (Tu et al. 2021). *S. brevidens* and *S. commersonii* were also utilized for introgression of freezing tolerance into cultivated potato (Cardi et al. 1993; Nyman and Waara 1997). Several researchers discussed the significance of *S. commersonii* in enhancing freezing tolerance. Dehydrins (*DHN2*) and chaperones (*CPN60b*) were expressed under cold acclimated *S. commersonii* plants (De Palma et al. 2008). The frost tolerance capability of *S. commersonii* is attributed to 126 cold related genes that lacked in *S. tuberosum. CBF* (C-repeat binding factor) genes such as, *CBF1* and *CBF3* originated from *S. commersonii* were most responsive to the acclimation response during cold stress. Likewise, cold responsive genes viz. *COR15a* and *COR413* were upregulated in *S. commersonii* in response to cold stress and confer cold acclimation. *COR* genes were regulated by ABA (Xiong et al. 2002). Other significant genes, which actively contribute to freezing tolerance in *S. commersonii*, includes transcription factor belonging to the *AP2/ERF* (apetala 2/ethylene-response element binding factor) domain, *CDPK7* (*Ca^{2+}-DEPENDENT PROTEIN KINASES 7*), *CDPK17*, *CDPK19*, *CIPK1* (*CBL INTERACTING PROTEIN KINASE 1*), *CIPK3*, *CIPK23*, and *GOLS1* (galactinol synthase). Genome sequencing along with candidate gene mining of *S. commersonii* provides new knowledge about the diversity which exists within *Solanum* genus combined with direction to translate this knowledge for the development of cold tolerant potato varieties (Aversano et al. 2015).

A differential response was observed between *S. tuberosum* and *S. phureja* accessions towards cold acclimation. The ability to tolerate cold and frost environment depends upon modifications in carbohydrate and amino acid metabolism, antioxidant defense mechanism, role of transcription factors, and various plant hormones involved in signal transduction pathways. For instance, sucrose synthase, arginine decarboxylase, histidine decarboxylase, methionine S-methyltransferase, S-adenosylmethionine decarboxylase (*SAMDC*) genes were mostly upregulated in *S. phureja* upon cold exposure and during cold acclimation (Oufir et al. 2008). *Solanum acaule* confer tolerance against frost due to the presence of *SaCBL1-like* (calcineurin B-like protein). This gene interacts with *SaCIPK3* to enhance cold acclimatization of potato (Chen et al. 2022). The cold acclimation potential of *S. acaule* was utilized for cloning and transformation of *SaADC1* (arginine decarboxylase) gene in *S. tuberosum,* that causes enhanced accumulation of putrescine, functioned in cold-acclimated freezing tolerance. Putrescine is responsible for improving membrane stability by scavenging free radicals and maintain ROS homeostasis through antioxidant defense system. Moreover, *SaADC1* increases the expression of highly responsive cold acclimation group of genes known as *CBF* genes in potato (Kou et al. 2018). Thus, our current understanding about allele mining for frost/cold tolerance in potato suggests that wild species of potatoes constitute a valuable resource for mining of genes/alleles associated with freezing tolerance. The identification of several genomic regions and genetic resources linked with cold tolerance will improve breeding of frost hardy potato cultivars.

2.3 BIOTIC STRESS RESISTANCE

Potato plants are affected by different sources of pathogens, which can lead to symptom or symptomless effects in the crops, yet eventually a significant number of these biotic stress factors are causing great yield losses in the production and threaten the food security with other environmental stress factors ex. abiotic stress. Plants have adapted a means of strategies for resistance to these biotic stress factors with two prominent resistance ways namely vertical resistance and horizontal resistance. In vertical resistance, the resistance is highly specific and can confer resistance to a single pathogen's race unlike horizontal resistance where the resistance is not specific and is for a group of pathogens. Vertical resistance are qualitative traits showing monogenic inheritance and ensures shorter and unstable resistance against pathogens while horizontal resistance are quantitative traits showing polygenic inheritance and provides durable resistance (Clair 2010; Spoel and Dong 2012; Keane 2012). Scientists exploit both approaches as combinatorial method to control and combat pathogen infection in plants (van der Plank 1966).

Resistance gene (*R*) in the host, directly or indirectly, trigger the signaling for ensuring the defense in plant against avirulence (*Avr*) genes in the pathogen and this is how these *R* genes are capturing the overwhelming interest of the scientists (Hammond-Kosack and Jones 1997). These genes play key roles to define specific alleles responsible during resistance to an infection. The resistance mechanism can be classified into two different groups, extreme resistance and hypersensitive response (Keane 2012). Extreme resistance (ER) appears to be mainly symptomless and characterized with restricted pathogen survival including the replication and movement while hypersensitive response (HR) is a preventive mechanism for pathogen attack/spread and it causes necrotic symptoms in a certain tissue to avoid further pathogen infection (Ross et al. 2021).

Allele mining is having a great impact on potato plant breeding and implementing modern techniques for sustaining the efficient agricultural productivity particularly for biotic stress (Figure 2.3). Several alleles and resistance genes (*R*) have been explored responsible of biotic stress with previous works and many studies are still ongoing to elucidate further alleles for potato (Wang et al. 2008). Many of these alleles have been already used for developing molecular markers to screen for resistance or susceptibility against especially for late blight (Tiwari et al. 2013), potato virus Y (PVY) (Kasai et al. 2000; Heldak et al. 2007), cyst nematode (Qin et al. 2000). Today, different platforms are publicly available such as a database named Sol*R*gene, using different approaches, including allele mining, provides information about *R* genes for a diverse group of pathogens but particularly for late blight disease in potato and in other tuberous crops and help in allele mining study efforts for biotic stress in potato (Butterbach et al. 2007; Vleeshouwers et al. 2011).

Advances in NGS technologies have contributed to allele mining studies in potato to a great extent both as a single main technique or a combined technique with another method such as NBS (nucleotide binding site) profiling and tagging. This is considered as sequence-based allele mining approach (Ashkani et al. 2015). Providing a cheaper and faster alternative to former sequencing techniques (ex. Sanger sequencing) enable a mass re-sequencing alternative for the crops particularly for allele mining works (Ashkani et al. 2015). An initial example to allele mining in plants using one of NGS techniques was in Arabidopsis for the *NBS-LRR* genes in 2011 (Guo et al. 2011).

NBS (nucleotide binding site–Leucine rich repeat) profiling is considered as one of the most essential and common allele mining approach in potato for biotic stress studies to find out *R* genes (van der Linden et al. 2004; Vossen et al. 2014; Brugmans et al. 2008). These highly conserved *NBS-LRR* regions in *R* genes have been effectively used for NBS profiling by sequencing in *Solanum* species to find out *R* gene analogues or fragments (RGA). Degenerate primers are used to target conserved motifs, *Ploop*, *Kinase 2*, and *GLPL*, within this region to identify homology between

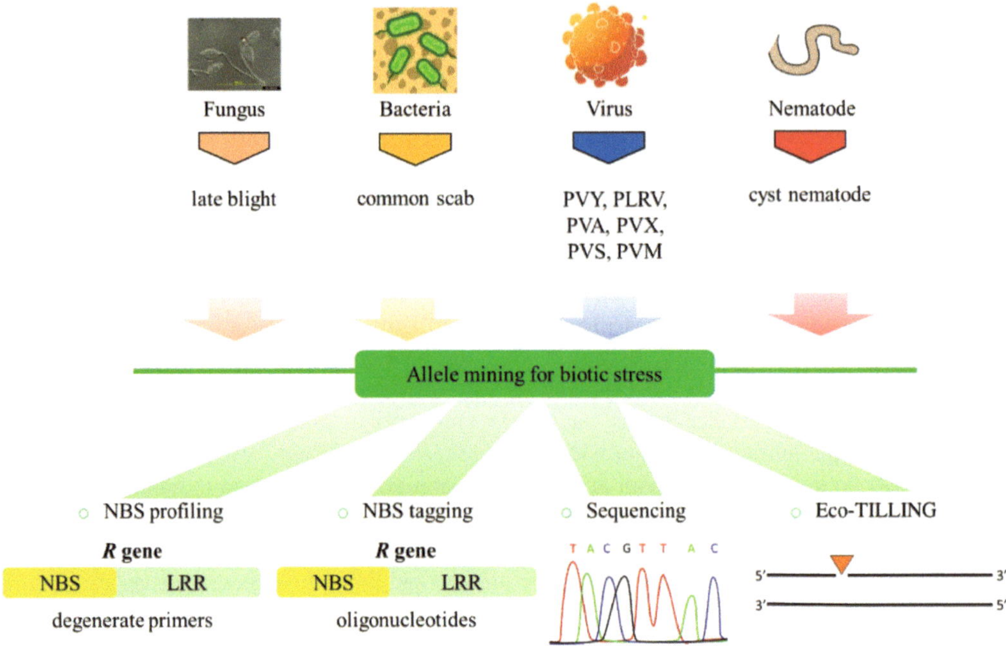

FIGURE 2.3 Allele mining for biotic stress in potato (*R* gene sketches modified from Prakash et al. 2020). (Permission for publication was taken from forestry images, Bruce Watt, University of Maine, Bugwood.org for fungus photo, upper left side.)

different sequences. Later, Prakash and the colleagues (2020) developed a system, which is called NBS tag. This method is close to methodology of NBS profiling however oligonucleotides are used unlike degenerate primers. The further steps in methodology also include an NGS technology, which is called Illumina HiSeq-2000 (Prakash et al. 2020). Scientists identified about 587 NBS domains using this NBS tagging approach, deposited in SolariX webpage (publicly available) (Prakash et al. 2020). These two NBS-based systems are widely applicable for *R* gene variant detection in the potato for a means of biotic stress factors.

Eco-TILLING is another strategy for determining the allelic variations in plants however, studies in potato are very less (Elias et al. 2009). It is simply based on reverse genetic approach (Bharathi Raja et al. 2016). In Eco-TILLING, the mutations because of a natural process are detected in a population unlike TILLING where a mutagen is applied externally to create mutant plants (Till et al. 2007; Pankin et al. 2011).

2.3.1 Allele Mining for Late Blight Resistance

The first studies in potato for allele mining started with identifying alleles for late blight resistance against *Phytophthora infestans* and many of the existing works in potato allele mining subject particularly focused on late blight in depth. Previous efforts showed that resistance could be achieved in cultivated potato plants by introgression of *Rpi* gene(s) from wild *Solanum* species however the resistance was observed to be not durable (Tiwari et al. 2015; Paluchowska et al. 2022). A total of 70 genes or QTL regions in 32 different *Solanum species* were identified so far however a very limited number of those genes, i.e., *R1, R2, R2-like, R3a, Rpiblb1*, *Rpiblb2, Rpi-blb3* and *Rpi-abpt, R2* were cloned (see detailed Table 2.1 in Paluchowska et al., 2022 review article) (Ballvora et al. 2002; Song et al. 2003; Huang et al. 2005; van der Vossen et al. 2005; Lokossou et al. 2009).

The studies are very few about allele mining in wild *Solanum* species for late blight resistance. There are several *R* genes, *R1, R2, R3, Rpi-blb1/RB, Rpi-blb2, Rpi-blb3, Rpi-vnt1, Rpi-abpt, Ph-3*, which have been characterized and studied well in late blight disease in potato (Ballvora et al. 2002; Park et al. 2005; Pel et al. 2009; Lokossou et al. 2009; Khavkin et al. 2010; Mosquera et al. 2016). Today, scientists are using some of these *R* genes for screening resistance in different populations (Trognitz and Trognitz 2007). *R* genes can have different alleles in several cases as for *R3* locus it has two *R3a* and *R3b* (Huang vd. 2005; Li vd. 2011). In an earlier work by Wang et al. (2008) found out *Rpi-blb1* homologous, *Rpi-sto1, Rpi-plt1, Rpi-pta1*, and *Rpi-pta2*, in *Solanum stoloniferum* and three main haplotypes were proposed having total six SNPs at different positions. This study also monitored *Rpi-blb2* homologous in the same species however the results did not show any homologous gene (Wang et al. 2008). In another research, for *Rpi-blb1* and *Rpi-blb2* genes, different *RXLR* effectors have been characterized using allele mining by PCR and sequencing. The study showed that 18 *RXLR* effectors have had amino acid changes (synonymous and non-synonymous substitutions, premature stop codon). *in planta* expression of *RXLR* effectors grouped *PexRD6/IpiO* for *Rpi-blb1* and *PexRD39/PexRD40* effectors for *Rpi-blb2* gene (Oh et al. 2009). A resistance gene in chromosome 8, *RB*, from *Solanum bulbocastanum* has been studied for the allelic variants using allele mining. The comparison of *RB* in resistant plants and *rb* in susceptible plants showed an earlier stop codon in Tyr-454 in susceptible plants. *RB* gene was cloned into potato cultivar, Katahdin. Transgenic plants showed resistance against late blight (Song et al. 2003). In a study, researchers used NBS profiling and *Tm-22*-based approach to characterize *Rpi* genes, *Rpi-vnt1.1*, and *Rpi-vnt1.3*, in *Solanum venturii*. The difference between these two alleles explained by an insertional mutation in N-terminal and LRR region (Pel et al. 2009). Association mapping work in different *Solanum tuberosum* group *phureja* accessions identified two candidate SNPs, a thylakoid protein and a glycoprotein, associated with late blight resistance and these two SNPs mapped on sixth chromosome along with *Rpi-blb2* (Alvarez et al. 2017). Another work focused on determining the *RB/Rpi-blb1* homologs where the study revealed that some *Solanum* species and different accessions had a unique locus-specific region, *RB-629* and *RB-226* (Pankin et al. 2011). The exonic regions were more conserved than intronic regions in these homologoues sequences. They showed that *Solanum* accessions clustered into four groups when the phylogenetic tree was constructed with these two alleles (Pankin et al. 2011). NBS profiling identified *R* genes, *Rpi-avl1* and *Rpi-cap1, Rpi-qum1*, in a diploid species, *Solanum avilesii*, and in *Solanum circaeifolium*, respectively, in chromosome 11 (Verzaux et al. 2011, 2012). Another study mapped *R8* gene in chromosome 9 in *Solanum demissum* with NBS profiling (Jo et al. 2011). Map-based cloning revealed a QTL region, *dPI09c*, which has been shown to have a similar resistance mechanism as *R8* gene in *Solanum demissum* (Jiang et al. 2018). Nine different SNPs in *R* gene against late blight disease have been explored in chromosome 9 with NBS tagging however a particular locus showing resistance against late blight is not observed. The same study focused on SNPs in two different *R* genes in chromosome 4 and 11 in a different population (Prakash et al. 2020). A GWAS work for late blight disease in potato revealed >80,000 SNPs of which a few (about 15) were present in organ-specific QTLs, including one *LRR* gene as well, such as mainly in leaf and stem (Rojas et al. 2019).

Besides allele mining in *R* genes for late blight resistance, there are a couple of studies in which several genes, unlike *R* gene, were involved in pathogen infection. Several studies are reporting novel QTLs as well (Collins vd. 1999; Rojas vd. 2019; Lindqvist-Kreuze vd. 2021). Scientists screened tetraploid potato cultivars for ~200 SNPs and ~150 alleles and they revealed that the natural variation in allene oxide synthase 2 gene (*StAOS2*) can mainly be responsible for late blight resistance in the field namely field resistance. Three different SNPs (*StAOS2_snp691, StAOS2_snp692*, and *StAOS2_snp727*) were found to be highly associated in this work (Pajerowska-Mukhtar et al. 2009). The same group further identified five different allelic variation of the *StAOS2* gene (*StAOS2-1, StAOS2-6, StAOS2-7, StAOS2-8*, and *StAOS2-12*) in 2008 yet in diploid potato population. They expressed these five different alleles in *aos* mutant *Arabidopsis* lines for complementation analysis and they observed resistance against late blight disease. These alleles were shown to have a key function also for a type of bacterial infection, *Erwinia carotovora ssp. atroseptica* (Pajerowska-Mukhtar et al. 2008). Genes involved in resistance against late blight could have a further role in other mechanisms as well, such as in abiotic stress. Pleitropic resistance genes, *StPDR1, StPDR2, StPDR3,* and *StPDR4*, are one example to this yet, also showed very high expression after *Phytophthora infestans* infection particularly with *StPDR1, StPDR2* (Ruocco et al. 2011). A GWAS work using 8.3k SolCAP SNP array and RNASeq in potato identified SNPs present in a number of genes, which are mainly involved in different metabolic pathways, jasmonate, terpene biosynthesis, and transcription factor upon *P. infestans* infection (Mosquera et al. 2016).

2.3.2 Allele Mining for Virus Resistance

Viruses are the most common disease factors in potato. Vegetative propagation spread this disease easily to next generations through seed tubers (Andrade et al. 2009). Today it is known that some of these viruses can be damaging for plant and for tuber as well. The tuber yield in infected plants can be decreased by almost 80% (Tsedaley 2015; Taye 2016). The degree of yield loss, absolutely, varies with virus type and other external parameters. There are >30 viruses, which can infect potato. Studies revealed that *Potato virus Y* (PVY), *Potato leafroll virus* (PLRV) are the most damaging types and the other most widespread viruses besides PVY and PLRV are *Potato virus A* (PVA), *Potato virus X* (PVX), *Potato virus S* (PVS), and *Potato virus M* (PVM) yet with comparably less effect on the yield loss, ranging 5–40% (Ahmadvand et al. 2012; Lacomme and Jacquot 2017). Some of these viruses can have serious symptoms and some may have symptomless effects in plants (Ahmadvand et al. 2012). These viruses can have different strains, i.e., PVY^O, PVY^C, PVY^N for PVY, which make the disease control more difficult (Ahmadvand et al. 2012). Breeding efforts introduced several PVY resistant cultivars, albeit they could not meet the market needs particularly for quality traits (Duan et al. 2012). Unfortunately, chemical application is also not an effective approach for avoiding or minimizing the infection or the symptoms of viral diseases (Boquel vd. 2015).

PVY having different strains (about seven) distributed in different geographical regions is the most destructive virus type in potato leading to severe necrotic lesions and three main alleles, *Ryadg, Rychc, Rysto*, from four different wild *Solanum* accessions are involved in PVY extreme resistance (ER) (Zoteyeva et al. 2014; Palma et al. 2015; Lacomme and Jacquot 2017). The hypersensitive response (HR), which makes the other resistance type against PVY, has been provided by *N* genes in potato yet, this type of resistance has limitations and is also not durable (Karasev and Gray 2013; Vidal et al. 2002). Albeit, the study conducted previously identified and highlighted SNPs in $Ny_{(o,n)}sto$ from *S. stoloniferum* against two particularly different PVY strains namely PVY^O and PVY^{NTN} using a GWAS-based approach (van Eck et al. 2017). The same study even further provided several SNPs in *Gpa5* locus for cyst nematode (van Eck et al. 2017). Moreover, some of the ER and HR genes have been reported that they could have epistatic effect (*Rysto*) even if they act independently (Grech-Baran et al. 2020). Introgression of these alleles from wild accession into cultivars could help to study resistance against PVY, however, this alone is not enough to minimize the infection and its spread in plants (Caruana et al. 2021). The extreme resistance genes have been extensively studied in potato. *Ryadg* locus has been mapped on chromosome 11 and 12 in *Solanum andigenum* (Hamalainen et al. 1997; Herrera et al. 2018). The resistance in *Ryadg* cultivars is not strain specific so that it can confer a broad-spectrum resistance against PVY infection (Sato et al. 2006). Several molecular markers have been developed such as *ADG2 BbvI, RYSC3, RYSC4, M6* to screen *Ryadg* alleles in different potato germplasms (Sorri et al. 1999; Kasai et al. 2000; Herrera et al. 2018; Slater et al. 2020; Tiwari et al. 2013). *Rychc* from *Solanum chacoense* has been mapped on chromosome 9. This resistance is making only 5% of extreme resistance to PVY and it is mostly introduced into and present in Japan cultivars, i.e., Konafubuki, Harimaru (Sato et al. 2006; Elison et al. 2021; Fujimatsu et al. 2018; Biryukova et al. 2019). Potato populations can be screened by

Ry186 and *Ry364* marker to find out *Rychc* dependent resistance in the plants and more molecular markers are available for this gene (Mori et al. 2011; Hosaka et al. 2001; Sato et al. 2006). *Rysto* and *Ry-fsto* alleles from *Solanum stoloniferum* is mapped on chromosome 12 for *Rysto* gene and is providing extreme resistance to PVY besides *Ryadg* and *Rychc* (Grech-Baran et al. 2020). This gene has been particularly introgressed into European cultivars (Valkonen et al. 2008). Besides *R* genes for ER to PVY, a number of genes, *eIF4E1, eIF4E1* variant *eva1, Pvr4* and *TPN1*, has been also shown to be involved in PVY resistance however *Pvr4* and *TPN1* genes have been characterized in only pepper yet, it could be promising for improving PVY resistance in potato as well (Duan et al. 2012; Lucioli et al. 2022; Grech-Baran et al. 2020). Plants have become resistant against certain strains such as PVY^{NTN} when *eIF4E1* gene (translation initiation factor) has been knocked out using CRISPR-Cas9 improving hypersensitive response in cv. Desiree (Lucioli et al. 2022). Yeast two-hybrid assays in wild pepper populations revealed substitutions in amino acid sequences for several alleles for *eIF4E1* cap region and these amino acid changes have been thought to play an important role for resistance in PVY and other potyviruses (Poulicard et al. 2016; Atarashi et al. 2020). It is estimated that potato is also likely to have two alleles of *eIF4E1* gene (Lucioli et al. 2022). A CAPS marker, chr09_11PagI, has been developed for PVY resistance screening using NBS tagging (Prakash et al. 2020). Allele mining for PVY has been thoroughly studied in other *Solanacea* family members as well (Rubio et al. 2009).

PLRV (potato leaf roll virus) is another virus type in potato and belongs to the *Luteoviridae* family and can cause more than 80% yield losses worldwide today (Abbas et al. 2016). Natural resistance can be observed in several wild *Solanum* accessions, such as *Solanum demissum, Solanum brevidens* (Mihovilovich et al. 1999). A few different loci, *PLRV.1, PLRV.2, PLRV.3,* and *PLRV.4,* has been reported in diploid potato clones particularly in chromosome 5, 6, and 11 using QTL mapping (Marczewski et al. 2001a; Marczewski et al. 2004). RIadg gene has been mapped on chromosome 5 in a dihaploid population (Velasquez et al. 2007). Besides, two different alleles, *UBC864600* and *UBC864800*, were explored for *PLRV.4* locus (Marczewski et al. 2004). An earlier work provided that mutant alleles of *pr17* gene (a phloem protein) (modifications in N- and C-termini) showed a reduced damage when they were infected with PLRV (Tacke et al. 1996). *R* gene in chromosome 4, *Rlretb*, from *Solanum etuberosum* has provided resistance against PLRV infection in introgressed cultivars (Kuhl et al. 2016). A number of markers, *Nl271164, E35M48192,* and *RGASC850*, particularly tightly linked to *Rladg* loci from LOP-868 accession, has been developed for PLRV screening in potato populations (Marczewski et al. 2001a; Velásquez et al. 2007; Mihovilovich et al. 2014; Byarugaba et al. 2021).

PVX, a member of Potexvirus, is another common potato virus yet, mildly affecting the yield, 10–40% however farmers can end up with more yield reductions in potato as a result of the combinatorial acting of PVX with PVY or PVA virus (Kreuze et al. 2020). There are four classified groups of PVX strains, *X1, X2, X3,* and *X4*, infecting potato (Cockerham 1955, 1970; Huang 2010). *Nb* gene in chromosome 5, as a common location with many other resistance genes, is responsible for hypersensitive response against PVX in potato and the same work suggested that *Rx2*, which is providing ER type resistance, and Nb genes are tightly linked (De Jong et al. 1997). *Rx1* and *Rx2* are two main genes, which are highly similar at amino acid level controlling ER type resistance to PVX in potato (Ritter et al. 1991; Bendahmane et al. 1995; Bendahmane et al. 2000; Ahmadvand et al. 2013). These two alleles are localized in chromosome 5 and 12, respectively (Ritter et al. 1991). Ahmadvand and the colleagues (2013) developed two markers, *5Rx1* and *106Rx2*, to differentiate the presence/absence of *Rx1* and *Rx2* genes in populations and identified several SNPs (Ahmadvand et al. 2013). These markers are confirmed for their selectivity in different populations (Shaikhaldein et al. 2018). Some other markers, *221R, 218R,* and *CP60*, have been developed (Kanyuka et al. 1999; Bendahmane et al. 1997). Farnham and Baulcombe (2006) developed several mutants to improve the recognition specificity of coat protein by introducing changes at two different amino acid positions for *LRR* (Leucine-rich repeat proteins) region in *Rx* gene by random mutagenesis.

The mutagenesis took place between residues 473 and 949 for *LRR* region. The mutations were *S516G, L706P, N796D,* and *N846D*. The mutated *Rx* alleles persisted ER resistance to PVX infection (Farnham and Baulcombe 2006).

PVS, a member of *Carlavirus* genus, is another potato virus causing symptomless effects although several studies are pointing the yield loss by PVS as negligible, yield reductions can reach up to 15% when potato plants are infected with this virus (Matousek et al. 2005). In some countries like Kenya, PVS is even making the most common viral incidence unlike PVY, PLRV (Were et al. 2013). There are two different strains of PVS with Andean origin (PVSA) and ordinary type (PVSO) (Daniela-Vallejo et al. 2016). The PVS resistance from *Solanum tuberosum andigena* has been provided with *Ns* gene presence that ensures HR type resistance in introgressed cultivars (Marczewski et al. 1998). This gene has been mapped on chromosome 8 (Marczewski et al. 2002). Several markers, SCG17321, SC811260, UBC811660, UBC811454, GP122718, have been developed to screen *Ns* gene for PVS resistance in cultivars (Marczewski et al. 2001b; Flis et al. 2005; Witek et al. 2006; Szajko et al. 2008).

PVM virus belongs to the same genus, *Carlavirus*, as the PVS virus, however, the yield loss can be almost five times higher (up to 75%) compared to PVS (Kostiw 2011; Brunt 2001). There are two major genes, *Rm* and *Gm*, that confer resistance against PVM infection. These two latter genes are originated from *Solanum megistacrolobum* and *Solanum gourlayi*, respectively. *Rm* is located on chromosome 9 while *Gm* is located on chromosome 11 (Marczewski et al. 2006). Several markers, GP250510, GP283320, and SC878885, have been developed and tested for screening PVM resistance for *Rm* and *Gm* genes in potato (Marczewski et al. 2006). A recent study showed that silencing of Arabidopsis Essential for potexvirus accumulation 1 (*EXA1*) gene homolog in *Nicotiana benthamiana* and in tomato has prevented viral accumulation, therefore infection by PVM. A similar response can be possible for potato as silenced *EXA1* gene function can be complemented with *EXA1* homolog from tomato and rice (Yusa et al. 2019).

PVA belongs to same genus with PVY, *Potyvirus* (Solomon-Blackburn and Barker 2001). PVA-infected plants can show mild to severe symptoms and yields are reduced by 40% in infected plants (Oruetxebarria et al. 2000; He et al. 2014). There are three to four strains of PVA (Valkonen et al. 1995; Rajamaki et al. 1998; Oruetxebarria et al. 2000). Extreme resistance can be achieved by *R* genes, *Rasto, Raadg, Rysto*, for PVA infection (Barker 1996; Hämäläinen et al. 1998). *Ryadg* confers HR type resistance against PVA whereas the same gene regulates ER type resistance for PVY (Kasai et al. 2000; Hämäläinen et el. 2000). *Naadg* and *Nydms* also provides HR for PVA (Cockerham et al. 1970; Hamalainen et al. 1998).

2.3.3 Allele Mining for Cyst Nematode Resistance

Cyst nematode is one of the most prominent pest diseases in potato leading 10–60% yield loss mostly by affecting the root parts and the field is quarantined after the infection (Karelov et al. 2013). There are two common species, *Globodera rostochiensis* and *Globodera pallidaa* spread worldwide, besides one recently discovered species, *Globodera ellingtonae* is also threatening the potato production particularly in the USA with an average 16–41% reduction in yield (Kort et al. 1977; Handoo et al. 2012; Dandurand et al. 2019). The resistance to different cyst nematode pathotypes (five different; *Ro1, Ro2, Ro3, Ro4,* and *Ro5*) has been simply achieved by wild *Solanum* accessions, *Solanum tuberosum* spp. *andigena, Solanum spegazzinii* and *Solanum vernei* (Jacobs et al. 1996; Khiutti et al. 2017). Partial resistance genes, particularly *Gro1.2, Gro1.3, Gro1.4, Grp1* and nine more, are involved in resistance against cyst nematode and all these genes are located in chromosome 3 (Kreike et al. 1993; Kreike et al. 1996; van der Voort et al. 1998; Finkers-Tomczak et al. 2011). Besides, *H1, GroV1, Gpa2,* and *Gro1* ensure absolute and durable resistance against cyst nematode (Bakker et al. 2004). Allele mining work exploited different *Gro1* alleles (approx. 10) in *LRR* domains of 16 wild *Solanum* accessions against the *Globodera rostochiensis Ro1* pathotype.

Sequencing of different *LLR* domains in these species revealed overall >400 SNPs (Nunziata et al. 2007). NBS profiling identified *SH5.2* (profiled RGA interval colocalize with *H1*) and *SH12.2* (profiled RGA interval colocalize with *Gpa2*) loci in chromosome 5 and 12, respectively (Brugmans et al. 2008). Although GWAS studies are very limited in cyst nematode resistance in potato, utilizing RenSeq (Resistance Gene Enrichment Sequencing) and GenSeq (Gene Enrichment Sequencing) approaches for identifying allelic variations for a couple of disease resistance genes against *Pa1* pathotype of *G. pallida* revealed several SNPs in *H2* locus in chromosome 5 (Strachan et al. 2019).

2.3.4 Allele Mining for Bacterial Disease Resistance

Common scab (*Streptomyces scabies*) is the most common bacterial diseases in cultivated potato (Gebhardt and Valkonen 2001). The infection by this pathogen has negative effect on tubers purposed for marketing and this situation raises concerns regarding tuber quality traits (Dees and Wanner 2012). Considering the severe lesions on tuber, it is very important to develop disease resistant potato cultivars. A wild *Solanum* species, *S. chacoense*, is naturally resistant to common scab and can be used for introgression (Dionne and Lawrence 1961). Allele mining works on this pathogen are limited but still a few studies are available in which particularly QTL regions have been detected. The earliest study about common scab resistance found out two dominant QTLs in breeding population (parents 12601 and Stirling) (Bradshaw et al. 2008). A recent GWAS work showed that three QTLs are responsible for common scab resistance in potato. Eight SNPs have been identified in three chromosomes, 2 (dominant), 4 (recessive), and 12 (recessive) with the same study (Yuan et al. 2020). More QTLs have been detected in chromosome 1 and 11 (Braun et al. 2017; Kaiser et al. 2020). A recessive QTL in chromosome 1 has been characterized in processing potato cultivars albeit the same study especially emphasize common scab is controlled by multiple genes so that existing studies are not adequate to explain the resistance mechanism behind common scab thoroughly (Koizumi et al. 2021). Besides QTL mapping research, genome wide identification of several transcription factor (TF) families, *MYB* (*PTSt00929.1* and *PTSt00907.1*) and *bHLH* (*PTSt01265.1*, *PTSt001238.1*, and *PTSt01245.1*), showed that these TFs are functioning for resistance against common scab disease in potato (Tai et al. 2013).

2.4 ALLELE MINING FOR IMPROVEMENT OF QUALITY TRAITS

Tuber flesh color in potato has been studied extensively particularly for yellow and white color determinants with the help of several previous works. Carotenoids and anthocyanins are major regulatory pigments for yellow and red-purple flesh and skin color, respectively, in potato (Brown 2006). Beta-carotene hydroxylase is the key gene involved in yellow flesh color and in this reaction this enzyme converts beta carotene into cryptoxanthin and later, zeaxanthin (Kim et al. 2012). Zeaxanthin is known to be responsible for yellow color formation and its accumulation not only affecting the flesh color in tuberous crops but also in other several plants. A locus named after yellow color formation in diploid hybrid population – *Y-locus* also called *Or* – in chromosome 3 has a role in flesh color formation, and studies have been reported that a dominant allele of *Y-locus* mainly determines the flesh color in potato tubers (Fruwirth 1912; Brown et al. 1993). Several alleles (~11) have been identified for beta-carotene hydroxylase 2 (*Chy2*) in different populations with different ploidy levels (monoploid, diploid, and tetraploid) using association mapping, however only one of these SNPs (named allele 3) performed its role in white to yellow color formation in these populations. However, the same study concluded this allele 3 as being not novel rather to be previously reported Y allele (Fruwith 1912; Wolters et al. 2010). A CASP marker has been developed to identify beta-carotene hydroxylase allele 3 and has been effectively used in different breeding studies (Wolters et al. 2010; Polivanova et al. 2021). However, it is estimated that there can be some other *Chy2* alleles, which can have roles in flesh color determination, not fully elucidated yet (Polivanova et al. 2021). A considerable number of studies are also available for the mechanism behind pigmented skin and flesh color in tubers. Anthocyanins have a role in pigmented color from red to purple in potato (Jong et al. 2003). These tubers are becoming popular and more visible in markets particularly for their high antioxidative activity. The *R* locus in chromosome 2 in potato encodes for dihydroflavonol 4-reductase, which is necessary for breakdown of dihydrokaempferol into leucopelorgonidin (precursor of pelorgonidin), and it is revealed that allelic variants of this enzyme have roles in red pigmentation in tuber skin color (Jong et al. 2003). In a diploid potato population, two alleles, showing polymorphism regarding restriction sites (*BamHI* and *DraI*), were found to play an important role for red skin color formation. However, it is known that *R* locus alone is not enough for triggering red pigmentation, rather its interplay with *I* locus (or *D* locus) in chromosome 10 controls red color formation in tubers (Dodds and Long 1955; Dodds and Long 1956; Jong et al. 2003). Allele present in *I* locus is not only involved in red pigmentation but also purple color formation in periderm (Jung et al. 2009). A *R2R3 MYB* transcription factor allele encoded by *I* locus, named *Stan2*^{816}, was present in pigmented tubers however the allele did not exist in white skin tubers (Jung et al. 2005, 2009). This allele, respective to another allele named *stan3*, had tandem repeat and duplication (Jung et al. 2009). *P* locus, like *I* and *R* loci, is another region controlling purple pigmentation in tuber skin (van Eck et al. 1994; Jung et al. 2005). A study in Potato GeneBank in Chile investigated the allelic variants of the genes encoded by these three loci and observed different alleles for each. Researchers found that SNP584 was a promising allele for distinguishing elevated anthocyanin levels compared to low levels for both flesh and skin color in this study (Solis et al. 2022). Comparative transcriptomics data using RNASeq analysis ascertained several SNPs and InDels in white and purple skin and flesh color potato cultivars and once again, previously attributed anthocyanin biosynthesis gene, anthocyanidin 3-O-glucosyltransferase and transcription factor, *R2R3 MYB*, have been shown to have critical roles for pigmentation in potato tubers. A SNP creating a restriction site in *3-O-glucosyltransferase* has been suggested to have a likely role in pigmentation mechanism as this allelic variation was present only in white cultivar. The same study identified several repressor and activator genes, which are involved in anthocyanin biosynthesis pathway (Liu et al. 2015).

2.5 ALLELE MINING IN OTHER TRAITS

Another important trait in potato is tuber dormancy and this was elaborately dissected genetically with QTL mapping studies in a backcross population and predominantly genes and their alleles present in chromosome 2 have been known to be responsible for controlling dormancy in potato although some more QTLs have been detected in other chromosomes (nine in total) unlike chromosome 6, 7, 11, and 12 (van den Berg et al. 1996; Ewing et al. 2004). The existence of T allele at *TG234* locus in chromosome 2 had shown to have a major effect in dormancy that tends to be shorter (average 29 days earlier) when T allele was present in population (van den Berg et al. 1996). Same allelic variation was also tested in another population showing an epistatic effect with *TG155* in chromosome 4 linked to abscisic acid level (Simko et al. 1997). The earliest study identified a protein marker, glutamate oxaloacetate transaminase 2 in chromosome 7 for delayed tuber dormancy (Freyre et al. 1994). More QTLs have been detected in seven different chromosomes for dormancy and for sprouting as well in a diploid population (Bisognin et al. 2018). These SNPs with main effect, i.e., c2_45308 in chromosome 2 has been related to a dormancy period (Bisognin et al. 2018).

2.6 FUTURE PROSPECTS

It is necessary to utilize the ever-increasing crop genebanks efficiently for the purpose of crop improvement programs. Crop wild relatives and landraces are reservoirs of useful alleles, hidden in their phenotype. Allele mining can be a game changer for molecular breeding. Its success depends upon various factors like, effective phenotyping platforms (high-throughput phenotyping), selection of diverse genetic material for screening, efficient and quick generation of allelic variation within the candidate genes, cost-effective sequencing techniques, and development of markers for MAS (Kumar et al. 2010). The availability of diversity in potato genebanks along with the sequence information for potato and its wild relative *S. commersonii* provided a successful platform for allele mining of complex traits. Sequence-based allele mining possesses the potential of identifying novel allelic variations within the candidate genes to construct haplotypes. Whereas next-generation sequencing (NGS)-based allele mining identified accessions and their relation with the variation in phenotype. It is particularly important for the mining of resistance/tolerant genes in potato. GWAS followed by validation approaches like KASP marker development or/and RNAi-based techniques could facilitate the identification of novel alleles and markers for MAS, useful for molecular potato breeding.

REFERENCES

Abbas A, Arif M, Ali A (2016) Use of hot water-thermotherapy to free potato tubers of potato leaf roll virus (PLRV). *Int J Life Sci Scienti Res* 2(2):155–162.

Ahmadvand R, Takács A, Taller J, Wolf I, Polgár Z (2012) Potato viruses and resistance genes in potato. *Acta Biol Hung* 60(3):283–298.

Ahmadvand R, Wolf I, Gorji AM, Polgár Z, Taller J (2013) Development of molecular tools for distinguishing between the highly similar *Rx1* and *Rx2* PVX extreme resistance genes in tetraploid potato. *Potato Res* 56(4):277–291.

Akbudak MA, Filiz E. (2020) Genome-wide investigation of proline transporter (ProT) gene family in tomato: Bioinformatics and expression analyses in response to drought stress. *Plant Physiol Biochem* 157: 13–22.

Aksoy E, Demirel U, Bakhsh A, Zia MAB, et al. (2021) Recent advances in potato (*Solanum tuberosum* L.) breeding. In: Al-Khayri JM, Jain SM, Johnson DV (eds), *Advances in Plant Breeding Strategies: Vegetable Crops*. Springer, Cham, Switzerland, pp 409–487.

Aliche EB, Oortwijn M, Theeuwen TP, Bachem CW, van Eck HJ, Visser RG, van der Linden CG (2019) Genetic mapping of tuber size distribution and marketable tuber yield under drought stress in potatoes. *Euphytica* 215(11): 1–9.

Álvarez MF, Angarita M, Delgado MC, García C, Jiménez-Gomez J, Gebhardt C, Mosquera T (2017) Identification of novel associations of candidate genes with resistance to late blight in *Solanum* tuberosum group *Phureja*. *Front Plant Sci* 8:1040.

Andrade CM, Pinto CA, de Paula Ribeiro SR, Peixouto LS, de Souza Vilela XM (2009) Potato clones with multiple copies of the Ryadg allele conferring resistance to PVY. *Crop Breed Appl Biotechnol* 9(4).

Ashkani S, Yusop MR, Shabanimofrad M, Azadi A, Ghasemzadeh A, Azizi P, Latif MA (2015) Allele mining strategies: principles and utilisation for blast resistance genes in rice (*Oryza sativa* L.). *Curr Iss Mol Biol* 17(1):57–74.

Asim A, Gökçe ZN, Bakhsh A, Çaylı İT, Aksoy E, Çalişkan S, Çalişkan ME, Demirel U (2021) Individual and combined effect of drought and heat stresses in contrasting potato cultivars overexpressing *miR172b-3p*. *Turk J Agric Forest* 45(5):651–68.

Atarashi H, Jayasinghe WH, Kwon J, Kim H, Taninaka Y, Igarashi M, Ito K, Yamada T, Masuta C, Nakahara KS (2020) Artificially edited alleles of the eukaryotic translation initiation factor 4E1 gene differentially reduce susceptibility to cucumber mosaic virus and potato virus Y in tomato. *Front Microbiol* 11:564310.

Aversano R, Contaldi F, Ercolano MR, Grosso V, Iorizzo M, Tatino F, Xumerle L, Dal Molin A, Avanzato C, Ferrarini A, Delledonne M (2015) The *Solanum commersonii* genome sequence provides insights into adaptation to stress conditions and genome evolution of wild potato relatives. *Plant Cell* 27(4):954–68.

Bakker E, Achenbach U, Bakker J, van Vliet J, Peleman J, Segers B, van der Heijden S, van der Linde P, Graveland R, Hutten R, van Eck H (2004) A high-resolution map of the *H1* locus harbouring resistance to the potato cyst nematode *Globodera rostochiensis*. *Theor Appl Genet* 109(1):146–52.

Ballvora A, Ercolano MR, Weiß J, Meksem K, Bormann CA, Oberhagemann P, ... Gebhardt C (2002) The *R1* gene for potato resistance to late blight (Phytophthora infestans) belongs to the leucine zipper/NBS/LRR class of plant resistance genes. *Plant J* 30(3):361–371.

Barker H (1996) Inheritance of resistance to potato viruses Y and A in progeny obtained from potato cultivars containing gene Ry: evidence for a new gene for extreme resistance to PVA. *Theor Appl Genet* 93(5):710–716.

Bendahmane A, Kanyuka K, Baulcombe DC (1997) High-resolution genetical and physical mapping of the Rx gene for extreme resistance to potato virus X in tetraploid potato. *Theor Appl Genet* 95(1):153–162.

Bendahmane A, Köhm BA, Dedi C, Baulcombe DC (1995) The coat protein of potato virus X is a strain-specific elicitor of *Rx1*-mediated virus resistance in potato. *Plant J* 8(6):933–941.

Bendahmane A, Querci M, Kanyuka K, Baulcombe DC (2000) Agrobacterium transient expression system as a tool for the isolation of disease resistance genes: application to the *Rx2* locus in potato. *Plant J* 21(1):73–81.

Bharathi Raja R, Agasimani S, Anusheela V, Thiruvengadam V, Chibbar RN, Ganesh Ram S. TILLING and EcoTILLING for discovery of induced and natural variations in sorghum genome. In: Rakshit S, Wang YH (eds), *The Sorghum Genome*. Springer, Cham, pp 257–267.

Biryukova VA, Shmiglya IV, Zharova VA, Beketova MP, Rogozina EV, Mityushkin AV, Meleshin AA (2019) Molecular markers of genes for extreme resistance to potato virus Y in *Solanum tuberosum* L. cultivars and hybrids. *Russ Agri Sci* 45(6):517–23.

Bisognin DA, Manrique-Carpintero NC, Douches DS (2018) QTL analysis of tuber dormancy and sprouting in potato. *Am J Potato Res* 95(4):374–382.

Blokhina O, Virolainen E, Fagerstedt KV (2003) Antioxidants, oxidative damage and oxygen deprivation stress: a review. *Ann Bot* 91(2):179–194.

Boquel S, Zhang J, Goyer C, Giguère MA, Clark C, Pelletier Y (2015) Effect of insecticide-treated potato plants on aphid behavior and potato virus Y acquisition. *Pest Manag Sci* 71(8):1106–12.

Bouaziz D, Charfeddine M, Jbir R, Saidi MN, Pirrello J, Charfeddine S,… Gargouri-Bouzid R (2015) Identification and functional characterization of ten *AP2/ERF* genes in potato. *Plant Cell Tiss Org Cult* 123(1):155–172.

Bouaziz D, Pirrello J, Amor HB, Hammami A, Charfeddine M, Dhieb A, ... Gargouri-Bouzid R (2012) Ectopic expression of dehydration responsive element binding proteins (*StDREB2*) confers higher tolerance to salt stress in potato. *Plant Physiol Biochem* 60:98–108.

Bradshaw JE, Hackett CA, Pande B, Waugh R, Bryan GJ (2008) QTL mapping of yield, agronomic and quality traits in tetraploid potato (*Solanum tuberosum* subsp. *tuberosum*). *Theor App Genet* 116(2):193–211.

Bradshaw JE, Ramsay G (2005) Utilisation of the Commonwealth Potato Collection in potato breeding. *Euphytica* 146:9–19.

Braun S, Gevens A, Charkowski A, Allen C, Jansky S (2017) Potato common scab: A review of the causal pathogens, management practices, varietal resistance screening methods, and host resistance. *Am J Potato Res* 94(4):283–96.

Brown CR, Edwards CG, Yang CP, Dean BB (1993) Orange flesh trait in potato: Inheritance and carotenoid content. *J Am Soc Hortic Sci* 118(1):145–150.

Brown CR, Kim TS, Ganga Z, Haynes K, De Jong D, Jahn M, Paran I, De Jong W (2006) Segregation of total carotenoid in high level potato germplasm and its relationship to *beta-carotene hydroxylase* polymorphism. *Am J Potato Res* 83(5):365–72.

Brugmans B, Wouters D, van Os H, Hutten R, van der Linden G, Visser RG, van Eck HJ, van der Vossen EA (2008) Genetic mapping and transcription analyses of resistance gene loci in potato using NBS profiling. *Theor App Genet* 117(8):1379–88.

Brunt AA (2001) Potato virus M (PVM; Genus *Carlavirus*). In: Loebenstein G, Berger PH, Brunt AA, Lawson RH (eds), *Virus and Virus-like Diseases of Potatoes and Production of Seed-Potatoes*. Springer, Dordrecht, pp 101–107.

Brush S, Kesseli R, Ortega R, Cisneros P, Zimmerer K, Quiros C (1995) Potato diversity in the Andean center of crop domestication. *Conserv Biol* 9(5):1189–1198.

Butterbach P (2007) *Molecular Evolution of the Disease Resistance Gene Rx in Solanum*. Wageningen University and Research, Copenhagen, Denmark.

Byarugaba AA, Mukasa SB, Barekye A, Rubaihayo PR (2021) Prevalence of R genes for resistance to potato viruses in Uganda germplasm. *Potato Res* 64(2):267–76.

Byun M, Kwon H, Park S (2007) Recent advances in genetic engineering of potato crops for drought and saline stress tolerance. In: Jenks MA, Hasegawa PM, Jain SM (eds), *Advances in Molecular Breeding Toward Drought and Salt Tolerant Crops*. Springer, pp 713–737.

Campbell R, Ducreux L, Cowan G, Young V, Chinoko G, Chitedze G, Kwendani S, Chiipanthenga M, Bita CE, Mwenye O, Were H (2022) Allelic variants of a potato *HEAT SHOCK COGNATE 70* gene confer improved tuber yield under a wide range of environmental conditions. *Food Energy Secur15*:e377.

Cardi T, D'Ambrosio E, Consoli D, Puite K, Ramulu KJ, Ramulu KS (1993) Production of somatic hybrids between frost-tolerant *Solanum commersonii* and *S. tuberosum*: characterization of hybrid plants. *Theor Appl Genet* 87:193–200.

Caruana BM, Rodoni BC, Constable F, Slater AT, Cogan NO (2021) Genome Enhanced Marker Improvement for Potato Virus Y Disease Resistance in Potato. *Agronomy* 11(5):832.

Chang DC, Sohn HB, Cho JH, Im JS, Jin YI, Do GR, ... Lee YB (2014) Freezing and frost damage of potato plants: a case study on growth recovery, yield response, and quality changes. *Potato Res* 57(2):99–110.

Chen L, Zhao H, Chen Y, Jiang F, Zhou F, Liu Q, Fan Y, Liu T, Tu W, Walther D, Song B (2022) Comparative transcriptomics analysis reveals a calcineurin B-like gene to positively regulate constitutive and acclimated freezing tolerance in potato. *Plant Cell Environ* 45(11):3305–3321.

Chen Y, Li C, Yi J, Yang Y, Lei C, Gong M (2019) Transcriptome response to drought, rehydration and re-dehydration in potato. *Int J Mol Sci* 21(1):159.

Chourasia KN, Lal MK, Tiwari RK, Dev D, Kardile HB, Patil VU, Kumar A, Vanishree G, Kumar D, Bhardwaj V, Meena JK (2021) Salinity stress in potato: Understanding physiological, biochemical and molecular responses. *Life* 11(6):545.

Cockerham G (1955) *Strains of Potato Virus X. In: Proceedings of the Second Conference on Potato Virus Diseases*, Lisse-Wageningen, June 25–29, 1954, pp. 89–92

Cockerham G (1970) Genetical studies on resistance to potato viruses X and Y. *Heredity* 25:309–48.

Collins A, Milbourne D, Ramsay L, Meyer R, Chatot-Balandras C, Oberhagemann P, De Jong W, Gebhardt C, Bonnel E, Waugh R (1999) QTL for field resistance to late blight in potato are strongly correlated with maturity and vigour. *Mol Breed* 5(5):387–98

Dandurand LM, Zasada IA, Wang X, Mimee B, De Jong W, Novy R, Whitworth J, Kuhl JC (2019) Current status of potato cyst nematodes in North America. *Annu Rev Phytopathol* 57:117–33.

De Jong W, Forsyth A, Leister D, Gebhardt C, Baulcombe DC (1997) A potato hypersensitive resistance gene against potato virus X maps to a resistance gene cluster on chromosome 5. *Theor Appl Genet* 95(1):246–52.

De Jong WS, De Jong DM, De Jong H, Kalazich J, Bodis M (2003) An allele of dihydroflavonol 4-reductase associated with the ability to produce red anthocyanin pigments in potato (*Solanum tuberosum* L.). *Theor Appl Genet* 107(8):1375–1383.

De Palma M, Grillo S, Massarelli I, Costa A, Balogh G, Vigh L, Leone A (2008) Regulation of desaturase gene expression, changes in membrane lipid composition and freezing tolerance in potato plants. *Mol Breed* 21(1):15–26.

Dees MW, Wanner LA (2012) In search of better management of potato common scab. *Potato Res* 55(3):249–68.

Díaz P, Sarmiento F, Mathew B, Ballvora A, Mosquera Vásquez T (2021) Genomic regions associated with physiological, biochemical and yield-related responses under water deficit in diploid potato at the tuber initiation stage revealed by GWAS. *PLoS One* 16(11):e0259690.

Dionne LA, Lawrence CH (1961) Early scab resistant derivatives of *Solanum chacoense* x*Solanum phureja*. *Am Potato J* 38(1):6–8.

Dodds ES, Long DH (1956) The inheritance oe colour in diploid potatoes II. A three-factor linkage group. *J Genet* 54(1):27–41.

Dodds KS, Long DH (1955) The inheritance of colour in diploid potatoes. *J Genet* 53(1):136–149.

Dong T, Xu ZY, Park Y, Kim DH, Lee Y, Hwang I (2014) Abscisic acid uridine diphosphate glucosyltransferases play a crucial role in abscisic acid homeostasis in Arabidopsis. *Plant Physiol* 165(1):277–289.

Duan H, Richael C, Rommens CM (2012) Overexpression of the wild potato *eIF4E-1* variant Eva1 elicits Potato virus Y resistance in plants silenced for native *eIF4E-1*. *Transgen Res* 21(5):929–38.

Dutt S, Manjul AS, Raigond P, Singh B, Siddappa S, Bhardwaj V, Kawar PG, Patil VU, Kardile HB (2017) Key players associated with tuberization in potato: potential candidates for genetic engineering. *Crit Rev Biotechnol* 37(7):942–57.

Elias R, Till BJ, Mba C, Al-Safadi B (2009) Optimizing TILLING and Ecotilling techniques for potato (*Solanum tuberosum* L). *BMC Res Notes* 2(1):1–5.

Elison GL, Novy RG, Whitworth JL (2021) Russet Potato Breeding Clones with Extreme Resistance to Potato Virus Y Conferred by Rychc as well as Resistance to Late Blight and Cold-Induced Sweetening. *Am J Potato Res* 98(5):411–419.

Ewing EE, Simko I, Omer EA, Davies PJ (2004) Polygene mapping as a tool to study the physiology of potato tuberization and dormancy. *Am J Potato Res* 81:281–289.

FAOSTAT (2022) Food and Agriculture Organization of the United Nations. www.fao.org/faost at/en/#data/QC/ (Accessed on 25 September 2022).

Farnham G, Baulcombe DC (2006) Artificial evolution extends the spectrum of viruses that are targeted by a disease-resistance gene from potato. *Pro Nat Acad Sci USA* 103(49):18828–18833.

Finkers-Tomczak A, Bakker E, de Boer J, van der Vossen E, Achenbach U, Golas T, Suryaningrat S, Smant G, Bakker J, Goverse A (2011) Comparative sequence analysis of the potato cyst nematode resistance locus H1 reveals a major lack of co-linearity between three haplotypes in potato (*Solanum tuberosum* ssp.). *Theor Appl Genet* 122(3):595–608.

Flis B, Hennig J, Strzelczyk-Żyta D, Gebhardt C, Marczewski W (2005) The *Ry-f sto* gene from *Solanum stoloniferum* for extreme resistant to Potato virus Y maps to potato chromosome XII and is diagnosed by PCR marker GP122718 in PVY resistant potato cultivars. *Mol Breed* 15(1):95–101.

Freyre R, Warnke S, Sosinski B, Douches DS (1994) Quantitative trait locus analysis of tuber dormancy in diploid potato (*Solanum* spp.). *Theor Appl Genet* 89(4):474–480.

Fruwirth C (1912) Zur Züchtung der Kartoffel. *Deutsche Landwirtschaftliche Presse* 39(551–552):565–567.

Fujimatsu M, Hashizume H, Fudan T, Koma Y, Sanetomo R, Ono S, Hosaka K (2018) Harimaru: a new potato variety for a local specialty. *Breed Sci* 68(2):284–288.

Gangadhar BH, Sajeesh K, Venkatesh J, Baskar V, Abhinandan K, Yu JW, ... Mishra RK (2016) Enhanced tolerance of transgenic potato plants over-expressing non-specific lipid transfer protein-1 (*StnsLTP1*) against multiple abiotic stresses. *Front Plant Sci* 7: 1228.

Gebhardt C, Valkonen JP (2001) Organization of genes controlling disease resistance in the potato genome. *Annu Rev Phytopathol* 39(1):79–102.

Grech-Baran M, Witek K, Szajko K, Witek AI, Morgiewicz K, Wasilewicz-Flis I, Jakuczun H, Marczewski W, Jones JD, Hennig J (2020) Extreme resistance to Potato virus Y in potato carrying the Rysto gene is mediated by a TIR-NLR immune receptor. *Plant Biotechnol J* 18(3):655–67.

Guedes ML, Haynes KG, Vinyard BT, Pinto CA (2019) Heat tolerance in diploid wild potato species in vitro. *Am J Potato Res* 96(3):294–302

Guo YL, Fitz J, Schneeberger K, Ossowski S, Cao J, Weigel D (2011) Genome-wide comparison of nucleotide-binding site-leucine-rich repeat-encoding genes in Arabidopsis. *Plant Physiol* 157(2):757–769.

Hämäläinen JH, Kekarainen T, Gebhardt C, Watanabe KN, Valkonen JP (2000) Recessive and dominant genes interfere with the vascular transport of Potato virus A in diploid potatoes. *Mol Plant-Microb Interact* 13(4):402–412.

Hämäläinen JH, Sorri VA, Watanabe KN, Gebhardt C, Valkonen JP (1998) Molecular examination of a chromosome region that controls resistance to potato Y and A potyviruses in potato. *Theor Appl Genet* 96(8):1036–1043.

Hämäläinen JH, Watanabe KN, Valkonen JP, Arihara A, Plaisted RL, Pehu E, Miller L, Slack SA (1997) Mapping and marker-assisted selection for a gene for extreme resistance to potato virus Y. *Theor Appl Genet* 94(2):192–197

Hammond-Kosack KE, Jones JD (1997) Plant disease resistance genes. *Ann Rev Plant Biol* 48(1):575–607.

Hancock RD, Morris WL, Ducreux LJ, Morris JA, Usman M, Verrall SR, Fuller J, Simpson CG, Zhang R, Hedley PE, Taylor MA (2014) Physiological, biochemical and molecular responses of the potato (*Solanum tuberosum* L.) plant to moderately elevated temperature. *Plant cell Environ* 37(2):439–50.

Handoo ZA, Carta LK, Skantar AM, Chitwood DJ (2012) Description of Globodera ellingtonae n. sp.(Nematoda: *Heteroderidae*) from Oregon. *J Nematol* 44(1):40.

Haussmann BIG, Parzies HK, Presterl T, Susic Z, Miedaner T (2004) Plant genetic resources in crop improvement. *Plant Genet Res* 2(1):3–21

He C, Zhang W, Hu X, Singh M, Xiong X, Nie X (2014) Molecular characterization of a Chinese isolate of potato virus A (PVA) and evidence of a genome recombination event between PVA variants at the 3′-proximal end of the genome. *Arch Virol* 159(9):2457–2462.

Heldák J, Bežo M, ŠtefúnoVá V, GAllikoVá A (2007) Selection of DNA markers for detection of extreme resistance to Potato Virus Y in tetraploid potato. *Czech J Genet Plant Breed* 43(4):125–134.

Herrera MD, Vidalon LJ, Montenegro JD, Riccio C, Guzman F, Bartolini I, Ghislain M (2018) Molecular and genetic characterization of the *Ryadg* locus on chromosome XI from Andigena potatoes conferring extreme resistance to potato virus Y. *Theor Appl Genet* 131(9):1925–1938.

Hijmans RJ, Jacobs M, Bamberg JB, Spooner DM (2003) Frost tolerance in wild potato species: Assessing the predictivity of taxonomic, geographic, and ecological factors. *Euphytica* 130(1):47–59

Hijmans RJ, Spooner DM (2001) Geographic distribution of wild potato species. *Am J Bot* 88(11):2101–2112

Hosaka K, Hosaka Y, Mori M, Maida T, Matsunaga H (2001) Detection of a simplex RAPD marker linked to resistance to potato virus Y in a tetraploid potato. *Am J Potato Res* 78(3):191–196.

Hu H, Dai M, Yao J, Xiao B, Li X, Zhang Q, Xiong L (2006) Overexpressing a NAM, ATAF, and CUC (NAC) transcription factor enhances drought resistance and salt tolerance in rice. *Proc Natl Acad Sci USA* 103(35):12987–12992.

Huamán Z, Ortiz R, Gómez R (2000) Selecting a *Solanum tuberosum* subsp. andigena core collection using morphological, geographical, disease and pest descriptors. *Am J Potato Res* 77(3):183–190

Huang S, Van Der Vossen EA, Kuang H, Vleeshouwers VG, Zhang N, Borm TJ, Van Eck HJ, Baker B, Jacobsen E, Visser RG (2005) Comparative genomics enabled the isolation of the *R3a* late blight resistance gene in potato. *Plant J* 42(2):251–261.

Huang Z, Phoolcharoen W, Lai H, Piensook K, Cardineau G, Zeitlin L, Whaley KJ, Arntzen CJ, Mason HS, Chen Q (2010) High-level rapid production of full-size monoclonal antibodies in plants by a single-vector DNA replicon system. *Biotechnol Bioengg* 106(1):9–17.

Jacobs JM, van Eck HJ, Horsman K, Arens PF, Verkerk-Bakker B, Jacobsen E, Pereira A, Stiekema WJ (1996) Mapping of resistance to the potato cyst nematode Globodera rostochiensis from the wild potato species *Solanum vernei. Mol Breed* 2(1):51–60.

Jansky SH, Simon R, Spooner DM (2009) A test of taxonomic predictivity: resistance to the Colorado potato beetle in wild relatives of cultivated potato. *J Econ Entomol* 102(1):422–431

Jiang R, Li J, Tian Z, Du J, Armstrong M, Baker K, Tze-Yin Lim J, Vossen JH, He H, Portal L, Zhou J (2018) Potato late blight field resistance from QTL dPI09c is conferred by the *NB-LRR* gene *R8. J Exp Bot* 69(7):1545–1555.

Jiang SC, Mei C, Liang S, Yu YT, Lu K, Wu Z, Wang XF, Zhang DP (2015) Crucial roles of the pentatricopeptide repeat protein *SOAR1* in *Arabidopsis* response to drought, salt and cold stresses. *Plant Mol Biol* 88(4): 369–85.

Jo KR, Arens M, Kim TY, Jongsma MA, Visser RG, Jacobsen E, Vossen JH (2011) Mapping of the *S. demissum* late blight resistance gene *R8* to a new locus on chromosome IX. *Theor Appl Genet* 123(8):1331–1340.

Jung CS, Griffiths HM, De Jong DM, Cheng S, Bodis M, De Jong WS (2005) The potato P locus codes for flavonoid 3′, 5′-hydroxylase. *Theor Appl Genet* 110(2):269–75

Jung CS, Griffiths HM, De Jong DM, Cheng S, Bodis M, Kim TS, De Jong WS (2009) The potato developer (D) locus encodes an *R2R3 MYB* transcription factor that regulates expression of multiple anthocyanin structural genes in tuber skin. *Theor Appl Genet* 120(1):45–57.

Juyo Rojas DK, Soto Sedano JC, Ballvora A, Léon J, Mosquera Vásquez T (2019) Novel organ-specific genetic factors for quantitative resistance to late blight in potato. *PLoS One* 14(7):e0213818.

Kaiser NR, Coombs JJ, Felcher KJ, Hammerschmidt R, Zuehlke ML, Buell CR, Douches DS (2020) Genome-wide association analysis of common scab resistance and expression profiling of tubers in response to thaxtomin A treatment underscore the complexity of common scab resistance in tetraploid potato. *Am J Potato Res* 97(5):513–522.

Kanyuka K, Bendahmane A, van der Voort JN, Van Der Vossen EA, Baulcombe DC (1999) Mapping of intra-locus duplications and introgressed DNA: aids to map-based cloning of genes from complex genomes illustrated by physical analysis of the *Rx* locus in tetraploid potato. *Theor Appl Genet* 98(5):679–89.

Karasev AV, Gray SM (2013) Continuous and emerging challenges of Potato virus Y in potato. *Ann Rev Phytopathol* 51:571–586

Karelov AV, Pylypenko LA, Kozub NO, Bondus RO, Borzykh OI, Sozinov IA, Blume YB, Sozinov AA (2013) Allelic state of the molecular marker for golden nematode (Globodera rostochiensis) resistance gene H1 among Ukrainian and world potato (*Solanum tuberosum* ssp. *tuberosum*) cultivars. *Cytol Genet* 47(5):294–297

Kasai K, Morikawa Y, Sorri VA, Valkonen JP, Gebhardt C, Watanabe KN (2000) Development of SCAR markers to the PVY resistance gene *Ry adg* based on a common feature of plant disease resistance genes. *Genome* 43(1):1–8.

Keane PJ (2012) *Horizontal or Generalized Resistance to Pathogens in Plants. Plant Pathology*. Rijeka, Croatia, InTech, pp 327–362.

Khavkin EE, Sokolova EA, Beketova MP, Pankin AA, Kuznetsova MA, Kozlovskaya IN, Spiglazova S, Statsyuk N, Yashina IM (2010) Potato resistance to late blight as related to the *R1* and *R3* genes introgressed from *Solanum demissum*. *PPO Spec Rep* 14:231–238.

Khiutti AV, Antonova OY, Mironenko NV, Gavrilenko TA, Afanasenko OS (2017) Potato resistance to quarantine diseases. *Russ J Genet: Appl Res* 7(8):833–844

Kim SH, Ahn YO, Ahn MJ, Lee HS, Kwak SS (2012) Down-regulation of β-carotene hydroxylase increases β-carotene and total carotenoids enhancing salt stress tolerance in transgenic cultured cells of sweetpotato. *Phytochemistry* 74:69–78

Koizumi E, Igarashi T, Tsuyama M, Ogawa K, Asano K, Kobayashi A, Sanetomo R, Hosaka K (2021) Association of genome-wide SNP markers with resistance to common scab of potato. *Am J Potato Res* 98(2):149–156.

Kort J, Ross H, Rumpenhorst HJ, Stone AR (1977) An international scheme for identifying and classifying pathotypes of potato cyst-nematodes *Globodera rostochiensis* and *G. pallida*. *Nematology* 23(3):333–339.

Kostiw, M. (2011) The occurrence of major potato viruses in Poland. *J Plant Prot Res* 51(3):204–209

Kou S, Chen L, Tu W, Scossa F, Wang Y, Liu J, Fernie AR, Song B, Xie C (2018) The arginine decarboxylase gene ADC 1, associated to the putrescine pathway, plays an important role in potato cold-acclimated freezing tolerance as revealed by transcriptome and metabolome analyses. *Plant J* 96(6):1283–1298.

Koyro H-W, Ahmad P, Geissler N (2012) Abiotic stress responses in plants: an overview. In: Ahmad P, Prasad MNV (eds), *Environmental Adaptations and Stress Tolerance of Plants in the Era of Climate Change*. Springer, New York, pp 1–28.

Krannich CT, Maletzki L, Kurowsky C, Horn R (2015) Network candidate genes in breeding for drought tolerant crops. *Int J Mol Sci* 16:16378–16400.

Kreike CM, De Koning JR, Vinke JH, Van Ooijen JW, Gebhardt C, Stiekema WJ (1993) Mapping of loci involved in quantitatively inherited resistance to the potato cyst-nematode *Globodera rostochiensis* pathotype *Ro1*. *Theor Appl Genet* 87(4):464–470.

Kreike CM, Kok-Westeneng AA, Vinke JH, Stiekema WJ (1996) Mapping of QTLs involved in nematode resistance, tuber yield and root development in *Solanum* sp. *Theor Appl Genet* 92(3):463–470.

Kreuze JF, Souza-Dias JA, Jeevalatha A, Figueira AR, Valkonen JP, Jones RA (2020) Viral diseases in potato. In: Campos H, Ortiz O (eds), *The Potato Crop*. Springer, Cham, pp 389–430.

Kuhl JC, Novy RG, Whitworth JL, Dibble MS, Schneider B, Hall D (2016) Development of molecular markers closely linked to the potato leafroll virus resistance gene, *Rlr etb*, for use in marker-assisted selection. *Am J Potato Res* 93(3):203–212.

Kumar GR, Sakthivel K, Sundaram RM, Neeraja CN, Balachandran SM, Rani NS, Viraktamath BC, Madhav MS (2010) Allele mining in crops: prospects and potentials. *Biotechnol Adv* 28(4):451–461.

Lacomme C, Jacquot E (2017) General characteristics of Potato virus Y (PVY) and its impact on potato production: an overview. In: Lacomme C, Glais L, Bellstedt D, Dupuis B, Karasev A, Jacquot E (eds), *Potato Virus Y: Biodiversity, Pathogenicity, Epidemiology and Management*. Springer, Cham, pp 1–9.

Lehretz GG, Sonnewald S, Lugassi N, Granot D, Sonnewald U (2021) Future-proofing potato for drought and heat tolerance by overexpression of hexokinase and SP6A. *Front Plant Sci* 11:614534.

Levy D, Veilleux RE (2007) Adaptation of potato to high temperatures and salinity-a review. *Am J Potato Res* 84(6):487–506.

Li G, Huang S, Guo X, Li Y, Yang Y, Guo Z, Kuang H, Rietman H, Bergervoet M, Vleeshouwers VG, van der Vossen EA (2011) Cloning and characterization of *R3b*; members of the R3 superfamily of late blight resistance genes show sequence and functional divergence. *Mol Plant-Microbe Interact* 24(10):1132–1142.

Lindqvist-Kreuze H, De Boeck B, Unger P, Gemenet D, Li X, Pan Z, Sui Q, Qin J, Woldegjorgis G, Negash K, Seid I (2021) Global multi-environment resistance QTL for foliar late blight resistance in tetraploid potato with tropical adaptation. *G3: Gene Genome Genet* 11(11):jkab251.

Liu MM, Li YL, Li GC, Dong TT, Liu SY, Pei LIU, Wang QG (2020) Overexpression of *StCYS1* gene enhances tolerance to salt stress in the transgenic potato (*Solanum tuberosum* L.) plant. *J Integrat Agric* 19(9):2239–2246

Liu Y, Lin-Wang K, Deng C, Warran B, Wang L, Yu B, Yang H, Wang J, Espley RV, Zhang J, Wang D (2015) Comparative transcriptome analysis of white and purple potato to identify genes involved in anthocyanin biosynthesis. *PLoS One* 10(6):e0129148.

Lokossou AA, Park TH, van Arkel G, Arens M, Ruyter-Spira C, Morales J, Whisson SC, Birch PRJ, Visser RGF, Jacobsen E, van der Vossen EA (2009) Exploiting knowledge of R/Avr genes to rapidly clone a new LZ-NBS-LRR family of late blight resistance genes from potato linkage group IV. *Mol Plant Microbe Interact* 22(6): 630–641.

Lu M, Ying S, Zhang DF, Shi YS, Song YC, Wang TY, Li Y (2012) A maize stress responsive NAC transcription factor, *ZmSNAC1*, confers enhanced tolerance to dehydration in transgenic Arabidopsis. *Plant Cell Rep* 31(9):1701–1711.

Lucioli A, Tavazza R, Baima S, Fatyol K, Burgyan J, Tavazza M (2022) CRISPR-Cas9 Targeting of the eIF4E1 Gene Extends the Potato Virus Y Resistance Spectrum of the *Solanum tuberosum* L. cv. Desirée. *Front Microbiol* 13: 1–12.

Luthra SK, Gopal J, Manivel P, Kumar V, Singh BP, Pandey SK (2007) Screening of wild and cultivated species of potato for frost tolerance in north-central plains of India. *Potato J* 34(1–2): 45–46.

Machida-Hirano R (2015) Diversity of potato genetic resources. *Breed Sci* 65(1):26–40

Marczewski W, Flis B, Syller J, Schäfer-Pregl R, Gebhardt C (2001a) A major QTL for resistance to Potato leafroll virus (PLRV) is located in a resistance hotspot on potato chromosome XI and is tightly linked to N-gene-like markers. *Mol Plant Microbe Interact* 14:1420–1425.

Marczewski W, Flis B, Syller J, Strzelczyk-Żyta D, Hennig J, Gebhardt C (2004) Two allelic or tightly linked genetic factors at the PLRV. 4 locus on potato chromosome XI control resistance to potato leafroll virus accumulation. *Theor Appl Genet* 109(8):1604–1609.

Marczewski W, Hennig J, Gebhardt C (2002) The Potato virus S resistance gene *Ns* maps to potato chromosome VIII. *Theor Appl Genet* 105(4):564–567.

Marczewski W, Ostrowska K, Zimnoch-Guzowska E (1998) Identification of RAPD markers linked to the Ns locus in potato. *Plant Breed* 117(1):88–90.

Marczewski W, Strzelczyk-Żyta D, Hennig J, Witek K, Gebhardt C (2006) Potato chromosomes IX and XI carry genes for resistance to potato virus M. *Theor Appl Genet* 112(7):1232–1238.

Marczewski W, Talarczyk A, Hennig J (2001b) Development of SCAR markers linked to the Ns locus in potato. *Plant Breed* 120(1):88–90.

Matoušek J, Schubert J, Ptáček J, Kozlová P, Dědič P (2005) Complete nucleotide sequence and molecular probing of potato virus S genome. *Acta Virol* 49:195–205.

Mayer MP (2013) *Hsp70* chaperone dynamics and molecular mechanism. *Trends Biochem Science* 38(10):507–514.

McCord PH, Sosinski BR, Haynes KG, Clough ME, Yencho GC (2011) QTL mapping of internal heat necrosis in tetraploid potato. *Theoret Appl Genet* 122(1):129–42.

Mihovilovich E, Alarcón L, Pérez A, Bonierbale M (1999) *Discovery and evaluation of the valuable new source of resistance to PLRV: Solanum tuberosum subsp. andigena. In: Scientist and ST Farmer· Partners in Research for the 21 Century*. International Potato Center, pp 1999–2000.

Mihovilovich E, Aponte M, Lindqvist-Kreuze H, Bonierbale M (2014) An RGA-derived SCAR marker linked to PLRV resistance from *Solanum tuberosum* ssp. *andigena*. *Plant Mol Biol Rep* 32(1):117–128.

Mori K, Sakamoto Y, Mukojima N, Tamiya S, Nakao T, Ishii T, Hosaka K (2011) Development of a multiplex PCR method for simultaneous detection of diagnostic DNA markers of five disease and pest resistance genes in potato. *Euphytica* 180(3):347–355.

Morris WL, Ducreux LJ, Morris J, Campbell R, Usman M, Hedley PE, Prat S, Taylor MA. (2019) Identification of *TIMING OF CAB EXPRESSION 1* as a temperature-sensitive negative regulator of tuberization in potato. *J Exp Bot* 70(20):5703–14.

Mosquera T, Alvarez MF, Jiménez-Gómez JM, Muktar MS, Paulo MJ, Steinemann S, Li J, Draffehn A, Hofmann A, Lübeck J, Strahwald J (2016) Targeted and untargeted approaches unravel novel candidate genes and diagnostic SNPs for quantitative resistance of the potato (*Solanum tuberosum* L.) to *Phytophthora infestans* causing the late blight disease. *PLoS One* 11(6):e0156254.

Munns R, James RA, Xu B, Athman A, Conn SJ, Jordans C, Byrt CS, Hare RA, Tyerman SD, Tester M, Plett D, Gilliham M (2012) Wheat grain yield on saline soils is improved by an ancestral Na^+ transporter gene. *Nat Biotechnol* 30:360–364.

Munns R, Tester M (2008) Mechanisms of salinity tolerance. *Annu Rev Plant Biol* 59:651–681.

Muthoni J, Kabira JN (2015) Potato production in the hot tropical areas of Africa: Progress made in breeding for heat tolerance. *J Agric Sci* 7(9):220.

Naeem M, Demirel U, Yousaf MF, Caliskan S, Caliskan ME (2021) Overview on domestication, breeding, genetic gain and improvement of tuber quality traits of potato using fast forwarding technique (GWAS): A review. *Plant Breed* 140(4):519–542.

Nunziata A, Ruggieri V, Frusciante L, Barone A (2007) Allele mining at the locus *Gro 1* in *Solanum* wild species. *Acta Hortic* 745:449.

Nyman M, Waara S (1997) Characterization of somatic hybrids between S. tuberosum and its frost-tolerant relative *S. commersonii. Theor Appl Genet* 95:1127–1132.

Oh SK, Young C, Lee M, Oliva R, Bozkurt TO, Cano LM, Win J, Bos JI, Liu HY, van Damme M, Morgan W (2009) In planta expression screens of Phytophthora infestans RXLR effectors reveal diverse phenotypes, including activation of the *Solanum bulbocastanum* disease resistance protein *Rpi-blb2. Plant Cell* 21(9):2928–2947.

Oruetxebarria I, Kekarainen T, Spetz C, Valkonen JP (2000) Molecular characterization of Potato virus V genomes from Europe indicates limited spatiotemporal strain differentiation. *Phytopathol* 90(4):437–444

Oufir M, Legay S, Nicot N, Van Moer K, Hoffmann L, Renaut J, Hausman JF, Evers D (2008) Gene expression in potato during cold exposure: changes in carbohydrate and polyamine metabolisms. *Plant Sci* 175(6):839–52.

Pajerowska-Mukhtar K, Stich B, Achenbach U, Ballvora A, Lübeck J, Strahwald J, Tacke E, Hofferbert HR, Ilarionova E, Bellin D, Walkemeier B (2009) Single nucleotide polymorphisms in the *Allene Oxide Synthase 2* gene are associated with field resistance to late blight in populations of tetraploid potato cultivars. *Genetics* 181(3):1115–1127.

Pajerowska-Mukhtar KM, Mukhtar MS, Guex N, Halim VA, Rosahl S, Somssich IE, Gebhardt C (2008) Natural variation of potato *allene oxide synthase 2* causes differential levels of jasmonates and pathogen resistance in Arabidopsis. *Planta* 228(2):293–306.

Palma ACF (2015) Management of Potato virus Y in potato. Doctoral dissertation, The University of Wisconsin-Madison, Madison, Wisconsin.

Paluchowska P, Śliwka J, Yin Z (2022) Late blight resistance genes in potato breeding. *Planta* 255(6):1–20.

Pankin A, Sokolova E, Rogozina E, Kuznetsova M, Deahl K, Jones R, Khavkin E (2011) Allele mining in the gene pool of wild *Solanum* species for homologues of late blight resistance gene *RB/Rpi-blb1. Plant Genet Res* 9(2):305–308.

Park HJ, Jung WY, Lee SS, Song JH, Kwon SY, Kim H, Kim C, Ahn JC, Cho HS (2013) Use of heat stress responsive gene expression levels for early selection of heat tolerant cabbage (*Brassica oleracea* L.). *Int J Mol Sci* 14(6):11871–94.

Park TH, Vleeshouwers VG, Huigen DJ, Van Der Vossen EA, Van Eck HJ, Visser RG (2005) Characterization and high-resolution mapping of a late blight resistance locus similar to *R2* in potato. *Theor Appl Genet* 111(3):591–597

Pel MA, Foster SJ, Park TH, Rietman H, van Arkel G, Jones JD, Van Eck HJ, Jacobsen E, Visser RG, Van der Vossen EA (2009) Mapping and cloning of late blight resistance genes from *Solanum venturii* using an interspecific candidate gene approach. *Mol Plant–Microbe Interact* 22(5):601–615.

Polivanova OB, Gins EM, Moskalev EA, Voinova MS, Koroleva AK, Semenov AZ, Sivolapova AB, Ivanova AS, Kazakov OG, Simakov EA, Chalaya NA (2021) Quality evaluation, phytochemical characteristics and estimation of *beta-carotene hydroxylase 2* (*Chy2*) alleles of interspecific potato hybrids. *Agronomy* 11(8):1619.

Potato Genome Sequencing Consortium (PGSC) (2011) Genome sequence and analysis of the tuber crop potato. *Nature* 475(7355):189–195.

Poulicard N, Pacios LF, Gallois JL, Piñero D, García-Arenal F (2016) Human management of a wild plant modulates the evolutionary dynamics of a gene determining recessive resistance to virus infection. *PLoS Genet* 12(8):e1006214.

Prakash C, Trognitz FC, Venhuizen P, von Haeseler A, Trognitz B (2020) A compendium of genome-wide sequence reads from NBS (nucleotide binding site) domains of resistance genes in the common potato. *Sci Rep* 10(1):1–21.

Premkumar A, Kumar M, Chauhan R (2020) Improvement of rice variety for biotic and abiotic stress resistance using the technique of allele mining. *J Pharmacogn Phytochem* 9(5):1888–1893.

Qin L, Overmars H, Helder J, Popeijus H, van der Voort JR, Groenink W, van Koert P, Schots A, Bakker J, Smant G (2000) An efficient cDNA-AFLP-based strategy for the identification of putative pathogenicity factors from the potato cyst nematode *Globodera rostochiensis*. *Mol Plant-Microbe Interact* 13(8):830–836.

Rajamäki M, Merits A, Rabenstein F, Andrejeva J, Paulin L, Kekarainen T, Kreuze JF, Forster RL, Valkonen JP (1998) Biological, serological, and molecular differences among isolates of potato A potyvirus. *Phytopathology* 88(4):311–321.

Ritter E, Debener T, Barone A, Salamini F, Gebhardt C (1991) RFLP mapping on potato chromosomes of two genes controlling extreme resistance to potato virus X (PVX). *Mol Gen Genet MGG* 227(1):81–85.

Ross BT, Zidack NK, Flenniken ML (2021) Extreme resistance to viruses in potato and soybean. *Front Plant Sci* 12:658981.

Rubio M, Nicolaï M, Caranta C, Palloix A (2009) Allele mining in the pepper gene pool provided new complementation effects between *pvr2-eIF4E* alleles and *pvr6-eIF* (iso) 4E for resistance to the Pepper veinal mottle virus. *J Gen Virol* 90(11):2808–2814.

Ruocco M, Ambrosino P, Lanzuise S, Woo SL, Lorito M, Scala F (2011) Four potato (*Solanum tuberosum*) ABCG transporters and their expression in response to abiotic factors and *Phytophthora infestans* infection. *J Plant Physiol* 168(18):2225–2233.

Sato M, Nishikawa K, Komura K, Hosaka K (2006) Potato virus Y resistance gene, *Ry chc*, mapped to the distal end of potato chromosome 9. *Euphytica* 149(3):367–372.

Schumacher C, Krannich CT, Maletzki L, Köhl K, Kopka J, Sprenger H, ... Horn R (2021) Unravelling differences in candidate genes for drought tolerance in potato (*Solanum tuberosum* L.) by use of new functional microsatellite markers. *Genes* 12(4):494.

Schumann MJ, Zeng ZB, Clough ME, Yencho GC (2017) Linkage map construction and QTL analysis for internal heat necrosis in autotetraploid potato. *Theor Appl Genet* 130(10):2045–56.

Shaikhaldein HO, Hoffmann B, Alaraidh IA, Aseel DG (2018) Evaluation of extreme resistance genes of Potato virus X (*Rx1* and *Rx2*) in different potato genotypes. *J Plant Dis Prot* 125(3):251–257.

Shaterian J, Waterer DR, de Jong H, Tanino KK (2008) Methodologies and traits for evaluating the salt tolerance in diploid potato clones. *Am J Potato Res* 85:93–100.

Shi J, Wang J, Wang N, Zhou H, Xu Q, Yan G (2019) Overexpression of *StGA2ox1* gene increases the tolerance to abiotic stress in transgenic potato (*Solanum tuberosum* L.) plants. *Appl Biochem Biotechnol* 187(4):1204–1219.

Simko I, McMurry S, Yang HM, Manschot A, Davies PJ, Ewing EE (1997) Evidence from polygene mapping for a causal relationship between potato tuber dormancy and abscisic acid content. *Plant Physiol* 115(4):1453–1459.

Singh A, Siddappa S, Bhardwaj V, Singh B, Kumar D, Singh BP (2015) Expression profiling of potato cultivars with contrasting tuberization at elevated temperature using microarray analysis. *Plant Physiol Biochem* 97:108–116.

Singh B, Kukreja S, Goutam U (2020) Impact of heat stress on potato (*Solanum tuberosum* L.): Present scenario and future opportunities. *J Hortic Sci Biotechnol* 95(4):407–24.

Slater AT, Schultz L, Lombardi M, Rodoni BC, Bottcher C, Cogan NO, Forster JW (2020) Screening for resistance to PVY in Australian potato germplasm. *Genes* 11(4):429.

Smýkal P, Nelson MN, Berger JD, Von Wettberg EJ (2018) The impact of genetic changes during crop domestication. *Agron* 8(7):119.

Solís JL, Muth J, Canales J, Lizana C, Pruefer D, Riegel R, Behn A (2022) Allelic diversity of three anthocyanin synthesis genes in accessions of native *Solanum tuberosum* L. ssp. *tuberosum* at the Potato Genebank of the Universidad Austral de Chile. *Genet Resour Crop Evol* 69(1):297–314.

Solomon-Blackburn RM, Barker H (2001) Breeding virus resistant potatoes (*Solanum tuberosum*): a review of traditional and molecular approaches. *Heredity* 86(1):17–35.

Song J, Bradeen JM, Naess SK, Raasch JA, Wielgus SM, Haberlach GT, Liu J, Kuang H, Austin-Phillips S, Buell CR, Helgeson JP (2003) Gene RB cloned from *Solanum bulbocastanum* confers broad spectrum resistance to potato late blight. *Pro NatlAcad Sci USA* 100(16):9128–9133.

Sorri VA, Watanabe KN, Valkonen JP (1999) Predicted kinase-3a motif of a resistance gene analogue as a unique marker for virus resistance. *Theor Appl Genet* 99(1):164–170.

Spoel SH, Dong X (2012) How do plants achieve immunity? Defence without specialized immune cells. *Nat Rev Immunol* 12(2):89–100.

Sprenger H, Kurowsky C, Horn R, Erban A, Seddig S, Rudack K, Fischer A, Walther D, Zuther E, Köhl K et al (2016) The drought response of potato reference cultivars with contrasting tolerance. *Plant Cell Environ* 39:2370–2389.

St. Clair DA (2010) Quantitative disease resistance and quantitative resistance loci in breeding. *Annual Rev Phytopathology* 48:247–268.

Strachan SM, Armstrong MR, Kaur A et al (2019) Mapping the H2 resistance effective against *Globodera pallida* pathotype *Pa1* in tetraploid potato. *Theor Appl Genet* 132(4): 1283–1294.

Sung DY, Guy CL. (2003) Physiological and molecular assessment of altered expression of *Hsc70-1* in Arabidopsis. Evidence for pleiotropic consequences. *Plant Physiol* 32(2):979–87.

Szajko K, Chrzanowska M, Witek K, Strzelczyk-Żyta D, Zagórska H, Gebhardt C, Hennig J, Marczewski W (2008) The novel gene *Ny-1* on potato chromosome IX confers hypersensitive resistance to Potato virus Y and is an alternative to *Ry* genes in potato breeding for PVY resistance. *Theor Appl Genet* 116(2):297–303.

Szalonek M, Sierpien B, Rymaszewski W, Gieczewska K, Garstka M, Lichocka M, ... Konopka-Postupolska D (2015) Potato annexin *STANN1* promotes drought tolerance and mitigates light stress in transgenic *Solanum tuberosum* L. plants. *PLoS One* 10(7):e0132683.

Tacke E, Salamini F, Rohde W (1996) Genetic engineering of potato for broad-spectrum protection against virus infection. *Nat Biotechnol* 14(11):1597–1601.

Tagliotti ME, Deperi SI, Bedogni MC, Huarte MA (2021) Genome-wide association analysis of agronomical and physiological traits linked to drought tolerance in a diverse potatoes (*Solanum tuberosum*) panel. *Plant Breed* 140(4):654–664.

Tai HH, Goyer C, Murphy AM (2013) Potato *MYB* and *bHLH* transcription factors associated with anthocyanin intensity and common scab resistance. *Botany* 91(10):722–730.

Taye MM (2016) *Determining the Type of PVY Extreme Resistance Genes (Ryadg and/or Rysto) in Genotypes Used in the Breeding Programs of the Potato Research Centre.*

Tian Z, He Q, Wang H, Liu Y, Zhang Y, Shao F, Xie C (2015) The potato ERF transcription factor *StERF3* negatively regulates resistance to *Phytophthora infestans* and salt tolerance in potato. *Plant Cell Physiol* 56(5):992–1005.

Tian ZD, Zhang Y, Liu J, Xie CH (2010) Novel potato *C2H2*-type zinc finger protein gene, *StZFP1*, which responds to biotic and abiotic stress, plays a role in salt tolerance. *Plant Biol* 12(5): 689–697.

Till BJ, Comai L, Henikoff S (2007) TILLING and EcoTILLING for crop improvement. In: Varshney RK, Tuberosa R (eds), *Genomics-Assisted Crop Improvement.* Springer, Dordrecht, pp 333–349.

Tiwari JK, Devi S, Sharma S, Chandel P, Rawat S, Singh BP (2015) Allele mining in *Solanum* germplasm: cloning and characterization of RB-homologous gene fragments from late blight resistant wild potato species. *Plant Mol Biol Rep* 33(5):1584–1598.

Tiwari JK, Kumar V, Devi S, Luthra SK, Chakrabarti SK, Rawat S, Nagesh M (2017) Genomics in management and genetic enhancement of potato germplasm. In: Chakrabarti SK, Xie C, Tiwari JK (eds), *The Potato Genome*. Springer, Cham, pp 123–142.

Tiwari JK, Siddappa S, Singh BP, Kaushik SK, Chakrabarti SK, Bhardwaj V, Chandel P (2013) Molecular markers for late blight resistance breeding of potato: an update. *Plant Breed* 132(3):237–245.

Trapero-Mozos A, Morris WL, Ducreux LJ, McLean K, Stephens J, Torrance L, Bryan GJ, Hancock RD, Taylor MA (2018) Engineering heat tolerance in potato by temperature-dependent expression of a specific allele of *HEAT-SHOCK COGNATE 70*. *Plant Biotechnol J* 16(1):197–207.

Trognitz BR, Trognitz FC (2007) Occurrence of the *R1* allele conferring resistance to late blight in potato *R*-gene differentials and commercial cultivars. *Plant Pathol* 56(1):150–155.

Tsedaley B (2015) A review paper on Potato virus Y (PVY) biology, economic importance and its managements. *J Biol Agric Healthcare* 5(9):110–126.

Tu W, Dong J, Zou Y, Zhao Q, Wang H, Ying J, Wu J, Du J, Cai X, Song B (2021) Interspecific potato somatic hybrids between *Solanum malmeanum* and *S. tuberosum* provide valuable resources for freezing-tolerance breeding. *Plant Cell Tiss Org Cult* 147(1):73–83.

Valkonen JP, Puurand U, Slack SA, Makinen K, Saarma M (1995) Three strain groups of potato A potyvirus based on hypersensitive responses in potato, serological properties, and coat protein sequences. *Plant Dis* 79(7):748–753.

Valkonen JP, Wiegmann K, Hämäläinen JH, Marczewski W, Watanabe KN (2008) Evidence for utility of the same PCR-based markers for selection of extreme resistance to Potato virus Y controlled by *Rysto* of *Solanum stoloniferum* derived from different sources. *Ann Appl Biol* 152(1):121–130.

Vallejo CD, Gutiérrez SPA, Marín MM (2016) Genome characterization of a Potato virus S (PVS) variant from tuber sprouts of *Solanum phureja* Juz. et Buk. *Agron Colomb* 34(1):51–60.

Van den Berg JH, Ewing EE, Plaisted RL, McMurry S, Bonierbale MW (1996) QTL analysis of potato tuber dormancy. *Theor Appl Genet* 93(3):317–324.

van der Linden CG, Wouters DC, Mihalka V, Kochieva EZ, Smulders MJ, Vosman B (2004) Efficient targeting of plant disease resistance loci using NBS profiling. *Theor Appl Genet* 109(2):384–393.

Van Der Plank JE (1966) Horizontal (polygenic) and vertical (oligogenic) resistance against blight. *Am Potato J* 43(2):43–52.

van der Voort JR, Lindeman W, Folkertsma R, Hutten R, Overmars H, Van der Vossen E, Jacobsen E, Bakker JA (1998) QTL for broad-spectrum resistance to cyst nematode species (*Globodera spp.*) maps to a resistance gene cluster in potato. *Theor Appl Genet* 96(5):654–661.

van der Vossen EA, Gros J, Sikkema A, Muskens M, Wouters D, Wolters P, Pereira A, Allefs S (2005) The *Rpi-blb2* gene from *Solanum bulbocastanum* is an Mi-1 gene homolog conferring broad-spectrum late blight resistance in potato. *Plant J* 44(2):208–222.

van Eck HJ, Jacobs JM, Van Den Berg PM, Stiekema WJ, Jacobsen E (1994) The inheritance of anthocyanin pigmentation in potato (*Solanum tuberosum* L.) and mapping of tuber skin colour loci using RFLPs. *Heredity* 73(4):410–421.

van Eck, HJ, Vos PG, Valkonen J, Uitdewilligen JG, Lensing H, de Vetten N, Visser RG (2017) Graphical genotyping as a method to map *Ny* (o, n) *sto* and *Gpa5* using a reference panel of tetraploid potato cultivars. *Theor Appl Genet* 130(3): 515–528.

Velásquez AC, Mihovilovich E, Bonierbale M (2007) Genetic characterization and mapping of major gene resistance to potato leafroll virus in *Solanum tuberosum* ssp. *andigena. Theor Appl Genet* 114(6):1051–1058.

Verzaux E, Budding D, de Vetten N, Niks RE, Vleeshouwers VG, van der Vossen EA, Jacobsen E, Visser RG (2011) High resolution mapping of a novel late blight resistance gene Rpi-avl1, from the wild Bolivian species *Solanum avilesii. Am J Potato Res* 88(6):511–519.

Verzaux E, van Arkel G, Vleeshouwers VG, van der Vossen EA, Niks RE, Jacobsen E, Vossen J, Visser RG (2012) High-resolution mapping of two broad-spectrum late blight resistance genes from two wild species of the *Solanum circaeifolium* group. *Potato Res* 55(2):109–123.

Vidal S, Cabrera H, Andersson RA, Fredriksson A, Valkonen JP (2002) Potato gene *Y-1* is an N gene homolog that confers cell death upon infection with potato virus Y. *Mol Plant–Microbe Interact* 15(7):717–727.

Vleeshouwers VG, Finkers R, Budding D, Visser M, Jacobs MM, van Berloo R, Pel M, Champouret N, Bakker E, Krenek P, Rietman H (2011) SolRgene: an online database to explore disease resistance genes in tuber-bearing *Solanum* species. *BMC Plant Biol* 11(1):1–9.

Vossen JH, Jo KR, Vosman B (2014) Mining the genus *Solanum* for increasing disease resistance. *Genom Plant Genet Res* 2014:27–46.

Wang M, Allefs S, van den Berg RG, Vleeshouwers VG, van der Vossen EA, Vosman B (2008) Allele mining in Solanum: conserved homologues of *Rpi-blb1* are identified in *Solanum stoloniferum. Theor Appl Genet* 116(7):933–943.

Were HK, Kabira JN, Kinyua ZM, Olubayo FM, Karinga JK, Aura J, Lees AK, Cowan GH, Torrance L (2013) Occurrence and distribution of potato pests and diseases in Kenya. *Potato Res* 56(4):325–342.

Witek K, Strzelczyk-Żyta D, Hennig J, Marczewski W (2006) A multiplex PCR approach to simultaneously genotype potato towards the resistance alleles *Ry-f sto* and *Ns. Mol Breed* 18(3):273–275.

Wolters AM, Uitdewilligen JG, Kloosterman BA, Hutten RC, Visser RG, van Eck HJ (2010) Identification of alleles of carotenoid pathway genes important for zeaxanthin accumulation in potato tubers. *Plant Mol Biol* 73(6):659–671.

Xiong L, Lee H, Ishitani M, Zhu JK (2002) Regulation of osmotic stress-responsive gene expression by the los6/aba1 locus in Arabidopsis. *J Biol Chem* 277(10):8588–96.

Xu Q, He Q, Li S, Tian Z (2014) Molecular characterization of *StNAC2* in potato and its overexpression confers drought and salt tolerance. *Acta Physiol Plant* 36(7):1841–1851.

Yuan J, Bizimungu B, De Koeyer D, Rosyara U, Wen Z, Lagüe M (2020) Genome-wide association study of resistance to potato common scab. *Potato Res* 63(2):253–266.

Yusa A, Neriya Y, Hashimoto M, Yoshida T, Fujimoto Y, Hosoe N, Keima T, Tokumaru K, Maejima K, Netsu O, Yamaji Y (2019) Functional conservation of EXA1 among diverse plant species for the infection by a family of plant viruses. *Sci Rep* 9(1):1–12.

Zhao P, Wang D, Wang R, Kong N, Zhang C, Yang C, . . . Chen Q (2018) Genome-wide analysis of the potato Hsp20 gene family: identification, genomic organization and expression profiles in response to heat stress. *BMC Genom* 19(1):1–13.

Zhou X, Zhang N, Yang J, Tang X, Wen Y, Si H (2018) Functional analysis of *StDWF4* gene in response to salt stress in potato. *Plant Physiol Biochem* 125:63–73.

Zoteyeva N, Meþaka I, Vilcane D, Carlson-Nilsson U, Skrabule I, Rostoks N (2014) Assessment of genes *R1* and *R3* conferring resistance to late blight and of gene *RYsto* conferring resistance to potato virus Y in two wild species accessions and their hybrid progenies/Izturîbas Pret Lakstu Puvi Gçnu R1 Un R3 Un Izturîbas Pret Kartupeïu Vîrusu Y Gçna Rysto Novçrtçjums Divu Savvaïas Sugu Lînijâs Un To Hibrîdos. *In Pro Lat Acad Sci De Gruyter Poland* 68(3–4):133.

3 Allele Mining in Tomato

Strategies and Applications in Gene Discovery for Food and Nutritional Security

Michael Kwabena Osei,[1] Joseph Adjebeng-Danquah,[2] Victor Acheampong Amankwaah,[1] Fredrick Justice Awuku,[2] Herbaud Phanuel Fortunat Zohoungbogbo,[3] Leander Dede Melomey,[4] Richard Yaw Agyare,[2] Doris Kanvenaa Puozaa,[2] Isaac Newton Boakye Mensah,[1] and Isaac Osei-Bonsu[1]
[1]CSIR-Crops Research Institute, Kumasi, Ghana
[2]CSIR-Savannah Agricultural Research Institute, Nyanpkala, Ghana
[3]Worldveg-West and Central Africa (WCA) - Coastal and Humid Regions, IITA Campus, Abomey-Calavi, Republic of Benin
[4]WACCI-University of Ghana, Legon, Ghana

3.1 INTRODUCTION

Innovation in plant breeding in terms of development of superior varieties of agricultural crops is promising by accumulation of beneficial alleles from large plant genetic resources present worldwide. Access to allelic variation that affects the plant phenotype is of maximum significance for the exploitation of genetic resources, such as in plant variety development. Considering the vast numbers of accessions that are held together by gene banks, genetic resource collections are considered to harbor a wealth of hidden allelic variants. Mutation is considered as an evolutionary driving force that underlies existing allelic diversity in any crop species. For creation of new alleles or causing variations in the existing allele and allelic combinations, mutations in the genic regions of the genome either as single nucleotide polymorphism (SNP) or as insertion and deletion (InDel) are important. The mutations in coding regions and or regulatory regions may have tremendous effect on the phenotype by altering the encoded protein structure and function. True allele mining includes analysis of non-coding and regulatory regions of the candidate genes in addition to analyzing sequence variations in the coding regions of genes. Perhaps, how to reveal this variation is the challenge. Enormous sequence information is available in public databases because of sequencing of diverse crop genomes (Kumar et al. 2010). It is important to use this genomic information for the identification and isolation of novel and superior alleles of agronomically important genes from crop gene pools to suitably deploy for the development of improved cultivars. Allele mining is a research field targeted at identifying or finding allelic variation of important traits within genetic resource collections (www.wur.nl/en/show/allele-mining.htm). It is a promising way to dissect naturally occurring allelic variants of candidate genes with essential agronomic qualities (Kumar et al. 2010). It helps in tracing the evolution of alleles, identification of new haplotypes and development of allele-specific markers for use in marker-assisted selection. Realizing the immense potential of

DOI: 10.1201/9781003376439-3

allele mining, concerted allele mining efforts are underway in many international crop research institutes. An allele is an alternative form of a gene found at a given locus on a chromosome. Mining an allele however, is searching the new alleles in say wild germplasm. Thus, allele mining involves searching of useful alleles of genes from a wide range of cultivars, related species, and even across species. Allele mining requires various sophisticated bio-informatic tools, viz., PLACE, plantCARE, TRANSFAC, JASPAR, MEME, Plantprom DB, DCPD, SCPD, BioEdit, ClustalW2, etc. These tools are useful for sequence alignment in order to compare the new sequence to reference genome, i.e., sequenced genome.

The concept of food security has moved beyond the quantity of food available to a population to include the nutritional component as well. For many decades tomato breeding efforts focused on the development of cultivars for higher yields and resistance to various biotic and abiotic stresses. Over time, consumers are increasingly demanding tasty and nutritious tomato (*Solanum lycopersicum* L.) fruits (Fortuny et al. 2021). This demand for better fruit quality is evidenced by the tail of the "water bombs" tomato cultivar that led to the decline in Dutch tomato export to Germany (Schouten et al. 2019). The challenge for plant breeders has been to continually provide tomato varieties with superior fruit qualities tailored to fit various market segments without compromising on important agronomic traits. This task has been made possible with the increasing genetic diversity observed in the tomato crop contrary to popular observations in other species where genetic diversity has been eroded because of extensive breeding efforts. The increase in genetic diversity is attributable to the use of wild relatives, landraces, modern cultivars, and mutants as sources of genetic variation for various traits coupled with advances in genomics. Today tomato cultivars come in numerous market classes and colors. These classes include the cherry, saladette, large round slicer types, heirloom, and plum types (good for processing). The fruits come in different colors and color combinations (Mckenzie and Zystro 2021). Many breeding programs are targeting flavors defined by sweetness, mainly determined by sugars; acidity, determined by the presence of organic acids; textural attributes, such as firmness and juiciness, and aroma (Roohanitaziani et al. 2020). This chapter examines the tomato genome, concept of allele mining, application of allele mining in tomato, mapping and validation of QTL in gene discovery, bioinformatics tools for allele mining, prospects, and challenges of allele mining in gene discovery.

3.2 TOMATO GENOME

3.2.1 The Tomato Genome: Current Status

The tomato genome was first sequenced through a concerted effort of scientists from ten different countries as the tomato genome project, which began in 2004 (Fernandez-Pozo et al. 2015). Combining high-coverage, shorter SOLiD and Illumina GAIIx reads with longer Sanger and 454/ Roche GS FLX reads allowed for the sequencing of the domesticated tomato Heinz 1706's genome in 2012 (Pietrella and Giuliano 2016). The sequences were put together into 91 scaffolds spanning about 84% of the 900 Mb genome and contained 34,727 putative protein-coding genes (Mueller et al. 2009; Ren 2012). The genomes of other economically valuable members of the Solanaceae family, including eggplant and pepper, and the tomato, have a high degree of synteny. In the Solanum lineage, two sequential triplication events were found using comparative genomic analysis. As a result, novel gene families, including transcription factors and enzymes required for ethylene biosynthesis and sensing, which mediate crucial fruit-specific functions were added (Ranjan 2012). The genome sequence of Heinz 1706 has since served as a reference sequence for most scientific studies. Improvement in the sequencing and assembly technique have improved the quality of the first Heinz 1706 genome. Using the PacBio sequencing platform (Hosmani et al. 2019) provided an updated reference genome (SL4.0) of Heinz 1706 (SL2.5) with fewer gaps and highly contiguous. The quest to refine the reference genome as advances in techniques keep emerging resulted in an update on SL4.0 by Su et al. (2021). Their reference genome STL1.0 was generated by comparing

TABLE 3.1
Summary of Evolution of Genome Assembly and Annotation of Tomato

	SLT1.0	SL4.0	SL3.0	SL2.5
Genome assembly (Mb)	799.09	782.52	828.08	823.94
Non-Nbases	797,955,212	782,475,302	746,357,470	737,636,348
Number of gaps	210	286	22,700	
Number of total contigs	1,615	504	-	
Longest contig length (Mb)	47.16	26.29	-	
N50 of contigs (Mb)	17.83	6.01	-	
Number of unplaced contigs	112	176	4,374	4,410
Unplaced contigs sequence length (Mb)	8.5	9.64	20.85	21.81
Number of genes	34,384	34,075	35,768	
percentage of gene length in genome (%)	16.21	15.56	17.33	
Mean gene length (bp)	3,766.53	3,572.44	4,011.09	
Gene density (per bp)	43.03	43.55	43.19	
Mean coding sequence length (bp)	223.02	228.01	219.97	
Mean exon length (bp)	310.11	275.03	308.36	
Mean intron length (bp)	270.41	606.69	632.38	
Masked repeat sequence length (Mb)	558.49	546.95	507.14	
Repeats percentage of genome size (%)	69.89	69.9	61.24	

the genomes of *S. pimpinellifolium* "LA2093", and *S. lycopersicum.* Table 3.1 gives a summarization of the genome evolution of tomato from SL2.5 to SLT1.0.

3.2.2 Sequencing of the Tomato Genome

The tomato genome sequence was an effort by a consortium of ten countries with each assigned a specific role and an overall aim to shed light on the diversification and adaptation of the *Solanaceae* family. The USA was assigned chromosome 1 and 10, Korea chromosome 2, China chromosome 3 and 11, UK chromosome 4, India chromosome 5, Netherlands chromosome 6, France chromosome 7, Japan chromosome 8, Spain chromosome 9, and Italy chromosome 12. Tomato was chosen as the model crop due to the large genetic resources of tomato available at the time, thus making it easier to start such an important project (Pietrella and Giuliano 2016; Ranjan et al. 2012).

The sequencing approach was initially based on the BAC-by-BAC sequencing method grounded on seed BAC genetic maps. Prior to the tomato genome project, the BAC-by-BAC method had been successful in sequencing the 125 Mb Arabidopsis reference genome, this method was less expensive than other Sanger methods available at the time considering the large availability of mapping population of lycopersicum and pennelli, and availabilities of BAC libraries (Pietrella and Giuliano 2016). Sequencing began in 2004 with focus on the euchromatic region of all 12-chromosome based on concentration of available markers at this segment of the chromosome compared to other regions, the region was also perceived to be gene rich and had 90% of all non-transposon genes. The final genetic map used in selection of seed BACs contained different DNA markers (SSR, AFLP, EST, etc.) obtained from the Tomato-EXPEN 2000 mapping population. Using the BAC libraries (Hind III, EcoR I, and Mbo I) and the fosmid library (Fosmid), marked progress was achieved in the sequencing of the genome in its fourth year (Pietrella and Giuliano 2016; Sato et al. 2012). Using the data on the sequences generated since the start of the project, a Snapshot physical map was generated as well as a WGP map, the summaries of these maps are presented in Table 3.2.

Though the BAC-by-BAC yielded significant results, there were unforeseen bottlenecks as the sequencing progressed. Again, advances in sequencing technology, such as the next-generation

TABLE 3.2
A Comparison of the WGP and Snapshot Map of Tomato

	WGP Map	SNAPshot Map
Total number of BACs	66,084	82,784
Number of contigs	2521	1835
Number of BACs in contigs	52,617	66,810
Number of singleton BACs	13,467	15,974
Coverage	953	859
Average Number of BACs/contig	21	36
N50 contig size (Number of BACs)	26	67
Average contig size (Mbp)	0.378	0.39
N50 contig size (Mbp)	0.563	0.697

sequencing (Roche/454, ABI, SoLiD, and illumine/solexa) provided better throughput. The assumption of only the euchromatin region of the chromosome being gene rich was not true. Hence, the whole genome sequencing approach was adapted to improve on the deficiencies of the BAC-by-BAC method. A combination of which resulted in the first tomato reference genome (SL2.5). Major additions and improvement have been made to the reference genome since the completion of the first genome.

3.3 TOMATO GENETICS AND GENOMICS

Plants are vital for maintaining the balance of the ecosystem in that they play other essential roles, such as balancing the oxygen and carbon dioxide in the atmosphere aside from their primary role of food, cloth, and shelter for animals and humans. Plant breeding plays an important role in maintaining this balance and providing for the health and nutritional needs of the growing population of humans. Since the beginning of agriculture, humankind learnt to select based on characters favorable to them, marking the beginning of crop genetics and its maintenance. The discovery of DNA and its manipulations with associated technologies revolutionized the genetics and genomics of crop plants (Abdurakhmonov 2016; Ahmad and Mukhtar 2017). The conservation of tomato germplasms helps prevent the erosion of genetic diversity and protects the gene pools that can be relied on in times of need. Without conversation, most novel genes usually stack in wild relative and less adapted cultivars and at times get lost to climate change and other natural disasters (Pathirana and Carimi 2022; Wambugu and Henry 2022). This will further reduce the already narrow genetic base of tomato. There are several centers for the collection and preservation of tomato germplasm across the world (Table 3.3), thus conserving the gene pools from wild and domesticated tomato.

Tomato characterization began with the use of morphological features before the advent of DNA technology (Ichihashi and Sinha 2014; Mata-Nicolás et al. 2020). Isozyme markers were used in the study of tomato genetics and together with morphological characters, linkage maps of tomato were constructed consisting of 12 linkage groups (Rick and Fobes 1975; Tanksley and Rick 1980). As tomato genetics advance, more dense and fine maps have been developed to understand in more details, tomato genetics using different marker systems (Foolad 2007; Gonda et al. 2019). This has resulted in the identification of several domestication related QTLs to help breeders produce much improved cultivars with traits of importance (Adhikari et al. 2020; Grandillo 1996; Rodríguez et al. 2011; Würschum 2012). Also, paving the way for marker-assisted breeding (MAB) approaches (Foolad and Panthee 2012, 2013) to reduce breeding time and have a targeted approach in tomato improvement targeting disease resistance (Barchenger et al. 2022; Bernal 2020; Lukyanenko 1991; Panthee 2017; Ramírez et al. 2020; Scott 2005), insect resistance

TABLE 3.3
Tomato Germplasm Conversation Centers

Acronym	Country	Number of Germplasms
AVRDC	Taiwan	8,259
NE9	USA	6,343
DHSNYST	USA	4,850
IPB-UPLB	Philippines	4,751
Campbell Inst. Agric. Res.	USA	4,572
IPK	Germany	4,063
GSLY	USA	3,395
VIR	Russian Federation	2,540
NIAS	Japan	2,428
PGRC	Canada	2,137
ICA/REGION 5	Colombia	2,018
BGUPV	Spain	1,928
NBPGR	India	1,788
INRAE	France	3410

(Nienhuis et al. 1987; Ricardo Zeist et al. 2019), nutrient use efficiencies, and increased yields. The availability of the tomato genome sequence has hastened our understanding of how genes function in metabolic and developmental processes and help map relevant tomato mutants to pinpoint crucial co-regulation mechanism steps. Additionally, numerous TILLING (Targeting Induced Local Lesions IN Genomes) resources with various backgrounds have been created for tomato functional genomics (Ranjan 2012). Several genetic databases have been developed for ease of access to tomato genetic and genomic resources, these include the tomato functional genomics database (TFGD) (http://ted.bti.cornell.edu/), tomato genomic resource database (TGRD) (Suresh et al. 2014), the Solanaceae Genomics Network also known as Sol genomic networks (SGN) (https://solgenomics.net/), Tomatonet (Kim et al. 2017), etc. Hence, making it easier to access these resources for breeding and research purposes.

3.4 RATIONALE, CONCEPTS, AND IMPORTANCE OF ALLELE MINING

3.4.1 What Is Allele Mining?

Crop improvement involves the accumulation of desirable genes present in the genome of a particular species. These genes can come from the same species or different species including wild relatives. However, some of these alleles are not always tapped from the genepool due to inability of breeders to identify these genes during the selection process (Kumar et al. 2010). Ability to identify and incorporate these desirable genes especially those from wild relatives into cultivated crops go a long way to improve their agronomic and economic performance including nutrition (McCouch et al. 2007). In recent times, molecular biology has made it possible to explore the genome of different crop species to provide information on the sequence of genes controlling important traits of economic importance. This will make it possible to develop allele-specific markers so that desirable genes in potential donors can be identified and isolated (Mardis 2008; Spooner et al. 2005).

3.4.2 The Concept of Allele Mining

Allele mining involves the process of identifying allelic variations or genomic regions controlling relevant traits within the genepool. It involves the identification of basic DNA sequences and their functions, which may be used to screen a collection of genetic resources including related species and even across species (Sarika and Amruta 2014). In allele mining, novel genomic tools are used to identify and isolate the alleles responsible for a given trait in a genotype and other variants of that same gene. The variations could be due to mutations in the genotype of the crop species. These mutations could be due to insertions or deletions of certain genomic regions that are necessary for the survival of the species (Singh 2005).

3.4.3 Current State of Allele Mining

Allele mining can take two main forms for the identification of desirable alleles of a particular gene (Kumar et al. 2010; Kumari et al. 2018; Thomson et al. 2022). These are as follows: (1) Target Induced Local Lesions in Genomes (TILLING) and (2) sequencing-based allele mining. TILLING involves identification of polymorphism that results from induced mutation (Till et al. 2004). In the case of allele mining, a special form of TILLING known as Eco-TILLING is used to target naturally occurring mutations in the genome of crops (Comai and Henikoff 2004). In Eco-TILLING, the single-stranded specific nucleases are used to cleave heteroduplexed DNA single nucleotide mismatch in DNA sequence between the text genotype and a reference sequence (Kumar et al. 2010). In sequencing-based allele mining, alleles in a diverse group of genotypes are amplified through polymerase chain reaction followed by DNA sequencing is used to identify nucleotide variation (Simsek and Kacar 2010). This technique helps to multiply specific segments of the genome and analyze the individual's haplotype structure and the detected diversity in order to determine the associations with other genotypes (Sharma et al. 2014).

New techniques such as next-generation sequencing provide new opportunities to reduce the cost of sequencing and provide more accurate and reliable data in a timely manner. These sequencing approaches are used for resequencing, alignment of sequencing information, and their comparison with the reference genotype. Comparison of the detected polymorphism with the reference genome can help detect new alleles that could be linked with traits of importance (Kumar et al. 2010). More so, identification of specific genes in identified genotypes and their relationships with desired traits of interest will result in marked progress in crop improvement. Following the identification and detection of new alleles, association mapping needs to be done in order to link these loci to specific traits (Kumari 2018). Association mapping using the identified markers enables breeders to associate specific regions of the genome with key phenotypic traits through linkage disequilibrium (Kumar et al. 2016).

3.4.4 The Role of Allele Mining in Gene Discovery

The role of allele mining in gene discovery cannot be over emphasized. Allele mining provides a unique opportunity for the identification and discovery of new alleles of genes. Through the use of bioinformatics tools, new alleles are discovered following sequence alignment of the reference genome with the newly sequenced genome (Kumari 2018). Sequence comparison has been used to identify syntenic relationships in rice and finger millet. Li et al. (2016) reported the discovery of QTLs of finger blast and neck blast in finger millet, which may be targeted for allele mining in order to identify genes for blast resistance. Allele mining has also been very useful in haplotype discovery in candidate genes for drought tolerance in barley (Al-Yassin and Khademian 2015). Allele mining also provides in-depth information concerning the molecular basis of unique trait variations that are associated with certain genotypes and identify potential changes that may be expected through the introgression of new superior alleles (Kumar et al. 2010). Allele mining has also been very helpful in unlocking the inherent genetic diversity among collections of genetic resources in gene banks (Kaur et al. 2008a). The construction of a pan-genome has facilitated a comprehensive exploration of the breeding history of tomato including domestication as well as complete characterization of the tomato gene functions. A pan-genome comprises all the useful genetic elements present in cultivated species and their wild progenitors (Gao et al. 2019). Some of these useful alleles are linked to fruit characteristics including flavor in tomatoes.

3.5 APPLICATION OF ALLELE MINING IN TOMATO

3.5.1 Approaches to Allele Mining in Tomato

There are two major approaches employed in the identification of polymorphisms for a given gene sequence within a population. These are modified TILLING, also known as Eco-TILLING and sequencing-based allele mining. There have been various arguments about which of the approaches is superior and cost-effective. However, each of them has its strengths and weaknesses (Table 3.4) the choice of the approach to use must be made after careful considerations such as the purpose, access to genotyping and phenotyping infrastructure and time. These approaches are briefly described below.

3.5.1.1 Modified TILLING *(Eco-TILLING)*

TILLING (Targeting Induced Local Lesions in Genomes) is the identification of polymorphisms, especially in point mutations, resulting from mutations induced in a target gene. It is a valuable and non-transgenic reverse-genetic used in mutagenized gene discovery and mapping and can be used in any organism that can be mutagenized (McCallum et al. 2000; Till et al. 2004; Slade et al. 2005; Barkley and Wang 2008). TILLING begins with the development of a mutagenized population. This is followed by DNA preparation and pooling, and mutation discovery. The mutations are induced in genomes by using chemical mutagens such as Ethylmethanesulfonate (EMS) to treat seeds. The M1 plants from the resulting mutagenized seed are self-fertilized and the M2 individuals subsequently obtained are then used to prepare DNA samples for screening for mutations (Kumari et al. 2018). Modified TILLING, also known as Eco-TILLING provides a means to determine the extent of natural

TABLE 3.4
A Comparison Between Eco-TILLING and Sequencing-Based Allele Mining

S. No.	Parameters	Eco-TILLING	Sequencing-Based Allele Mining
1.	Technical expertise	Requires high technical expertise starting from DNA pooling to detection of cleavage	Require less expertise with direct sequencing of heteroduplexes of PCR products
2.	Complexity	More complex	Less complex
3.	Efficiency	Less efficient due to high chances of false positives, non-specific cleavage and chances of non-detection in DNA pools	Highly efficient as it involves one step direct identification of sequence
4.	Utility	Proposed as effective in detection of SNPs rather than InDels	Effective in detection of any type of nucleotide
5.	Cost per data point	Comparatively high as it needs the intervention of confirmatory sequencing step	Comparatively less cost is involved
6.	Time	Requires more time especially for sample preparation	Comparatively less time is required
7.	Throughput	Associated complexity reduces the throughput only less sample can be processed	Throughput and sample size increases with and massively parallel sequencing platforms

Source: Gokidi et al. 2017.

variation in selected genes. That is, in Eco-TILLING (Comai et al. 2004), the mutations are from naturally occurring alleles in primary and secondary gene pools. Eco-TILLING, like TILLING, relies on enzymatic cleavage of heteroduplexed DNA formed from single nucleotide mismatch in sequence between the reference and test genotype, with a single-strand-specific nuclease. This is followed by detection through Li-Cor genotypers. The CEL I-cleaved products generated during TILLING are commonly detected by polyacrylamide gel electrophoresis using a high-throughput genotyping platform and the SNPs are determined by sequencing. For increased throughput and economy of scale, multiple thermal cyclers and analyzers can be used (Till 2006). As an alternative to the use of polyacrylamide gel (Raghavan et al. 2007) explored the use of conventional agarose gels and were successful in detecting SNPs in DNA templates with throughput comparable to the Li-Cor genotyper. They also showed that the throughput could be increased by analyzing larger amplicons (Raghavan et al. 2007). Eco-TILLING is useful when screening many individuals at one locus, discovering rare haplotypes and good when working on heterozygous populations (Kumar et al. 2014). Eco-TILLING/TILLING has been successfully applied including in maize, rice, wheat, barley, tomato, and soybean.

3.5.1.2 Sequencing-Based Allele Mining

The sequencing-based allele mining approach involves the amplification of alleles of different genotypes using polymerase chain reaction and subsequent identification of nucleotide variation via DNA sequencing. The main procedures include (i) sample collection; (ii) extraction of DNA; (iii) designing of the primer; (iv) PCR amplification; and (v) sequencing and identifying the allelic variations (Gokidi et al. 2017).

Next-generation sequencing for allele mining. DNA sequencing enables the understanding of the arrangement of nucleotides in DNA molecules. The development of advanced sequencing techniques initiated after the completion of the human genome draft led to the emergence of high-throughput sequencing, known as next-generation sequencing (NGS) technologies (Gokidi et al. 2017). Over time, “massively parallel” methods were developed leading to the development of “next-generation” sequencing platforms with high throughput and accuracy (Kumar et al. 2010). These methods are used for resequencing and aligning the sequence data and making comparisons with the reference genome. They include the 454 FLX, which marked the first commercialization of this technology and relied on pyrosequencing while eliminating the need for cloning (Gokidi et al. 2017; Kumari et al. 2018). Other sequencers are HiSeq 2000, SOLiDv4, and Sanger 3730xl. They vary in their sequencing mechanism, read length, accuracy, output as well as run time and hence differ in cost and applicability (Kumar et al. 2010; Liu et al. 2012). The progress made in the development of sequence technologies has increased the sequencing capacities and whole-genome resequencing achievable by individual laboratories. Using these technologies, candidate genes for thousands of individuals, entire transcriptomes or entire plant genomes more efficiently and economically and in greater depth than using more technologies that are traditional.

3.5.2 Exploration of Allele Mining for Enhanced Seed Quality, Fruit Quality and Shelf Life

Seed sets the upper limit for every other production input and determines the success of germination, growth, and yield of the crop. It is one of the most prominent tools by which the gains made in crop improvement efforts are carried over time and space. Seed quality establishes the level of performance of the seed when planted. It also describes the physiological soundness and health status of seed. Ghana, like other countries all over the world, has set minimum quality standards for seed of food crops that are termed seed certification standards. A good seed lot must meet the minimum quality standards defined for the crop and seed class. Seed quality can be measured by the genetic purity, physical purity, germination, and vigor. A good seed lot must be free from diseases, have acceptable moisture content, and be free from other crop seeds, weed seeds, shriveled seeds, plant debris, and other foreign materials. Other characteristics are seed size, weight, specific gravity, and seed color. It is vital for all individual seeds in the lot to be as uniform in these parameters as possible (Sarika and Amruta 2014). Depending on the purpose of the seed the physical quality parameters may not be very important to the user. Seeds of many tomato cultivars are clean and bright in color, plump, and have yellowish seed coats.

In tomato, the quality of the seed is associated with a complex of physiological and genetic traits, unlike other characteristics in the crop. Most plant processes are controlled by small to large effect gene action that follows the Mendelian inheritance. In an earlier study, Kazmi et al. (2012) found that seed quality is primarily quantitative and genetically complex. In a recombinant inbred line population of *Solanum lycopersicum* × *Solanum pimpinellifolium*, QTLs influencing seed quality under abiotic stress conditions, including salt, cold, and high temperature stresses were identified. A total of 42 seed quality traits were analyzed and yielded 120 QTLs under different conditions. Phenotypic correlations among germination traits under both optimal and stress conditions were significant. One or more QTLs were identified for each trait with some being co-locating, a possible indication of the presence of pleiotropic effects on multiple traits resulting from common mechanistic basis (Kazmi et al. 2012). Also, many QTLs dissected seed quality into separate components, which suggest different physiological mechanisms and signaling pathways for the quality parameters. Similar outcomes were obtained by Khan et al. 2012) in a RIL population. In this study, seed weight had a strong effect on seedling vigor. These results, backed by extensive phenotyping are essential for breeding superior seed and seedling traits.

Like other crop seeds, tomato seeds acquire several physiological traits during maturation (Kazmi et al. 2012; Khan et al. 2012b; Bizouerne et al. 2021). Temporal RNA-seq analyses were undertaken to explore transcriptome signatures associated with desiccation tolerance, seed longevity, dormancy, and germination. Over 15,000 genes differentially expressed were detected, forming a gene network of 21 expression modules. Three out of the 21 were specific to seed coat and embryo while five were specific to the endosperm. Gene-trait significance measure identified a common gene module between endosperm and embryo associated with desiccation tolerance and conserved with non-endospermic seeds. LEA and HSP, genes normally involved in protection, as well as ABA response, antioxidants, and repair genes, were included. Seed dormancy was released in tandem with an increase in longevity throughout fruit ripening until 14 days after the red ripe fruit stage. This was paralleled by increase in SIDOG1-2 and PROCERA transcripts. Other genes such as AB14 and CHOTTO1, and ABA/GA response genes were found to be associated with various seed quality attributes (Bizouerne et al. 2021).

Molecular basis of the various fruit and fruit quality traits has been studied and led to the identification of genes affecting these traits. Barrantes et al. (2016) identified chromosomal regions associated with vegetative and fruit (morphology, organoleptic quality, color) characteristics. They found QTLs related to fruit weight, soluble solids' content, and fruit internal color. The QTLs differ in stability and the stability is dependent on the trait (Barrantes et al. 2016) of interest. Furthermore, sequenced cultivated accessions showed the presence of mutations and sequence variation in key genes. These genes underlie important domestication and agronomic traits, including inflorescence architecture, fruit pedicel abscission, fruit number, size and shape, fruit color, and soluble solid content. These results yielded valuable information for selecting genotypes for further forward studies. The information generated also demonstrates the possibility of using the sequenced genomes to mine alleles in genes of interest (Roohanitaziani et al. 2020). Allele mining has also been employed in breeding for heat tolerance as a way to mitigate the negative effects of high temperatures (HT) (Jesús 2018) on crop yield and yield quality characteristics and this shows great promise improving tomato yield and yield quality.

3.5.3 Influence of Allele Mining on Disease Resistance in Tomato

Disease resistance has been treated as a must-have trait during crop improvement for many years. Conventional breeding methods have played a crucial role in developing disease resistant cultivars. With the advent of molecular technologies, there have been more efficient methods in the selection of resistant genotypes in an environment-dependent manner (Maurya 2020). The wild tomato relatives often serve as sources of resistance. Resistance alleles for four important tomato diseases, tomato leaf curl, tomato root knot, tomato late blight, and tomato mosaic have been identified. These were the Ty3 resistance allele (for leaf curl disease), Mi1.2 resistance allele (for root knot disease), the Ph3 resistance allele (for late blight disease), and the Tm2 resistance gene (for tomato mosaic disease). There is the opportunity to develop varieties with single gene resistance, also known as vertical resistance, for certain races of leaf mold, Septoria, late blight, early blight, Stemphylium, and Verticillium and Fusarium wilts disease of tomatoes. Dominant resistance to tobacco mosaic virus and nematode resistance is also available and other forms of complex, polygenic (multi-gene) resistance, also known as horizontal resistance, exist. These are extremely beneficial to a robust tomato-breeding program with the objective of breeding for disease resistance.

3.5.3.1 Characterization of Allelic Diversity in Tomato Genetic Resources

Allelic diversity exists in naturally occurring populations, gene banks or accessions and mutant populations. Discovering allelic diversity involves the characterization of genetic diversity and allelic diversity of a candidate gene or genes among a collection of accessions to establish the genetic relationships_among them. TILLING/Eco-TILLING or the sequence-based approaches are approached used in characterizing the allelic diversity of a candidate gene in a collection of accessions or mutant populations. The World Vegetable Centre, Taiwan and many National institutions and Universities maintain tomato germplasm for continuous characterization of allelic diversity for important traits and accumulation in many tomato genotypes (Kumar et al. 2010). In recent studies, novel diversity may be mined from databases, but functional characterization remains a limiting step in identifying new alleles. Complementing classical breeding approaches with marker-assisted selection (MAS) provide means to quickly assess the function of a new variation in a coding or regulatory region to modulate traits (Orchard et al. 2021).

3.5.3.2 Identification of New Haplotypes

Haplotype is a pair or group of genes within an individual or organism that was inherited together from one parent on a chromosome. Haplotypes can also refer to inheritance of a cluster of SNPs, which are variations at single positions (nucleotide variations) in the DNA sequences among individuals. The nucleotide variations in the DNA sequence among the individuals could lead to observed phenotypic differences. Allelic mining can be deployed in the identification of the nucleotide variation in the DNA sequence (candidate gene) associated with phenotypic variation of the gene (Kumar et al. 2010). The nucleotide variation in the DNA sequence could be a novel or superior allele. Information generated from haplotype frequencies and their changes within population could form the basis for organization of association mapping research (Amaranatha Reddy 2014).

3.5.3.3 Development of Allele-Specific2 Markers for MAS

Once a superior or novel allele is identified through allele mining, an allelic-specific marker assay can be developed. This marker-specific assay could be utilized in introgression superior or novel alleles into a suitable genetic background through marker-assisted selection (MAS). Several studies have demonstrated sequence variations in key genes in many crops and the utilization of these variations to develop allele-specific molecular markers for MAS (Kumar et al. 2010).

3.5.4 Utilization of Allele Mining in Tomato Breeding Programs

The application of allele mapping in tomato breeding programs has been diverse, ranging from mapping and cloning of *R* gene identification of novel genetic resources to more scientific applications like *R* gene geographic distribution and evolution (Vossen 2014). In tomato (Rose 2005) investigated the extent to which variation in *Pto gene* is responsible for natural occurring variation in resistance to *Pst* (*Pseudomonas syringae pv. tomato*) by sequencing and amplifying alleles of *Pto* in 49 genotypes and testing the functions of a subset. This led to identification of 11 alleles responsible for hypersensitive resistance (HR) in the presence of the pathogen avirulence protein. Allelic variations of Tomato Yellow Leaf Curl Virus resistant Ty-1/Ty-3 gene was studied in few *Solanum chilense* to identify a number of allele-specific polymorphisms. Nine polymorphisms representing three unique alleles were identified (Yan et al. 2018). Orchard et al. (2021) characterized variation in the Cyc-B gene (B) that modulate levels of β-carotene in ripe tomato from 84 sequences recovered from public databases and additional 29 high β-carotene tomato. Thirteen unique haplotypes were identified with 11 occurring in high β-carotene accessions when sequenced. Association analysis suggested two single nucleotide polymorphisms (SNPs) are the most likely causes of the high β-carotene. A marker-assisted backcross-breeding scheme leveraged SNPs for background genome selection to rapidly develop germplasm resources containing different alleles of *B* in a genetic background. Pereira et al. (2021) exploited natural genetic diversity in tomato flavor genes including *LIN5, ALMT9, AAT1, CXE1, and LoxC* to identify any association of these genes with trait variation and to describe the genetic diversity of these loci in the red-fruit tomato clade. High genetic diversity was observed at the five loci, including novel haplotypes that could be utilized in breeding programs. The expression patterns of the genes increased during ripening but diverged between accessions without any clear relationship between expression and metabolic levels.

3.6 MAPPING POPULATIONS AND VALIDATION OF QTLS FOR GENE DISCOVERY IN TOMATO

Quantitative trait loci (QTL) mapping refers to finding an association between a genetic marker and a trait for population developed from a controlled cross or a breeding population (Mulualem and Bekeko 2016). In tomato, breeding, mapping populations are used for gene discovery. These mapping populations are very diverse and demonstrate specific and complementary features (Rothan et al. 2018). Plant progenies obtained from experimental crosses are used for genetic dissection of complex traits and support modern plant breeding. Different types of mapping populations have been used for QTL detection for food and nutritional quality in tomato. Examples include bi-parental mapping populations, recombinant inbred lines, multi-parent advanced generation inter-cross (MAGIC), and diverse breeding populations.

The genome-wide inference of the relationship between genotype at various genomic locations and phenotype for a set of quantitative traits in terms of the number, genomic positions, and effects is termed quantitative trait loci. Principal step in plant breeding activities is QTL mapping in which genetic linkage maps are employed to provide basis for mapping of QTLs (Mulualem and Bekeko 2016). For the purposes of developing a consensus genetic map that can be a common framework for mapping studies, a complete genetic map that covers all chromosomes within a species genome is important. This presupposes that accurate and high-resolution genetic maps are critical for crop improvement (Bhandari and Lee 2021). A prerequisite for the construction of a high-density genetic linkage map is the availability of a bulk number of molecular markers. Several molecular markers, such as express sequence tag (EST), single sequence repeats (SSR), amplified fragment length polymorphism (AFLP), randomly amplified fragment length polymorphism (AFLP), randomly amplified polymorphic DNA (RAPD), and single nucleotide polymorphism (SNP) are employed in the construction of a linkage map. In recent times, SNPs that are the most abundant has taken the center stage for linkage map construction, QTL discovery and validation (Lindblad-Toh et al. 2000; Kho et al. 2021).

Important target traits for breeding and development of new cultivars of tomato (*Solanum lycopersicum* L.) include fruit nutritional and flavor components (Kimbara et al. 2018). These same traits are also important to consumers. Previous studies have discovered QTLs for traits such as soluble sugars, titratable acidity and pH that are important indices of eating and processing quality have been dissected (Ashrafi et al. 2012; Kimbara et al. 2018). Also important traits including beta-carotene and lycopene that are important antioxidants and of enormous health benefit have been studied by many research groups (Ashrafi et al. 2012; Ruggieri et al. 2014). Many QTLs associated with fruit size, shape, color intensity, firmness, and composition have been mapped (Rothan et al. 2018).

Bi-parental mapping population is developed from a cross of two diverse breeding lines as parents to develop F_1 hybrid. The F_1 hybrid is then self-pollinated for the formation of segregating F_2 population that are further selfed using the single descent breeding method to form recombinant inbred lines (RIL). Each resulting inbred line has a unique genetic structure that distinguishes it from the parental lines in both genotypic and phenotypic characteristics that provides opportunities for phenotypic evaluation under different conditions for various traits contrasting in the two parental lines (Gonda et al. 2019). Adhikari et al. (2020) developed the mapping population NC 10204 from a cross of the plum tomato breeding line NC 30P and the grape tomato breeding line NC 22L-1(2008) at Mountain horticultural crops research and extension center (MHCREC), North Carolina State University, Mills River, NC consisting of 110 recombinant inbred lines. This mapping population was used for the detection of quantitative trait loci (QTL) associated with tomato fruit shape and size. The two parents were diverse in characteristics. While NC 30P is large and highly elongated fruits with 2–3 locules with fruit weight of 115–140 g per fruit, NC 22L-1(228) is small elongated fruit with 2–3 locules that is dark red with and smooth with average fruit size of 15–20 g per fruit. For the detection of novel QTL for increased lycopene content and other fruit quality traits, NCEBR-1, an advanced and *S. pimpinellifolium* accession LA2093 were selected as parental lines to develop a RIL population. NCEBR-1, which is advanced fresh-market breeding line with multiple disease resistance whilst LA2093 is a self-incompatible, inbred accession with numerous desirable characteristics that includes very high lycopene content (Ashrafi et al. 2011).

Advantages of using bi-parental mapping population for QTL mapping include the relative simplicity of their construction, their high power for QTL detection because allele frequencies are typically close to the optimal value of 50% as well as low rate of linkage equilibrium decay within chromosomes. In spite of these advantages, there is a lack of mapping precision as a result of limited effective recombination during population development and low genetic diversity sometimes due to genetic bottle necks caused by choice of founders that could affect the number of QTLs detected (Scott et al. 2020). Precision of mapping of QTLs have been reported and depends on the extent of recombination in chromosomal region harboring the genetic variant and on the size of the population studied (Rothan et al. 2018). Consequently, due to these demerits, two most popular multi-parent populations evolved, which are nested association mapping (NAM) and multi-parent advanced generation inter-cross (MAGIC) populations. Bi-parental populations all linked to a common parent is called NAM. MAGIC, F_2 generation is further crossed and several founders are intercrossed over multiple generations before selfing to generate inbred lines (Scott et al. 2020). Multi-parental population was developed in tomato with eight parental lines belonging to the cultivated clade *S. lycopersicum var lycopersicum*. With regards to the genetic architecture of fruit quality traits, this population demonstrated its potential for the better understanding of these traits (Pascual et al. 2015; Rothan et al. 2018). Well-characterized tomato lines at different molecular and physiological levels (Pascual et al. 2013; Pascual et al. 2015) as founders of an eight way MAGIC population for detection of QTLs for fruit weight. Founders used included four *S. lycopersicum* (Levovil, Stupicle PR, LA0147 and Ferum) and four *S. lycopersicum* var cerasiforme (Cervil, Criollo, Plovoliv24A and LA01420) lines. The chosen lines were to maximize the genetic diversity (Pascual et al. 2015).

After QTL detection using a particular mapping population, they are cross-validated in different mapping population and in different environment to ensure its stability. Sequences of markers of QTLs are then blasted using different blasting tools to discover genes associated with the QTLs. Sequencing technologies for gene discovery keeps on increasing and have triggered new developments such as genome-wide association studies (Zhu et al. 2018), QTL-seq (Illa-Berenguer

2015) and mapping-by-sequencing (MBS) (Garcia et al. 2016). These new technologies facilitate the exploitation of natural and induced genetic diversity, thus accelerating gene discovery (Rothan et al. 2018). Genes discovered are very important for marker development to assist in marker-assisted breeding and genome editing techniques

3.7 BIOINFORMATICS TOOLS FOR ALLELE MINING IN TOMATO

3.7.1 Bioinformatics in General and the Tools Used

Biologists are increasingly generating new data on organisms, specifically genomes, biomolecules, as well as their interactions and evolution. In this context, there is a growing need for computational approaches for the manipulation, storage, visualization, and analysis of often very complex data (Aslam et al. 2017). Bioinformatics is a multi-disciplinary research field that brings together all the concepts and techniques necessary for the computer interpretation of biological information. So, it represents a growing area of science that uses computational approaches to answer biological questions, ranging from structural biology, genomics to gene expression studies (Aslam et al. 2017; Baxevanis et al. 2020). Bioinformatics is now almost essential for an effective study of genetic and genomic data, regardless of the living organisms targeted (Barnes and Gray 2003).

The next table (Table 3.5) that is mainly inspired by the one made by (Luscombe et al. 2001), present the sources of data used in bioinformatics, the sub-disciplines of bioinformatics involved, and bioinformatics topics linked to this data

Bioinformatics has a great impact in the agricultural sector. Its benefits are countless and contribute to the achievement of the objectives of food security in the world. More specifically, bioinformatics has fostered the development of software, algorithms, databases, and tools of data analysis that facilitate studies in agronomy (Iquebal et al. 2015; Singh et al. 2011; Xue et al. 2007). In fact, using computational tools to quickly find new genes promotes the development of nutritional genomics, improves seed quality, develops effective disease and pest control programs, improves nutritional quality and growth in the poorest soils and allows the development of plants adapted to climate change (Aslam 2017; Batley and Edwards 2016; Dixit et al. 2022). On the other hand, recent advances in computational software and computer tools, as well as modern bioinformatics and genome analysis, are proving effective in reducing the effects of harmful agriculture and creating a better, sustainable agro atmosphere (Mishra et al. 2022). In short, the applications of bioinformatics to agriculture (Khalid et al. 2021; Kushwaha et al. 2017) are as follows:

- Insect resistance.
- Pedigree analysis.
- Nutritional quality.
- Forecasting models.
- Biometrical analysis.
- Plant disease management.
- Varietal information system.
- Designing of gRNAs for CRISPR/CAS.
- Mining quantitative trait loci studies (QTL).
- Applying machine learning to crop breeding.
- Integrated crop databases (like Genbank, Phytozome, PlantGBD…).
- Identification of breeding targets by Genome-wide association studies (GWAS).

Sequence assembly and similarity search tools form the basis of various software applications for the examination, analysis, and interpretation of crop genome information (Aslam 2017). Likewise, since genome data has grown exponentially, it was useful to find a way to store it and make it

TABLE 3.5
Sources of Data Used in Bioinformatics, the Sub-Disciplines of Bioinformatics Involved, and Bioinformatics Topics Linked to This Data

Sources of Data used in Bioinformatics	Sub-Disciplines of Bioinformatics	Bioinformatics Subject Areas
DNA sequence	Sequence Bioinformatics (concepts, data-structures and algorithms for sequence analysis)	Separating coding and non-coding regions Identification of introns and exons Genome comparison promoter and gene prediction Forensic analysis
Protein sequence	Sequence Bioinformatics	Sequence comparison algorithms Multiple sequence alignment algorithms Identification of conserved sequence motifs Prediction of nucleic acid and protein structure
Macromolecular structure	Structural bioinformatics (reconstruction, prediction or analysis of the 3D structure or folding of biological macromolecules)	Secondary, tertiary structure prediction 3D structural alignment algorithms Protein geometry measurements Surface and volume shape calculations Intermolecular interactions Molecular simulations
Genomes	Network bioinformatics (interactions between genes, proteins, cells, organisms)	Characterisation of repeats Structural assignments to genes Phylogenetic analysis Genomic-scale censuses (characterisation of protein content, metabolic pathways and their simulation) Linkage analysis relating specific genes to diseases
Gene expression	Network bioinformatics	Correlating expression patterns Mapping expression data to sequence, structural and biochemical data
Population	Population bioinformatics (computer science applied to the analysis of a population taken as a whole on the biological level)	Identification of the genetic and molecular mechanisms underlying variation in the human mitochondrial transcriptome, and how perturbations in these processes modulate disease risk Assessing next-generation sequencing performance Population structure analysis Population modeling
Statistical data	Statistical bioinformatics (identify significant changes in biological processes)	Counting and manipulating k-mer Using genomic information to distinguish good or bad prognosis Clustering to get a data-driven structure
Literature		Web-based genome browsers Gene Ontology

accessible to researchers. This is why analysis tools and databases are available on the Web in the form of user-friendly interfaces, called Genome browser.

Genome browser (genome browsers with multiple species and species-specific genome browsers) provides a graphical interface for users to browse, search, retrieve, and analyze genomic sequence and

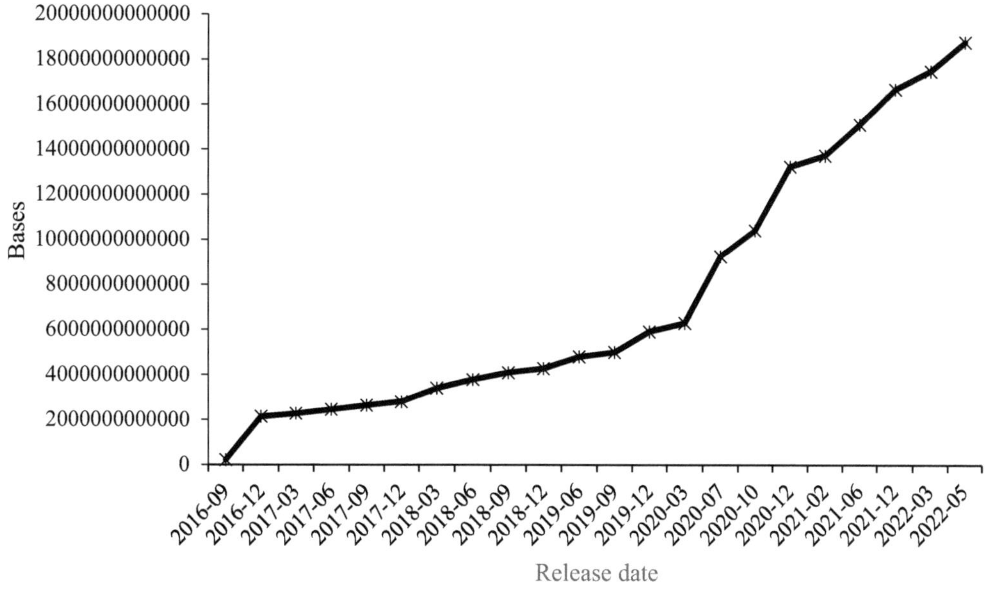

FIGURE 3.1 Evolution of bases from 2016 to 2022.

Source: DDBJ/ENA/GenBank database growth, www.ddbj.nig.ac.jp/statistics/index-e.html#ddbj-release.

annotation data (Wang et al. 2013). For example, National Agricultural Biotechnology Information Center (NABIC) is an integrated Web service and genome database for agricultural plants with biotechnology information (Kim et al. 2015; Seol et al. 2016). Another interesting example is Ensembl Plants (https://plants.ensembl.org/index.html). However, Rice-Map (www.ricemap.org/) and MaizeDB (http://gbrowse.maizegdb.org/) are species-specific genome browsers respectively for *Oryza sativa* japonica/indica (Rice) and *Zea mays* (Maize). Three major institutions are responsible for archiving data: the NCBI in the United States, the EBI (European bioinformatics institute) in Europe and the DDBJ (DNA Data Bank of Japan). These notably manage large nucleotide sequence databases such as GenBank or the EMBL database, and protein sequence databases such as UniProt or TrEMBL (Figure 3.1).

3.7.1.1 Bioinformatics Tools for Sequence Alignment and Analysis

As shown previously, sequence analysis plays an important role in biological studies. Since the invention of DNA sequencing by Frederick Sanger, researchers have been working on sequencing the genome of living organisms and therefore plants. As a result, the volume of available DNA sequences has grown exponentially (Figure 3.2), and with the creation of databases, it is possible to store them. The development of sequence analysis tools aims to facilitate the determination of their properties. In the sub-discipline of sequence bioinformatics, sequential alignment or sequence alignment is a way of representing two or more sequences of biological macromolecules (DNA, RNA or proteins) under each other, in order to highlight homologous or similar regions (Aslam 2017; Daugelaite 2013). The search for sequences in a database from another sequence or from a fragment of a sequence is also often carried out in sequential bioinformatics.

The ability to produce increased amounts of paired read sequence data, combined with customized bioinformatics tools, provides the basis for sequencing large and complex plant genomes (Lee et al. 2012). It should be specified that it is also possible to carry out a global multiple sequence

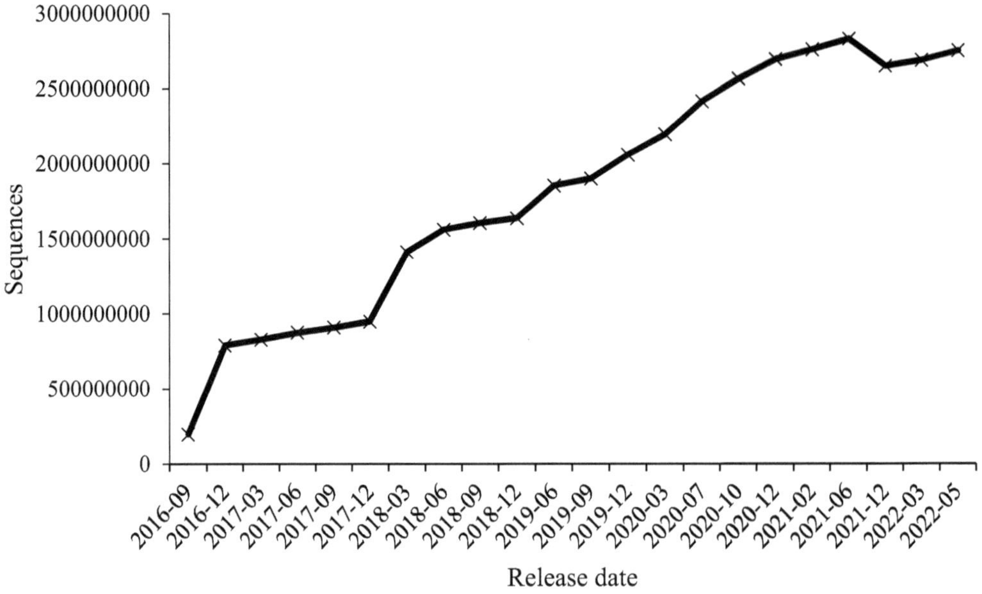

FIGURE 3.2 Evolution of sequences from 2016 to 2022.

Source: DDBJ/ENA/GenBank database growth, www.ddbj.nig.ac.jp/statistics/index-e.html#ddbj-release.

alignment (MSA), which aims to organize the motifs of conserved sequences of smaller size among the sequences of all the queries.

After performing genome assembly with the appropriate algorithms (e.g. VCAKE, ALLPATHS, Edena, Velvet, ABySS; EULER SR; SOAPdenovo, SaSSY, or PE Assembler), the next step could involve sequence alignment to perform a comparison (Lee et al. 2012). To do this, the most frequently used software is from the BLAST (Basic Local Alignment Search Tool) family (Aslam et al. 2017; Lee et al. 2012; Xue et al. 2007). In fact, many genomics labs provide a BLAST Web interface to their sequence databases to allow scientists to easily identify homologs of an input sequence of interest (Xue et al. 2007). Sequence alignment tools typically work by subdividing the alignment problem into two steps: (1) heuristically searching for candidate alignment locations (hits) by indexing either the sequences or the reference genome; (2) performing the actual alignment (Lee et al. 2012). In the following table (Table 3.6), we have presented some examples of tools used in sequence bioinformatics.

Sequence redundancy removal is an essential preprocess in several bioinformatics tasks. There are software tools that can be used to deal with redundant sequences efficiently and accurately such as CD-HIT, VSEARCH, and USEARCH (Zou et al. 2020). In the same sense, there are tools that promote the processing of readings and quality control, such as iSeqQC, MultiQC, NGS QC, FASTQC, and FASTX-Toolkit (Etukuri et al. 2022).

3.7.1.2 Bioinformatics Tools Developed for Allele Mining in Tomato

Allele mining is a field of research aimed at identifying allelic variation in relevant traits in genetic resource collections. Allele mining is a promising approach to dissect naturally occurring allelic variation at candidate genes controlling key agronomic traits, which has potential applications in crop improvement programs (Kumar et al. 2010). It helps to trace allele evolution, identify new

TABLE 3.6
Some Popular and Common Tools Used in Sequence Alignment

Tools for Sequence Alignment	About	Source
BLAST-like alignment tool BLAT	An alignment tool that is designed for the ultra-rapid alignment of sequences to a reference genome. Facilitates the rapid search for a location in a genome by sequence similarity	(Kent 2002).
MAQ	This tool introduces the concept of mapping quality, which is a Phred-like measure of alignment confidence	(Li 2008)
BFAST (BLAT-like Fast Accurate Search Tool)	In particular, this tool seeks sensitivity and alignment accuracy at the expense of alignment speed	(Homer et al. 2009)
Burrows-Wheeler aligner (BWA)	Burrows-Wheeler aligner (BWA) consists of three algorithms: BWA-backtrack, BWA-SW, and BWA-MEM. The first is designed for reads up to 100 bp, while the other two for longer sequences.	(Abuín et al. 2015)
Bowtie and Bowtie 2	Bowtie implements Burrows–Wheeler indexing for the rapid alignment of short nucleotide sequences to genomic sequence Bowtie 2 combines the strengths of the full-text minute index with the flexibility and speed of hardware-accelerated dynamic programming algorithms to achieve a combination of high speed, sensitivity, and accuracy.	(Langmead et al. 2009; Langmead and Salzberg 2012)
SOAP2 (a redesign of the original SOAP: Short Oligonucleotide Analysis Package application)	This tool implements Burrows–Wheeler indexing to dramatically improve performance and memory usage over its predecessor	(Li et al. 2009; Li et al. 2008)
The Short-Read Mapping Package (SHRiMP)	It is a set of algorithms and methods to map short reads to a genome, even in the presence of a large amount of polymorphism	(Rumble et al. 2009)
PuffAligner	A fast, accurate and versatile aligner built on top of the Pufferfish index. It is able to produce highly sensitive alignments, similar to those of Bowtie2.	(Almodaresi et al. 2021)
STAR	This tool ultrafast universal RNA-seq aligner	(Dobin et al. 2013)
HAlign	This tool is a fast multiple similar DNA/RNA sequence alignment based on the centre star strategy	(Zou et al. 2015)
HISAT	HISAT is a fast spliced aligner with low memory requirements	(Kim et al. 2015)
HTSeq	HTSeq is a Python framework to work with high-throughput sequencing data	(Anders et al. 2015)
DiAlign	DiAlign is a software tool for multiple sequence alignment by combining global and local alignment features	(Morgenstern 2014, 2004)
Avid	A global alignment method practical for sequence alignments of large genomic regions up to megabases long	(Bray et al. 2003)

haplotypes and develop allele-specific markers for use in marker-assisted selection. Allele mining exploits the deoxyribonucleic acid (DNA) sequence of one genotype to isolate useful alleles from related genotypes. Note that, bioinformatics prediction can increase the efficiency of SNP discovery.

Tomato (*Solanum lycopersicum L.*) is the most intensively investigated Solanaceous species both in genetic and genomics studies (Barone et al. 2008). The completion of the complete tomato genome sequence has opened new perspectives for the development of innovative tools that strengthen the position of tomato as a reference species for Solanaceae and fleshy fruits (Sato et al. 2012). Thus, some tools have been developed to facilitate allele mining, always with the aim of facilitating the production of this crop that occupies an important place in food systems.

Genotyping by sequencing (GBS) is a high-throughput approach based on the genome complexity reduction method and is cost-effective for genotyping large numbers of samples. Thus, GBS has become a powerful tool for several applications in crop species, including high-resolution genetic mapping, genetic diversity analysis, QTL mapping, and genomic selection (Phan et al. 2019). Different protocols are used to prepare GBS libraries. Here is an example of using genome-wide single nucleotide polymorphisms to identify loci associated with fruit traits in a core collection of tomato (Phan et al. 2019). First genomic DNA was extracted from seedlings. The DNA samples were after digested using an appropriate enzyme (here *Ape*KI enzyme). After that, the DNA fragments were ligated to the adapters with different barcodes that were assigned to eachplant sample. The resulting DNA samples were then pooled and amplified by PCR to generate GBS libraries. The Agilent Bioanalyzer was used to check the library quality. The GBS libraries were sequenced using the HiSeq 2000 and the pair-end method. The raw sequencing reads were demultiplexed based on the individual barcode. The adapter and barcode sequences were then removed. As stated previously, it is imported to check the sequence quality. The authors used the SolxaQA v1.13 package to do it. Then, the Burrows-wheeler aligner (BWA) was used to map the cleaned reads to the tomato reference genome. The mapped reads were extracted using SAM tools v 0.1.16 and a SNP matrix between the tomato accessions was generated using an appropriate script. SEAN is also a SNP prediction and display program that utilize expressed sequence tags EST sequence clusters (Huntley et al. 2006).

There are many analytical tools for tomato. Here are some interesting examples:

- Tomato expression dwatabase (TED) is a suite of data presentation and analysis tools for tomato (Fei et al. 2006).
- Tomato Functional Genomics Database is a comprehensive resource and analysis package for tomato functional genomics (Fei et al. 2011).
- NexGenEx-Tom Web-based platform that offers processed NGS transcriptome collections of the tomato and enables immediate analyses of the results (Bostan and Chiusano2015).
- TomExpress, which is a unified tomato RNA-Seq platform for visualization of expression data, clustering and correlation networks (Zouine et al. 2017).

3.7.1.3 Bioinformatics Analysis of a Gene from a Database: An Example on the Tomato

Table 3.7 summarizes a practical case of the applications of bioinformatics in genetic research in agriculture. The crop in question is tomato and the gene involved is AMT1 related to Ammonium transporters (AMTs) (Filiz and Akbudak 2020).

TABLE 3.7
A Bioinformatics Application Case on the Study of the AMT1 Gene in Tomato

Analysis	Bioinformatics Tools
Database mining and identification of AMT1 genes	UniProtKB database BLASTP analysis on tomato genome in Phytozome v12 database
Analysis of SlAMT1 gene sequences including estimates of genetic diversity and segregating (polymorphic) sites (S)	DnaSP 6.12
Determination of Physicochemical details of AMT1 proteins	ProtParam tool (http://web.expasy.org/protparam/)
Verification of all identified AMT1 protein sequencesby a conserved domain search	Pfam 32.0 (https://pfam.xfam.org/)
Displaying gene structures	Phytozome v12 database
Identification of the predicted subcellular localizations of AMT1 proteins	WoLF PSORT (https://wolfpsort.hgc.jp/)
Prediction of transmembrane helices in AMT1 proteins	TMHMM Server v.2.0 (www.cbs.dtu.dk/services/TMHMM-2.0/)
Location of predicted serine, threonine or tyrosine phosphorylation sites	NetPhos 3.1 server (www.cbs.dtu.dk/services/NetPhos/)
Investigation of the Protein sequence identities	blastP analyses in NCBI (https://blast.ncbi.nlm.nih.gov/Blast.cgi?PAGE=Proteins)
Conserved motif analysis	MEME (Multiple Em for Motif Elicitation) server v 5.0.5 (http://meme-suite.org/tools/meme)
Multiple sequence alignment on the full-length amino acid sequences of the tomato	BioEdit v7.2.5
Phylogenetic analyses	MEGA7
Detection of transcription factor binding sites (TFBSs) and corresponding TFs	PlantPAN (Plant Promoter Analysis Navigator) 2.0 database (http://plantpan2.itps.ncku.edu.tw/)
Analyze the expression patterns of SlAMT1 genes	Genevestigator database (https://genevestigator.com/gv/doc/intro_plant.jsp)
Synteny analyses	Circoletto platform (http://tools.bat.infspire.org/circoletto/)
Gene ontology (GO) analysis	PANNZER (Protein ANNotation with Z-scoRE) server (http://ekhidna2.biocenter.helsinki.fi/sanspanz/)
Prediction of Ligand binding site of SlAMT1 proteins	GalaxyWEB server (http://galaxy.seoklab.org/cgi-bin/submit.cgi?type=SITE)
Construction of the Protein-protein interaction (PPI) network	STRING v11 server (https://string-db.org/)
Analyses of secondary structures such as helix, beta, coil and turn	VADAR (Volume, Area, Dihedral Angle Reporter) server (http://vadar.wishartlab.com/)
Predicted 3D structures	PHYRE (Protein Homology/analogY Recognition Engine V 2.0) protein fold recognition server (www.sbg.bio.ic.ac.uk/phyre2/html/page.cgi?id=index)
Structure overlap of AMT1 proteins calculation	Pairwise alignment by CLICK server (http://cospi.iiserpune.ac.in/click/),
Finding pockets and cavities in SlAMT1 proteins	CASTp v3.0 (Computed Atlas of Surface Topography of proteins)(http://sts.bioe.uic.edu/castp/calculation.html)

Source: Filiz and Akbudak 2020.

3.8 PROSPECTS AND CHALLENGES OF ALLELE MINING IN TOMATO FOR GENE DISCOVERY

The importance of allele mining for gene discovery in tomato cannot be overemphasized. Demand-led tomato breeding programs can benefit massively from allele mining technology that involves the dissection of allelic variations naturally occurring at several loci in the genome (Sharma 2014). Allele mining has become significant in plant breeding due to the availability of sequence information on the genome of many important crops including tomato (Kumar et al. 2010). Graner (2006) indicated that availability of sequence information on candidate genes of agronomically important traits would facilitate gene-based use of genetic resources. Apart from the availability of sequence information on the genome of crops, many software tools have been developed for the deployment of allele mining technology in agriculture (Kumar 2010). The application of allele mining technology in tomato breeding involves but is not limited to the following:

1. Unlocking the genetic potential among accessions in gene banks.
2. Development of allele specific molecular markers.
3. Establishing allelic synteny and evolutionary relationships.
4. Identification of new haplotypes.

3.8.1 Unlocking the Genetic Potential among Accessions in Gene Banks

This application is based on assessment of genetic (allelic) diversity among tomato accessions held in the gene banks. Assessment of genetic diversity within a crop is a very important pre-breeding activity to establish the genetic relationships and worthiness among the accessions with regards their usefulness for crop improvement (Kumar et al. 2010). This will not only help in the identification of distinct individuals for selection and hybridization purposes, it will also lead to the identification of duplicates, which can be discarded to reduce the cost of germplasm conservation in gene banks. Even though unlocking the potential of allelic variability among accessions in gene banks have not been fully exploited for most important crops (Kaur et al. 2008a), superior alleles can be discovered among genetically diverse germplasm.

3.8.1.1 Development of Allele Specific Molecular Markers

Molecular markers have been used in plant breeding to aid in the selection of desirable traits in crops. The use of molecular markers has improved the efficiency of breeding programs through selection of desirable individuals and shortening the breeding cycle. This not only saves time but also the cost of accurate phenotyping of the trait. Allele mining technology can be exploited to develop allele-specific molecular markers for marker-assisted breeding in crops (Graner 2006). This will enable the introgression of superior alleles into desirable tomato genetic backgrounds using marker-assisted backcrossing and other genetic engineering tools (Mikami et al. 2008; Bao et al. 2006).

3.8.1.2 Establishing Evolutionary Relationships and Allelic Synteny

Most domesticated crops evolved from their wild or ancestral progenitors and may share certain portions of their genome together. This phenomenon of sharing portions of the genome between related species is referred to as syntenic relationship. Syntenic relationship among related species can be harnessed through allele mining to trace variability in nucleotide sequences in traits of agronomic importance. Fontecha et al. (2007) reported that allele-mining studies could reveal genomic regions with syntenic relationships among related species. Allele mining can be a useful tool in assessing allelic variability among individuals of a species or related species. Evolutionary forces, such as mutation, genetic drift, gene flow, and natural selection act on crop species causing allelic variability, which will lead to alterations in gene functions and allele frequency (Singh 2005). Even though, most of the allelic variations may not result in changes in the phenotype, some of the allelic variability may result in better adaptation and improved performance of crops. Such allelic variability can be used in studying evolutionary relationships among the species. Rose (2005) reported changes in nucleotide sequences in the Pto gene responsible for pathogen recognition and hypersensitive reaction in tomato and indicated natural variation in resistance to *Pseudomonas* strains as the cause of the changes.

3.8.1.3 Identification of New Haplotypes

Allele mining studies can uncover the occurrence of new haplotypes and reveal the observable phenotypic variation depending on the kind and extent of new haplotype in crops (Kumar et al. 2010). For example, the *Pseudomonas* resistance locus (Pto gene) in tomato was identified through allele mining. Information on the existence of haplotype blocks is imperative in studies in association mapping in crops.

Notwithstanding the numerous prospects and opportunities that allele mining offers, there are challenges that need to be overcome in order to benefit fully from allele mining technology in crops. Some of the challenges of allele mining in crops have been discussed below.

3.8.1.4 Diversity Panel Selection

To select diverse tomato genotypes to form a core collection that constitute the total genetic variability (Brown 1989) within the crop pose a challenge to allele mining. To overcome this, diverse tomato accessions obtained from gene banks should be phenotyped using reliable protocols under conducive environments to establish the genetic variability among the collection. The extent of variability among the accessions would determine the degree of allelic variability, which could be exploited through allele mining.

3.8.1.5 Prioritizing Agronomically Important Traits

Prioritizing agronomically important traits in tomato poses a challenge to allele mining. Breeding programs may have specific product profiles that seeks to address different constraints pertaining to their environment. This makes it difficult to prioritize the list of important constraints that need to be addressed in tomato using allele mining technology. In addition, many agronomically important traits are controlled by complex quantitative trait loci (QTLs), which makes it challenging to prioritize them.

Other challenges associated with allele mining include the high cost of sequencing, efficient bioinformatics tools for identification of variations in nucleotides (handling genomic resources, characterization of regulatory sites, demarcation of promoter regions, etc.) and marker construction technologies (Kumar et al. 2010).

3.9 CONCLUSION

An allele is an alternative form of a gene found at a given locus on a chromosome. Allele mining is a research field that aims to detect or find allelic variation of essential traits within genetic resource collections. This involves searching for useful alleles of genes from a wide range of cultivars, related species, and even across species. Advances in development of molecular biology tools and availability of sequencing information for more crops make allele mining a promising tool for crop improvement. This has been explored in this chapter with special emphasis on tomatoes. The recent consumer demands and challenges associated with cultivation of tomato has necessitated the exploitation of different sources of genes of desirable traits for improvement of the crop. Traditional breeding has its limits, some of which can be overcome using molecular breeding techniques that also require knowledge of genes controlling the traits of interest. In this regard, allele mining techniques are being utilized to facilitate achievement of modern breeding goals.

In order to enable effective mining of alleles, the genome of the crop of interest or closely related species need to be sequenced. Fortunately, tomato as a model crop has its genome sequenced. The initial sequencing project used the BAC-by-BAC method but the current reference genome for tomato was completed using a combination of next-generation sequencing technologies following the whole genome sequencing approach. The two main techniques used in allele mining, the modified TILLING (Eco-TILLING) and sequencing-based approaches each has its own strengths and limitations. Consequently, the choice of approach to be used is guided by the purpose, time and access to phenotyping and genotyping infrastructure. The latter is necessary because the use of allele mining in improvement of tomato is facilitated by the development of many desktop and web-based bioinformatics tools such as sequence assembly and similarity search tools as well as genome browsers. Specific examples include Tomato Expression Database, Tomato Functional Genomics Database, TomExpress, and NexGenEx-Tom.

The application of allele mining for gene discovery and tomato improvement includes but is not limited to characterization of allelic diversity in tomato genetic resources (which unlocks the genetic potential among accessions within gene banks), identification of new haplotypes, development of allele-specific molecular markers for MAS and establishment of allelic synteny and evolutionary relationships. The later enables identification and utilization of genes of desirable traits from closely related species and wild relatives of tomato. The potential for allele mining to speed up gene discovery for improvement of tomato is enormous, although, this potential has not been fully exploited. With the development and improvement of new molecular biology and bioinformatics tools, it will become increasingly easier and relatively cheaper to access the full potential and utility of allele mining, consequently fast-tracking improvement of tomato for food and nutritional security.

REFERENCES

Abdurakhmonov IY(2016) Genomics era for plants and crop species–advances made and needed tasks ahead. *Plant Genom*. https://doi.org/10.5772/62083

Abuín JM, Pichel JC, Pena TF, Amigo J (2015) BigBWA: Approaching the Burrows–Wheeler aligner to Big Data technologies. *Bioinformatics* 31: 4003–4005.

Adhikari P, McNellie J, Panthee DR (2020) Detection of quantitative trait loci (Qtl) associated with the fruit morphology of tomato. *Genes* 11(10): 1117. https://doi.org/10.3390/genes11101117

Ahmad N, Mukhtar Z (2017) Genetic manipulations in crops: Challenges and opportunities. *Genomics* 109(5–6):494–505. https://doi.org/10.1016/j.ygeno.2017.07.007

Almodaresi F, Zakeri M, Patro R (2021) PuffAligner: A fast, efficient and accurate aligner based on the Pufferfish index. *Bioinformatics* 37, 4048–4055.

Al-Yassin A. and Khademian R. (2015). Allelic variation of salinity tolerance genes in barley ecotypes (natural populations) using ecotilling. *Am-Eur Agric & Environ S.* 15: 563–572.

Amaranatha Reddy M, Gambhire VB, Rajeswar Reddy T (2014) Allele Mining in Crop Improvement. *Int J Dev Research*. 4(2): 300–305.

Anders S, Pyl PT, Huber W (2015) HTSeq – A Python framework to work with high-throughput sequencing data. *Bioinformatics* 31, 166–169.

Ashrafi H, Hill T, Stoffel K, Kozik A, Yao JQ, Chin-Wo SR, van Deynze A (2012) De novo assembly of the pepper transcriptome (*Capsicum annuum*): A benchmark for in silico discovery of SNPs, SSRs and candidate genes. *BMC Genom* 13(1), 1–15 https://doi.org/10.1186/1471-2164-13-571

Ashrafi H, Kinkade MP, Merk HL, Foolad MR (2011) Identification of novel quantitative trait loci for increased lycopene content and other fruit quality traits in a tomato recombinant inbred line population. *Mol Breed* 30(1), 549–567. https://doi.org/10.1007/s11032-011-9643-1

Aslam Z, Khattak JZK, Ahmed M, Asif M (2017) A role of bioinformatics in agriculture. In: Ahmed M, Stockle CO (eds), *Quantification of Climate Variability, Adaptation and Mitigation for Agricultural Sustainability*. Springer International Publishing, Cham, pp 413–434. https://doi.org/10.1007/978-3-319-32059-5_17

Bao JS, Corke H, Sun M (2006) Nucleotide diversity in starch synthase IIa and validation of single nucleotide polymorphisms in relation to starch gelatinization temperature and other physicochemical properties in rice (*Oryza sativa* L.). *Theor Appl Genet.* 113:1171–1183.

Barchenger DW, Hsu Y, Ou J, Lin Y, Lin Y, Balendres MAO, Hsu Y, Schafleitner R, Hanson P (2022) Whole genome resequencing and complementation tests reveal candidate loci contributing to bacterial wilt (*Ralstonia sp.*) resistance in tomato. Sci Rep. 12(1), 1–12. https://doi.org/10.1038/s41598-022-12326-x

Barkley NA, Wang ML (2008) Application of TILLING and EcoTILLING as reverse genetic approaches to elucidate the function of genes in plants and animals. *Curr Genom.* 9(4), 212. https://doi:10.2174/138920208784533656

Barnes MR, Gray IC (2003) *Bioinformatics for Geneticists*. John Wiley & Sons, New York.

Barone A, Chiusano ML, Ercolano R, Giuliano G, Grandillo S, Frusciante L (2008) Structural and functional genomics of tomato. *Int J Plant Genomics* 2008:820274. doi: 10.1155/2008/820274

Barrantes W, López-Casado G, García-Martínez S, Alonso A, Rubio F, Ruiz JJ, Fernández-Muñoz R, Granell A, Monforte AJ (2016) Exploring new alleles involved in tomato fruit quality in an introgression line library of *Solanum pimpinellifolium. Front Plant Sci. 7(AUG2016).* https://doi:10.3389/FPLS.2016.01172

Batley J, Edwards D (2016) The application of genomics and bioinformatics to accelerate crop improvement in a changing climate. *Curr Opin Plant Biol.* 30, 78–81.

Baxevanis AD, Bader GD, Wishart DS (2020) *Bioinformatics*. John Wiley & Sons, New York.

Bernal E, Liabeuf D, Francis DM (2020) Evaluating quantitative trait locus resistance in tomato to multiple Xanthomonas spp. *Plant Dis.* 104(2), 423–429. https://doi.org/10.1094/PDIS-03-19-0669-RE/ASSET/IMAGES/LARGE/PDIS-03-19- 0669-RE_F3.JPEG

Bhandari P, Lee TG (2021) A genetic map and linkage panel for the large-fruited fresh-market tomato. *J Am Soc Hortic Sci.* 146(2), 125–131. https://doi.org/10.21273/JASHS04999-20

Bizouerne E, Buitink J, Vu BL, Vu JL, Esteban E, Pasha A, Provart N, Verdier J, Leprince O (2021) Gene co-expression analysis of tomato seed maturation reveals tissue-specific regulatory networks and hubs associated with the acquisition of desiccation tolerance and seed vigour. *BMC Plant Biol.* 21(1), 1–23 https://doi:10.1186/S12870-021-02889- 8/FIGURES/9

Bostan H, Chiusano M. (2015) NexGenEx-Tom: A gene expression platform to investigate the functionalities of the tomato genome. *BMC Plant Biol.* 15, 48. https://doi.org/10.1186/s12870-014-0412-2

Bray N, Dubchak I, Pachter L (2003) AVID: A global alignment program. *Genome Res.* 13, 97–102. https://doi.org/10.1101/gr.789803

Brown AHD (1989) Core collections: A practical approach to genetic resource management. *Genome* 31, 818–824.

Comai L, Young K, Till BJ, Reynolds SH, Greene EA, Codomo CA, Enns LC, Johnson JE, Burtner C, Odden AR (2004) Efficient discovery of DNA polymorphisms in natural populations by Ecotilling. *Plant J* 37(5), 778–786. https://doi:10.1111/j.0960-7412.2003.01999.x

Daugelaite J, O'Driscoll A, Sleator RD (2013) An overview of multiple sequence alignments and cloud computing in bioinformatics. *Int Sch Res Note* 2013(615630): 1–14.

Dixit R, Kapoor N, Singh NP, Kumar A (2022) Bioinformatics resources for sustainable agriculture. In Sengar, Rakesh S, Chaudhary R, Bhadauria H (Eds.): *Handbook of Research on Green Technologies for Sustainable Management of Agricultural Resources*. IGI Global, pp 461–471.

Dobin A, Davis CA, Schlesinger F, Drenkow J, Zaleski C, Jha S, Batut P, Chaisson M, Gingeras TR (2013) STAR: ultrafast universal RNA-seq aligner. *Bioinformatics* 29, 15–21.

Etukuri SP, Anche VC, Ayubov MS, Walker LT, Sripathi VR (2022) *Transcriptome Analysis Using RNA Sequencing for Finding Genes Related to Fiber in Cotton: A Review.* Intechopen, London, UK.

Fei Z, Joung J.G, Tang X, Zheng Y, Huang M, Lee JM, McQuinn R, Tieman DM, Alba R, Klee HJ, Giovannoni JJ (2011) Tomato Functional Genomics Database: A comprehensive resource and analysis package for tomato functional genomics. *Nucleic Acids Res.* 39, D1156–D1163. https://doi.org/10.1093/nar/gkq991

Fei Z, Tang X, Alba R, Giovannoni J (2006) Tomato Expression Database (TED): A suite of data presentation and analysis tools. *Nucleic Acids Res.* 34, D766–D770. https://doi.org/10.1093/nar/gkj110

Fernandez-Pozo N, Menda N, Edwards JD, Saha S, Tecle IY, Strickler SR., Bombarely A, Fisher-York T, Pujar A, Foerster H, Yan A, Mueller LA (2015) The Sol Genomics Network (SGN)-from genotype to phenotype to breeding. *Nucleic Acids Res* 43(D1), D1036–D1041. https://doi.org/10.1093/nar/gku1195

Filiz E, Akbudak MA (2020) Ammonium transporter 1 (AMT1) gene family in tomato (*Solanum lycopersicum* L.): Bioinformatics, physiological and expression analyses under drought and salt stresses. *Genomics* 112, 3773–3782. https://doi.org/10.1016/j.ygeno.2020.04.009

Fontecha G, Silva-Navas J, Benito C, Mestres MA, Espino FJ, Hernandez-Riquer MV (2007) Candidate gene identification of an aluminium-activated organic acid transporter gene at the Alt4 locus for aluminium tolerance in rye (*Secale cereale* L.). *Theor Appl Genet* 114, 249–60.

Foolad MR (2007) Genome mapping and molecular breeding of tomato. *Int J Plant Genomics* 2007(64358): 1–152. https://doi.org/10.1155/2007/64358

Foolad MR, Panthee DR (2012) Marker-assisted selection in tomato breeding. Crit Rev Plant Sci. 31(2), 93–123. https://doi.org/10.1080/07352689.2011.616057

Foolad MR, Panthee DR (2013) Erratum to Marker-Assisted Selection in Tomato Breeding. *Crit Rev Plant Sci* (2012), 31, 2 (93–123), 10.1080/07352689.2011.616057). CRC Crit Rev Plant Sci. 32(6), 483 https://doi.org/10.1080/07352689.2013.795451

Fortuny AP, Bueno RA, Pereira Da Costa JH, Zanor MI, Rodríguez GR (2021) Tomato fruit quality traits and metabolite content are affected by reciprocal crosses and heterosis. *J Exp Bot.* 72(15), 5407–5425. doi:10.1093/JXB/ERAB222.

Gao L, Gonda I, Sun H, Ma Q, Bao K, Denise M, Tieman DT, Burzynski-Chang EA, Fish TL, Stromberg KA, Sacks GL, Thannhauser TW, Foolad MR, Diez MJ, Blanca J, Canizares J, Xu Y, van der Knaap E, Huang S, Klee HJ, Giovannoni JJ, Fei Z (2019). The tomato pan-genome uncovers new genes and a rare allele regulating fruit flavor. *Nat Genet* 51, 1044–1051.

Garcia V, Bres C, Just D, Fernandez L, Tai FWJ, Mauxion JP, le Paslier MC Bérard A, Brunel D, Aoki K, Alseekh S, Fernie AR, Fraser PD, Rothan C (2016) Rapid identification of causal mutations in tomato EMS populations via mapping-by-sequencing. *Nat Protoc* 11(12), 2401–2418. https://doi.org/10.1038/nprot.2016.143

Gokidi Y, Bhanu AN, Chandra K, Singh MN, Hemantaranjan A (2017) Allele mining-an approach to discover allelic variation in crops. *J Plant Sci Res* 33(2), 167–180.

Gonda I, Ashrafi H, Lyon DA, Strickler SR, Hulse-Kemp, AM, Ma Q, Sun H, Stoffel K, Powell AF, Futrell S, Thannhauser TW, Fei Z, Deynze AE, Van Mueller A, Giovannoni JJ, Foolad MR (2019) Sequencing-based Bin map construction of a tomato mapping population, facilitating high-resolution quantitative trait loci detection. *Plant Genome,* 12(1), 180010. https://doi.org/10.3835/PLANTGENOME2018.02.0010

Grandillo S, Ku HM Tanksley SD (1996) Characterization of fs8.1, a major QTL influencing fruit shape in tomato. *Mol Breed* 2(3), 251–260 https://doi.org/10.1007/BF00564202

Graner A (2006) Barley research at IPK. In: de Vicente MC, Glaszmann JC (eds) *Molecular Markers for Allele Mining, IPGRI,* p 25.

Homer N, Merriman B, Nelson SF (2009) BFAST: An alignment tool for large scale genome resequencing. *PLoS One* 4, e7767.

Hosmani PS, Flores-Gonzalez M, Geest H, van de, Maumus F, Bakker LV, Schijlen E, Haarst J van, Cordewener J, Sanchez-Perez G, Peters S, Fei Z, Giovannoni J.J, Mueller LA, Saha S (2019) An improved de novo assembly and annotation of the tomato reference genome using single-molecule sequencing, Hi-C proximity ligation and optical maps. *BioRxiv2012*, 767764. https://doi.org/10.1101/767764

Huntley D, Baldo A, Johri S, Sergot M (2006) SEAN: SNP prediction and display program utilizing EST sequence clusters. *Bioinformatics* 22, 495–496.

Ichihashi Y, Sinha NR (2014) From genome to phenome and back in tomato. *Curr Opin Plant Biol* 18(1), 9–15. https://doi.org/10.1016/j.pbi.2013.12.004

Illa-berenguer E, Houten J, Huang Z, Knaap E (2015) Rapid and reliable identification of tomato fruit weight and locule number loci by QTL–seq. *Theor Appl Genet* 128, 1329–1342. https://doi.org/10.1007/s00122-015-2509-x

Iquebal MA, Jaiswal S, Mukhopadhyay CS, Sarkar C, Rai A, Kumar D (2015) Applications of Bioinformatics in Plant and Agriculture. In: Barh D, Khan MS, Davies E (eds), *PlantOmics: The Omics of Plant Science.* Springer India, New Delhi, pp 755–789. https://doi.org/10.1007/978-81-322-2172-2_27

Jesús J (2018) Seed quality of three native tomato varieties and a commercial one produced under high temperatures. *Revista Chapingo Serie Horticultura* 24(3):215–227. https://doi:10.5154/r.rchsh.2018.04.009

Kaur N, Street K, Mackay M, Yahiaoui N, Keller B (2008a) *Allele mining and sequence diversity at the wheat powdery mildew resistance locus Pm3.* In: *Proceedings of the 11th International Wheat Genetics Symposium.* Sydney, pp 1–3.

Kazmi R, Khan N, Willems L, Heusden A van, Ligterink W (2012) Complex genetics controls natural variation among seed quality phenotypes in a recombinant inbred population of an interspecific cross between *Solanum lycopersicum* x Solanum *pimpinellifolium. Plant Cell Environ.* 35, 929–951.

Kent WJ (2002) BLAT – the BLAST-like alignment tool. *Genome Res.* 12, 656–664.

Khalid MN, Abdullah A, Ijaz Z, Naheed N, Hamad A, Sheir MA, Shabir F, Parveen K, Khan MD (2021) Application and potential use of advanced bioinformatics techniques in agriculture and animal sciences. *Ind J Pure App Biosci* 9, 237–246.

Khan N, Kazmi RH, Willems LAJ, van Heusden AW, Ligterink W, Hilhorst HWM (2012b) Exploring the natural variation for seedling traits and their link with seed dimensions in tomato. *PLoS One* 7(8), e43991. https://doi:10.1371/journal.pone.0043991

Kho KH, Sukhan ZP, Hossen S, Cho Y, Kim SC, Sharker MR, Jung HJ, Nou IS (2021) Construction of a genetic linkage map based on SNP markers, QTL mapping and detection of candidate genes of growth-related traits in pacific abalone using genotyping-by-sequencing. *Front Mar Sci.* 8(August), 1–13. https://doi.org/10.3389/fmars.2021.713783

Kim D, Langmead B, Salzberg SL (2015) HISAT: A fast spliced aligner with low memory requirements. *Nat Methods* 12, 357–360.

Kim H, Kim BS, Shim JE, Hwang S, Yang S, Kim E, Iyer-Pascuzzi AS, Lee I (2017) TomatoNet: A genome-wide co-functional network for unveiling complex traits of tomato, a model crop for fleshy fruits. *Mol Plant.* 10(4), 652–655. https://doi.org/10.1016/j.molp.2016.11.010

Kimbara J, Ohyama A, Chikano H, Ito H, Hosoi K, Negoro S, Miyatake K, Yamaguchi H, Nunome T, Fukuoka H, Hayashi T (2018) QTL mapping of fruit nutritional and flavor components in tomato (*Solanum lycopersicum*) using genome-wide SSR markers and recombinant inbred lines (RILs) from an intra-specific cross. *Euphytica* 214(11), 1–12. https://doi.org/10.1007/s10681-018-2295-z

Kumar A, Babu BK, Yadav S, Agrawal PK (2016) Allele mining for resistance gene analogs (RGAs) in crop plants: A special emphasis on blast resistance in finger millet (*Eleusine coracana* L.). *Indian J Genet.* 76: 1–9

Kumar GR, Sakthivel K, Sundaram RM, Neeraja CN, Balachandran SM, Rani NS, Viraktamath BC, Madhav MS (2010) Allele mining in crops: Prospects and potentials. *Biotechnol Adv.* 28(4), 451–461. https://doi.org/10.1016/j.biotechadv.2010.02.007

Kumar M, Rama SS, Singh S, Sardar RSS, Patel V (2014) Allele Mining and its Applications Study on molecular basis of salt tolerance of chickpea genotypes. *Int J Biol Sci* 4(1): 34–37. https://doi:10.5958/2322-0996.2014.00016.7

Kumari R, Prasad R, Sharma VK, Kumar H (2018) Allele mining for crop improvement Characterization and micropropagation of *Physalis minima* L. in Bihar View project. *Int J Pure App Biosci.* 6(1):1456–1465. doi:10.18782/2320-7051.6073.

Kushwaha UKS, Deo I, Jaiswal JP, Prasad B (2017) Role of bioinformatics in crop improvement. *Glob J Sci Front Res Agric Vet.* 17, 13–24.

Langmead B, Salzberg SL (2012) Fast gapped-read alignment with Bowtie 2. *Nat Methods* 9, 357–359.

Langmead B, Trapnell C, Pop M, Salzberg SL (2009 Ultrafast and memory-efficient alignment of short DNA sequences to the human genome. *Genome Biol.* 10, 1–10.

Lee HC, Lai K, Lorenc MT, Imelfort M, Duran C, Edwards D (2012) Bioinformatics tools and databases for analysis of next-generation sequence data. *Brief Funct Genomics* 11, 12–24. https://doi.org/10.1093/bfgp/elr037

Li C., Zhang J., Hu G., Fu Y. and Wang Q. (2016). Association mapping and favorable allele mining for node of first fruiting/sympodial branch and its height in Upland cotton (*Gossypium hirsutum* L.). *Euphytica* 210, 57–68

Li R, Li Y, Kristiansen K, Wang J (2008) SOAP: Short oligonucleotide alignment program. *Bioinformatics* 24, 713–714.

Li R, Yu C, Li Y, ultrafast tool for short read Lam TW, Yiu SM, Kristiansen K, Wang J (2009) SOAP2: An improved alignment. *Bioinformatics* 25, 1966–1967.

Lindblad-Toh K, Tanenbaum DM, Daly MJ, Winchester E, Lui WO, Villapakkam A, Stanton SE, Larsson C, Hudson TJ, Johnson BE, Lander ES, Meyerson M (2000) Loss-of-heterozygosity analysis of small-cell lung carcinomas using single-nucleotide polymorphism arrays. *Nat Biotechnol.* 18(9), 1001–1005. https://doi.org/10.1038/79269

Liu L, Li Y, Li S, Hu N, He Y, Pong R, Lin D, Lu L, Law M (2012) Comparison of next-generation sequencing systems. *J Biomed Biotechnol.* 2012, 11. https://doi:10.1155/2012/251364

Lukyanenko AN (1991) Disease Resistance in Tomato. In Kalloo G (Ed.): Genetic Improvement of Tomato,. Springer Berlin, Heidelberg, pp 99–119. https://doi.org/10.1007/978-3-642-84275-7_9

Luscombe NM, Greenbaum D, Gerstein M (2001) What is bioinformatics? An introduction and overview. *Yearb Med Inform.* 10, 83–100.

Mardis E.R. (2008) Next-generation DNA sequencing methods. *Ann Rev Genom Hum Genet.* 9, 387–402.

Mata-Nicolás E, Montero-Pau J, Gimeno-Paez E, Garcia-Carpintero V, Ziarsolo P, Menda N, Mueller L A, Blanca J, Cañizares J, van der Knaap E, Díez MJ (2020) Exploiting the diversity of tomato: The development of a phenotypically and genetically detailed germplasm collection. *Hortic Res* 7(1), 1–14. https://doi.org/10.1038/s41438-020-0291-7

Maurya DB, Akhtar S, Chattopadhyay T (2020) Exploring allelic status of selected disease resistance genes in a set of tomato genotypes using gene-linked molecular markers. *J Crop Weed.* 16(1), 236–241. https://doi:10.22271/09746315.2020.V16.I1.1299

McCallum CM, Comai L, Greene EA, Henikoff S. 2000. Targeted screening for induced mutations. *Nat Biotechnol.* 18(4), 455–457. https://doi:10.1038/74542

McCouch SR, Sweeney M, Li J, Jiang H, Thomson M, Septiningsih E et al. (2007) Through the genetic bottleneck: *O. rufipogon* as a source of trait-enhancing alleles for *O. sativa. Euphytica* 154, 317–339.

Mckenzie L, Zystro J (2021) *Tomato Seed Production Guide-Organic Seed Alliance seedalliance.org/publications/tomato-seed-production-guide.* Table of Contents Tomato Background, Lifecycle, and Basic Biology.

Mikami I, Uwatoko N, Ikeda Y, Yamaguchi J, Hirano HY, Suzuki Y, Sano, Y (2008) Allelic diversification at the locus in landraces of Asian rice. Theor Appl Genet, 116(7), 979–989.

Mishra DP, Badajena JC, Nayak SK, Baliyarsingh B (2022) Bioinformatics: A tool for sustainable agriculture. In: Nayak SK, Baliyarsingh B, Singh A, Mannazzu I, Mishra BB (eds), *Advances in Agricultural and Industrial Microbiology: Volume-2: Applications of Microbes for Sustainable Agriculture and in-Silico Strategies.* Springer Nature, Singapore, pp 233–246. https://doi.org/10.1007/978-981-16-9682-4_13

Morgenstern B (2004) DIALIGN: Multiple DNA and protein sequence alignment at BiBiServ. *Nucleic Acids Res.* 32, W33–W36.

Morgenstern B (2014) Multiple sequence alignment with DIALIGN. In Russell D (Ed.): *Multiple Sequence Alignment Methods.* Humana Press, Totowa, NJ. https://doi.org/10.1007/978-1-62703-646-7_12

Mueller LA, Lankhorst RK, Tanksley SD, Giovannoni JJ, White R, Vrebalov J, Fei Z, Eck J, Buels R, Mills AA, Menda N, Tecle IY, Bombarely A, Stack S, Royer SM, Chang S, Shearer LA, Kim BD, Jo S, Stiekema W (2009) A snapshot of the emerging tomato genome sequence. *The Plant Genome* 2(1): 78–92. https://doi.org/10.3835/plantgenome2008.08.0005

Mulualem T, Bekeko Z (2016) Advances in quantitative trait loci, mapping and importance of markers assisted selection in plant breeding research. *Int J Plant Breed Genet.* 10(2), 58–68 https://doi.org/10.3923/ijpbg.2016.58.68

Nienhuis J, Helentjaris T, Slocum M, Ruggero B, Schaefer A (1987) Restriction fragment length polymorphism analysis of loci associated with insect resistance in Tomato 1. *Crop Sci.* 27(4), 797–803. https://doi.org/10.2135/CROPSCI1987.0011183X002700040039X

Orchard, CJ, Cooperstone JL, Gas-Pascual E, Andrade MC, Abud G, Schwartz SJ, Francis DM (2021) Identification and assessment of alleles in the promoter of the Cyc-B gene that modulate levels of β-carotene in ripe tomato fruit. *Plant Genome* 14(1), 1–17. https://doi.org/10.1002/tpg2.20085

Panthee DR, Piotrowski A, Ibrahem R (2017) Mapping quantitative trait loci (QTL) for resistance to late blight in tomato. *Int J Mol Sci.* 18(7): 1–14. https://doi.org/10.3390/IJMS18071589

Pascual L, Desplat N, Huang BE, Desgroux A, Bruguier L, Bouchet JP, Le QH, Chauchard B, Verschave P, Causse M (2015) Potential of a tomato MAGIC population to decipher the genetic control of quantitative traits and detect causal variants in the resequencing era *J Plant Biotechnol.* 3(4), 565–577. https://doi.org/10.1111/pbi.12282

Pascual L, Xu J, Biais B, Maucourt M, Ballias P, Bernillon S, Deborde C, Jacob D, Desgroux A, Faurobert M, Bouchet JP, Gibon Y, Moing A, Causse M (2013) Deciphering genetic diversity and inheritance of tomato fruit weight and composition through a systems biology approach. *J Exp Bot.* 64(18), 5737–5752. https://doi.org/10.1093/jxb/ert349

Pathirana R, Carimi F (2022) Management and utilization of plant genetic resources for a sustainable agriculture. *Plants* 11(2038). https://doi.org/https://doi.org/10.3390/ plants11152038

Pereira L, Sapkota M, Alonge M, Zheng Y, Zhang Y, Razifard H, Taitano NK, Schatz MC, Fernie AR, Wang Y, Fei Z, Caicedo AL, Tieman DM, van der Knaap E (2021) Natural genetic diversity in tomato flavor genes. *Front Plant Sci* 12(June), 1–23. https://doi.org/10.3389/fpls.2021.642828

Phan NT, Trinh LT, Rho MY, Park TS, Kim OR, Zhao J, Kim HM, Sim SC (2019) Identification of loci associated with fruit traits using genome-wide single nucleotide polymorphisms in a core collection of tomato (*Solanum lycopersicum* L.). *Sci Hortic.* 243, 567–574. https://doi.org/10.1016/j.scienta.2018.09.003

Pietrella M, Giuliano G (2016) The sequencing: How it was done and what it produced. In: Causse M, Giovannoni J, Bouzayen M, Zouine M (eds), *The Tomato Genome*. Compendium of Plant Genomes. Springer, Berlin, Heidelberg. https://doi.org/10.1007/978-3-662-53389-5_6

Raghavan C, Naredo MEB, Wang H, Atienza G, Liu B, Qiu F, McNally KL, Leung H (2007) Rapid method for detecting SNPs on agarose gels and its application in candidate gene mapping. *Mol Breed.* 19(2), 87–101. https://doi:10.1007/S11032-006-9046-X

Ramírez M, Moncada RN, Villegas-Escobar V, Jackson RW, Ramírez CA (2020) Phylogenetic and pathogenic variability of strains of Ralstonia solanacearum causing moko disease in Colombia. *Plant Pathol.* 69(2), 360–369. https://doi.org/10.1111/ppa.13121

Ranjan A, Ichihashi Y, Sinha NR (2012) The tomato genome: implications for plant breeding, genomics and evolution. *Genome Biol.* 13(8), 167. https://doi.org/10.1186/gb4037

Ren J, Li C, Li C (2012) Tomato genome gets fully decoded – paves way to tastier and healthier fruits. *J Genet Genomics = Yi Chuan Xue Bao*, 39(7), 303–305. https://doi.org/10.1016/j.jgg.2012.06.001

Ricardo Zeist A, Antônio da Silva A, Tadeu Vilela de Resende J, Roberto Maluf, W, Gabriel A, Zanin DS, Perez Guerra E (2019) Tomato breeding for insect-pest resistance. *Recent Advances in Tomato Breeding and Production*. https://doi.org/10.5772/INTECHOPEN.75978

Rick CM, Fobes JF(1975). Allozyme variation in the cultivated tomato and closely related species. *Bull Torre Bot Club* 102(6), 376–384. https://doi.org/10.2307/2484764

Rodríguez GR, Muños S, Anderson C, Sim SC, Michel A, Causse M, McSpadden Gardener BB, Francis D, van der Knaap E (2011) Distribution of SUN, OVATE, LC, and FAS in the tomato germplasm and the relationship to fruit shape diversity. *Plant Physiol* 156(1), 275–285. https://doi.org/10.1104/PP.110.167577

Roohanitaziani R, de Maagd RA, Lammers M, Molthoff J, Meijer-Dekens F, van Kaauwen MPW, Finkers R, Tikunov Y, Visser RGF, Bovy AG (2020) Exploration of a resequenced tomato core collection for phenotypic and genotypic variation in plant growth and fruit quality traits. *Genes* 11(11), 1278. https://doi:10.3390/GENES11111278

Rose LE, Langley CH, Bernal AJ, Michelmore RW (2005) Natural variation in the Pto pathogen resistance gene within species of wild tomato (*Lycopersicon*). I. Functional analysis of Pto alleles. *Genetics* 171(1), 345–357. https://doi.org/10.1534/genetics.104.039339

Rothan C, Diouf I, Causse M (2018). Trait discovery and editing in tomato. *Plant J.* 97(1), 73–90. https://doi.org/10.1111/tpj.14152

Ruggieri V, Francese G, Sacco A, D'Alessandro A, Rigano MM, Parisi M, Milone M, Cardi T, Mennella G, Barone A (2014) An association mapping approach to identify favourable alleles for tomato fruit quality breeding. *BMC Plant Biol.* 14(1), 1–15. https://doi.org/10.1186/s12870-014-0337-9

Rumble SM, Lacroute P, Dalca AV, Fiume M, Sidow A, Brudno M (2009) SHRIMP: accurate mapping of short color-space reads. *PLoS Comput. Biol.* 5, e1000386.

Sarika GU, Amruta N (2014) Allele mining and its importance in seed quality. *Int J of Dev Res.* 4(4), 819–824.

Sato S, Tabata S, Hirakawa H, Asamizu E, Shirasawa K, Isobe S, Kaneko T, Nakamura Y, Shibata D, Aoki K, Egholm M, Knight J, Bogden R, Li C, Shuang Y, Xu X, Pan S, Cheng S, Liu X, Gianese G (2012) The tomato genome sequence provides insights into fleshy fruit evolution. *Nature* 485(7400), 635–641. https://doi.org/10.1038/nature11119

Schouten HJ, Tikunov Y, Verkerke W, Finkers R, Bovy A, Bai Y, Visser RGF (2019) Breeding has increased the diversity of cultivated tomato in The Netherlands. *Front Plant Sci* 10, 1606. https://doi:10.3389/FPLS.2019.01606/BIBTEX

Scott JW, Wang JF, Hanson PM (2005) Breeding tomato for resistance to bacterial wilt, a global view. *Acta Hortic* 695, 161–172. https://doi.org/10.17660/actahortic.2005.695.18

Scott MF, Ladejobi O, Amer S, Bentley AR, Biernaskie J, Boden SA, Clark M, Dell'Acqua M, Dixon LE, Filippi CV, Fradgley N, Gardner KA, Mackay IJ, O'Sullivan D, Percival-Alwyn L, Roorkiwal M, Singh RK, Thudi M, Varshney RK, Mott R (2020) Multi-parent populations in crops: A toolbox integrating genomics and genetic mapping with breeding. *Heredity* 125(6), 396–416. https://doi.org/10.1038/s41437-020-0336-6

Seol YJ, Lee TH, Park DS, Kim CK (2016) NABIC: A new access portal to search, visualize, and share agricultural genomics data. *Evol Bioinformat.* 12, EBO.S34493. https://doi.org/10.4137/EBO.S34493

Sharma MK, Singh S, Sengar, RS (2014) Allele mining and its applications. *Biotechnol Today*, 4(1), 34–37.
Simsek O. and Kacar Y.A. (2010). Discovery of mutations with tilling and ecotilling in plant genomes. *Sci Res Essays*. 5, 3799–3802.
Singh BD (2005). *A Text Book of Plant Breeding*. Kalyani Publishers, New Delhi.
Singh VK, Singh AK, Chand R, Kushwaha C (2011) Role of bioinformatics in agriculture and sustainable development. *Int J Bioinformat Res*. 3, 221–226.
Slade AJ, Fuerstenberg SI, Loeffler D, Steine MN, Facciotti D (2005) A reverse genetic, nontransgenic approach to wheat crop improvement by TILLING. *Nat Biotechnol*. 23(1), 75–81. https://doi:10.1038/nbt1043
Spooner D., vanTreuren R. anddeVicente M.C. (2005). *Molecular markers for genebank management*. International Plant Genetic Resources Institute, Rome, Italy. IPGRI Technical Bulletin No. 10.
Su X, Wang B, Geng X, Du Y, Yang Q, Liang B, Meng G, Gao Q, Yang W, Zhu Y, Lin T (2021) A high-continuity and annotated tomato reference genome. *BMC Genomics* 22(1), 1–12. https://doi.org/10.1186/s12864-021-08212-x
Suresh BV, Roy R, Sahu K, Misra G, Chattopadhyay D (2014) Tomato genomic resources database: An integrated repository of useful tomato genomic information for basic and applied research. *PLoS One*. 9(1), 1–9. https://doi.org/10.1371/journal.pone.0086387
Tanksley SD, Rick CM (1980) Isozymic gene linkage map of the tomato: Applications in genetics and breeding. *TAG. Theor Appl Genet Theoretische Und Angewandte Genet*. 58(2), 161–170. https://doi.org/10.1007/BF00279708
Thomson MJ, Biswas S, Tsakirpaloglou N, Septiningsih EM (2022) Functional allele validation by gene editing to leverage the wealth of genetic resources for crop improvement. *Int J Mol Sci*. 23(6565), 1–24 https://doi.org/10.3390/ijms23126565
Till BJ, Reynolds SH, Weil C, Springer N, Burtner C, Young K, Bowers E, Codomo CA, Enns LC, Odden AR, et al. (2004) Discovery of induced point mutations in maize genes by TILLING. *BMC Plant Biol*. 4, 12. https://doi:10.1186/1471-2229-4-12
Till BJ, Zerr T, Comai L, Henikoff S (2006) A protocol for TILLING and Ecotilling in plants and animals. *Nat Protoc*. 1(5), 2465–2477. https://doi:10.1038/nprot.2006.329
Vossen JH, Kwang-Ryong Jo, B Vosman B (2014) Chapter 2 mining the genus solanum for increasing disease resistance. In: Tuberosa R, Graner A, Frison E (eds), *Genomics of Plant Genetic Resources*. Springer, Dordrecht, Netherlands, Vol. 2, pp 1–46. https://doi.org/10.1007/978-94-007-7575-6-2
Wambugu PW, Henry R (2022) Supporting in situ conservation of the genetic diversity of crop wild relatives using genomic technologies. *Mol Ecol* 31(8), 2207–2222. https://doi.org/10.1111/mec.16402
Wang J, Kong L, Gao G, Luo J (2013) A brief introduction to web-based genome browsers. *Brief Bioinform*. 14, 131–143. https://doi.org/10.1093/bib/bbs029
Würschum, T. (2012) Mapping QTL for agronomic traits in breeding populations. *Theor Appl Genet*. 125(2), 201–210. https://doi.org/10.1007/S00122-012- 1887-6
Xue J, Zhao S, Liang Y, Hou C, Wang J (2007) *Bioinformatics and Its Applications in Agriculture. In: International Conference on Computer and Computing Technologies in Agriculture*. Springer, pp 977–982.
Yan Z, Pérez-de-Castro A, Díez MJ, Hutton SF, Visser RGF, Wolters AMA, Bai Y, Li J (2018) Resistance to tomato yellow leaf curl virus in tomato germplasm. *Frontiers in Plant Sci*. 9(August), 1–14. https://doi.org/10.3389/fpls.2018.01198
Zhu G, Wang S, Huang Z, Zhang S, Liao Q, Zhang C, Lin T, Qin M, Peng M, Yang C, Cao X, Han X, Wang X, van der Knaap E, Zhang Z, Cui X, Klee H, Fernie AR, Luo J, Huang S (2018) Rewiring of the fruit metabolome in tomato breeding. *Cell* 172(1–2), 249–261.e12. https://doi.org/10.1016/j.cell.2017.12.019
Zou Q, Hu Q, Guo M, Wang G (2015) HAlign: Fast multiple similar DNA/RNA sequence alignment based on the centre star strategy. *Bioinformatics* 31, 2475–2481.
Zou Q, Lin G, Jiang X, Liu X, Zeng X (2020) Sequence clustering in bioinformatics: An empirical study. *Brief. Bioinform*. 21, 1–10. https://doi.org/10.1093/bib/bby090
Zouine M, Maza E, Djari A, Lauvernier M, Frasse P, Smouni A, Pirrello J, Bouzayen M (2017) TomExpress, a unified tomato RNA-Seq platform for visualization of expression data, clustering and correlation networks. *Plant J*. 92, 727–735. https://doi.org/10.1111/tpj.13711

4 High-Throughput Genomics for Allele Mining in Tomato

Archana Bhardwaj,[1,2] *Priti Prasad,*[1,3] *and Sumit Kumar Bag*[1,*]
[1]Molecular Biology and Biotechnology Division, CSIR-National Botanical Research Institute, Rana Pratap Marg, Lucknow, India
[2]Institute of Clinical Molecular Biology, Christian-Albrechts-University Kiel, Germany
[3]Redcliffe Lifesciences Pvt. Ltd., Noida, Uttar Pradesh, India
*Correspondence author: sumit.bag@nbri.res.in

4.1 INTRODUCTION

Tomatoes are significant crops in food security and considered as a rich source of vitamin C. Scientific studies have shown importance of tomato juice in diabetics condition (Tarigan et al. 2019). One of the essential uses of crop genetics and genomics is to identify all allelic variations in trait-associated genes. Before we attempt to discuss allele mining in tomato, it is important to discuss various platforms and computational workflows used in previous studies to identify different types of alleles in form of single nucleotide polymorphism (SNP). Advancements in sequencing technology have led to identifying alleles in the form of SNPs within multiple plant species (Sim et al. 2012). SNPs are one of the predominantly occurring variations in plant species (Haas et al. 2020). In tomato, SNPs were discovered using hybridization to oligonucleotide arrays; expressed sequencing tags (ESTs), and amplicon sequencing approaches (Sim et al. 2009). High-throughput SNP discovery from the array-based method has facilitated high-density genetic mapping and genome-wide association analysis (Sim et al. 2012). On the other hand, SNP discovery from sequencing datasets (Bulk-RNASeq and EST) provides an essential source of variation information, mainly focusing on the coding region of the plant genome. Depending on the "technology usage" and "traits relevant allele in tomatoes" we have classified this chapter into three sections: (1) existing approaches and analytical workflows for allele mining; (2) functional association of alleles in tomato; and (3) multi-omics with allele information for tomato phenotypic traits.

4.2 HIGH-THROUGHPUT APPROACHES AND ANALYTICAL WORKFLOWS FOR ALLELE MINING

Until now, various approaches exist that highlight the usage of various high-throughput sequencing technology for allele mining. This section focuses on using different technologies for allele mining in tomato.

4.2.1 Array-Based SNP Genotyping

The array-based method consists of probe sequences that hybridize with DNA sequences, allows SNP analysis (Min et al. 2020). SNPs are filtered based on the missing allele frequency (MAF) value. Filtered data used for the PCA analysis helps to identify sub-population, mainly to understand the genetic relationship between the individuals. The following analytical step, i.e., linkage

DOI: 10.1201/9781003376439-4

disequilibrium (LD), helps to identify the genes associated with attractive traits (Figure 4.1: refers to array-based section). GWAS is highly useful in identifying the biologically relevant alleles of traits in plants. Based on the GWAS protocol, Kim et al.'s group (Kim et al. 2021) identified various eQTL for eight fruit traits, a valuable resource for the molecular biology. Phan et al. (2019) identified 41 loci associated with the fruit traits in tomato. In another study of GWAS, the authors carried out an investigation into LD patterns at the genome-wide scale using the MLMM approach (Sauvage et al. 2014) whereas using 174 tomato accessions for the GWAS analysis, Zhang et al. (2015) identified 125 significant associations ($P < 0.005$) with multiple SSR markers. In one of the meta-analysis studies, SNPs were identified from multiple GWAS panels using >700 tomato accessions (Zhao et al. 2019). It led to the discovery of >300 significant associations for flavor-related volatiles.

4.2.2 High-Throughput Sequencing Approaches

Identifying alleles is always critical for plant studies (Wu et al. 2019). Various methods have been developed for allele mining and tested mainly in human and plant studies. The choice of the variant calling method affects the precision depending on the genome complexity and sequence diversity (Wu et al. 2019). This section focuses on various sequencing-based methods and their use in marker identification.

NGS sequencing data undergoes many filtration steps, such as trimming adapter sequences and removing low-quality reads (Bhardwaj et al. 2016). High-quality data further follows by mapping to tomato reference genome v.2.50 using different mapping tools (Bowtie2 v.2.3.2 or BWA) (Langmead et al. 2009; Li and Durbin 2009). Finally, mapping data after removing duplicated reads with Picard mark-duplicates used for allele calling from multiple software such as samtools, GATK, Varscan, etc. (Figure 4.1) (Koboldt et al. 2009; Li et al. 2009; McKenna et al. 2010). Samtools is one of the well-known variant calling methods. Researchers used Samtools and its modified version for the variant analysis from the Illumina read with improved sensitivity (Kim et al. 2014; Schmidt et al. 2017). GATK, along with other variant calling methods in NGS data, highlighted the importance of multiple variant detection tools to reduce the false positives in the mutation discovery rate (Gupta et al. 2017). In another study, cryptic variations in tomato were identified using GATK's Unified Genotyper with default parameters (Soyk et al. 2019). In the bulk-RNASeq sequencing data, Scarano et al. (2017) used the Simply Unified Pair-End Read (S.U.P.E.R.) approach for the identification of sequence variations and alternative splicing variants (Scarano et al. 2017). This workflow was developed to analyze the variation data from the resequencing data. Zhou et al. (2019) identified various SNPs in tomato transcriptomes. Müller et al. (2016) applied SNP analysis of cultivated tomato (*Solanum lycopersicum*) and identified that gene *EID1* regulating phase delay (Müller et al. 2016). Multiple studies have reported the usage of GATK in tomato for allele mining (Barchi et al. 2019; Wu et al. 2019). Freebayes algorithm was used for the SNP identification among multiple wild and cultivated tomato species (Blanca et al. 2022). Due to the availability of various methods, it is mandatory to study the impact of various methods on the allele outcome.

4.2.3 Use of Machine Learning in Tomato

Due to the advancement of high-throughput sequencing data and data visualization in plant sciences, the deep learning (DL) role in plant science research is undeniable. Any form of visual input can be used for DL analysis. Artificial intelligence could be classified into three forms that use SNPs. In the first form, raw imaging data (coming from pileup form) could be used for the SNP calling (Figure 4.1) with the CNN model. DL-based CNN models are ideal for high-dimensional data such as spectral image inputs. Deep learning approaches helps to capture complex interactions with high accuracy (González-Camacho et al. 2018). A deep convolutional neural network identifies genetic variation in next-generation sequencing read data (Poplin et al. 2018).

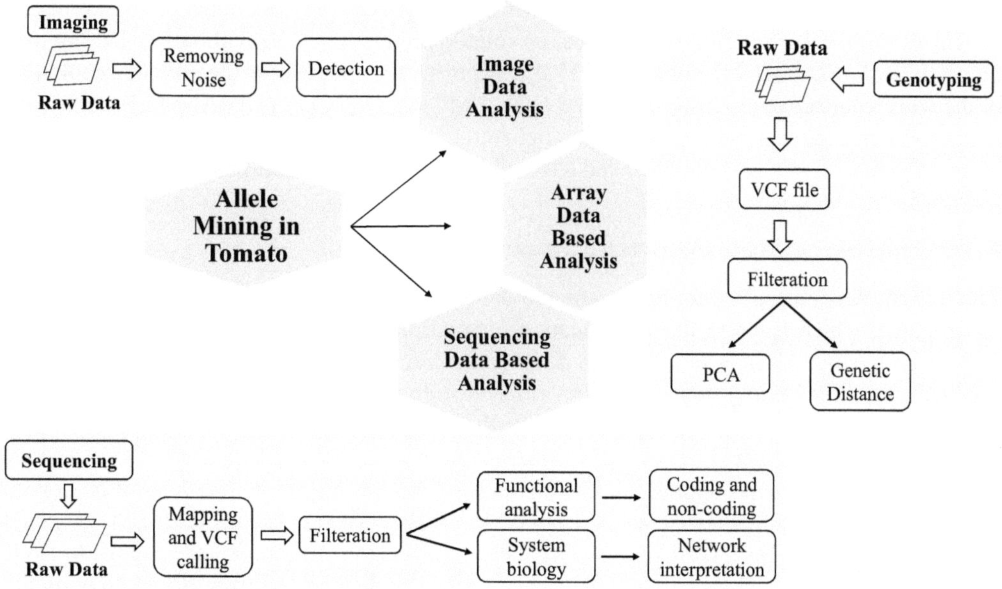

FIGURE 4.1 Multiple approaches with step-wise analysis for allele mining. Each methodology generates multiple intermediate format files for SNP mining. Both sequencing data and array-based methods generate VCF as intermediate files, while image analysis deals with "pixel" information of given input data for SNP prediction.

Various studies have used the DL for classification model development and predicting stable yields in tomato species (Alajrami and Abu-naser 2020). Shi et al. (2019) applied CNNs to segment various 2D and 3D tomato images (Shi et al. 2019). A deep neural network approach was used to locate the disease symptoms in tomato (Fuentes et al. 2017). DeepVar (https://github.com/google/deepvariant) is the method that makes use of pileup image data (Poplin et al. 2018). The second form of AI works directly on the genomic data in SNPs and is used as input in the CNN model for the final output (Figure 4.2). Researchers proposed the GenNet, a deep learning method for predicting phenotypes (van Hilten et al. 2021). This approach could be used to obtain deeper insights into the genetic architecture of tomato. In the third model, SNP features could be used, and the model could be developed with the help of SVM, and RF as illustrated in Bhardwaj and Bag's (2019) study.

4.3 FUNCTIONAL ALLELES ASSOCIATED WITH VARIOUS TRAITS

Various studies have reported the functional role of an allele associated with tomato phenotypic behaviors. In this section, we have discussed tomato phenotypic traits with various alleles.

4.3.1 Alleles Regulating Tomato Fruit Flavor

Tomato flavor is one of the desired qualities, which basically depends on different volatile and sugar molecules. Its flavor is directly associated with the large number of volatile substances that are obtained from lipid precursors during fruit ripening (Chen et al. 2020). Pangenome analysis identifies the promoter allele for the lipoxygenase gene (TomLoxC) that has a crucial role in volatile biosynthesis and apocarotenoid production. This apocarotenoid is positively linked to tomato fruit flavor and aroma at the cost of minimal metabolic processes (Tieman et al. 2017). The 4 kb substitution at the promoter region of TomLoxC gene contributes to the fruit flavor in tomatoes (Shen et al.

2014). It is located on chromosome 1 along with many shared QTL in a range of 153 kb intervals, which are responsible for 28 different types of volatile production (Gao et al. 2019). During tomato domestication, *FLORAL4* (Solyc04g063350) was positively selected for the volatile accumulation in the fruit (Martina et al. 2021). Multiple NON-SMOKY GLYCOSYLTRANSFERASE1 (*nsgt1*) mutant analysis suggests their participation in natural variations of smoky flavor due to the accumulation of guaiacol, methyl salicylate, and eugenol in mature tomato fruit (Alonge et al. 2020).

4.3.2 Mining of Stress-Associated Tomato Alleles

Through the breeding program in tomato and using Bulk RNAseq, many tomato lines have been interrogated, which leads to the identification of multiple alleles for different beneficial traits in *S. lycopersicoides* (Canady et al. 2005). *S. lycopersicoides* is a sexually compatible wild tomato producing healthy anthocyanin. It also has resistance ability for different biotic and abiotic stress (Li et al. 2011; Zhao et al. 2005). For instance, the *Ptr1* locus mediate the recognition of type III bacterial effector *AvrRpt2* (Li et al. 2021; Mazo-Molina et al. 2019, 2020; Zhao et al. 2005) while *Aubergine* (*Abg*) locus producing the healthy anthocyanin during the domestication process. The purple fruit phenotype of *Abg* lines *S. lycopersicoides* are very much similar to the *S. chilense* (Powell et al. 2022). *Ty1, Ty2, TY3,* and *Ty5* allelic genes also confer resistance against the Yellow Leaf Curl virus disease in tomato (Rothan et al. 2019). The different cultivar of tomato participates in different type of trait for domestication purpose. For example, *Solanum pimpinellifolium* (*SP*) alleles contribute to disease resistance, abiotic stress tolerance and different fruit quality-related traits in modern tomato cultivars (Wang et al. 2020).

4.3.3 Allele Regulating the Fruit Color in Tomato

Domestication and the crop improvement of tomato lead to significant morphological changes, such as fruit size, shape, and color. Basically, tomato fruit color is decided by the accumulation of lycopene that was governed by the upstream and downstream transcriptional regulation of genes, including Phytoene synthase (PSY1), ζ-carotene isomerase (ZISO), CRTISO, and lycopene ß-cyclase 2 CRTL-B2, etc. (Ronen et al. 2000). Some mutational studies also confirmed the involvement of different allele in defining fruit color, for instance, yellow-flesh, old gold and yellow (y) allele participate in carotenoids biosynthesis (Roohanitaziani et al. 2020) or in degradation of chlorophyll. The yellow-colored fruit in different tomato accessions is due to the deletion in *PHYTOENE SYNTHASE 1 (PSY1)* gene, and the orange-colored fruit is caused by the *tangerine* mutation, which disrupted the *CAROTENE ISOMERASE* activity near the 5' open reading frame (Domínguez et al. 2020; Isaacson et al. 2002). PSY1 controls the first rate-limiting step for the carotenoid biosynthesis that catalyzes two molecules of GGPP to form the colorless phytoene. Other mutations named *green flesh* maintain the chlorophyll content during the ripening stages (Barry and Pandey 2009). Mutation in *the y* allele results in pink-colored fruit that have the 603 bp deletion in promoter region of MYB12 in different tomato cultivars (Ballester et al. 2010; Fernandez-Moreno 2016). The resulting pink color is due to the reduction in epidermal anthocyanin accumulation, directly and indirectly affecting the 152 metabolites (Giovannoni 2018).

4.3.4 Alleles Associated with Plant Growth and Fruit Quality

The resequencing efforts of whole genome assembly of different cotton accessions identified the different allele that plays a significant role in the domestication of modern cultivars. Multiple variations were identified, which transformed the small berries tomatoes into largely cultivated fruits (Tanksley 2004; Van der Knaap and Tanksley 2003). For instance, fruit weight, locule number, ovate, and sun that determine the fruit shape, size and weight in tomato (Chakrabarti et al. 2013;

Xiao et al. 2008). The *fruit weight* loci, i.e., *fw2.2* is the most vital locus to determine the fruit size for tomato domestication. The *fw3.2* (KLUH) locus controls the fruit shape and weight (Brewer et al. 2007). The gain of function of this truncated *CSR* gene encodes a FAF domain protein to increase the fruit weight by regulating the mesocarp cell size and the enhancement of pericarp areas (Gao et al. 2019; Mu et al. 2017). Other alleles like *fascinated* and *locule number* are closely linked to two SNPs regulating fruit size and shape (Cong et al. 2008; Muños et al. 2011; Xu et al. 2015). The carpel number in the flower determines the increase in fruit size due to the expansion of the WUS expression domain under the effect of these two alleles (Chu et al. 2019; Cong et al. 2008). However, the ~294 Kb inversion at the *fas* locus is more effective in increasing the locule number and fruit size compared to the gain of *lc* allele function (Chu et al. 2019; Roohanitaziani et al. 2020; Xu et al. 2015). The 85 bp deletion in the promoter of the EXCESSIVE NUMBER OF FLORAL ORGANS *(ENO)* gene is also responsible for fruit size enhancement through the floral meristem activity and thus positively selected during tomato domestication (Yuste-Lisbona et al. 2020). Floral meristem size is regulated by two highly conserved regulators *SlWUSCHEL* and *CLV3,* that act in a negative feedback loop (Pereira et al. 2021). The two other alleles, namely, *ovate* (chromosome 2) and *sun* (chromosome 7), cause the elongated fruit shape but have no or minimal effect on fruit size and weight. *SUN* regulates the fruit shape by increasing the longitudinal cell division (Wang et al. 2019). *OVATE* protein negatively regulates the transcription of their target genes by affecting cell division (Wu et al. 2018). Along with these, *Solanum lycopersicum* elongated fruit1 *(Slef1)* elevate the proximal region of the ovary in order to get the elongated-shaped fruits (Chusreeaeom et al. 2014; Liu et al. 2002; Sun et al. 2015). Other than this genome-wide association study identifies novel loci like *SlEZ1, SlIAA17, fab2,* and *SlMYB33* to determine the fruit size in tomato (Pons et al. 2022).

4.3.5 Allele Regulating the Fruit Ripening Process

Fruit ripening is one of the critical ecological and necessary evolutionary processes in different fleshy fruit. In tomato, many mutants were identified that have a crucial role in fruit ripening, such as *ripening inhibitor (rin), Never ripe (Nr), nonripening (nor),* and *Colorless nonripening (Cnr)* (Zhu et al. 2022). The ripening delaying mutant, *rin*, causes the lacking of a significant amount of ethylene burst through the transcription regulatory network, *Never ripe(Nr)* mutant becomes insensitive to the ethylene signal transduction (Lanahan et al. 1994; Vrebalov et al. 2002). Two FRUITFUL (*FUL*) genes, namely, *FUL1* and *FUL2*, control fruit ripening through the interaction with RIPENING INHIBITOR (*RIN*) gene (Bemer et al. 2012) that targets the ethylene biosynthesis and cell wall metabolism (Fujisawa et al. 2013). *FUL* gene is consistently expressed at a higher level in modern tomato for specific development functions and thus suggesting their positive selection during the domestication of tomato (Bemer et al. 2012; Wang et al. 2014; Wang et al. 2020). *Cnr* locus is another key component in the tomato regulatory network, which causes non-ripening fruit (Manning et al. 2006). The *non-ripening* allele is 2 bp spontaneous deletion in the third exon of *NON-RIPENINNG (NAC-NOR)* transcription factor (TF), resulting in a truncated protein and thus reducing the ethylene production and lycopene biosynthesis (Wang et al. 2020). *NOR-LIKE1* mutant analysis also confirmed their direct role in fruit ripening after direct recruitment on the promoters of various fruit color and ethylene synthesis-related genes (Gao et al. 2018). Transcription factors (TFs) families like MADS-box (TAGL1, TAGL2, RIN, SLCMB1) and helix loop helix also act as positive regulators for fruit ripening (Waseem et al. 2019), which is recruited and stabilized by the *Sl*WD40 complex during the anthocyanin biosynthesis (Zhu et al. 2022). Other TFs, such as the ABA stress ripening 3 (ASR3) gene, positively regulate the fruit ripening through intermediating the cross signaling between ABA and sucrose (Jia et al. 2016), whereas the activating enhancer-binding protein 2 alpha *(AP2a),* negatively regulates the ethylene biosynthesis for reduced ethylene production in mature tomato fruit (Chung et al. 2010). Golden 2-like transcription factor *(SlGLK2, u)* governs uniform fruit ripening by regulating chloroplast development and fruit sugar content (Powell et al. 2012). This uniform fruit ripening is the most desired trait in modern-day tomato cultivation.

Fruit ripening is epigenetically controlled by various epigenetic regulators (Gallusci et al. 2016). DNA methylation profiling of *Solanum lycopersicum* DNA demethylases 2 (*sldml2*) mutant line suggested an essential role of DNA demethylation during the fruit ripening. Similarly, the mutant line of CHROMOMETHYLASE3 (*SlCMT3*) and of METHYLTRANSFERASE1 *(SlMET1)* reveals the negative correlation that caused the tomato *Cnr* and *RIN* based fruit ripening (Liu et al. 2022; Wang 2020). Another demethylase, *SlJMJ6* and SlJMJ7, regulates the ethylene biosynthesis genes and transcription regulation (*NOR*) by the removal of H3K27me3 marker and thus accelerating the fruit ripening (Ding et al. 2022; Li et al. 2020).

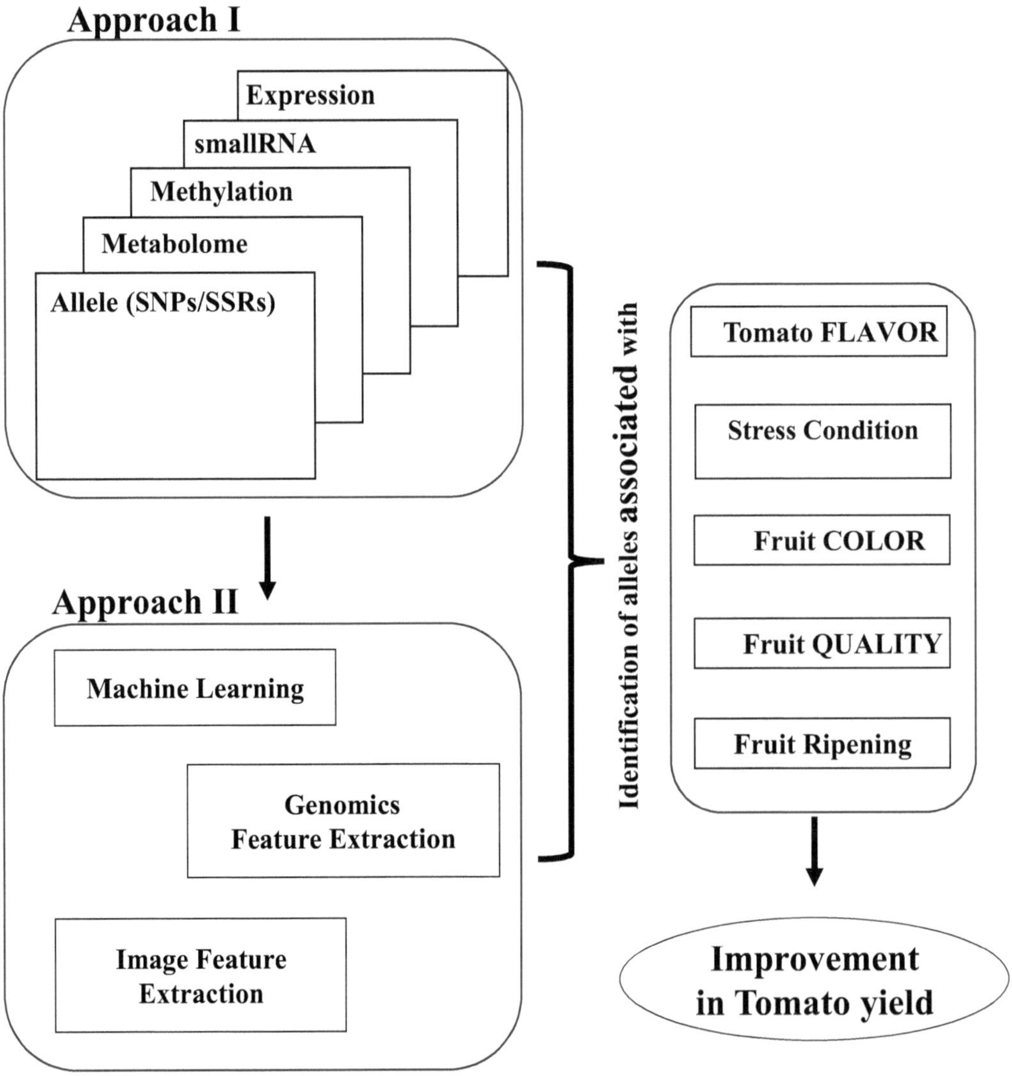

FIGURE 4.2 Two systematic omics-based approaches for cultivating tomato species. Approach I start with omics data and SNP (allele) profile for trail association. Approach II explains machine learning approaches over multiple omics data for trait-linked allele prediction.

4.4 USE OF MULTI-OMICS TOGETHER WITH ALLELE INFORMATION

Advancement in the high-throughput sequencing methods leads to the developing of various omics technologies, such as methylation, proteomics, expression, and metabolomics. Integration of these omics' profiles or data types, such as genomics, epigenomics, transcriptomics, proteomics, and metabolomics, refers to multi-omics integrative analysis. The "conceptual" integration refers to each omics data set analyzed separately and matched without statistics (Cavill et al. 2016). Therefore, omics provide valuable knowledge of plant organizational functionality. Due to the large data usually provided by omics studies methodologies, DL became an inseparable tool of omics data processing and reasoning. Each type of omics data provides helpful information associated with plant cultivation (Figure 4.2). Now, one can integrate SNPs with expression data and identify the eQTL. However, integrating different omics profiles is always beneficial to identify the potential changes linked to specific traits and tested in molecular studies. This section mainly focuses on using the "ALLELE" or "SNP" profile with multiple types of omics data applied in various plant studies. Depending on the study's goal, multi-omics approaches are grouped into the genome, phenotype, and environment-based analysis (Hasin et al. 2017). In the context of allele mining in tomato, it would be interesting to highlight the potential use of various approach that helps to identify the GWAS loci contributing to trait improvement explained in Figure 4.2.

Allele-centred integration with omics layers helps to identify SNPs and genes which could further explain the association of tomato phenotypic behavior and diseases. GWAS study identified *GREEN STRIPE* (*GS*) locus, regulating carotenoid accumulation (Liu et al. 2020). Association of DNA methylation with the ripening-induced/repressed genes highlighted in the fruits of tomato *sldml2* mutants (Liu et al. 2022). Balcke et al. (2017) used multiple omics profiles (comparative metabolomics, transcriptomics, proteomics) to study tomato accessions.

ACKNOWLEDGEMENTS

The manuscript number is provided by the institute is **CSIR-NBRI_MS/2022/11/07**.

REFERENCES

Alajrami, Mahmoud A. & Abu-Naser, Samy S. Type of Tomato Classification Using Deep Learning. *International Journal of Academic Pedagogical Research* 2020 (IJAPR) 3 (12):21–25.

Alonge M, Wang X, Benoit M, Soyk S, Pereira L, Zhang L, Suresh H, Ramakrishnan S, Maumus F, Ciren D, Levy Y. Major impacts of widespread structural variation on gene expression and crop improvement in tomato. *Cell.* 2020 Jul 9;182(1):145–61.

Balcke GU, Bennewitz S, Bergau N, Athmer B, Henning A, Majovsky P, Jiménez-Gómez JM, Hoehenwarter W, Tissier A. Multi-omics of tomato glandular trichomes reveals distinct features of central carbon metabolism supporting high productivity of specialized metabolites. *The Plant Cell.* 2017 May;29(5):960–83.

Ballester AR, Molthoff J, de Vos R, Hekkert BT, Orzaez D, Fernaǒndez-Moreno JP, Tripodi P, Grandillo S, Martin C, Heldens J, Ykema M. Biochemical and molecular analysis of pink tomatoes: Deregulated expression of the gene encoding transcription factor SlMYB12 leads to pink tomato fruit color. *Plant Physiology.* 2010 Jan;152(1):71–84.

Barchi L, Acquadro A, Alonso D, Aprea G, Bassolino L, Demurtas O, Ferrante P, Gramazio P, Mini P, Portis E, Scaglione D. Single Primer Enrichment Technology (SPET) for high-throughput genotyping in tomato and eggplant germplasm. *Frontiers in Plant Science.* 2019 Aug 7;10:1005.

Barry CS, Pandey P. A survey of cultivated heirloom tomato varieties identifies four new mutant alleles at the green-flesh locus. *Molecular Breeding.* 2009 Oct; 24(3):269–76.

Bemer M, Karlova R, Ballester AR, Tikunov YM, Bovy AG, Wolters-Arts M, Rossetto PD, Angenent GC, de Maagd RA. The tomato FRUITFULL homologs TDR4/FUL1 and MBP7/FUL2 regulate ethylene-independent aspects of fruit ripening. *The Plant Cell.* 2012 Nov; 24(11):4437–51.

Bhardwaj A, Bag SK. PLANET-SNP pipeline: PLants based ANnotation and Establishment of True SNP pipeline. *Genomics.* 2019 Sep 1;111(5):1066–77.

Bhardwaj A, Dhar YV, Asif MH, Bag SK. In silico identification of SNP diversity in cultivated and wild tomato species: insight from molecular simulations. *Scientific Reports*. 2016 Dec 8;6(1):1–3.

Blanca J, Sanchez-Matarredona D, Ziarsolo P, Montero-Pau J, van der Knaap E, Díez MJ, Cañizares J. Haplotype analyses reveal novel insights into tomato history and domestication driven by long-distance migrations and latitudinal adaptations. *Horticulture Research*. 2022 Jan 5;9.

Brewer MT, Moyseenko JB, Monforte AJ, van der Knaap E. Morphological variation in tomato: a comprehensive study of quantitative trait loci controlling fruit shape and development. *Journal of Experimental Botany*. 2007 Apr 1;58(6):1339–49.

Canady MA, Meglic V, Chetelat RT. A library of Solanum lycopersicoides introgression lines in cultivated tomato. *Genome*. 2005 Aug 1;48(4):685–97.

Cavill R, Jennen D, Kleinjans J, Briedé JJ. Transcriptomic and metabolomic data integration. *Briefings in Bioinformatics*. 2016 Sep 1;17(5):891–901.

Chakrabarti M, Zhang NA, Sauvage C, Muños S, Blanca J, Cañizares J, Diez MJ, Schneider R, Mazourek M, McClead J, Causse M. A cytochrome P450 regulates a domestication trait in cultivated tomato. *Proceedings of the National Academy of Sciences*. 2013 Oct 15;110(42):17125–30.

Chen T, Qin G, Tian S. Regulatory network of fruit ripening: current understanding and future challenges. *New Phytologist*. 2020 Nov;228(4):1219–26.

Chu YH, Jang JC, Huang Z, van der Knaap E. Tomato locule number and fruit size controlled by natural alleles of lc and fas. *Plant Direct*. 2019 Jul; 3(7):e00142.

Chung MY, Vrebalov J, Alba R, Lee J, McQuinn R, Chung JD, Klein P, Giovannoni J. A tomato (Solanum lycopersicum) APETALA2/ERF gene, SlAP2a, is a negative regulator of fruit ripening. *The Plant Journal*. 2010 Dec;64(6):936–47.

Chusreeaeom K, Ariizumi T, Asamizu E, Okabe Y, Shirasawa K, Ezura H. A novel tomato mutant, Solanum lycopersicum elongated fruit1 (Slelf1), exhibits an elongated fruit shape caused by increased cell layers in the proximal region of the ovary. *Molecular Genetics and Genomics*. 2014 Jun;289(3):399–409.

Cong B, Barrero LS, Tanksley SD. Regulatory change in YABBY-like transcription factor led to evolution of extreme fruit size during tomato domestication. *Nature Genetics*. 2008 Jun;40(6):800–804.

Ding X, Liu X, Jiang G, Li Z, Song Y, Zhang D, Jiang Y, Duan X. SlJMJ7 orchestrates tomato fruit ripening via crosstalk between H3K4me3 and DML2-mediated DNA demethylation. *New Phytologist*. 2022 Feb;233(3):1202–19.

Domínguez M, Dugas E, Benchouaia M, Leduque B, Jiménez-Gómez JM, Colot V, Quadrana L. The impact of transposable elements on tomato diversity. *Nature Communications*. 2020 Aug 13;11(1):1–1.

Fernandez-Moreno JP, Tzfadia O, Forment J, Presa S, Rogachev I, Meir S, Orzaez D, Aharoni A, Granell A. Characterization of a new pink-fruited tomato mutant results in the identification of a null allele of the SlMYB12 transcription factor. *Plant Physiology*. 2016 Jul 1;171(3):1821–1836.

Fernandez-Moreno JP, Tzfadia O, Forment J, Presa S, Rogachev I, Meir S, Orzaez D, Aharoni A, Granell A. Characterization of a new pink-fruited tomato mutant results in the identification of a null allele of the SlMYB12 transcription factor. *Plant Physiology*. 2016 Jul;171(3):1821–36.

Fuentes A, Yoon S, Kim SC, Park DS. A robust deep-learning-based detector for real-time tomato plant diseases and pests recognition. *Sensors*. 2017 Sep 4;17(9):2022.

Fujisawa M, Nakano T, Shima Y, Ito Y. A large-scale identification of direct targets of the tomato MADS box transcription factor RIPENING INHIBITOR reveals the regulation of fruit ripening. *The Plant Cell*. 2013 Feb;25(2):371–86.

Gallusci P, Hodgman C, Teyssier E, Seymour GB. DNA methylation and chromatin regulation during fleshy fruit development and ripening. *Frontiers in Plant Science*. 2016 Jun 14;7:807.

Gao L, Gonda I, Sun H, Ma Q, Bao K, Tieman DM, Burzynski-Chang EA, Fish TL, Stromberg KA, Sacks GL, Thannhauser TW. The tomato pan-genome uncovers new genes and a rare allele regulating fruit flavor. *Nature Genetics*. 2019 Jun;51(6):1044–51.

Gao Y, Wei W, Zhao X, Tan X, Fan Z, Zhang Y, Jing Y, Meng L, Zhu B, Zhu H, Chen J. A NAC transcription factor, NOR-like1, is a new positive regulator of tomato fruit ripening. *Horticulture Research*. 2018 Dec 1;5.

Giovannoni J. Tomato multiomics reveals consequences of crop domestication and improvement. *Cell*. 2018 Jan 11;172(1–2):6–8.

González-Camacho JM, Ornella L, Pérez-Rodríguez P, Gianola D, Dreisigacker S, Crossa J. Applications of machine learning methods to genomic selection in breeding wheat for rust resistance. The Plant Genome. 2018 Jul;11(2):170104.

Gupta P, Reddaiah B, Salava H, Upadhyaya P, Tyagi K, Sarma S, Datta S, Malhotra B, Thomas S, Sunkum A, Devulapalli S. Next-generation sequencing (NGS)-based identification of induced mutations in a doubly mutagenized tomato (Solanum lycopersicum) population. The Plant Journal. 2017 Nov;92(3):495–508.

Haas FB, Fernandez-Pozo N, Meyberg R, Perroud PF, Göttig M, Stingl N, Saint-Marcoux D, Langdale JA, Rensing SA. Single nucleotide polymorphism charting of P. patens reveals accumulation of somatic mutations during in vitro culture on the scale of natural variation by selfing. *Frontiers in Plant Science*. 2020 Jul 7;11:813.

Hasin Y, Seldin M, Lusis A. Multi-omics approaches to disease. *Genome Biology*. 2017 Dec;18(1):1–5.

Isaacson T, Ronen G, Zamir D, Hirschberg J. Cloning of tangerine from tomato reveals a carotenoid isomerase essential for the production of β-carotene and xanthophylls in plants. *The Plant Cell*. 2002 Feb;14(2):333–42.

Jia H, Jiu S, Zhang C, Wang C, Tariq P, Liu Z, Wang B, Cui L, Fang J. Abscisic acid and sucrose regulate tomato and strawberry fruit ripening through the abscisic acid-stress-ripening transcription factor. *Plant Biotechnology Journal*. 2016 Oct;14(10):2045–65.

Kim JE, Oh SK, Lee JH, Lee BM, Jo SH. Genome-wide SNP calling using next generation sequencing data in tomato. *Molecules and Cells*. 2014 Jan 1;37(1):36.

Kim M, Nguyen TT, Ahn JH, Kim GJ, Sim SC. Genome-wide association study identifies QTL for eight fruit traits in cultivated tomato (*Solanum lycopersicum* L.). *Horticulture Research*. 2021 Dec 1;8.

Koboldt DC, Chen K, Wylie T, Larson DE, McLellan MD, Mardis ER, Weinstock GM, Wilson RK, Ding L. VarScan: variant detection in massively parallel sequencing of individual and pooled samples. *Bioinformatics*. 2009 Sep 1;25(17):2283–5.

Lanahan MB, Yen HC, Giovannoni JJ, Klee HJ. The never ripe mutation blocks ethylene perception in tomato. *The Plant Cell*. 1994 Apr;6(4):521–30.

Langmead B, Trapnell C, Pop M, Salzberg SL. Ultrafast and memory-efficient alignment of short DNA sequences to the human genome. *Genome Biology*. 2009 Mar;10(3):1.

Li G, Wang L, Yang J, He H, Jin H, Li X, Ren T, Ren Z, Li F, Han X, Zhao X. A high-quality genome assembly highlights rye genomic characteristics and agronomically important genes. *Nature Genetics*. 2021 Apr;53(4):574–84.

Li H, Durbin R. Fast and accurate short read alignment with Burrows–Wheeler transform. *Bioinformatics*. 2009 Jul 15;25(14):1754–60.

Li H, Handsaker B, Wysoker A, Fennell T, Ruan J, Homer N, Marth G, Abecasis G, Durbin R. The sequence alignment/map format and SAMtools. *Bioinformatics*. 2009 Aug 15;25(16):2078–9.

Li J, Liu L, Bai Y, Zhang P, Finkers R, Du Y, Visser RG, van Heusden AW. Seedling salt tolerance in tomato. *Euphytica*. 2011 Apr;178(3):403–14.

Li Z, Jiang G, Liu X, Ding X, Zhang D, Wang X, Zhou Y, Yan H, Li T, Wu K, Jiang Y. Histone demethylase SlJMJ6 promotes fruit ripening by removing H3K27 methylation of ripening-related genes in tomato. *New Phytologist*. 2020 Aug;227(4):1138–56.

Liu G, Li C, Yu H, Tao P, Yuan L, Ye J, Chen W, Wang Y, Ge P, Zhang J, Zhou G. GREEN STRIPE, encoding methylated TOMATO AGAMOUS-LIKE 1, regulates chloroplast development and Chl synthesis in fruit. *New Phytologist*. 2020 Oct;228(1):302–17.

Liu J, Van Eck J, Cong B, Tanksley SD. A new class of regulatory genes underlying the cause of pear-shaped tomato fruit. *Proceedings of the National Academy of Sciences*. 2002 Oct 1;99(20):13302–6.

Liu Z, Wu X, Liu H, Zhang M, Liao W. DNA methylation in tomato fruit ripening. *Physiologia Plantarum*. 2022 Jan;174(1):e13627.

Manning K, Tör M, Poole M, Hong Y, Thompson AJ, King GJ, Giovannoni JJ, Seymour GB. A naturally occurring epigenetic mutation in a gene encoding an SBP-box transcription factor inhibits tomato fruit ripening. *Nature Genetics*. 2006 Aug;38(8):948–52.

Martina M, Tikunov Y, Portis E, Bovy AG. The genetic basis of tomato aroma. *Genes*. 2021 Feb 4;12(2):226.

Mazo-Molina C, Mainiero S, Haefner BJ, Bednarek R, Zhang J, Feder A, Shi K, Strickler SR, Martin GB. Ptr1 evolved convergently with RPS2 and Mr5 to mediate recognition of AvrRpt2 in diverse solanaceous species. *The Plant Journal*. 2020 Aug;103(4):1433–45.

Mazo-Molina C, Mainiero S, Hind SR, Kraus CM, Vachev M, Maviane-Macia F, Lindeberg M, Saha S, Strickler SR, Feder A, Giovannoni JJ. The Ptr1 locus of Solanum lycopersicoides confers resistance to race 1 strains of Pseudomonas syringae pv. tomato and to Ralstonia pseudosolanacearum by recognizing the type III effectors AvrRpt2 and RipBN. *Molecular Plant–Microbe Interactions*. 2019 Aug 12;32(8):949–60.

McKenna A, Hanna M, Banks E, Sivachenko A, Cibulskis K, Kernytsky A, Garimella K, Altshuler D, Gabriel S, Daly M, DePristo MA. The Genome Analysis Toolkit: a MapReduce framework for analyzing next-generation DNA sequencing data. *Genome Research*. 2010 Sep 1;20(9):1297–303.

Min CY, Wu LQ, Qian TT, Ain NU, Liu WJ, Wu XN, Zhang C, Chen Z, Xie HP. Typing and determination of SNP functional gene based on highly selective and signal-amplified fluorescence double-probe with the help of ExoIII nuclease and magnetic bead. *Journal of Pharmaceutical and Biomedical Analysis*. 2020 Feb 5;179:112917.

Mu QI, Huang Z, Chakrabarti M, Illa-Berenguer E, Liu X, Wang Y, Ramos A, van der Knaap E. Fruit weight is controlled by Cell Size Regulator encoding a novel protein that is expressed in maturing tomato fruits. *PLoS Genetics*. 2017 Aug 17;13(8):e1006930.

Müller NA, Wijnen CL, Srinivasan A, Ryngajllo M, Ofner I, Lin T, Ranjan A, West D, Maloof JN, Sinha NR, Huang S. Domestication selected for deceleration of the circadian clock in cultivated tomato. *Nature Genetics*. 2016 Jan;48(1):89–93.

Muños S, Ranc N, Botton E, Bérard A, Rolland S, Duffé P, Carretero Y, Le Paslier MC, Delalande C, Bouzayen M, Brunel D. Increase in tomato locule number is controlled by two single-nucleotide polymorphisms located near WUSCHEL. *Plant Physiology*. 2011 Aug;156(4):2244–54.

Pereira L, Zhang L, Sapkota M, Ramos A, Razifard H, Caicedo AL, van Der Knaap E. Unraveling the genetics of tomato fruit weight during crop domestication and diversification. *Theoretical and Applied Genetics*. 2021 Oct;134(10):3363–78.

Phan NT, Trinh LT, Rho MY, Park TS, Kim OR, Zhao J, Kim HM, Sim SC. Identification of loci associated with fruit traits using genome-wide single nucleotide polymorphisms in a core collection of tomato (Solanum lycopersicum L.). *Scientia Horticulturae*. 2019 Jan 3;243:567–74.

Pons C, Casals J, Palombieri S, Fontanet L, Riccini A, Rambla JL, Ruggiero A, Figás MD, Plazas M, Koukounaras A, Picarella ME. Atlas of phenotypic, genotypic and geographical diversity present in the European traditional tomato. *Horticulture Research*. 2022 Jan 5;9: uhac112.

Poplin R, Chang PC, Alexander D, Schwartz S, Colthurst T, Ku A, Newburger D, Dijamco J, Nguyen N, Afshar PT, Gross SS. A universal SNP and small-indel variant caller using deep neural networks. *Nature Biotechnology*. 2018 Nov;36(10):983–7.

Powell AF, Feder A, Li J, Schmidt MH, Courtney L, Alseekh S, Jobson EM, Vogel A, Xu Y, Lyon D, Dumschott K. A Solanum lycopersicoides reference genome facilitates insights into tomato specialized metabolism and immunity. *The Plant Journal*. 2022 Apr 12; 110(6): 1791–1810.

Powell AL, Nguyen CV, Hill T, Cheng KL, Figueroa-Balderas R, Aktas H, Ashrafi H, Pons C, Fernández-Muñoz R, Vicente A, Lopez-Baltazar J. Uniform ripening encodes a Golden 2-like transcription factor regulating tomato fruit chloroplast development. *Science*. 2012 Jun 29;336(6089):1711–5.

Ronen G, Carmel-Goren L, Zamir D, Hirschberg J. An alternative pathway to β-carotene formation in plant chromoplasts discovered by map-based cloning of Beta and old-gold color mutations in tomato. *Proceedings of the National Academy of Sciences*. 2000 Sep 26;97(20):11102–7.

Roohanitaziani R, de Maagd RA, Lammers M, Molthoff J, Meijer-Dekens F, van Kaauwen MP, Finkers R, Tikunov Y, Visser RG, Bovy AG. Exploration of a resequenced tomato core collection for phenotypic and genotypic variation in plant growth and fruit quality traits. *Genes*. 2020 Oct 29;11(11):1278.

Rothan C, Diouf I, Causse M. Trait discovery and editing in tomato. *The Plant Journal*. 2019 Jan;97(1):73–90.

Sauvage C, Segura V, Bauchet G, Stevens R, Do PT, Nikoloski Z, Fernie AR, Causse M. Genome-wide association in tomato reveals 44 candidate loci for fruit metabolic traits. *Plant Physiology*. 2014 Jul;165(3):1120–32.

Scarano D, Rao R, Corrado G. In Silico identification and annotation of non-coding RNAs by RNA-seq and De Novo assembly of the transcriptome of Tomato Fruits. *PloS One*. 2017 Feb 10;12(2):e0171504.

Schmidt MH, Vogel A, Denton AK, Istace B, Wormit A, van de Geest H, Bolger ME, Alseekh S, Maß J, Pfaff C, Schurr U. De novo assembly of a new Solanum pennellii accession using nanopore sequencing. *The Plant Cell*. 2017 Oct;29(10):2336–48.

Shen J, Tieman D, Jones JB, Taylor MG, Schmelz E, Huffaker A, Bies D, Chen K, Klee HJ. A 13-lipoxygenase, TomloxC, is essential for synthesis of C5 flavour volatiles in tomato. *Journal of Experimental Botany*. 2014 Feb 1;65(2):419–28.

Shi W, van de Zedde R, Jiang H, Kootstra G. Plant-part segmentation using deep learning and multi-view vision. *Biosystems Engineering*. 2019 Nov 1;187:81–95.

Sim SC, Durstewitz G, Plieske J, Wieseke R, Ganal MW, Van Deynze A, Hamilton JP, Buell CR, Causse M, Wijeratne S, Francis DM. Development of a large SNP genotyping array and generation of high-density genetic maps in tomato. *PloS One*. 2012 Jul 10;7(7):e40563.

Sim SC, Robbins MD, Chilcott C, Zhu T, Francis DM. Oligonucleotide array discovery of polymorphisms in cultivated tomato (Solanum lycopersicum L.) reveals patterns of SNP variation associated with breeding. *Bmc Genomics*. 2009 Dec;10(1):1–0.

Soyk S, Lemmon ZH, Sedlazeck FJ, Jiménez-Gómez JM, Alonge M, Hutton SF, Van Eck J, Schatz MC, Lippman ZB. Duplication of a domestication locus neutralized a cryptic variant that caused a breeding barrier in tomato. *Nature Plants*. 2019 May;5(5):471–9.

Sun L, Rodriguez GR, Clevenger JP, Illa-Berenguer E, Lin J, Blakeslee JJ, Liu W, Fei Z, Wijeratne A, Meulia T, van der Knaap E. Candidate gene selection and detailed morphological evaluations of fs8. 1, a quantitative trait locus controlling tomato fruit shape. *Journal of Experimental Botany*. 2015 Sep 1;66(20):6471–82.

Tanksley SD. The genetic, developmental, and molecular bases of fruit size and shape variation in tomato. *The Plant Cell*. 2004 Jun;16(suppl_1):S181–9.

Tarigan H, Marlina S, Natalia K, Barus DT. The impact of tomato juice on reducing blood sugar levels at diabetes mellitus patients. In Proceedings of the International Conference on Health Informatics and Medical Application Technology (ICHIMAT 2019), 2019 Nov, pp. 6–8.

Tieman D, Zhu G, Resende Jr MF, Lin T, Nguyen C, Bies D, Rambla JL, Beltran KS, Taylor M, Zhang B, Ikeda H. A chemical genetic roadmap to improved tomato flavor. *Science*. 2017 Jan 27;355(6323):391–4.

Van der Knaap E, Tanksley SD. The making of a bell pepper-shaped tomato fruit: Identification of loci controlling fruit morphology in Yellow Stuffer tomato. *Theoretical and Applied Genetics*. 2003 Jun;107(1):139–47.

van Hilten A, Kushner SA, Kayser M, Ikram MA, Adams HH, Klaver CC, Niessen WJ, Roshchupkin GV. GenNet framework: Interpretable deep learning for predicting phenotypes from genetic data. *Communications Biology*. 2021 Sep 17;4(1):1–9.

Vrebalov J, Ruezinsky D, Padmanabhan V, White R, Medrano D, Drake R, Schuch W, Giovannoni J. A MADS-box gene necessary for fruit ripening at the tomato ripening-inhibitor (rin) locus. *Science*. 2002 Apr 12;296(5566):343–6.

Wang R, Lammers M, Tikunov Y, Bovy AG, Angenent GC, de Maagd RA. The rin, nor and Cnr spontaneous mutations inhibit tomato fruit ripening in additive and epistatic manners. *Plant Science*. 2020 May 1;294:110436.

Wang S, Lu G, Hou Z, Luo Z, Wang T, Li H, Zhang J, Ye Z. Members of the tomato FRUITFULL MADS-box family regulate style abscission and fruit ripening. *Journal of Experimental Botany*. 2014 Jul 1;65(12):3005–14.

Wang X, Gao L, Jiao C, Stravoravdis S, Hosmani PS, Saha S, Zhang J, Mainiero S, Strickler SR, Catala C, Martin GB. Genome of *Solanum pimpinellifolium* provides insights into structural variants during tomato breeding. *Nature Communications*. 2020 Nov 16;11(1):1–1.

Wang Y, Clevenger JP, Illa-Berenguer E, Meulia T, van der Knaap E, Sun L. A comparison of sun, ovate, fs8. 1 and auxin application on tomato fruit shape and gene expression. *Plant and Cell Physiology*. 2019 May 1;60(5):1067–81.

Waseem M, Li N, Su D, Chen J, Li Z. Overexpression of a basic helix–loop–helix transcription factor gene, SlbHLH22, promotes early flowering and accelerates fruit ripening in tomato (*Solanum lycopersicum* L.). *Planta*. 2019 Jul;250(1):173–85.

Wu S, Zhang B, Keyhaninejad N, Rodríguez GR, Kim HJ, Chakrabarti M, Illa-Berenguer E, Taitano NK, Gonzalo MJ, Díaz A, Pan Y. A common genetic mechanism underlies morphological diversity in fruits and other plant organs. *Nature Communications*. 2018 Nov 9;9(1):1–2.

Wu X, Heffelfinger C, Zhao H, Dellaporta SL. Benchmarking variant identification tools for plant diversity discovery. *BMC Genomics*. 2019 Dec;20(1):1–5.

Xiao H, Radovich C, Welty N, Hsu J, Li D, Meulia T, Van Der Knaap E. Integration of tomato reproductive developmental landmarks and expression profiles, and the effect of SUN on fruit shape. *BMC Plant Biology*. 2009 Dec;9(1):1–21

Xu C, Liberatore KL, MacAlister CA, Huang Z, Chu YH, Jiang K, Brooks C, Ogawa-Ohnishi M, Xiong G, Pauly M, Van Eck J. A cascade of arabinosyltransferases controls shoot meristem size in tomato. *Nature Genetics*. 2015 Jul;47(7):784–92.

Yuste-Lisbona FJ, Fernández-Lozano A, Pineda B, Bretones S, Ortíz-Atienza A, García-Sogo B, Müller NA, Angosto T, Capel J, Moreno V, Jiménez-Gómez JM. ENO regulates tomato fruit size through the floral meristem development network. *Proceedings of the National Academy of Sciences*. 2020 Apr 7;117(14):8187–95.

Zhang J, Zhao J, Xu Y, Liang J, Chang P, Yan F, Li M, Liang Y, Zou Z. Genome-wide association mapping for tomato volatiles positively contributing to tomato flavor. *Frontiers in Plant Science*. 2015 Nov 27;6:1042.

Zhao J, Sauvage C, Zhao J, Bitton F, Bauchet G, Liu D, Huang S, Tieman DM, Klee HJ, Causse M. Meta-analysis of genome-wide association studies provides insights into genetic control of tomato flavor. *Nature Communications*. 2019 Apr 4;10(1):1–2.

Zhao L, Qiu C, Li J, Chai Y, Kai G, Li Z, Sun X, Tang KX. Investigation of disease resistance and cold tolerance of Solanum lycopersicoides for tomato improvement. *HortScience*. 2005 Feb 1;40(1):43–6.

Zhou Y, Cho WK, Byun HS, Kil EJ, Bak SI, Moon DH, Chavan V, Park TS, Lee S, Hong SW. Transcriptome profiles of tomato plants after neutron irradiation and infection with TYLCV. *Physiologia Pantarum*. 2019 Feb;165(2):427–41.

Zhu F, Jadhav SS, Tohge T, Salem MA, Lee JM, Giovannoni JJ, Cheng Y, Alseekh S, Fernie AR. A comparative transcriptomics and eQTL approach identifies SlWD40 as a tomato fruit ripening regulator. *Plant Physiology*. 2022 May 4; 190(1): 250–266.

5 Allele Mining in *Alliums*

Anil Khar[1*] *and Yogesh P. Khade*[2]
[1]Division of Vegetable Science, ICAR-IARI, Pusa Campus, New Delhi, Delhi, India
[2]ICAR-Directorate of Onion and Garlic Research, Pune, Maharashtra, India
[*]Corresponding author: anil.khar@gmail.com

5.1 INTRODUCTION

Globally, *Allium* crops are being cultivated for their multiple uses as a vegetable, spice, condiment, and medicinal purposes. Researchers perceive *Alliums* to be extremely challenging crops, both in terms of their cultivation and omics. They belong to the family *Amaryllidaceae* having bulb, stem, and leaf as edible parts. These crops are biennial, developing bulbs in the first year, then resting for 4–6 months before producing seeds the following year. From seed to seed, this lifetime takes 18–22 months (bulb to bulb). Consequently, the duration of breeding and cultural practices for this particular vegetable will be extended by two seasons, aligning with the typical seasonal nature of most other vegetables. This hinders the genetic advance by extending the breeding cycle. Furthermore, most of the *Alliums* reproduce asexually, hence lacking natural sexual reproduction and the inherent variability in desired qualities necessary for selective breeding in the progeny of a specific plant or group of plants. In contrast, *Allium* species that engage in sexual reproduction have a high degree of open pollination, heterozygous in nature and diversity. Moreover, self-pollination can harm fitness in these species, affecting the breeding goals. The issue of inbreeding depression is a significant matter of concern among onion breeders, prompting the exploration of the doubled haploid onion concept by prominent research groups on a global scale. Moreover, *Alliums* possess genomes characterized by their substantial size, extensive repetition, intricate structure, and heterozygosity. Thus, they are presently positioned among those awaiting complete genome sequencing toward the end of the line.

The production of better and high-yielding varieties of *Allium* has advanced thanks to the accumulation of advantageous alleles from extensive plant genetic resources around the world. Nevertheless, a sizable number of these superior/beneficial genes went unused since they were lost during domestication and evolution. Crop plants' landraces and wild cousins contain untapped genetic variety that could be profitably used to create cultivars with improved agronomic traits. Novel alleles from *Allium* families in the wild have been introduced into cultivated varieties, and studies on this phenomenon (De Vicente and Tanksley 1993; Xiao et al. 1996) have shown that doing so can circumvent the genetic bottlenecks that prevented the introduction of these alleles outside of cultivars. To enhance the genetic potential of *Allium* varieties for commercially significant traits, it is imperative to revisit the vast germplasm resources to identify new alleles. The amount of sequencing data being deposited into gene banks has increased exponentially over the past 15 years (Chan 2005).

The discovery and annotation of new genes in various genomic databases has enabled the development of allele-specific markers (Spooner et al. 2005). PCR techniques are developed to isolate beneficial alleles of genes from a wide range of species based on gene and genome sequences (Latha et al. 2004). This capability allows for direct access to important alleles that resist biotic and abiotic stressors, increase nutrient use efficiency, increase yield and improve quality. Similar alleles

DOI: 10.1201/9781003376439-5

TABLE 5.1
Most Commonly Cultivated *Allium* Wild Species

Sr. No.	Common Name	Botanical Name	Somatic Chromosome Number	Ploidy
1	Common onion	*Allium cepa* L.	16	2X
2	Garlic	*A. sativum* var. *sativum*	16	2X
3	Multiplier onion/ potato onion group	*A. cepa* L. *aggregatum*	16	2X
4	Chives	*A. schoenoprasum* L.	24 and 32	2X, 3X and 4X
5	Shallot	*A. cepa var. ascalonicum* L.	16	2X
6	Japanese bunching Onion/ Welsh onion	*A. fistulosum* L.	16	2X

underlying a particular trait and their variations in other genotypes can be found using modern genomic methods. Scientists often use the term "allele mining" to refer to the dissection of naturally occurring variation at candidate genes or loci. In addition to supplying fresh germplasm for introducing novel alleles to improve desired traits, the identification of allelic variants from germplasm collections helps classify entries for germplasm conservation (Table 5.1). We examine the idea of allele mining and its methodologies, as well as the prospects and obstacles, after realizing the significance of allele mining in the era of plant breeding based on genomics.

5.2 ALLELE MINING, WHAT DOES THIS IMPLY?

In light of these, Table 5.2 lists the requirements for allele mining. Any crop with sequence information, or a related family member, can undergo allele mining using flowing parameters.

5.2.1 True Allele Mining

To analyze functional variations in genes, allele mining should include non-coding and regulatory regions along with coding regions (Latha et al. 2004). The identification of genetic variation in regulatory regions has led to the selection of gene promoters as a potential target for allele mining, also known as "promoter mining". Due to the anticipated impact on protein structure and function, allele mining initially focused on identifying SNPs or InDels in the gene's coding sequences or exons. Numerous instances show how such sequence changes in genomic regions might change phenotypes. Recent studies have shown that nucleotide alterations in non-coding regions (including the promoter, introns, and 3′ UTR) significantly affect transcript synthesis and accumulation, which in turn impacts trait expression

5.2.2 Has There Been an Increase in Novel Alleles?

The mutation is regarded as a significant evolutionary mechanism underpinning allelic variation in various crop species. The introduction of new genetic variations through mutations like SNPs and InDels is crucial. Mutations occurring in coding areas and regulatory regions can significantly impact the phenotype by modifying the structure and function of the encoded protein. Conversely, mutations that arise in non-coding sections of a gene frequently remain inconsequential, resulting in a quiet phenotype devoid of any noticeable effects. While most mutations are detrimental, it is generally observed that a small proportion, approximately 0.1%, of mutations significantly impact gene function, potentially playing a crucial role in the survival of plants (Tan 2005). Numerous

TABLE 5.2
Consideration for Allele Mining Activities

S. No.	Steps	Selection Criteria
1	Species naming	Reference collection with the greatest genetic diversity in the fewest number of individuals.
2	*Allium* species	The presence of a genetic map, segregating groups, ESTs, or gene/genome sequence. Mini-core collections, consisting of 200–300 reference accessions selected from all varietal groups, are available
3	Trait of target	Trait of interest
4	Species identification	Using effective procedures in the precise phenotyping, right environmental conditions
5	Potential destinations for genes	A prioritised gene list taking into account the desired characteristics

occurrences of this nature are commonly found across various crop species. The color of bulbs on onions is a significant trait (Table 5.3). The onion germplasm exhibits diverse bulb colors, including the main hues of white, yellow, and red, and an extensive spectrum of other shades like chartreuse and gold. Onions possess a diverse range of flavonoids, comprising a total of 54 distinct types, contributing to the bulb's characteristic pigmentation. As Shirley (1996) stated, flavonoids play a crucial role as antioxidants in protecting against ultraviolet radiation, plant fertilization, and pigmentation.

5.3 ALLELE MINING FOR IMPORTANT TRAITS IN *ALLIUMS*

5.3.1 Bulb Color

Onions with a gold color are found to have lower levels of quercetin, the most common flavonoid in onions. According to Kim et al. (2016), critical mutations in the chalcone isomerase (CHI) gene cause the development of gold onions. The difference in color between yellow and red onions is due to the inactivation of the DFR enzyme in the anthocyanin synthesis pathway, while the pink color of the bulbs is caused by two recessive alleles of the ANS gene (Kim et al. 2005). Kim et al. (2009) identified two previously unknown inactive alleles of *DFR-A* in Korean and Japanese yellow onion cultivars and breeding lines. A simple PCR-based molecular marker for *DFR-APS* allele selection was created by exploiting a 20-bp deletion in the allele promoter region of the simple sequence repeat. Additionally, they could not find any *DFR-A* transcript in yellow F_2 individuals through RT-PCR analysis. Furthermore, Park et al. (2013b) developed functional genetic markers, called cleaved amplified polymorphic sequence (CAPS) markers, to identify inactive DFR-A alleles responsible for anthocyanin production failure in onions. *DFR-APS* is the more common of the two alleles in yellow onion cultivars.

5.3.2 Taste

Since ancient times, people have prized onions and other *Alliums* for their intense aroma and flavor. The traditional wisdom that flavor-enhancing organosulfur compounds have significant positive effects on human health has been supported by modern science. The taste of onions comes from sulphoxides of 1-propenyl, propyl, and methyl cysteine (Randle and Lancaster 2002). Galmarini et al. (2001) developed a genetic map to analyze the impact of QTLs on the phenotypic correlations between soluble solids content (SSC), total dry matter, pungency, and onion-induced *in vitro* antiplatelet activity. In the interspecific population of *A. cepa* × *A. roylei*. McCallum et al. (2006)

discovered a polymorphic SSR marker with high disequilibrium with bulb fructan content on chromosome 8. A QTL analysis of total bulb fructan content revealed one significant QTL that may be responsible for differences in carbohydrate content between onion cultivars. McCallum et al. (2007) conducted a study on the genetic mapping of sulfur assimilation genes, revealing QTL's presence for bulb pungency in onions. A PCR-based molecular marker map, which includes candidate genes for sulfur assimilation, was used to identify genomic regions affecting pungency in the cross W202A × Texas Grano 438. Genes for plastidic ferredoxin-sulfite reductase (SiR) and plastidic ATP sulfurylase (ATPS) are located closely on chromosome 3 as shown by linkage mapping (within 1–2 cM range). QTL analysis found significant links between pungency and bulb soluble solids content, linked to specific marker regions on chromosomes 3 and 5. These chromosome segments have been previously recognized for influencing diverse impacts on bulb carbohydrate composition. The non-structural dry matter content of onion bulbs consists mainly of fructose, glucose, sucrose, and fructans. Raines et al. (2007) built a cDNA subtraction library to distinguish between high and low fructan accumulation backgrounds.

5.3.3 Photosynthesis

The genes responsible for chloroplast ferredoxin-sulfite reductase and ATPS are located near each other on chromosome 3 with a genetic distance of 1–2 centimorgans (cM). According to QTL analysis, there were significant correlations between marker intervals on chromosomes 3 and 5 and both pungency and bulb-soluble solids content (McCallum et al. 2007). Onion bulbs' non-structural dry matter composition consists mostly of fructose, glucose, sucrose, and fructans. The researchers utilized the restriction fragment length polymorphism (RFLP) method to determine the cytoplasmic male sterility (CMS) genotypes in onions. This was achieved by employing specific probes targeting mitochondrial genes, such as *atp*A, *atp*6, *atp*9, *cob*, *cox*1, *nad*3, *nad*4, and *nad*6, as described by Szklarczyk et al. (2002). These genetic markers are located within an amplicon of the chloroplast *psb*A gene, which distinguishes between male fertile (N) and male-sterile (S) cytoplasm, as demonstrated by Bang et al. (2013).

5.4 Mapping of Genetic Factors Responsible for Significant Traits in *Alliums*

Gene mapping involves determining the specific location of genes on an organism's genome to understand its organization and behavior during recombination. A genetic map can be constructed in two ways: a physical map based on the position of genes on the genome or a linkage map based on gene recombination frequency. Only a few *Alliums* have seen significant studies in the genetic map creation over the past 15 years utilizing intra- and interspecific crosses. These maps provide details regarding the organism's genetic structure, sequencing, linkage, and recombination.

5.4.1 Post-Harvest Traits

The breeding strategy involves hybridization or self-fertilization, both assist breeders in predicting the results of recombination. In the past 20 years, essential genes in the *Allium* genus have mostly been identified and assigned to specific positions on chromosomes in the wild species *Allium roylei* Stearn, Japanese bunching onion, and onion. These include traits related to quality such as pungency (sulfur content) (McCallum 2007), dry matter content (Martin 2005), potential benefits of antioxidants and medicinal properties (Tsukazaki 2012), bulb pigments (Martin 2005), plant physiological disorders (McCallum 2016), and pathogen resistance (Scholten 2016).

Marker-assisted breeding and genetic transformation enable the rapid development of specialized cultivars for agronomic and industrial uses, including food processing, pharmaceuticals,

TABLE 5.3
Mapping of Genetic Factors Responsible for Important Traits in *Alliums*

Gene/QTL/Trait	Type of Marker	Location	*Allium* Species	Involved Methodology	Reference
Downy mildew resistance	Functional (named DMR1)	Chr. 3	*Allium cepa L.*	cDNA sequencing, indel polymorphism study, primer pair development	Kim et al. 2016
Nuclear male sterility (*Ms/ms*)	SSCP and SNE	Chr. 2	*Allium cepa* L.	RFLP and AFLP analysis	Gokce et al. 2002
Cytoplasmic male sterility (N/S)	SCAR and RAPD (named SCS13 and	mtDNA	*Allium fistulosum* L.		Gai and Meng, 2004
Male fertility restorer locus (*Rf*)	-	Chr. 5F	*Allium fistulosum L.*	Backcrossing and GISH	Gai and Meng, 2002
Botrytis squamosa resistance (Bs1)	SNP	Chr. 6	*Allium roylei* L.	Transcriptome sequencing followed by QTL mapping	Scholten et al. 2016
I (locus for white bulb color)	SSR 21 cM from	ACM006	*Allium cepa* L.	-	Yamashita et al. 2005
B (seed coat color)	SSR	Chr. 1	*Allium cepa* L.	-	McCallum et al. 2006
Frc (bulb dry matter	SSR	Chr. 6	*Allium cepa* L.	-	McCallum et al. 2007
Bulb color		Chr. 8	*Allium cepa* L.	-	Masuzaki et al. 2006
Low pungency	Functional (named ATPS and SiR)	Chr. 3	*Allium cepa* L.	-	McCallum et al. 2016
Structural genes for bulb color	SCAR (named CHS-A,CHS-B, CHI, F3H, DFR, ANS)	Chr. 2A, 4A, 3A, 3A, 7A, 4A	*Allium cepa* L. *Aggregatum group*	Alien monosomic additions and PCR-based marker analysis	Masuzaki et al. 2006

and cosmeceuticals. Onion breeders are less reliant on phenotypic scoring, which is frequently deceptive, due to improved understanding of inheritance mechanisms and connections between features. The benefits of gene mapping are illustrated by the association study of low pungency and high solid content in onions using two molecular markers – ATPS and SiR. Most onion breeders strive to develop mild onion cultivars that have high soluble solid content, as these are preferred in Western countries and can be used to make powder, chips, flakes, and rings (McCallum 2007). Because of this, onions must have a lot of soluble solids. However, achieving the target through phenotypic evaluation is challenging due to the strong correlation between low pungency (mildness) and low soluble solid content. Now, it is possible to choose low-pungency onion types without linkage drag thanks to the mapping of ATPS and SiR markers in the onion genome. These types also have low soluble solid content in bulbs (Table 5.3 and Table 5.4). This will conserve resources and improve the selection process to create mild onions with high soluble solids suitable for processing.

TABLE 5.4
Molecular Markers in *Alliums* for Breeding Objectives

Type of Marker	Purpose	*Alliums* Involved	References
Microsatellite, EST-SSR, ISSR	Inter-specific taxonomic analysis	*Allium cepa* L., *Allium fistulosum* L, *Allium sativum* L. *Allium galanthum, Allium roylei, Allium vavilovii, Allium altaicum*	Malik et al. 2020
	Intra- and interspecific relatedness within subgenus *Rhizirideum*	*Allium sativum L.*	Khar et al. 2022
	Genetic diversity studies, allele mining, mapping and associative studies	*Allium sativum* L.	
	Diversity analysis, genotype identification, assessment of population structure	*Allium sativum* L.	Gokce et al. 2002
	EST-SSRs for genetic diversity, mapping and association studies	*Allium sativum* L.	Gai and Meng 2004
	SSRs and ISSRs for genetic diversity analysis	*Allium sativum* L.	Gai and Meng 2002
	EST-SSRs for diversity studies, mapping, association studies and fingerprinting	*Allium sativum* L.	Yamashita et al. 2005
	Construction of genetic map and phylogenetic relationship studies with other *Alliums*	*Allium fistulosum* L.	Scholten et al. 2016
ILP	To assess genetic separateness among Czechoslovakian garlic varieties	*Allium sativum* L.	Martin et al. 2005
SNP	Genetic variability and diversity studies	*Allium cepa* L.	McCallum et al. 2006
Functional (named DFR-PS)	Marker assisted selection (MAS) of yellow	*Allium cepa* L.	McCallum et al. 2007
Functional (MS2 gene-based)	MAS of fertility in garlic	*Allium sativum* L.	Masuzaki et al. 2006
SNP, RFLP (named ACP052 and API66C-E5	MAS for low levels of lachrymatory factor	*Allium cepa* L.	McCallum et al. 2016

The incorporation of advantageous alleles from wild species improved the phenotypic performance of rice (Xiao et al. 1996) and tomato (Devicente and Tanksley 1993). Preserving key genes in different wild and cultivated varieties of a crop depends on their role in plant adaptation through natural or artificial selection. Hence, investigating inherent variability in crucial genes is vital in contemporary agricultural enhancement initiatives.

5.5 ALLELE MINING FOR STRESS TOLERANCE

Globally, crop production faces major challenges and constraints due to abiotic stresses, such as water stress (drought and waterlogging), thermal stress (low and high temperature), and salinity stress, which leads to a 50% reduction in crop yield (Atkinson and Urwin, 2012). Yiu et al. (2009)

and Sta-Baba et al. (2010) reported that flooding, waterlogging, and salinity can significantly reduce the yield of *Allium* crops while Gedam et al. (2021) found that drought is the major abiotic constraint for crop production.

5.5.1 Water Stress

Frequent and extended drought, attributed to climate change, has emerged as a significant limitation resulting in a substantial reduction of approximately 30% in global onion production. Water scarcity considerably constraints crop production in regions with continental climates. Consequently, there is an ongoing need for genetically advanced crops and vegetables that exhibit enhanced resistance to drought conditions (Csiszar et al. 2007). The proliferation of this environmental hazard has become a global issue in various agricultural regions, particularly in the cultivation of *Alliums*, where comparable circumstances have facilitated its dissemination. According to Pelter et al. (2004), the impact of drought stress on the onion varied in morpho-physiological characteristics, yield attributes, and the plant's growth stage and genotypic sensitivity. Onion crops are vulnerable to drought stress due to a shallow root system, with a depth of 0.18 meters, and low water levels (Drinkwater and Janes 1955). Furthermore, it has been observed that it exerts a substantial impact on the process of photosynthesis as well as the plant pigments responsible for reducing the pace of photosynthesis, the mechanism by which light is harvested, and the distribution of assimilates inside plants (Cornic and Massacci 1996). In their study, Pelter et al. (2004) examined the impact of soil-water stress during specific growth stages on the yield of onion bulbs. The yield of long-day yellow onions decreased when exposed to stress at any growth stage, as discovered by researchers. Withholding irrigation during the 5-leaf, 7-leaf, 3- and 7-leaf stage caused the most significant reduction in output. According to the findings of Shock et al. (1998), optimal onion yield is achieved when the soil is consistently kept moist, and it is recommended to discontinue irrigation 2 weeks before harvest to mitigate the risk of rotting and sprouting during storage.

In their study, Abdelrahman et al. (2015) studied the variability of target metabolome and transcriptome profiling in doubled haploid onions. The researchers aimed to identify crucial metabolites and genes in dihaploid *Allium cepa* and dihaploid *Allium cepa* var. *aggregatum* (DHA) under normal settings and stress conditions. The researchers also observed an elevation in specific metabolites and the upregulation of pertinent genes in both the DHA and F_1 genotypes. These findings contribute to a more comprehensive comprehension of the molecular genetic factors that influence the variation of adaptive traits. Moreover, the presence of DHA suggests its potential significance as a valuable resource in an onion-breeding initiative aimed at enhancing tolerance to environmental stress. In their study, Han et al. (2016) investigated to identify the specific genes that exhibit differential expression throughout cold acclimation in several onion genotypes. Their research aimed to elucidate the underlying mechanisms responsible for developing freezing tolerance in onion crops. A comparison study was conducted on expression profiles, revealing a substantial quantity of differentially expressed genes that respond to cold and snowy conditions. In addition, the researchers conducted pathway analysis on the enriched 491 differentially expressed genes (DEGs), leading to the identification of significant metabolic processes associated with freezing conditions.

The study conducted by Zhang et al. (2018) employed transcriptome sequencing and metabolism analysis techniques to investigate the involvement of cyanidin metabolism in the dark red pigmentation of onions. The researchers identified 39 CepNAC transcription factors (TFs) belonging to the NAC family of genes, specifically the NAC-IV and NAC-V groups. These TFs were found to be potentially associated with the onion's stress response mechanisms. According to Dufoo-Hurtado et al. (2013), the exposure of seed cloves to low temperatures in garlic leads to the overexpression of genes involved in phenolic metabolism. The initial documentation of the transcriptome study of drought tolerance in distinct onion genotypes was presented by Ghodke et al. (2020) in India.

Overall, 1100 differentially expressed genes (DEGs) were discovered from the differing genotypes in response to drought stress. The functional annotation of these genes was performed utilizing various commonly used bioinformatics programs. The tolerant genotype (Acc. 1656) showed increased expression of drought-responsive genes, including those that encode transcription factors (TFs), cytochrome 450, and membrane transporters. Furthermore, it was found that genes responsible for carbohydrate metabolism and flavonoid biosynthesis were upregulated. According to Currah (2002), shallots have a higher tolerance to abiotic stressors than onions, among other *Allium* species.

5.5.2 Thermal Stress

The temperature range affects biological processes like seed germination, growth, and metabolism. Therefore, it is a significant concern in the cultivation of *Allium* crops within the context of climate change. Galsurker et al. (2018) examined the different reactions of outer and inner onion bulb scales to thermal stress. The researchers successfully identified the genes responsible for heat tolerance in the internal scales and browning development in the outer scale. Additionally, they provided evidence to support the notion that the discoloration of the external layer after applying heat results from an enzymatic mechanism linked to elevated levels of oxidation-related constituents, including peroxidase and quercetin glucoside. During the heating process, genes associated with lipid metabolism, oxidation pathways, and cell wall alteration on the outer scale were significantly upregulated. After being exposed to heat, the inner scale primarily induced defence-related genes. These genes produce heat-shock proteins, antioxidative stress defence mechanisms, and metabolites that protect against osmotic stress. The issue of flowering is a significant challenge in garlic cultivation throughout numerous tropical and subtropical regions. According to Takagi (1990), subjecting garlic to extended periods of low temperature induces the flower scape's development of floral components and elongation. In their study, Son et al. (2012) found that 15 genes were upregulated and four genes were downregulated in response to low temperature, all of which were classified as cold-responsive (CR) genes. The CR genes can be modified to reduce the harmful effects of cold or frost stress in the field.

5.6 ALLELE MINING FOR MALE STERILITY

The molecular-level characterization of male sterility in *Alliums* still needs to be understood. Various molecular markers have been established to select onion cytoplasmic male sterility (CMS) genotypes. In their study, Kim et al. (2015) employed BSA (bulked segregant analysis) and RNA-seq (RNA sequencing) methodologies to discern potential genes that could be involved in the restoration of fertility. There are certain single nucleotide polymorphisms (SNPs) that can differentiate between male-sterile and male-fertile bulks. Additionally, 14 contigs exhibited a complete correlation with the *Ms* locus. Among the contigs under consideration, *AcPMS1*, a contig involved in DNA repair, is the most promising gene and marker for selecting the restorer line. Moreover, an analysis was conducted during the tetrad stage, and the transcriptome of male-sterile and maintenance lines' anthers were compared. Yuan et al. (2018) discovered a set of five genes linked with male sterility in onions. Two of these genes were located in the cytoplasm, specifically *atp9* and *cox1*. The remaining three genes, *SERK1*, *AG*, and *AMS*, were determined to be encoded by the nucleus. An investigation was conducted using the transcriptome approach to compare differential gene expression between male-sterile and maintainer lines of *Allium fistulosum*. The analysis revealed a total of over 4000 unigenes that exhibited differential expression. Several unigenes related to male sterility were identified, including F-type *ATPase*, *NADH dehydrogenase*, and *cytochrome c oxidase*. These genes have previously been recognized for their involvement in this biological process (Liu et al. 2016). The investigations above have elucidated that CMS in *Alliums* is regulated by a combination of mitochondrial and nuclear genes and their interplay. The garlic genotype exhibiting male

TABLE 5.5
Molecular Markers Related to Male Sterility in *Alliums*

Trait/Gene	Species	Marker Name	Reference
N/S/T-cytoplasm	*Allium cepa L.*	(a) *orf* 501 (b) *cob* (c) *orf*725 and *cox*1	Engelke et al. 2003
	Allium fistulosum L.	(a) S327and N1412 (b) SCAR1	Gai and Meng 2004
Ms/ms or Rf/rf	*Allium cepa L.*	(a) Isotig 34671_610 (b) Isotig 30856_1351 (c) Isotig 29186_1830 (d) DNF 566 and RNS 357 (e) Jnurf13 (f) Acms 1100 (g) Jnurf 05, Jnurf 17 (h) AcSKP1	Havey 2013 Yang et al. 2013 Kim and Kim 2015 Bang et al. 2013 Park et al. 2013a, Huo et al. 2015
Male fertility transcripts	*Allium cepa L.*	Transcript contigs 2,71,665 in number (CUDH2107)	Khosa et al. 2016

sterility demonstrated tapetal enlargement, contributing to its sterility. Integrating transcriptome and proteome studies facilitated the identification of numerous genes and proteins that showed differential expression between male-sterile and fertile genotypes. The genes *AP3*, *ms2*, *mmd1*, *gpat2*, *nad7*, *ccmC*, *cox2*, *18S rRNA*, *flavanol synthase*, and *sod* are linked to male sterility in garlic. The findings of Shemesh-Mayer et al. (2015) suggest that uncontrolled programmed cell death (PCD) or restrictions in respiration can cause low energy levels that result in sterile pollens. (Table 5.5).

5.7 FERTILITY OF GARLIC

It has been suggested in garlic that genetic variation may result from differences in the quantity and morphology of somatic cell chromosomes. This is a sign that heritable variability may be obtained in other Allium crops, which do not produce true seeds and are therefore challenging to propagate. Cloves, bulbils, and bulb-lets, which are bulky propagating materials, demand a lot of room and storage, which raises the cost to farmers. Additionally, they serve as a major origin for viral load, infecting the harvested crop. True seeds, on the other hand, propagate plants without passing on diseases, are cheap to store, and do not transfer viruses. Garlic's apical meristem undergoes a one-way shift from producing flowers to bulbils in bolting varieties. This supports the theory that, it has undergone a gradual transformation from sexual to asexual reproduction (Konvica 1972). Considering both of these hypotheses, it is possible that the sexual reproduction of garlic progenitors led to the wide range of varieties we observe in garlic today. Applying this knowledge, numerous studies were conducted over the past three decades in Japan, the United States, and Israel. These studies shed light on garlic's fertility and restoration, resulting in the creation of multiple true seed-producing lines. Extensive molecular research has also been conducted to understand garlic's sterility and fertility processes better. Garlic genes involved in various aspects of fertility control, including tapetum development, male meiosis, pollen development, flower differentiation, exine formation, and tapetum viability. Some genes, such as *AP3*, *MMD1*, *MS2*, and *GPAT2* have established roles in these processes in other species and were found to be differentially expressed in fertile versus sterile garlic strains, suggesting their involvement in garlic fertility as well (Shemesh-Mayer 2015) and were proposed to control fertility in garlic. This study suggests that the gene *MS2* is a potential marker for fertility in garlic due to its strong expression in fertile garlic (Figure 5.1).

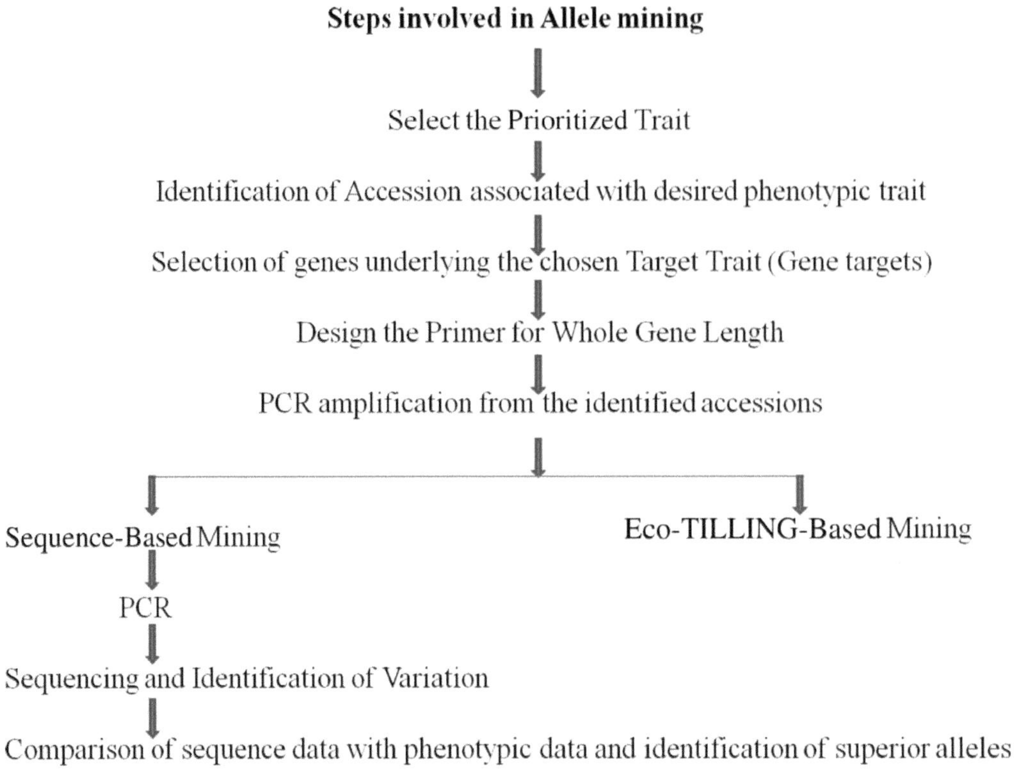

FIGURE 5.1 Steps involved in allele mining.

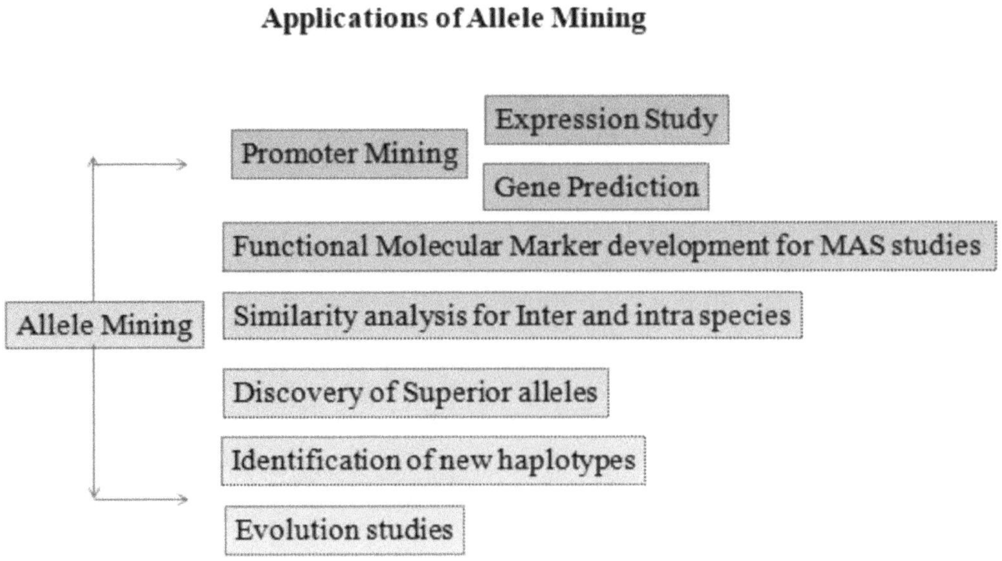

FIGURE 5.2 Applications of allele mining.

5.8 APPLICATIONS OF ALLELE MINING IN *ALLIUMS*

Although there has been extensive research on Allium species functionality and modes of action, particularly garlic and onion, progress in genetic enhancement for functional enrichment and quantitative improvement of these bioactive is nascent. Among all the medically significant *Allium* species constituents, allinase, the enzyme that produces allicin, has undergone the most research. To understand the *Allium* species phylogenetic relationships and evolutionary history concerning alliinase, studies on this enzyme involved analysis of the encoding genes (Figure 5.2). A cluster of closely related genes encodes alliinases across this genus was assumed after it was shown that the cDNA made from garlic alliinase shared a high degree of similarity throughout other species.

REFERENCES

Abdelrahman M, Sawada Y, Nakabayashi R, Sato S, Hirakawa H, El-Sayed M, Hirai MY, Saito K, Yamauchi N, Shigyo M (2015) Integrating transcriptome and target metabolome variability in doubled haploids of Allium cepa for abiotic stress protection. *Mol Breed*, 35, 1–11.

Atkinson NJ, Urwin PE (2012) The Interaction of Plant Biotic and Abiotic Stresses: From Genes to the Field. *J Exp Bot*, 63: 3523–3543. https://doi.org/10.1093/jxb/ers100

Bang H, Kim S, Park SO, Yoo KS, Patil BS (2013) development of a codominant CAPS marker linked to the *Ms* locus controlling fertility restoration in onion (*Allium cepa* L.). *Sci Horti* 153(4): 42–49. DOI: 10.1016/j.scienta.2013.01.020

Chan EY (2005) Advances in sequencing technology. *Mut Res/Fun and Mol Mech of Mut* 573(1–2): 13–40.

Cornic G, Massacci A (1996) Leaf Photosynthesis under Drought Stress. In: Baker, N.R., Ed., *Photosynthesis and the Environment*, Kluwer Academic Publishers, New York, pp. 347–366.

Csiszar J, Lantos E, Tari I, Madossa E, Wodala B, Vashegyi A, Horvath F, Pecsvaradi A, Szabo M, Bartha B, Galle A, Lazar A, Coradini G, Staicu M, Postelnicu S, Mihacea S, Nedelea G, Erdei L (2007) Antioxidant enzyme activities in *Allium* species and their cultivars under water stress. *Plant Soil Environ* 53(12): 517–523.

Currah L (2002) Onions in the tropics: cultivars and country reports. In: *Allium crop science: Recent advances.* (Rabinowitch, H. D. and Currah, L., Eds). CABI Publishing, Wallingford, UK, pp. 379–407.

De Vicente MC, Tanksley SD (1993) QTL analysis of transgressive segregation in an interspecific tomato cross. *Genetics* 134: 585–596.

Drinkwater WO, Janes BE (1955) Effects of irrigation and soil water on maturity, yield and storage of two onion hybrids. *Proc Am Soc Hortic Sci* 66: 267–279.

Dufoo-Hurtado MD, Huerta-Ocampo JÁ, Barrera-Pacheco A, Barba de la Rosa AP, Mercado-Silva EM (2015) Low temperature conditioning of garlic (Allium sativum L.) "seed" cloves induces alterations in sprouts proteome. *Front Plant Sci* 13;6:332.

Engelke T, Terefe D, Tatlioglu T (2003) A PCR-based marker system monitoring CMS-(S), CMS-(T) and (N)-cytoplasm in the onion (*Allium cepa* L.). *Theor Appl Genet* 107: 162–167.

Gai SP, Meng XD (2002) Development and identification of RAPD markers linked to cytoplasmic male sterility in Welsh onion (*Allium fistulosum* L.). *J Agric Biotechnol* 10: 94–97.

Gai SP, Meng XD (2004) Study on conversion of RAPD markers linked to cytoplasmic male sterile loci to SCAR markers in Welsh onion (*Allium fistulosum* L.). *J Laiyang Agric College* 21: 189–192.

Galmarini CM, Mackey JR, Dumontet C (2001) Nucleoside analogues: mechanisms of drug resistance and reversal strategies. *Leukemia* 15(6): 875–890.

Galsurker O, Doron-Faigenboim A, Teper-Bamnolker P, Daus A, Lers A, Eshel D (2018) Differential response to heat stress in outer and inner onion bulb scales. *J Exp Bot* 69(16): 4047–4064. DOI: 10.1093/jxb/ery189

Gedam PA, Thangasamy A, Shirsat DV, Ghosh S, Bhagat KP, Sogam OA, Singh M (2021) Screening of onion (*Allium cepa* L.) genotypes for drought tolerance using physiological and yield based indices through multivariate analysis. *Front Plant Sci*, 12: 1–16.

Ghodke P, Khandagale K, Thangasamy A, Kulkarni A, Narwade N, Shirsat D, Singh M (2020) Comparative transcriptome analyses in contrasting onion (*Allium cepa* L.) genotypes for drought stress. *PLoS One*, 15(8): 1–21.

Ghodke PH, Andhale PS, Gijare UM, Thangasamy A, Khade YP, Mahajan V, Singh M (2018) Physiological and biochemical responses in onion crop to drought stress. *Int J Curr Microbiol App Sci* 7(1): 2054–2062.

Gokce AF, McCallum J, Sato Y, Havey MJ (2002) Molecular Tagging of the Ms Locus in Onion. *J Am Soc Hortic Sci* 127(4): 576–582. DOI: 10.21273/JASHS.127.4.576

Han J, Thamilarasan SK, Natarajan S, Park JI, Chung MY, Nou IS (2016). De novo assembly and transcriptome analysis of bulb onion (Allium cepa L.) during cold acclimation using contrasting genotypes. *PloS One* 11(9):e0161987.

Havey MJ (2013) Single nucleotide polymorphisms in linkage disequilibrium with the male-fertility restoration (*Ms*) locus in open-pollinated and inbred populations of onion. *J Am Soc Hortic Sci* 138(4): 306–309. DOI: 10.21273/JASHS.138.4.306

Huo YM, Liu BJ, Yang, YY, Miao J, Gao LM, Kong SP, Wang ZB, Kitano H, Wu X (2015) *ACSKP1,* a Multiplex PCR based co-dominant marker in complete linkage disequilibrium with the male fertility restoration (Ms) locus and its application in open pollinated populations of onion. *Euphytica* 204(3): 711–722. DOI: 10.1007/s10681-015- 1374-7

Khar A, Singh H, Verma P (2022) Mitigating abiotic stresses in *Allium* under changing climatic scenario. In: Kole C (ed) *Genomic Designing for Abiotic Stress Resistant Vegetable Crops.* Springer International Publishing, Cham, Switzerland, pp 253–278.

Khosa JS, Lee R, Brauning S, Lord J, Pither-Joyce M, McCallum J, Macknight RC (2016) Doubled haploid 'CUDH2107'as a reference for bulb onion (*Allium cepa* L.) research: development of a transcriptome catalogue and identification of transcripts associated with male fertility. *PLoS One* 11(11): 1–17.

Kim S, Baek D, Cho DY, Lee ET, Yoon MK (2009) Identification of two novel inactive DFR-A alleles responsible for failure to produce anthocyanin and development of a simple PCR-based molecular marker for bulb color selection in onion (*Allium cepa* L.). *Theo and App Gen* 118: 1391–1399.

Kim S, Kim CW, Choi MS, Kim S (2016) Development of a simple PCR marker tagging the *Allium roylei* fragment harboring resistance to downy mildew (*Peronospora destructor*) in onion (*Allium cepa* L.). *Euphytica* 208: 561–569.

Kim S, Kim S (2015) Application of the molecular marker in linkage disequilibrium with *ms*, a restorer-of-fertility locus for improvement of onion breeding efficiency. *Kor J Hortic Sci* 33: 550–558.

Kim S, Yoo KS, Pike LM (2005) The basic color factor, the C locus, encodes a regulatory gene controlling transcription of chalcone synthase genes in onions (*Allium cepa*). *Euphytica* 142: 273–282.

Konvicka O, Levan A (1972) Chromosome Studies in *Allium sativum. Hereditas* 72: 129–148. DOI: 10.1111/j.1601-5223.1972.tb01035.x

Latha R, Rubia L, Bennett J, Swaminathan MS (2004) Allele mining for stress tolerance genes in Oryza species and related germplasm. *Mol Biotechnol* 27: 101–108.

Liu Q, Lan Y, Wen C, Zhao H, Wang J et al. (2016) Transcriptome sequencing analyses between the cytoplasmic male sterile line and its maintainer line in Welsh onion (*Allium fistulosum* L.). *Int J Mol Sci* 17(7): 1058. DOI: 10.3390/ijms17071058

Malik G, Dhatt AS, Malik AA (2020) A review of genetic understanding and amelioration of edible *Allium* species. *Food Rev Int* 37(4): 415–446.

Martin WJ, McCallum J, Shigyo M, Jakse J, Kuhl JC, Yamane N, Pither-Joyce M, Gokce AF, et al. (2005) Genetic mapping of expressed sequences in onion and in silico comparisons with rice show scant colinearity. *Mol Genet Genom* 274(3): 197–204.

Masuzaki S, Shigyo M, Yamauchi N (2006) Direct comparison between genomic constitution and flavonoid contents in *Allium* multiple alien addition lines reveals chromosomal locations of genes related to biosynthesis from dihydrokaempferol to quercetin glucosides in scaly leaf of shallot (*Allium cepa* L.). *Theo Appl Genet* 112: 607–617.

McCallum J, Clarke A, Pither-Joyce M, Shaw M, Butler R, Brash D, Scheffer J, Sims I, et al. (2006) Genetic mapping of a major gene affecting onion bulb fructan content. *Theor Appl Genet* 112(5): 958–967.

McCallum J, Khosa JS, Lee R, Brauning S, Lord J, Pither-Joyce M, Macknight R (2016) Doubled haploid 'CUDH2107' as a reference for bulb onion (*Allium cepa* L.) Research: Development of a transcriptome catalogue and identification of transcripts associated with male fertility. *PLoS One* 11(11): 1–17. DOI: 10.1371/journal. pone.0166568

McCallum J, Pither-Joyce M, Shaw M, Kenel F, Davis S, Butler R, Havey MJ (2007) Genetic mapping of sulfur assimilation genes reveals a QTL for onion bulb pungency. *Theo Appl Genet* 114: 815–822.

McCallum, J, Baldwin S, Thomson S, Pither-Joyce M, Kenel F, et al. (2016) Molecular Genetics analysis of onion (*Allium cepa* L.) adaptive physiology of bulb. *Acta Hortic* 1110: 71–76. DOI: 10.17660/ActaHortic.2016.1110.11

Park J, Bang H, Cho DY, Yoon MK, Patil BS, Kim S (2013a) Construction of high-resolution linkage map of the *ms* locus, a restorer-of-fertility gene in onion (*Allium cepa* L.). *Euphytica* 192(2): 267–278. DOI: 10.1007/s10681-012-0851-5

Park J, Cho DY, Moon JS, Yoon MK, Kim S (2013b) Development of functional markers for detection of inactive *DFR-A* alleles responsible for failure of anthocyanin production in onions (*Allium cepa* L.). *Hort Sci Technol* 31(1): 72–79.

Pelter GQ, Mittelstadt R, Leib BG, Redulla CA (2004) Effects of water stress at specific growth stages on onion bulb yield and quality. *Agric Water Manag* 68(2): 107–115.

Raines S, Henson C, Havey MJ (2007) Genetic analyses of soluble carbohydrate concentrations in onion bulbs. *J Am Soc Hortic Sci* 134(6): 618–623.

Randle, W. M., and Lancaster, J. E. (2002). Sulphur compounds in alliums in relation to flavour quality. *Allium Crop Sci* 329–356. DOI: https://doi.org/10.1079/9780851995106.0329

Scholten OE, van Kaauwen MP, Shahin A, Hendrickx PM, Keizer LC, Burger K, Vosman B (2016). SNP-markers in *Allium* species to facilitate introgression breeding in onion. *BMC Plant Biol* 16(1): 1–10. DOI: 10.1186/s12870-016-0879-0

Shemesh-Mayer E, Ben-Michael T, Rotem N, Rabinowitch HD, Doron-Faigenboim A, Kosmala A, Perlikowski D, Sherman A, Kamenetsky R (2015) Garlic (*Allium sativum* L.) Fertility: transcriptome and proteome analyses provide insight into flower and pollen development. *Front Plant Sci* 6: 1–17. DOI: 10.3389/fpls.2015.00271

Shirley BW (1996) Flavonoid biosynthesis: 'new' functions for an 'old' pathway. *Trends Plant Sci* 1(11): 377–382.

Shock CC, Feibert EB, Saunders M (1998) Irrigation Management for drip-irrigated onions. *Oregon State University Agricultural Experiment Station, Special Report.* 988:42–48.

Son JH, Park KC, Lee SI, Kim HH, Kim JH, Kim SH, Kim NS (2012). Isolation of cold-responsive genes from garlic, Allium sativum. *Genes Genom* 34:93–101.

Spooner DM, McLean K, Ramsay G, Waugh R, Bryan GJ (2005) A single domestication for potato based on multilocus amplified fragment length polymorphism genotyping. *Proc Natl Acad Sci USA* 102(41): 14694–14699.

Sta-Baba, R., Bkheder, M., Chaar, H., Harrabi, M. (2005). Identification des critères de sélection pour l'adaptation variétale d'oignon (*Allium cepa* l. *Rev. Inst. Nat. Agr. Tunisie* 20, 141–159.

Szklarczyk M, Simlat M, Jagosz B, Ba G (2002) The use of cytoplasmic markers in onion hybrid breeding. *Cell Mol Biol Lett* 7(2B): 625–634.

Takagi S (1990) Garlic (*Allium sativum* L.) In: Brewster JL, Rabinowitch HD (eds) *Onions and Allied Crops: Biochemistry, Food Science and Minor Crops.* CRC Press, Boca Raton, Florida, 3, pp 109–158.

Tan S, Evans RR, Dahmer ML, Singh BK, Shaner DL (2005) Imidazolinone-tolerant crops: history, current status and future. *Pest Manag Sci* 61(3): 246–257.

Tsukazaki H, Yaguchi S, Yamashita K, Kojima A, Wako T, Sato S, Shigyo M (2012) QTL mapping for bulb pigmentation in F_2 population between bulb onion and shallot. In: *VI International Symposium on Edible Alliaceae*, 969, pp 85–89.

Xiao X, Li J, Samulski RJ (1996) Efficient long-term gene transfer into muscle tissue of immunocompetent mice by adeno-associated virus vector. *J Virol* 70(11): 8098–8108.

Yamashita KI, Takatori Y, Tashiro Y (2005) Chromosomal location of a pollen fertility-restoring gene, *Rf*, for CMS in Japanese bunching onion (*Allium fistulosum* L.) possessing the cytoplasm of *A. galanthum* Kar. et Kir. revealed by genomic in situ hybridization. *Theor Appl Genet* 111: 15–22.

Yang YY, Meng Y, Miao HJ, Liu BJ, Kong SP, Gao LM, Liu C, Wang ZB, Tahara Y, et al. (2013) Identification of two SCAR markers co-segregated with the dominant *Ms* and recessive *Ms* alleles in onion *(Allium cepa* L.). *Euphytica* 190(2): 267–277.

Yiu, J.C., Juang, L.D., Fang, D.Y.T., Liu, C.W. and Wu, S.J., 2009. Exogenous putrescine reduces flooding-induced oxidative damage by increasing the antioxidant properties of Welsh onion. *Sci Hort* 120: 306–314.

Yuan Q, Song C, Gao L, Zhang H, Yang C, Sheng J, Ren J, Chen D, Wang Y (2018) Transcriptome de novo assembly and analysis of differentially expressed genes related to cytoplasmic male sterility in onion. *Pl Phy and Bioch* 125: 35–44.

Zhang C, Li X, Zhan Z, Cao L, Zeng A, Chang G, Liang Y (2018). Transcriptome sequencing and metabolism analysis reveals the role of cyanidin metabolism in dark-red onion (Allium cepa L.) bulbs. *Sci Rep* 20;8(1):14109.

6 Current Status of Viral Resistance Molecular Breeding in Cucumbers (*Cucumis sativus* L.)

Mahdi Badri Anarjan,[1] *Haisu Li,*[1] *and Sanghyeob Lee*[1,2]
[1]Plant Genomics Laboratory, Department of Bio-resource Engineering, College of Life Sciences, Sejong University, 209 Neungdong-ro, Gwanjing-gu, Seoul, Republic of Korea
[2]Plant Engineering Research Institute, Sejong University, 209 Neungdong-ro, Gwanjing-gu, Seoul, Republic of Korea

6.1 GENERAL INTRODUCTION

Cucumber has been cultivated in many places worldwide. Due to the world's love for crunchy cucumber, this agricultural product has contributed greatly to the economies of many countries. Most people consume fresh or processed cucumber (Staub and Bacher 1997; Staub et al. 2008). Various cultivation practices depend on the fruit (market) type, growing location, and cultivar type (Ranjan et al. 2008). To obtain high production and good quality, many efforts, such as the development of sophisticated cultivation practices and pesticide spray methods, and disease-free cultivar(s), have been devoted.

It is estimated that four of the world's top 12 vegetable crops come from the Cucurbitaceae family (FAOSTAT 2018; www.fao.org/faostat/en/#data/QCL). This family comprises 965 species in ~118 genera (Christenhusz and Byng 2016; Pandey and Kujur 2022). Among them, the three genera, *Citrullus* [watermelon (*Citrullus lannatus* and *Citrullus colocynthis*)], *Cucumis* [melon (*Cucumis melo*) and cucumber (*Cucumis sativus*)], and *Cucurbita* [gourd, pumpkin, squash, and zucchini (*Cucurbita* spp.)], are significantly important for humans (Rolnik and Olas 2020). However, others are also widely grown worldwide because they are consumed as common or human health-promoting foods (Rolnik and Olas 2020).

It has been considered that cucumber was domesticated from *C. sativus* var. *hardwickii* ~3,000 years ago in India (Qi et al. 2013; Pandey and Kujur 2022). India has been considered its primary center of origin, and cucumber has spread quickly in the south and east Himalayas (Whitaker and Davis 1962; Pandey and Kujur 2022). China has been considered a secondary diversity center as it spread to China ~2,000 years ago (Keng 1974; Paris et al. 2012; Pandey and Kujur 2022). Now, cucumber is globally grown and consumed. Up to this point, China (79.9%) produces the most cucumbers worldwide (FAOSTAT 2018; Table 6.1), followed by Iran (3.2%), Turkey (2.6%), and Russia (2.3%) (Figure 6.1; FAOSTAT 2018).

Reductions in the quality and quantity of yield are usually considered crop losses and occur at any stage of plant development (Zadoks and Schei 1979; Oerke and Dehne 2004). It has been considered that the 20% to 40% loss of global crop productivity could be caused by pathogens, animals, and weeds (Oerke 1994; Oerke and Dehne 2004). A significant crop loss occurs each year

DOI: 10.1201/9781003376439-6

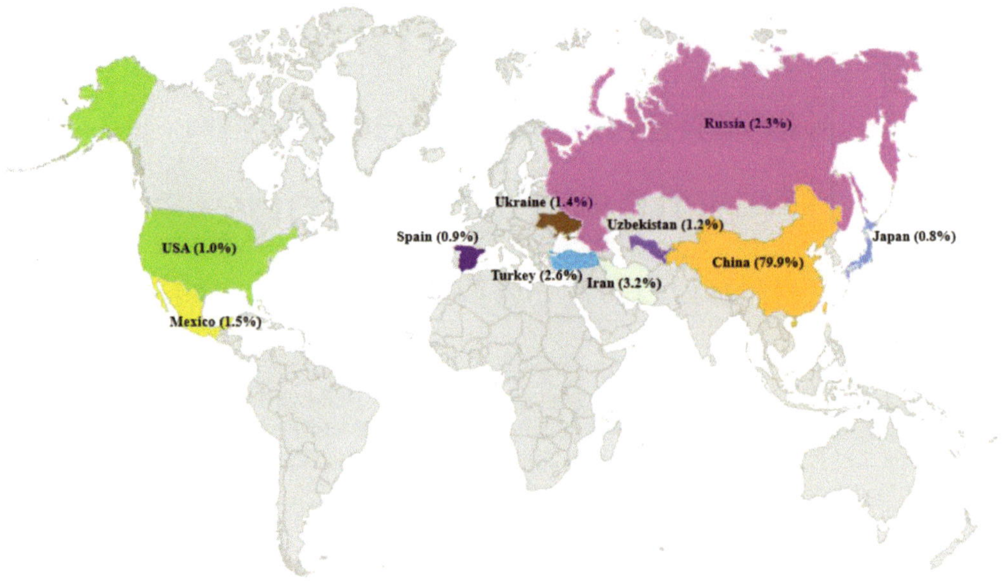

FIGURE 6.1 Top ten cucumber cultivation in the world. Percentages in parentheses indicate the country's share of global cucumber production.

Source: www.fao.org/faostat/en/#data/QCL

due to plant viruses (Sastry and Zitter 2014). Furthermore, it has been predicted to have a greater economic impact on the current global warming situation (Jones 2016). Cucurbitaceae vegetables are usually susceptible to many viruses easily spread by mechanically damaging cells or transmitting vectors, such as insects (Lecoq and Desbiez 2012; Velasco et al. 2020).

Cucumoviruses [*Cucumber Mosaic Virus* (CMV)], poleoviruses [*Cucurbit Aphid-Borne Yellows Virus* (CABYV)], and potyviruses [*Moroccan Watermelon Mosaic Virus* (MWMV), *Papaya Ringspot Virus* (PRSV), *Watermelon Mosaic Virus* (WMV), and *Zucchini Yellow Mosaic Virus* (ZYMV)], are usually spread by aphids and cause severe epidemics, leading to the severe breakout of viral diseases in key Cucurbitaceae vegetable production (Desbiez 2020; Martin-Hernandez and Pico 2021; Radouane et al. 2021). In addition, criniviruses [*Beet Pseudo-Yellows Virus* (BPYV), *Cucurbit Chlorotic Yellows Virus* (CCYV), and *Cucurbit Yellow Stunting Disorder Virus* (CYSDV)] and ipomoviruses [*Cucumber Vein Yellowing Virus* (CVYV)] are also transmitted by whitefly (*Trialeurodes vaporariorum*) and/or tobacco whitefly (*Bemisia tabaci*) (Abrahamian and Abou-Jawdah 2014; Desbiez 2020; Martin-Hernandez and Pico 2021). Mechanically spreading viruses, such as tobamoviruses [*Cucumber Green Mottle Mosaic Virus* (CGMMV) and *Zucchini Green Mottle Mosaic Virus* (ZGMMV)] are widely disseminated worldwide by seed transmission (Lewandowski 2008; Desbiez 2020). Although numerous viruses can damage Cucurbitaceae vegetables, the host range of the virus determines the limit of the range of damaged species and/or genera (Singh et al. 2003).

The spraying of chemicals for controlling viral diseases is unavoidable because viruses are obligate intracellular parasites. However, applying the chemical spray to control viral- and insect-originated diseases could be harmful to growers, consumers, and environments. For this reason, the utilization of genetic resistance from resistance germplasm is the most efficient method to lessen yield losses. Now, the reference genome sequences of most major cucurbits are available, and it has greatly advanced the study of virus and insect resistance (Huang et al. 2009; Garcia-Mas et al. 2012;

TABLE 6.1
List of Virus-Resistant Cucumber Accession

Genus	Virus	Resistant Accession	Reference
Criniviruses	*CYSDV*	Ames 13334, Ames 3950, PI 177364, PI 211589, PI 605923, PI 250147, A1 & A2	Aguilar et al. 2006; Eid et al. 2006; Wang et al. 2020
Cucumoviruses	*CMV*	Chinese Long, Santo, Tokyo Long Green, Wis. SMR 12, Wis. SMR 15, Wis. SMR 18, TMG-1, Inbred line 02245, IC-277048 (var. Hardwickii), Kyoto-3-Feet, Srinagar Local-I, Srinagar Local-II, Paprola Local, F-3	Igarashi et al. 1986; Kooistra 1969; Porter 1929; Shifriss and Myers. 1942; Wasuwat and Walker, 1961; Munger and Newhall 1953; Provvidenti 1985; Shi et al. 2018; Munshi et al. 2008; Gil-Salas et al. 2009; Dhillon 1992; Huang 2007; Wang 2010
Ipomoviruses	*CVYV*	C.sat-10, Kyoto-3-Feet, CE0749, TMG-1	Pico et al. 2003; Pujol et al. 2019; Pico et al. 2008; Gil-Salas et al. 2009; Nitzany 1970; Nitzany 1975; Monnot et al. 2022
Potyviruses	*MWMV*	TMG-1	Kabelka and Grumet, 1997
	PRSV	TMG-1, Inbred line 02245, Dina-1, Surinam	Wal et al. 1997; Tian et al. 2015; Grumet et al. 2000; Kabelka and Grumet 1997
	WMV	TMG-1, Inbred line 02245, Dina-1, Kyoto 3 feet	Wal et al. 1997; Wal and Grumet 1995; Tian et al. 2016; Grumet et al. 2000; Cohen et al. 1971
	ZYFV	TMG-1	Gilbert-Albertini et al. 1995
	ZYMV	Dina-1, TMG-1, Formosa, A192-18, G22, A202-18, R10, S93-18, Sweet-Slice	Wal et al. 1997; Cardoso et al. 2010; Provvidenti 1985; Ramirez-Madera and Havey 2017; Hayja and Al-Shahwan 1991; Kabelka et al. 1997; Svoboda et al. 2013; Amano et al. 2013; Wang et al. 2020
Tobamoviruses	*CGMMV*	BGV001358, CGN19818, TMG-1, C. sativus var. hardwickii	Crespo et al. 2018; Carnide and Barroso 2006; Monnot et al. 2022
Tospovirus	*MYSV*	27028930, Yamakyuri-1	Sugiyama et al. 2009; Sugiyama et al. 2015

Source: Modified and updated from the table of Martin-Hernandez and Pico (2021).

Yang et al. 2012; Guo et al. 2013; Qi et al. 2013; Zhang et al. 2015a; Sun et al. 2017; Wu et al. 2017; Montero-Pau et al. 2018; Guo et al. 2019; Li et al. 2019; Xie et al. 2019; Ma et al. 2020; Osipowski et al. 2020; Barrera-Redondo et al. 2021; Qin et al. 2021). In the past, resistant sources were chosen through field or in-house screening. However, resistance sources are currently being thoroughly examined with the assistance of genetic and genomic tools.

The plant has two types of resistance against pathogens classified as race-specific resistance and broad resistance (non-host resistance). The former follows Mendelian genetics and explains the interaction between single resistance genes (R-genes) of plants and *Avirulence* (*Avr*) genes of specific races of pathogens (Gabriel and Rolfe 1990; Keen 1990; De-Wit 1977; de Ronde et al. 2014). The latter represents a more complex and robust system against many pathogens (Heath 2000). Of all known plant viruses, only a few viruses are capable of causing viral diseases, and their effects are limited to very few crops (de Ronde et al. 2014). Generally, this phenomenon is called a natural host range. Most of the serious diseases in Cucurbitaceae vegetables originated from the virus involved in this case. Therefore, finding R-genes in Cucurbitaceae vegetables against various viruses is an emerging task for a plant breeder.

To achieve this objective, researchers should first find the accessions of Cucurbitaceae vegetables that showed no symptoms of a specific virus. The resistance gene involved in this specific interaction should be identified. Finally, the resistance gene should be successfully introgressed into the elite lines used for the breeding program. Thus far, many cutting-edge technologies are being developed to support these processes. Although improvement in technologies is being processed, molecular markers representing the specific traits of plants are widely accepted and used in breeding programs. Therefore, many efforts have been devoted to finding suitable markers for marker-assisted selection (MAS) and the genes responsible for the resistance. Depending on the species and/or genus variability, the source of the resistances, and the cross-ability relationships between various taxa, different crop strategies are used (Martin-Hernandez and Pico 2021). Therefore, it is necessary to expand the genetic resistance to respond to new viruses (Wang et al. 2018; Gonzalo et al. 2019; Wu et al. 2019; Xanthopoulou et al. 2019). This chapter reviewed the studies done in the past to understand the inheritance and genetic variability in cucumber resistance to viruses.

6.2 TAXONOMY OF THE GENUS *CUCUMIS*

The genus *Cucumis* contains ~66 species (Weng 2021). There are four varieties of *C. sativus* that are cross-compatible with each other; *C. sativus* var. *hardwickii* (wild-type cucumber), *C. sativus* var. *xishuangbannesis* (semi-wild-type Xishuangbanna cucumber), *C. sativus* var. *sikkimensis* (Sikkim-type cucumber), and *C. sativus* var. *sativus* (cultivated cucumber) (Pandey and Kujur 2022). Since *C. sativus* var. *hardwickii* showed the largest variability among *C. sativus* accessions, it has a better chance of creating genetic diversity in cultivated cucumbers (Horst and Lower 1978; Staub et al. 1992; Dijkhuizen et al. 1996). In contrast, *Cucumis hystrix*, a sister species of cucumber, has $2n = 2x = 24$ chromosomes (Weng et al. 2021). Due to unique genetic characteristics, the taxonomic identification of the wild species *C. hystrix* is difficult (Chen et al. 1995, 1997a, b). *C. melo*, another distinct species of *Cucumis* also has chromosome number $2n = 2x = 24$ (Kirkbride 1993). It has been proposed that the cucumber evolved from its extinct ancestor ($2n = 2x = 24$) via dysploid chromosome reduction during the evolution of the *Cucumis* species (Weng 2021).

The initial step in any crop improvement project is collecting and conserving numerous germplasms and their usage in breeding programs. It has been proposed that there are three types of the gene pool (GP) [primary GP (GP-I), secondary GP (GP-II), and tertiary GP (GP-III)] based on the cross-ability conduct and interconnection in the middle of a crop species (Figure 6.2; Raamsdonk et al. 1989; Den-Nijs and Custers 1990; Bates et al. 1995; Das et al. 2022). In GP-I of cucumbers, there are two species: *C. sativus* var. *sativus* and *C. sativus* var. *hardwickii*; in contrast, GP-II is composed of wild African cucumber species, such as *Cucumis dipsaceus* and *Cucumis myriocarpus*, and also *C. hystrix* (Naegele and Wehner 2016; Das et al. 2022). GP-III comprises *Cucumis* and *Cucurbita* genera that do not share a sexual relationship with others (Das et al. 2022).

The wild cucumber *C. sativus* var. *hardwickii* in GP-I has been a source of resistance to several biotic diseases, such as CGMMV and scab disease (Horst and Lower 1978; Staub and Palmer 1987; Sharma and Hore 1996). Because it is cross-compatible with cultivated cultivar (*C. sativus* var. *sativus*), the investigation of cucumber breeding programs for various biotic stress resistances, such as viral diseases and insect pests, has been pursued (Horst and Lower 1978). Two additional varieties of cultivated *C. sativus* var. *sativus* were also attracted to the biotic stress breeding program. One is *C. sativus* var. *xishuangbannanesis*, which is considered a selection of cultivated cucumber (Bo et al. 2015). The other is *C. sativus* var. *sikkimensis,* called Sikkim-type cucumber, and is considered an ecotype cucumber (Wan et al. 2013). Interestingly, the partial chromosome of the wild cucumber, *C. sativus* var. *hardwickii*, was introgressed into the genome of var. *xishuangbannanesis* and var. *sikkimensis* considered via chromosomal inversions and translocations (Yang et al. 2012; Grumet et al. 2021).

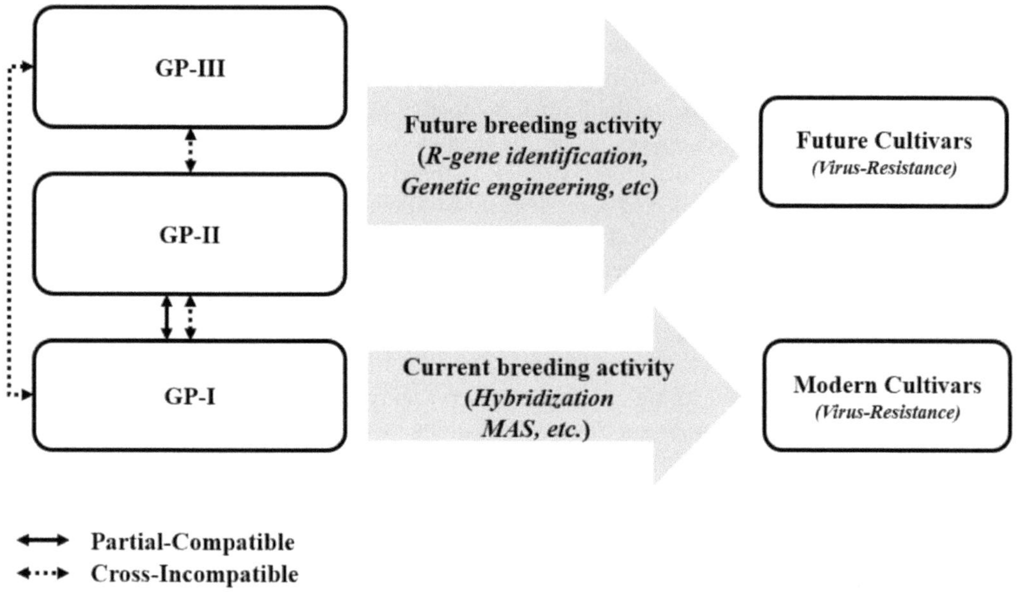

FIGURE 6.2 Utilization of cucumber gene pools (GP-I, GP-II, and GP-III) for current and future virus-resistance cucumber breeding program.

Current biotic stress resistance cultivars, including virus resistance, have been developed from the germplasm of GP-I. However, cucumbers included in GP-I showed low genetic diversity despite the apparent morphological diversity (Staub et al. 1997, 1999; Horejsi and Staub 1999; Munshi et al. 2007; Aydemir 2009; Lv et al. 2012; Zhang et al. 2012; Innark et al. 2013; Pandey et al. 2013). Therefore, utilization of GP-I for biotic stress resistance cucumber cultivar development has been facing the limitation of breeding materials. However, the wild African and *C. sativus* var. *hystrix* are cross-incompatible and partially compatible with *C. sativus* var. *sativus*, respectively (Kho et al. 1980; Chen et al. 1995; Chen et al. 1997a, b; Chen et al. 1998; Chen and Kirkbride 2000; Naegele and Wehner 2016). This makes it hard to utilize the accession in GP-II for the cucumber breeding program. It has been believed that wild cucumber species included in GP-III store numerous genes related to various biotic stress R-genes. However, this hinders the utilization of the germplasm in GP-III for the cucumber breeding program. The cross-incompatibility of accession in GP-II and III could be solved by applying modern genetic engineering tools. Therefore, many efforts should be focused on selecting virus-resistance sources and identifying R-genes. This would be a key factor in activating future virus resistance cucumber breeding programs.

6.3 GERMPLASM COLLECTION AND GENETICS RESOURCE OF CUCUMBER

Plant breeding requires divergent genetic resources, and a successful breeding program is based on genetic diversity in all accessions of a crop species (Sugiyama 2013; Pandey and Kujur 2022). Although the evaluation of morphology-based diversity was traditionally used in the breeding program, it has limitations because most of them are influenced by divergent factors, not only genetic factors. It means that morphology-based diversity evaluation could not reflect the real diversity because the morphology can be easily affected by a factor(s) other than the genotype. Accordingly, it is crucial to evaluate genetic diversity using current molecular techniques for modern breeding programs (Pandey and Kujur 2022).

Various biotic and abiotic stressors affect the yield and quality of cucumber production. Recently, insect-mediated viral diseases are getting attention because the damage rapidly increases with increasing global warming. Although the generation of disease-resistant cultivars is efficient, it would be challenging because of the limited number of resistant genes available. Furthermore, the emergence of new viruses makes it hard to generate new resistance cultivars because of the lack of resistance accessions to the new virus. It is therefore necessary to screen disease resistance accessions from newly collected accessions. In addition, rapid, accurate, and high-throughput screening techniques should be developed for this purpose. Therefore, current phenotype-based screening should be replaced by other techniques. Genotype-based screening systems should be needed shortly and would stimulate cucumber disease resistance cultivar breeding programs. For these new screening techniques, the collection and maintenance of divergent cucumber accessions are still important.

Currently, various gene banks are working to preserve and maintain cucumber germplasm worldwide, including the Crop Germplasm Resources Institute of the Chinese Academy of Agricultural Sciences (www.cgris.net/cgris_english.html), the National Agriculture and Food Research Organization of Japan (www.naro.affrc.go.jp/archive/nias/eng/genresources/index.html), the International Plant Genetic Resources Institute of the European Union (www.agent-project.eu/partners/ipgri), N.I. Vavilov All-Russian Institute of Plant Genetics Resources of Russia (www.vir.nw.ru), and the National Agrobiodiversity Center of Korea (http://genebank.rda.go.kr/eng/uat/uia/actionMain.do). In addition, the GeneSys system maintains records of the cucurbit germplasm donated from the gene bank worldwide (www.genesys-pgr.org). The National Plant Germplasm System (www.ars-grin.gov) of the USA collected and maintained ~1,552 cucumber collections (2022.09.10 status).

Because the importance of sequences in a molecular breeding program is increasing, many efforts to reveal the whole genome sequences (WGS) and expressed gene sequences have been trying, and the results are being deposited into the public database. The National Center for Biotechnology Information (www.ncbi.nlm.nih.gov/) and European Bioinformatics Institute (www.ebi.ac.uk/) are the representative databases of all sequence-related information. In the case of cucurbit vegetables, the Cucurbit Genomics Database (http://cucurbitgenomics.org/v2/) is the most important sequence-related database.

6.4 GENOMIC RESOURCES IN CUCUMBERS

Today, the WGS of commercially important vegetable crops, including cucurbit crops, are available (Chen et al. 2019; Ma et al. 2022). With the development of next-generation sequencing (NGS) analysis and bioinformatics tools, the WGS could serve as a key technology for molecular breeding programs (Koboldt et al. 2013; Pawelkowicz et al. 2016; Osipowski et al. 2020; Das et al 2022). In 2009, the first reference genome of cucumber using the northern Chinese fresh market type 9930 (*C. sativus* L. var. 9930) was revealed (Huang et al. 2009). The genome sequence of the North American pickling type Gy14 (*C. sativus* L. var. Gy14) followed (Yang et al. 2012). The genome sequence of B10 (*C. sativus* L. var. B10) and PI 183967 (*C. sativus* var. *hardwickii*) used for European inbred line and wild cucumber, respectively, followed (Qi et al. 2013; Osipowski et al. 2020). Recently, the genome sequence of wild cucumber *C. hystrix* was also revealed (Qin et al. 2021).

After getting a reference genome, many efforts have been made to deeply understand the genetic diversity of cucumber germplasm and domestication. For example, deep re-sequencing of a cucumber core collection (115 lines) from 3,342 accessions worldwide confirmed the existence of four geographical groups in cucumber germplasm and identified >100 selective sweeps associated with domestication (Qi et al. 2013). Lv et al. (2012) used a mega-collection consisting of all available cucumber accessions from China, the Netherlands, and the United States and fingerprinted with 23 simple sequence repeat markers to identify genetic diversity and population structure.

More recently, high-throughput genotyping-by-sequencing (GBS) technology using 1,234 NPGC collections was applied to understand genetic diversity and population structure and the development of a core collection (Wang et al. 2018).

Genome assembly and the determination of quantitative trait loci (QTLs) linked to several important traits are required for highly saturated genetic linkage maps in cucumber breeding (Zhu et al. 2016). For this purpose, many analysis methods, including cytogenetic maps, restriction site-associated DNA (RAD) sequencing, GBS, specific length amplified fragment-sequencing (SLAF-seq), and NGS, were applied (Sun et al. 2013a, b; Wei et al. 2014; Xu et al. 2015a, b; Zhu et al. 2016).

In addition, due to the availability of the complete genome sequence of wild and cultivated cucumbers, researchers have performed an in-depth analysis of the resistance gene family (Wan et al. 2013). However, there are still many missing sequences in the primary genome assemblies of wild and cultivated cucumbers, and it is still difficult to isolate individual genes and identify variations because most disease resistance-related loci could be composed of multiple copies of R-genes (Li et al. 2022). Therefore, it would be necessary to employ cutting-edge technologies, such as single-molecule real-time sequencing, to improve the quality of reference genome sequences. However, there are still some restrictions on using these high-throughput sequencing (HTS) technologies to fully capture genomic variation, especially structural variants (SVs) (Golicz et al. 2016). Recently, a pan-genome and graph-based pan-genome study on plants including cucumber have revealed genome-wide biodiversity, and SVs and sequence-based variations linked to several agronomic traits (Zhang et al. 2015b; Gao et al. 2019; Tao et al. 2019; Bayer et al. 2020; Lei et al. 2021; Li et al. 2022; Oren et al. 2022; Sun et al. 2022b; Wang et al. 2022). Currently, rapidly and significantly gathered genomic information on cucumbers is being used for this purpose.

6.5 DIFFICULTY OF VIRUS RESISTANCE BREEDING IN CUCUMBERS

Biotic and other environmental stresses in this era of global warming act as roadblocks that are counterproductive to achieving stable food production (Das et al. 2022). Vegetable crops, such as cucumber, are susceptible to biotic stresses, including bacteria, viruses, fungi, and insects (Hashemi et al. 2020). Cucumber is incredibly sensitive and susceptible to potyvirus-, cucumovirus-, tobamovirus-, ipomovirus-, crinivirus-, begomovirus-, and tospovirus-type (Moriones et al. 2017; Martin-Hernandez and Pico 2021).

Most viral diseases are usually transmitted via transmitting vectors, such as insects or pests. Despite being easily controlled by toxic pesticides, they easily develop resistance due to the repeated application of the same pesticide (Crow 1957). Furthermore, the benefits of being chemical-free; the lower cost of production; and the improved export quality of vegetables, such as cucumber, have motivated scientists to work on developing virus- and pest-resistant varieties (Dash et al. 2022). Developing insect- or pest-resistant varieties is a more fundamental strategy than the virus resistance itself. To create varieties tolerant to insects, breeders have been working continuously to introduce R-genes. Finally, the number of R-genes and/or candidate genes for virus and insect stresses in cucumbers has been discovered in a small number of studies (Das et al. 2022). In contrast to other disease resistance, breeding for insect resistance is more difficult because insects easily adapt to stress more quickly. Despite being known to act separately in other Cucurbitaceae plants, antibiosis and non-preference work in tandem to provide insect resistance in cucumbers. The cucumber gains its bitterness from Cucurbitacin and other tetracyclic terpenoid compounds, which also serve as poisonous insect pest deterrents (Da-Costa and Jones 1971; Kaushik et al. 2015). There is only one dominant gene, *Bi*, located on chromosome 6, which regulates bitterness in the entire cucumber plant (Da-Costa and Jones 1971).

Introducing target genes via hybridization and marker-assisted introgression has revolutionized cucumber breeding by producing new varieties that are early-maturing, high-yielding, and disease-resistant (tolerant to viruses and insects; Wang et al. 2020; Nerkar et al. 2022). Molecular markers could be used to speed up the breeding program, select desirable traits, and overcome obstacles in traditional breeding. Cucumber high-density genetic maps have been created for various traits to find QTLs for genetic improvement in various market classes of cucumber (Wang et al. 2020; Jat et al. 2021). In the last few centuries, cucumber breeders have developed several genetic stocks resistant to pests and diseases (Wang et al. 2020). Future cucumber breeding programs should focus on introducing more R-genes from available sources to develop multiple disease resistance varieties, including virus-resistant varieties. Because several F_1 hybrid cucumbers have been developed through hybridization across the globe, an understanding of the genetic inheritance of these traits would be necessary and will be provided with recently updated information. The analysis of QTLs and genome-wide association studies (GWAS) will speed up the cucumber breeding program (Jat et al. 2021).

6.6 BREEDING SOURCES FOR RESISTANCE TO VIRUSES IN CUCUMBERS

Virus-resistant cultivars are becoming more frequently in demand due to rapidly changing agricultural environments. Finding resistance sources of viral diseases is the initial step in the breeding program. A screening process using all available collections has been widely applied to determine, which sources might be suitable for the breeding program. Virus resistance cucumber accessions can be obtained from these efforts, as shown in Table 6.1.

Cucumber accessions PI 177364, 211589, 605923, 250147, Ames 13334, 3950 have reported resistance to CYSDV (Aguilar et al. 2006; Eid et al. 2006; Wang et al. 2020). Chinese Long, Santo, Tokyo Long Green, Kyoto-3-feet, IC 277048 (var. *hardwickii*), 02245 (inbred line), Srinagar-1 and Srinagar-2 (landrace), Paprola (landrace), Taichung Mou Gua-1 (TMG-1), and Wis. SMR 12, 15, and 18 have indicated CMV resistance (Porter 1929; Shifriss and Myers 1942; Munger and Newhall 1953; Wasuwat and Walker 1961; Kooistra 1969; Provvidenti 1985; Igarashi et al. 1986; Dhillon 1992; Munshi et al. 2008; Gil-Salas et al. 2009; Shi et al. 2018). Cucumber accessions C.sat-10, CE0749, and Kyoto-3-feet have reported resistance to CVYV (Pico et al. 2003, 2008; Gil-Salas et al. 2009; Pujol et al. 2019). Only one cucumber accession (TMG-1) showed resistance to MWMV (Kabelka and Grumet 1997), and 27028930 and Yamakyuri-1 showed resistance to MYSV (Sugiyama et al. 2009, 2015). Cucumber accessions TMG-1, Dina-1, Surinam, and 02245 showed resistance to PRSV (Wal et al. 1997; Kabelka and Grumet 1997; Grumet et al. 2000; Tian et al. 2015). Cucumber accessions TMG-1, Dina-1, 02245, and Kyoto-3-feet showed resistance to WMV (Cohen et al. 1971; Wal and Grumet 1995; Wal et al. 1997; Grumet et al. 2000; Tian et al. 2016). Only one cucumber accession (TMG-1) showed resistance to *Zucchini Yellow Fleck Virus* (ZYFV; Gilbert-Albertini et al. 1995). Cucumber accessions TMG-1, Formosa, A192-18, G22, A202-18, R10, S93-18, and Dina-1 showed resistance to ZYMV (Provvidenti 1985; Hayja and Al-Shahwan 1991; Kabelka et al. 1997; Wal et al. 1997; Cardoso et al. 2010; Svoboda et al. 2013; Amano et al. 2013; Ramirez-Madera and Havey 2017). Cucumber accessions BGV001358 and CGN19818 showed CGMMV resistance (Crespo et al. 2018).

When the origin of virus resistance accessions was considered, most resistance sources were derived from the Asian region. Resistance to CGMMV, CMV, MWMV, WMV, and ZYFV was only derived from Asian countries, China, Japan, and India. Other virus resistance, ZYMV, RPSV, CYSDV, and CVYV, were also extensively derived from Asian countries, with some exceptions (Spain, Brazil, and The Netherlands). The fact could be easily explained by considering the origin of cucumbers. Unlike the accession with resistance to bacteria and fungi, the cucumber accession showed multi-viral resistance (Table 6.2). The inbred line 02445 showed resistance to CMV, PRSV, and WMV. Kyoto-3-feet also showed resistance to CMV, CVYV, and WMV. Dina-1 showed resistance to PRSV and WMV. The var. *hardwickii* showed resistance to CGMMV and CMV. The most dramatic example of multi-viral resistance is shown in TMG-1. Cucumber accession TMG-1 showed

TABLE 6.2
Multi-Virus-Resistant Cucumber Germplasms

Cucumber Accession	Virus
var. *hardwickii*	*CGMMV, CMV*
02445	*CMV, PRSV, WMV*
Dina-1	*PRSV, WMV*
Kyoto-3-feet	*CMV, CVYV, WMV*
TMG-1	*CMV, MWMV, PRSV, WMV, ZYFV, ZYMV*

resistance to CMV, MWMV, PRSV, WMV, ZYFV, and ZYMV. From a practical point of view, utilizing these multi-virus resistance accessions in the cucumber breeding program is a very effective way of making long-lasting, durable virus-resistant cultivars.

In Tables 6.1 and 6.2, numerous viral disease resistance sources exist, and these traits have been introduced into the current commercial cultivars. Some of these genetic traits are simple to pass down from generation to generation, such as the $zym^{A192\text{-}18}$, $zym^{Dina\text{-}1}$, and $zym^{TMG\text{-}1}$ for ZYMV resistance (Provvidenti 1985; Hayja and Al-Shahwan 1991; Kabelka et al. 1997; Svoboda et al. 2013; Amano et al. 2013; Ramirez-Madera and Havey 2017), *wmv* and *Wmv* for WMV resistance (Cohen et al. 1971; Wal and Grumet 1995; Grumet et al. 2000; Tian et al. 2016), *zyf* for ZYFV resistance (Gilbert-Albertini et al. 1995), *prsv-1* and *prsv-2* for PRSV resistance (Wal et al. 1997; Tian et al. 2015), and *mvm* for MWMV resistance (Kabelka et al. 1997). However, some of the resistance is intermediate or followed by polygenic inheritance (Table 6.3). Resistance to CGMMV, CMV, CYSDV, and WMV may encompass this category. A recent study suggests that Chinese long has stronger resistance to six viruses (CMV, CVYV, CGMMV, WMV, ZYMV, and PRSV) than TMG-1, an interesting result that requires further research (Monnot et al. 2022). However, it is challenging to pyramid the genes causing multi-virus resistance. Furthermore, inferior traits related to resistant genes could not be ruled out. Breeding virus-resistant cultivars using polygenic resistance inheritance requires MAS to address these issues (Huang et al. 2009; Ren et al. 2009; Diaz et al. 2011; Garcia-Mas et al. 2012; Sugiyama 2013).

6.7 IDENTIFIED GENES AND QTLS FOR VIRAL DISEASE RESISTANCE

Because most viruses usually show a broad host range, many viruses affect all main Cucurbitaceae vegetables. However, some viruses showed a limited host range, infecting a small portion of Cucurbitaceae vegetables. The severity of disease symptoms differs from species to species. For example, potyviruses can infect all major Cucurbitaceae vegetables (Martin-Hernandez and Pico 2021). However, all are hosts for CMV, but melon and cucumber infections are more significant, whereas watermelon infections are less severe (Seminis 2015). CYSDV also significantly impacts melon and cucumber, whereas ToLCNDV has a particularly negative impact on zucchini and pumpkin (Romay et al. 2019; Velasco et al. 2020). It appears that cucumber is susceptible to CGMMV and CVYV but least affected by begomoviruses (Lecoq and Desbiez 2012; Seminis 2015; Romay et al. 2019). In addition, viruses either affect all developmental stages or specific developmental stages of plants. Some viruses are more common at particular developmental stages. Cucumber has been the subject of recent studies focusing on mapping QTLs for disease resistance, and several viral disease resistances are governed by one or two resistant genes (Sun et al. 2022a). Several research groups have published many studies on the genetic inheritance of key yield-related traits and biotic stresses (Dey et al. 2022). Table 6.3 shows cucumber accession resistance to various viruses and their map positions in genomic regions.

6.7.1 Resistance to *Cucurbit Yellow Stunting Disorder Virus* (CYSDV)

CYSDV is a member of the *Closteroviridae* family and is usually transmitted by whitefly to burst severe damage to cucurbit in many areas worldwide (Martelli et al. 2000, 2002). In the 1990s, CYSDV was first identified in the United Arab Emirates and then spread throughout the Mediterranean Basin (Duffus 1995; Kao et al. 2000). From extensive screening using several hundreds of cucumber accessions composed of divergent collected origin and stages of breeding status, several accessions showed CYSDV resistance (Eid et al. 2006; Aguilar et al. 2006). Aguilar et al. (2006) identified two cucumber accessions (A1 and A2) from a screening of 300 accessions with CYSDV partial resistance. Their analysis showed that more than one recessive gene was needed for CYSDV resistance. Furthermore, the introgression of A2 into the commercial cultivars should be followed, because A2 accession showed non-preference to the VYSDV vector (Aguilar et al. 2006). In addition, seven PI lines in USDA GRIN, including PI 177364, PI 211589, PI 293432, PI 279807, PI 605923, Ames 13334, and NLS 5746 showed partial resistance against CYSDV with delayed infection and less severe symptoms (Eid et al. 2006). Among them, PI 211589 (Afghanistan), PI 605923 (India), and Ames 13334 (Spain) showed relatively higher tolerance to CYSDV than those of others. Considering the origin of these lines, the sources of CYSDV resistance could be considered to be Asia and Spain. Furthermore, de Ruiter et al. (2008) identified a major CYSDV resistance QTL (*QTL1*) derived from PI 250147 (Pakistan) explaining 42% of CYSDV resistance. Later, *cysdv5.1* followed single genetic inheritance and was identified to control CDSDV resistance in the PI 250147 (Faber et al. 2008; Wang et al. 2020; Pandey and Kujur 2022).

6.7.2 Resistance to *Cucumber Mosaic Virus* (CMV)

CMV is a member of the *Cucumovirus* genus (Bromoviridae) and was first described more than a century ago (Doolittle 1916; Jagger 1916). Although CMV resistance of cucurbit was mostly controlled by recessive genetic inheritance, resistance to CMV in cucumbers is inconsistent (Risser et al. 1977; Marathe et al. 2004; Wang et al. 2006; Huang 2007; Shi et al. 2018). With the different sources of cucumber accessions and CMVs, CMV resistance of cucumber followed divergent patterns, dominant, or recessive genetic inheritance and was mostly controlled by a single gene (Wasuwat and Walker 1961; Kooistra 1969; Marathe et al. 2004; Wang et al. 2006; Huang 2007; Wang et al. 2010; Shi et al. 2018). Several cucumber accessions, Chinese Long, TGM-1, 02245, Tokyo Long Green, Kyoto-3-feet, *Hardwicki*, Chinese F-3, Wis. SMR 12, Srinagar Local-I and -II, and Paprola Local showed CMV resistance (Table 6.3). Chinese Long used the CMV resistance sources to create the first CMV resistance cucumber variety "Shamrock" (Porter 1932; Munger and Newhall 1953; Schaible 1954; Wasuwat and Walker 1961). CMV resistance of Chinese Long was first suggested that they followed a single dominant genetic inheritance in pickling cultivars (Wasuwat and Walker 1961). Later, Havey (1997) reported that CMV resistance of Chinese Long followed polygenic inheritance. In addition, Tokyo Long Green showed CMV resistance and used the resistance source of Ohio 31, the first CMV-resistance pickling cucumber (Doolittle et al. 1939; Munger and Newhall 1953; Havey 1997). In addition, Tokyo Long Green showed CMV resistance, followed by three incomplete dominant genetic inheritances (Kooistra 1969). TMG-1 and the Himalayan landraces (Srinagar Local-I and -II and Paprola Local) showed single dominant gene control the CMV resistance (Provvidenti 1985; Dhillon 1992). The allelism test among the three Himalayan landraces revealed that the genes that control CMV resistance of Srinagar Local-I and -II were allelic, but Paprola Local was nonallelic (Dhillon 1992). The Chinese inbred line 02245 showed polygenic CMV resistance inheritance with recessive resistance (Shi et al. 2018). The recessive CMV resistance locus spanning 624 kb is located on chromosome 6 and is named *cmv6.1* (Shi et al. 2018). From this locus, nine disease resistance-related genes were identified; finally, *Csa6M133680* was considered a CMV resistance candidate gene (Shi et al. 2018). *C. sativus* var. *hardwickii* also showed CMV resistance and followed single recessive genetic inheritance (Munshi et al. 2008).

TABLE 6.3
The Origin, Locus, Genetic Inheritance, and Candidate Genes of Cucumber Virus-Resistance Accessions

Genus	Virus	Resistant Source	Origin	Genetic Inheritance	Locus Name	Candidate Gene	Mapping Population	Chromosome	Reference
Criniviruses	*CYSDV*	PI 250147	Pakistan	Partial and Quantitative	*cysdv5.1* *QTL-1* *QTL-2*	-	$F_{2:3}$	5 4	Wang et al. 2020 De Ruiter et al. 2008
		PI 211589	Afghanistan	-	-	-	124 Accessions	-	Eid et al. 2006
		PI 605923	India	-	-	-	124 Accessions	-	Eid et al. 2006
		PI 177364	India	-	-	-	124 Accessions		Eid et al. 2006
		A1 & A2	-	-	-	-	300 Accessions		Aguilar et al. 2006
		Ames 13334	Spain	-	-	-	124 Accessions	-	Eid et al. 2006
Cucumoviruses	*CMV*	Chinese Long	China	Polygenic (dominant or recessive/ singles or polygenic)	-	-	-	-	Porter 1932; Shifriss et al., 1942; Munger and Newhall 1953; Wasuwat and Walker 1961; Kooister 1969; Havey 1997
		TMG-1	China	Single Dominant	-	-	-	-	Provvidenti 1985
		Srinagar Local-I, Srinagar Local-II, Paprola Local	Southern Himalaya (India)	Single Dominant	-	-	-	-	Dhillon 1992
		Inbred line 02245	China	Polygenic Recessive	*cmv6.1*	*Csa6M133680*	RIL	6	Shi et al. 2018
		Tokyo Long Green	Japan	Polygenic Recessive	-	3 Genes	F_1	-	Kooistra 1969
		Kyoto-3-Feet	Japan	-	-	-	-	-	Gil-Salas et al. 2009

(continued)

TABLE 6.3 (Continued)
The Origin, Locus, Genetic Inheritance, and Candidate Genes of Cucumber Virus-Resistance Accessions

Genus	Virus	Resistant Source	Origin	Genetic Inheritance	Locus Name	Candidate Gene	Mapping Population	Chromosome	Reference
		IC-277048 (var. *hardwickii*)	India	Single Recessive	-	-	F_2& 31 Indian Accessions	-	Munshi et al. 2008
		F-3	China	Polygenic	*cmv* 1 ~ *cmv* 17	-	F_2	-	Huang 2007; Wang 2010
		Cultivar Wis. SMR12/ Wis.SMR15/ Wis.SMR18/	China	Single Dominant	-	-	Cultivar	-	Shifriss et al., 1942; Wasuwat and Walker 1961; Kooistra 1969
Ipomoviruses	*CVYV*	C.sat-10	Spain	Single Dominant	-	-	Spanish Collection		Pico et al. 2003; Pico et al. 2008
		CE0749	Dutch	Single Recessive	*CsCvy-1*	24 Genes	$F_{2:3}$ (BSA)	5	Pujol et al. 2019
		Kyoto-3-Feet	Japan	-	-	-	-	-	Gil-Salas et al. 2009; Nitzany 1970; Nitzany 1975
		TMG-1	China	Single Recessive	$QTL_{CvyvCH5}$	-	1219 Accessions	5	Monnot et al. 2022
Potyviruses	*MWMV*	TMG-1	China	Single Recessive	*mwm*	-	$F_{2:3}$	-	Kabelka and Grumet 1997
	PRSV	Inbred line 02245	China	Single Recessive	$prsv^{02245}$	-	$F_{2:3}$	6	Tian et al. 2015
		TMG-1	China	Single Dominant	*Prsv-2*	-	$F_{2:3}$	-	Grumet et al. 2000; Kabelka and Grumet 1997
		Dina-1	Dutch	Single Recessive	-	-	$F_{2:3}$	-	Grumet et al. 2000; Kabelka and Grumet 1997
		Surinam	South America	Single Recessive	*prsv-1*	-	$F_{2:3}$	-	Wal et al. 1997
	WMV	Kyoto 3 feet	Japan	Single Dominant	*Wmv*	-	-	-	Cohen et al. 1971
		TMG-1	China	Polygenic	*wmv-2, wmv-3, wmv-4*	-	F_2 & backcross	-	Wal and Grumet 1995; Wal et al. 1997
		Dina-1	Dutch	Polygenic	-	-	$F_{2:3}$	-	Grumet et al. 2000

		Inbred line 02245	China	Single Recessive	*wmv*^{02245}	21 Genes	$F_{2:3}$	6	Tian et al. 2016
		Surinam	South America	Single Recessive	*wmv-1-1*		$F_{2:3}$		Wang et al. 1984
	ZYFV	TMG-1	China	Single Recessive	*zyf*	-	F_2 & NIL	-	Gilbert-Albertini et al. 1995
	ZYMV	TMG-1	China	Single Recessive	*zym*$^{TMG\text{-}1}$	*CsVPS4*	$F_{2:3}$	6	Provvidenti 1985; Ramirez-Madera and Havey 2017
		Formosa	Brazil	Single Recessive	-	-	F_2& F_1BC_1	-	Cardoso et al. 2010
		A192-18	Japan	Single Recessive	*zym*$^{A192\text{-}18}$	*CsVPS4*	$F_{2:3}$	6	Ramirez-Madera and Havey 2017; Svoboda et al. 2013; Amano et al. 2013
		G22	Japan	-	-	-	F_1	-	Svoboda et al. 2013
		A202-18	Japan	-	-	-	F_1	-	Svoboda et al. 2013
		R10	Japan	-	-	-	F_1	-	Svoboda et al. 2013
		S93-18	Japan	-	-	-	F_1	-	Svoboda et al. 2013
		Dina-1	Dutch	Single Recessive	*zym*$^{Dina\text{-}1}$	*CsVPS4*	$F_{2:3}$	6	Ramirez-Madera and Havey 2017; Hayja and Al-Shahwan 1991; Kabelka et al. 1997
		Sweet-Slice	Japan & Europe	Single Recessive	*zym*$^{A192\text{-}18,}$ *zym*$^{Dina\text{-}1}$	*CsVPS4*	$F_{2:3}$	6	Wang et al. 2020
Tobamoviruses	*CGMMV*	*C. sativus* var. *hardwickii*	India	Single Recessive	-	-	-	-	Carnide and Barroso 2006
		BGV001358& CGN19818	India	Partial	-	-	-	-	Crespo et al. 2018
		TMG-1	China	Quantitative	$QTL_{CgmmvCH2}$ $QTL_{CgmmvCH5}$	-	1219 Accessions	2 5	Monnot et al. 2022
Tospovirus	*MYSV*	27028930	Thailand	Quantitative	*Swf-1* *Swf-2* *Swf-3* *Swf-4* *Sws*	-	F_2	1 3 4 7 3	Sugiyama et al. 2009; Sugiyama et al. 2015

Source: Modified and updated from the table of Martin-Hernandez and Pico (2021).

6.7.3 Resistance to *Cucumber Vein Yellowing Virus* (CVYV)

CVYV is a member of the *Potyviridae* family and is transmitted by the whitefly *B. tabaci* (Pujol et al. 2019). CVYV was first reported in Israel in the 1960s and later expanded to eastern and western Mediterranean countries (Cohen and Nitzany 1960; Navas-Castillo et al. 2014). Currently, two isolates, CVYV-Isr (Israel) and CVYV-Jor (Jordan), have been identified (Lecoq et al. 2000).

Several cucumber accessions, including CE0749, C. sat-10, C.sat-11, C.sat-14, C.sat-40, Kyoto-3-feet, and TMG-1, showed partial resistance to CVYV (Nitzany 1970, 1975; Pico et al. 2003, 2008; Gil-Salas et al. 2009; Pujol et al. 2019). CVYV resistance was first identified in a Japanese cucumber Kyoto-3-feet (Nitzany 1970, 1975). Later, four accessions (C.sat-10, C.sat-11, C.sat-14, and C.sat-40) were identified that showed fewer symptoms and accumulated CVYV with partial resistance to genetic inheritance (Pico et al. 2003). Additional investigation revealed that CVYV resistance of C.sat-10 followed a single dominant genetic inheritance (Pico et al. 2008). Gil-Salas et al. (2009) screened ten Spanish cultivars supplied by the seed company and identified that CVYV resistance could fluctuate by the seasonal effect, especially CVYV concentration. Long Dutch-type cucumber accession CE0749 harbored a monogenic and incomplete dominant resistance on chromosome 5, named *Cscvy-1*, encompassing 625 kb, containing 24 annotated genes, and explains > 80% of CVYV resistance (Pujol et al. 2019).

6.7.4 Resistance to *Moroccan Watermelon Mosaic Virus* (MWMV)

MWMV is a member of potyviruses and was first reported in Morocco in 1974 (Fischer and Lockhart 1974). Progeny analyses using F_1, F_2, and backcross progenies represented that TMG-1 is resistant to MWMV and governed by the single recessive gene named *mwm* (Kabelka and Grumet 1997). In addition, analysis of F_3 population revealed that *mwm* also controls the ZYMV resistance or is tightly linked to the ZYMV resistance gene (Kabelka and Grumet 1997). In addition to cucumber, the resistance sources of MWMV were identified in *Citrullus ecirrhosus*, *Cocdnia sessilifoli*, *Cucumis metuliferus*, and *Luffa cylindrica* (Van der Meer and Garnett 1987). When we consider the low level of cross-compatibility of cucumber to those outside of the *Cucumis* genus, extensive efforts should be followed to find MWMV resistance resources from cross-compatible Cucumis genus.

6.7.5 Resistance to *Papaya Ringspot Virus* (PRSV)

PRSV is a member of prtyvirus (Bateson et al. 1994). Two major biotypes of PRSV (PRSV-P and PRSV-W) have been identified as both could infect Cucurbit crops (Yeh and Gonsalves 1984; Bateson et al. 1994; Gonsalves 1998; Attasart et al. 2002). Among them, PRSV-W causes severe damage to cucumber production (Tian et al. 2015). PRSV-W resistance has been identified in Surinam, TMG-1, Dina-1, and 02245 (Tian et al. 2015). Surinam Local is a South American cultivar, and its resistance is governed by a single recessive locus, *Prsv-1* (Wang et al. 1984; Provvidenti 1985; Wal et al. 1997; Grumet et al. 2000). *Prsv-2*, a single dominant locus, appears responsible for conferring PRSV resistance in TMG-1 (Wal and Grumet 1995; Kabelka and Grumet 1997). PRSV resistance of Dina-1 appeared to be controlled by dominant or incompletely dominant inheritance, and the resistance locus cosegregated with *Prsv-2* (Wal and Grumet 1995; Grumet et al. 2000). Park et al. (2000) reported that *Prsv-1* is allelic to *Prsv-2*. The resistance of the Chinese inbred line 02245 was governed by a single recessive locus called $prsv^{02245}$ and mapped on chromosome 6 (Tian et al. 2015).

6.7.6 Resistance to *Watermelon Mosaic Virus* (WMV)

Similar to PRSV, WMV is also a member of prtyvirus (Lecoq and Desbiez 2008). It can infect outside of cucurbits and be transmitted by aphids in a non-persistent manner (Lecoq and Desbiez 2008). Cucumber has several different sources of resistance to WMV, following different genetic inheritance. For example, one single dominant locus, *Wmv*, was considered to control WMV resistance in the Japanese cultivar Kyoto-3-feet (Cohen et al. 1971). However, the Chinese inbred line 02245 showed *WMV* resistance, and the resistance was controlled by a single recessive locus (*wmv*^{02245}) mapped on chromosome 6 (Tian et al. 2016). The progeny of TMG-1 showed complex resistance inheritance (Wal and Grumet 1995; Wal et al. 1997). The first resistance is controlled by *wmv-2* (single recessive locus) located on chromosome 6, expressed throughout the entire plant, including the cotyledons (Wal and Grumet 1995; Wal et al. 1997). The second resistance is controlled by epistatic interaction between *wmv-2 (*second recessive locus), and either *wmv-3* (single recessive locus) or *Wmv-4* (single dominant locus) inherited from the susceptible line (Wal and Grumet 1995; Wal et al. 1997). The second resistance is not expressed in the seedling stage but only expressed true leaf stage (Wal and Grumet 1995; Wal et al. 1997).

6.7.7 Resistance to *Zucchini Yellow Fleck Virus* (ZYFV)

ZYFV is a member of a potyvirus and was first identified in Italy in 1981 and later detected in the Mediterranean basin (Vovlas et al. 1981; Avgelis 1985; Katul and Makkouk 1987). Similar to MWMV resistance, ZYFV resistance was also identified in TMG-1 (Gilbert-Albertini et al. 1995). It has been identified that the ZYFV resistance was controlled by a single recessive gene, named *zyf* (Gilbert-Albertini et al. 1995; Kabelka and Grumet 1997). Interestingly, *mwm* and *zyf* have been considered either a single gene to control either trait or both genes are tightly linked (Kabelka and Grumet 1997).

6.7.8 Resistance to *Zucchini Yellow Mosaic Virus* (ZYMV)

ZYMV is usually transmitted by aphid and affect the fruit yield and quality of cucumber worldwide (Desbiez and Lecoq 1997; Gal-on 2007; Paris 2008). Several cucumber accessions have been identified as resistant to ZYMV (Table 6.3). The genetic inheritance of the ZYMV resistance trait has been well characterized by several inbred lines. First, an inbred line derived from the Taiwanese cucumber variety TMG-1 was found to have resistance controlled by a single recessive allele, *zym*$^{TMG\text{-}1}$ (Provvidenti 1985, 1987). Dina-1, an inbred line that evolved from the Dina (Dutch hybrid), also has a single recessive locus *zym*$^{Dina\text{-}1}$ (Kabelka et al. 1997). The Brazilian cultivar Formosa was also identified as a single recessive genetic inheritance (Cardoso et al. 2010). The five slicing-type cucumber inbred lines (A202-18, R10, S93-18, G22, and A192-18) showed varying resistance levels of ZYMV (Svoboda et al. 2013). The *zym* locus (ZYMV resistance locus) is especially interesting because it showed resistance to all 173 ZYMV isolates worldwide (Lecoq et al. 2000). Since *zym* locus (ZYMV resistance locus) showed resistance to all 173 ZYMV isolates worldwide, it is receiving lots of attention (Lecoq et al. 2000). However, the melon ZYMV resistance line PI 414723 lost resistance to some of these isolates (Lecoq et al. 2000). According to Amano et al. (2013), *zym*$^{A192\text{-}18}$ is a single recessive locus located on chromosome 6, encompassing six candidate genes from the Japanese inbred line A192-18. The gene encoding the putative vacuolar protein sorting 4 protein was identified as a candidate gene for the *zym-1* locus based on single-nucleotide polymorphism (SNP) comparisons between susceptible and resistant parent lines (Amano et al. 2013). The sweet slice landrace showed ZYMV resistance, followed by single recessive genetic inheritance, and was considered to be controlled by *zym*$^{A192\text{-}18}$ or *zym*$^{Dina\text{-}1}$ (Wang et al. 2020). Furthermore, allele tests using these alleles (*zym*$^{TGM\text{-}1}$, *zym*$^{Dina\text{-}1}$, and *zym*$^{A192\text{-}18}$) were applied. Results suggested that they are strongly linked together (Hayja and Al-Shahwan 1991; Kabelka et al. 1997; Amano et al 2013; Martin-Hernandez and Pico 2021).

6.7.9 Resistance to *Cucumber Green Mottle Mosaic Virus* (CGMMV)

CGMMV is a member of the *Tobamovirus* genus and *Virgaviridae* family (Adams et al. 2009). CGMMV induces deformation and mosaic symptoms in cucumbers and is transmitted mechanically by pollens and seeds (Van Koot and Van Dorst 1959; Liu et al. 2014). The first CGMMV-infected cucumber was identified in England and rapidly distributed all worldwide following the international seed trade (Ainsworth 1935; Dombrovsky et al. 2017). CGMMV resistance resources were not found in cucumbers, but only in melons and wild *Cucumis* species (Rajamony et al. 1987, 1990a, 1990b; Pan and More 1996; Pitrat 2016; Sugiyama et al. 2006, 2007). Therefore, several efforts to find CGMMV resistance in cucumbers have been followed. One of the results, *C. sativus* var. *hardwickii* showed resistance to CGMMV and has been used to introduce CGMMV resistance into the breeding lines (Carnide and Barroso 2006). Another study reported that *C. sativus* var. *sikkimensis* showed intermediate symptoms, and two Indian-originated *C. sativus* accessions (BGV001358 and CGN19818) showed mild symptoms in both Asian and European strains of CGMMV (Crespo et al. 2018). Unlikely cucumber, wild relatives of the *Cucumis* genus, *Cucumis anguria*, and *Cucumis ficifolius* showed no symptoms in both CGMMV strains (Ruiz et al. 2021). Extensive efforts to find resources of CGMMV resistance in cucumber should be paralleled with efforts to introduce CGMMY resistances of relative *Cucumis* genus into cucumber breeding program.

6.7.10 Resistance to *Melon Yellows Spot Virus* (MYSV)

MYSV is a member of the *Tospovirus* genus and was first isolated from cucumber in Kochi Prefecture, Japan (Takeuchi et al. 2001). Melon thrips (*Thripspalmi* Karny) usually transmit MYSV persistently (Sugiyama et al. 2009). Sugiyama et al. (2009) used 398 cucumber accessions from 26 different geographical origins and evaluated MYSV resistance by mechanical inoculation. Thirteen accessions showed mild disease symptoms, and finally 10 of them (27028923, 27028930, 27028932, 27028933, C4-6-5, C6-14, Cucumber-125, Yamakyuri-1, Yamakyuri-2, and Wa-19) showed no disease symptoms. The double-antibody sandwich enzyme-linked immunosorbent assay (DAS-ELISA) revealed that accessions 27028930 and Yamakyuri-1 showed resistance to a melon isolate (MYSV-S) and moderate resistance to a cucumber isolate (FuCu05P), respectively (Sugiyama et al. 2009). QTL analysis using accession 27028930 identified four QTLs involved in *MYSV*-FuCu05P resistance and located on chromosomes 1, 3, 4, and 7 (Sugiyama et al. 2015). QTLs on chromosome 1 (*Swf-1*) and chromosome 3 (*Swf-2*) identified as major QTLs and one minor QTL (*Swf-4*) on chromosome 7 were derived from the MYSV resistance accession 27028930. However, one minor QTL (*Swf-3*) on chromosome 4 derived from the MYSV susceptible accession Tokiwa. Another QTL analysis using melon isolate (MYSV-S) revealed that the MYSV-S resistance gene (*Sws*) was mapped on chromosome 3 (Sugiyama et al. 2015). These results indicate that the utilization of tightly linked markers to *Swf-1*, *Swf-2*, and *Sws* could be useful for MAS program of MNSV resistance.

6.8 NEWLY IDENTIFIED VIRAL RESISTANCE QTLS BY GENOME-WIDE ASSOCIATION STUDY (GWAS)

Although many efforts have been tried, a small number of QTLs harbor viral resistance and were successfully mapped on divergent cucumber accessions (Table 6.3). In summary, resistance to ZYMV, PRSV, and WMV was mapped to chromosome 6 (Tian et al. 2015, 2016; Ramirez-Madera and Havey 2017). CVYV and CGMMV resistances were mapped to chromosome 5 (Leibman et al. 2011, 2018). Resistance to CMV, CYSDV, and MYSV followed polygenic inheritance instead of monogenic inheritance (de Ruiter et al. 2008; Sugiyama et al. 2015; Shi et al. 2018; Wang et al. 2020; Martin-Hernandez and Pico 2021). QTL analysis of monogenic viral resistances is relatively easy, whereas polygenic resistances are more difficult; however, the resistance of the latter is more durable than the former. In addition, traditional QTL analysis usually uses specific cucumber accessions with resistant and susceptible to specific viral diseases. Using the limited number of cucumber diversity is also the weak point of a traditional QTL study. However, GWAS could conduct large divergent cucumber accession panels and overcome this limitation.

Recently, Monnot et al. (2022) used 240, 241, 243, 245, and 250 accessions (elite and landrace lines) to conduct GWAS analysis for resistance to CVYV, PRSV, ZYMV, CGMMV, WMV, and CMV. In this analysis, 18 QTLs were identified on chromosomes 2, 3, 5, and 6. In detail, resistance to CMV was represented by three QTLs located on chromosome 2 ($QTL_{cmvCHR2}$), chromosome 5 ($QTL_{cmvCHR5}$), and chromosome 6 ($QTL_{cmvCHR6}$). Resistance to CVYV followed monogenic inheritance and was located on chromosome 5 ($QTL_{cvyvCHR5}$). Resistance to WMV also followed monogenic inheritance and was located on chromosome 5 ($QTL_{wmvCHR5}$). $QTL_{wmvCHR5}$ is tightly located on $QTL_{cmvCHR5}$ and $QTL_{CVYVCHR5}$. Resistance to CGMMV followed polygenic inheritance and was located on chromosome 2 ($QTL_{cgmmvCHR2}$) and chromosome 5 ($QTL_{cgmmvCHR5}$). Interestingly, $QTL_{cmvCHR5}$ and $QTL_{cvyvCHR5}$ are positioned at the same position, and this position is tightly linked to $QTL_{wmvCHR5}$ and $QTL_{cgmmvCHR5}$, leading to the conclusion that this locus could control CMV, CVYV, WMV, and CGMMV resistance. These colocalized QTLs in chromosome 5 are usually defined as resistance hot spots and related to broad-spectrum resistance (Figure 6.3). Resistance to PRSV was governed by three overlapping QTLs mapped on chromosome 6 ($QTL_{prsvCHR6-1}$, $QTL_{prsvCHR6-2}$, and $QTL_{prsvCHR6-1,}$). Eight QTLs for ZYMV resistance were mapped on chromosome 2 ($QTL_{zymvCHR2}$), chromosome 3 ($QTL_{zymvCHR3-1}$ and $QTL_{zymvCHR3-2}$), and chromosome 6 ($QTL_{zymvCHR6-1}$, $QTL_{zymvCHR6-2}$, $QTL_{zymvCHR6-3}$, $QTL_{zymvCHR6-4}$, and $QTL_{zymvCHR6-5}$). Similar to PRSV QTLs on chromosome 6, five QTLs on chromosome 6 for ZYMV resistance are also indistinguishable. Therefore, it proposed two possibilities: they are real QTLs or simply one big QTL that needs to be verified. The second resistance hot spots reside on chromosome 6 (Figure 6.3). The first hotspot (chromosome 5) contained 127 annotated genes. Because ~50 genes have SNP in linkage disequilibrium (LD), they are considered resistance candidate genes. The second spot (chromosome 6) contained 414 annotated genes. Because ~140 genes have SNP in LD, they are considered resistance candidate genes. In conclusion, all currently identified virus resistance QTLs and their physical positions are listed in Figure 6.3.

6.9 FUTURE RESEARCH NEEDED AND CONCLUDING REMARKS

Viral infections are more challenging to treat with conventional chemical treatment than other biotic infections, and they are almost impossible to lessen these threats. To successfully establish disease in host plants, viruses frequently affect their epitope profiles or delay the onset of visible symptoms. In contrast, limited genetic diversity and resistance sources and a lack of compatibility with sexual reproduction severely limit the conventional breeding of cucumbers.

Applying molecular breeding technology to develop virus-resistant cucumber cultivars is advantageous because many genetic and genomic resources are accumulated and available. The small-sized genome of cucumber and the advancement of HTS technologies enable to analyze the genetic diversity, and population structure analysis was successfully applied to various cucumber accessions (Zhu et al. 2016). To promote the molecular breeding of virus-resistant cucumber cultivars, several activities should be followed by developing genome-assisted technologies.

First, we need to extend viral resistance to cucumber accessions. New screening methods are used to select virus resistance cucumber accessions to achieve this objective. The screening method should have an accurate, fast, high-throughput scale. Eventually, morphology-based screening should be replaced by genotype-based screening. Second, mutagenesis should be extensively applied to the cucumber accession to generate and extend new virus resistance cucumber accessions. To achieve this goal, many virus R-genes should be identified. Third, enough molecular markers should be developed based on all available virus resistance cucumber accessions. Fourth, traditional QTL analysis is an inappropriate method to develop many molecular markers in a short period. Although GWAS could be an alternative method for this purpose, more advanced and sophisticated methods

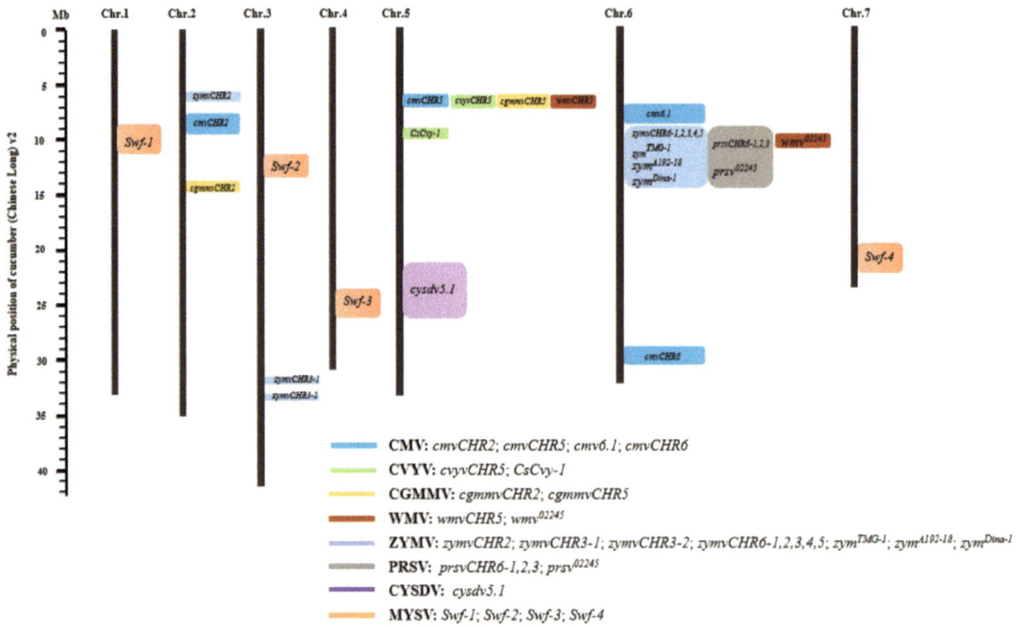

FIGURE 6.3 Physical locations of virus resistance QTLs. Cucumber chromosomes are represented by seven vertical black bars, and the physical position of each QTL is shown on the left scale bar. The virus resistance loci are represented on the right of each cucumber chromosome.

should be followed with the assistance of solid statistics enough to reveal the complexity of viral resistance genetic inheritances. Finally, the new MAS system with high-throughput applicable and low cost should be developed. The new MAS will promote the pyramiding of various virus R-genes into one elite line.

Ideally, molecular breeding would be done at the whole genome level. In addition, a machine learning prediction model and a breeding design scheme that intelligently combines natural variation, artificial variation, and excellent alleles for QTL would be applied. Furthermore, phenotyping technology would be replaced by automated genotyping.

ACKNOWLEDGMENT

This work was supported under the framework of an international cooperation program (2022K2A9A2A06037703) and the basic science research Program (RS-2023-00208020) managed by the National Research Foundation of Korea (NRF).

REFERENCES

Abrahamian PE, Abou-Jawdah Y (2014) Whitefly-transmitted criniviruses of cucurbits: current status and future prospects. *Virusdisease* 25:26–38.

Adams MJ, Antoniw JM, Kreuze J (2009) Virgaviridae: a new family of rod-shaped plant viruses. *Arch* 154(12):1967–72.

Aguilar JM, Abad J, Aranda MA (2006) Resistance to *Cucurbit yellow stunting disorder virus* in cucumber. *Plant Dis* 90(5):583–586.

Ainsworth GC (1935) Mosaic disease of cucumber. *Ann Appl Biol* 22(1):55–67.

Amano M, Mochizuki A, Kawagoe Y, Iwahori K, Niwa K, Svoboda J, Maeda T, Imura Y (2013) High-resolution mapping of *zym*, a recessive gene for *Zucchini yellow mosaic virus* resistance in cucumber. *Theor Appl Genet* 126(12):2983–93.

Attasart P, Charoensilp G, Kertbundit S, Panyim S, Juricek M (2002) Nucleotide sequence of a Thai isolate of *Papaya ringspot virus* type W. *Acta Virol* 46(4):241–6.

Avgelis AD (1985) Epidemiological studies of Zucchini yellow fleck virus in Crete. *Phytopath Mediterr* 24:208–210.

Aydemir I (2009) Determination of genetic diversity in cucumber (*Cucumis sativus* L.) germplasms. Graduate School of Engineering and Sciences of Izmir Institute of Technology, Izmir, Turkey. Thesis.

Barrera-Redondo J, Sánchez-de la Vega G, Aguirre-Liguori JA, Castellanos-Morales G, Gutiérrez-Guerrero YT, Aguirre-Dugua X, Aguirre-Planter E, Tenaillon MI, Lira-Saade R, Eguiarte LE (2021) The domestication of *Cucurbita argyrosperma* as revealed by the genome of its wild relative. *Hortic Res* 8:109.

Bates DM, Merrick LC, Robinson RW (1995) Minor cucurbits. In: Smartt J, Simmonds NW (eds) *Evolution of Crop Plants*, 2nd ed. Longman Scientific, Harlow, Essex, UK. pp 105–111.

Bateson MF, Henderson J, Chaleeprom W, Gibbs AJ, Dale JL (1994) *Papaya ringspot potyvirus*: isolate variability and the origin of PRSV type P (Australia). *J Gen Virol* 75(12):3547–53.

Bayer PE, Golicz AA, Scheben A, Batley J, Edwards D (2020) Plant pan-genomes are the new reference. *Nat Plants* 6(8):914–20.

Bo K, Ma Z, Chen J, Weng Y (2015) Molecular mapping reveals structural rearrangements and quantitative trait loci underlying traits with local adaptation in semi-wild Xishuangbanna cucumber (*Cucumis sativus* L. *var. xishuangbannanesis*). *Theor Appl Genet* 128(1):25–39.

Cardoso AII, Agenor P, Krause-Sakate R, Fattori K (2010) Inheritance of cucumber tolerance to *Zucchini yellow mosaic virus*. *J Plant Pathol* 92:245–248.

Carnide V, Barroso MDR (2006) Las cucurbitaceas: bases para su mejora genetica. *Horti International* 53:16–21.

Chen F, Song Y, Li X, Chen J, Mo L, Zhang X, Lin Z, Zhang L (2019) Genome sequences of horticultural plants: past, present, and future. *Hortic Res* 6:112.

Chen J, Adelberg JW, Staub JE, Skorupska HT, Rhodes BB (1998) A new synthetic amphidiploid in *Cucumis* from *C. sativus* x *C. hytrix*F1 interspecific hybrid. *Cucurbit Genet Coop Rep* 21:336–339.

Chen JF, Isshiki S, Tashiro Y, Miyazaki S (1995) Studies on a wild cucumber from China (*Cucumis hystrix* Chakr.). I. Genetic distances between *C. hystrix* and two cultivated *Cucumis* species (*C. sativus* L. and *C. melo* L.) based on isozyme analysis. *J Jpn Soci Horti Scie* 64:264–265.

Chen JF, Isshiki S, Tashiro Y, Miyazaki S (1997a) Biochemical affinities between *Cucumis hystrix* Chakr. and two cultivated *Cucumis* species (*C. sativus* L. and *C. melo. L.) based on isozyme analysis. Euphytica* 97(2):139–41.

Chen JF, Kirkbride JH (2000) A new synthetic species *Cucumis* (Cucurbitaceae) from interspecific hybridization and chromosome doubling. 52(4):315–9.

Chen JF, Staub JE, Tashiro Y, Isshiki S, Miyazaki S (1997b) Successful interspecific hybridization between *Cucumis sativus* L. and *Cucumis hystrix* Chakr. *Euphytica* 96(3):413–9.

Christenhusz MJM, Byng JW (2016) The number of known plants species in the world and its annual increase. *Phytotaxa* 261(3):201–17.

Cohen D, Gertnan E, Kedar N (1971) Inheritance of resistance to *Watermelon mosaic virus* in cucumber. *Phytopathology* 61:253–255.

Cohen S, Nitzany FE (1960) A whitefly-transmitted virus of cucurbits in Israel. *Phytopathol Mediterr* 1(1):44–46.

Crespo O, Janssen D, Robles C, Ruiz L (2018) Resistance to *Cucumber green mottle mosaic virus* in *Cucumis sativus* L. *Euphytica* 214(11):1–1.

Crow JF (1957) Genetics of insect resistance to chemicals. *Annu Rev Entomol* 2(1):227–46.

Da-Costa CP, Jones CM (1971) Cucumber beetle resistance and mite susceptibility controlled by the bitter gene in *Cucumis sativus* L. *Science* 172(3988):1145–1146.

Das A, Singh S, Islam Z, Munshi AD, Behera TK, Dutta S, Weng Y, Dey SS (2022) Current progress in genetic and genomics-aided breeding for stress resistance in cucumber (*Cucumis sativus* L.). *Sci Hortic* 300:111059.

Dash L, Tripathy B, Rout S (2022) Breeding for pest resistance in vegetable crops: A Review. *Int J of Mech Engi* 7:0974–5823.
de Ronde D, Butterbach P, Kormelink R (2014) Dominant resistance against plant viruses. *Front Plant Sci* 5:307.
de Ruiter W, Hofstede R, de Vries J, van den Heuvel H (2008) Combining QTL for resistance to CYSDV and powdery mildew in a single cucumber line. In: Pitrat M (ed) *Proceedings of 9th EUCARPIA Meeting on Genetics and Breeding of INRA*. Avignon, France, pp 181–188.
Den-Nijs APM, Custers JBM (1990) Introducing resistances into the cucumber by interspecific hybridization. In: Bates DM, Robinson RW, Jeffrey C (eds) *Biology and Utilization of the Cucurbitaceae*. Comstock Publishing Associates, Ithaca, NY, pp 382–396.
Desbiez C (2020) The never-ending story of cucurbits and viruses. Acta Hortic 1294:173–192.
Desbiez C, Lecoq H (1997) *Zucchini yellow mosaic virus*. *Plant Pathol* 46(6):809–29.
De-Wit PJGM (1977) A light and scanning-electron microscopic study of infection of tomato plants by virulent and *avirulent* races of Cladosporium fulvum. *Neth J Plant Pathol* 83(3):109–22.
Dey SS, Singh S, Munshi AD, Behera TK (2022) Classical Genetics and Traditional Breeding. In: Pandey S, Weng Y, Behera TK, Bo K (eds) *The Cucumber Genome. Compendium of Plant Genomes*. Springer, Cham, Switzerland, pp 159–83.
Dhillon NPS (1992) New sources of *Cucumber Mosaic Virus* tolerant land races of cucumber identified at the centre of origin (India). *Plant Breed* 109(2):172–4.
Diaz A, Fergany M, Formisano G, Ziarsolo P, Blanca J, Fei Z, Staub JE, Zalapa JE, Cuevas HE, Dace G, Oliver M (2011) A consensus linkage map for molecular markers and Quantitative Trait Loci associated with economically important traits in melon (*Cucumis melo* L.). *BMC Plant Biol* 11(1):1–14.
Dijkhuizen A, Kennard WC, Havey MJ, Staub JE (1996) RFLP variability and genetic relationships in cultivated cucumber. *Euphytica* 90(1):79–87.
Dombrovsky A, Tran-Nguyen LT, Jones RA (2017) *Cucumber green mottle mosaic virus*: rapidly increasing global distribution, etiology, epidemiology, and management. *Ann Rev Phytopathol* 4; 55:231–56.
Doolittle SP (1916) A new infectious mosaic disease of cucumber. *Phytopathology* 6:145–147.
Doolittle SP, Beecher FS, Porte WS (1939) A hybrid cucumber resistant to bacterial wilt. *Phytopathol* 29:996–998.
Duffus JE (1995) Whitefly-transmitted yellowing viruses of the Cucurbitaceae. In: Lester GE, Dunlap Jr (eds) *Cucurbitaceae 94: Evaluation and Enhancement of Cucurbit Germplasm.* 1–4 Nov., 1994. South Padre Island, Texas. Gateway Printing, Edinburg, TX, pp 12–16.
Eid S, Abou-Jawdah Y, El-Mohtar C, Sobh H, Havey M (2006) Tolerance in cucumber to Cucurbit yellow stunting disorder virus. *Plant Dis* 90(5):645–9.
Faber NM, Torres LM, Sanchez LO (2008) Method of Breeding *Cysdv*-Resistant Cucumber Plants (European Patent No. #EP09775344A) European Patent Office. www.epo.org
FAOSTAT (2018) www.fao.org/faostat/en/#data/QCL
Fischer HU, Lockhart BEL (1974) Serious losses in cucurbits caused by *watermelon mosaic virus* in Morocco. *Plant Dis Rep* 58:143–146.
Gabriel DW, Rolfe BG (1990) Working models of specific recognition in plant–microbe interactions. *Annu Rev Phytopathol* 28:365–391.
Gal-On A (2007) *Zucchini yellow mosaic virus*: insect transmission and pathogenicity-the tails of two proteins. *Mol Plant Pathol* 8:139–150.
Gao L, Gonda I, Sun H, Ma Q, Bao K, Tieman DM, Burzynski-Chang EA, Fish TL, Stromberg KA, Sacks GL, Thannhauser TW (2019) The tomato pan-genome uncovers new genes and a rare allele regulating fruit flavor. *Nat Genet* 51(6):1044–1051.
Garcia-Mas J, Benjak A, Sanseverino W, Bourgeois M, Mir G, Gonzalez VM, Henaff E, Camara F, Cozzuto L, Lowy E (2012) The genome of melon (*Cucumis melo* L.). *Proc Natl Acad Sci USA* 109:11872–11877.
Gilbert-Albertini F, Pitrat M, Leco H (1995) Inheritance of Resistance to *Zucchini Yellow Fleck Virus* in *Cucumis sativus* L. *HortScience* 30(2):336–337.
Gil-Salas FM, Colyer A, Boonham N, Cuadrado IM, Janssen D (2009) Resistance screening against *Cucumber vein yellowing virus* using a real-time (Taqman®) RT-PCR assay in cucumber (*Cucumis sativus* L.). *Crop Protect* 28(1):109–112.
Golicz AA, Batley J, Edwards D (2016) Towards plant pangenomics. *Plant Biotechnol J* 14(4):1099–1105.
Gonsalves D (1998) Control of Papaya ringspot virus in papaya: a case study. *Ann Rev Phytopathol* 36:415–437.

Gonzalo MJ, Diaz A, Dhillon NPS, Reddy UK, Pico B, Monforte AJ (2019) Re-evaluation of the role of Indian germplasm as center of melon diversification based on genotyping-by-sequencing analysis. *BMC Gen* 20:448.

Grumet R, Kabelka E, McQueen S, Wal T, Humphrey R (2000) Characterization of sources of resistance to the watermelon strain of *Papaya ringspot virus* in cucumber: Allelism and co-segregation with other potyvirus resistances. *Theor Appl Genet* 101(3):463–472.

Grumet R, McCreight JD, McGregor C, Weng Y, Mazourek M, Reitsma K, Labate J, Davis A, Fei Z (2021) Genetic Resources and Vulnerabilities of Major Cucurbit Crops. *Genes (Basel)* 12(8):1222.

Guo S, Zhang J, Sun H, Salse J, Lucas WJ, Zhang H, Zheng Y, Mao L, Ren Y, Wang Z (2013) The draft genome of watermelon (*Citrullus lanatus*) and re-sequencing of 20 diverse accessions. *Nat Genet* 45:51–58.

Guo S, Zhao S, Sun H, Wang X, Wu S, Lin T Ren Y, Gao L, Deng Y, Zhang J, Lu X, Zhang H, Shang J, Gong G, Wen C, He N, Tian S, Li M, Liu J, Wang Y, Zhu Y, Jarret R, Levi A, Zhang X, Huang S, Fei Z, Liu W, Xu Y (2019) Resequencing of 414 cultivated and wild watermelon accessions identifies selection for fruit quality traits. *Nat Genet* 51:1616–1623.

Hashemi L, Golparvar AR, Nasr-Esfahani M, Golabadi M (2020) Expression analysis of defense-related genes in cucumber (*Cucumis sativus* L.) against Phytophthora melonis. *Mol Biol Rep* 47(7):4933–4944.

Havey MJ (1997) *CMV* resistance in cucumber. A correction. *Cucurbit Genet Coop Rep* 20:9.

Hayja AZ, Al-Shahwan IM (1991) Inheritance of resistance to *zucchini yellow mosaic virus* in cucumber. *Zeitschrift für Pflanzenkrankheiten und Pflanzenschutz* 98(3):301–4.

Heath MC (2000) Nonhost resistance and nonspecific plant defenses. *Curr Opin Plant Biol* 3(4):315–319.

Horejsi T, Staub JE (1999) Genetic variation in cucumber (*Cucumis sativus* L.) as assessed by random amplified polymorphic DNA. *Genet Resour Crop Evol* 46(4):337–350.

Horst EK, Lower RL (1978) *Cucumis hardwickii*, a source of germplasm for the cucumber breeder. *Cucurbit Genet Coop Rep* 1:5.

Huang HH (2007) Studies on Molecular Maker of *Cucumber Mosaic Virus* Resistance-Related Gene in Cucumber (*Cucumis Sativus* L.). Northwest A&F University, Xianyang, Shaanxi, China. Thesis.

Huang S, Li R, Zhang Z, Li LI, Gu X, Fan W, Lucas WJ, Wang X, Xie B, Ni P, Ren Y (2009) The genome of the cucumber, *Cucumis sativus* L. *Nat Genet* 41(12):1275–1281.

Igarashi I, Iwanaga Y, Kawaide T (1986) Studies on the breeding of cucumber cultivars resistance to *Cucumber mosaic virus*. II. Variental differences and resistance of F1 hybrids. Bull Veg Ornament. *Crops Res Stn A* 14:35–41.

Innark P, Khanobdee C, Samipak S, Jantasuriyarat C (2013) Evaluation of genetic diversity in cucumber (*Cucumis sativus* L.) germplasm using agro-economic traits and microsatellite markers. *Sci Hortic* 162:278–84.

Jagger IC (1916) Experiments with the cucumber mosaic disease. *Phytopathology* 6:148–151.

Jat GS, Behera TK, Lata S, Kumar S (2021) Classical Genetics and Traditional Breeding in Cucumber (*Cucumis sativus* L.). In: Wang H (ed) *Cucumber Economic Values and Its Cultivation and Breeding*. IntechOpen, London, UK, pp 201–215.

Jones RA (2016) Future Scenarios for Plant Virus Pathogens as Climate Change Progresses. *Adv Virus Res* 95:87–147.

Kabelka E, Grumet R (1997) Inheritance of resistance to the *Moroccan watermelon mosaic virus* in the cucumber line TMG-1 and cosegregation with *Zucchini yellow mosaic virus resistance*. *Euphytica* 95(2):237–242.

Kabelka E, Ullah Z, Grumet R (1997) Multiple alleles for *zucchini yellow mosaic virus resistance* at the *zym* locus in cucumber. *Theor Appl Genet* 95(5):997–1004.

Kao J, Jia L, Tian T, Rubio L, Falk BW (2000) First report of *Cucurbit yellow stunting disorder virus* (genus Crinivirus) in *North America Plant Dis* 84(1):101.

Katul L, Makkouk KM (1987) Occurrence and serological relatedness of five cucurbit potyviruses in Lebanon and Syria. *Phytopath Mediterr* 26:36–42.

Kaushik U, Aeri V, Mir SR (2015) Cucurbitacins-An insight into medicinal leads from nature. *Pharmacogn Rev* 9(17):12–18.

Keen NT (1990) Gene-for-gene complementarity in plant–pathogen interactions. *Annu Rev Genet* 24:447–63.

Keng H (1974) Economic plants of ancient north China as mentioned in Shih Ching (Book of Poetry). *Econom Botany* 28(4):391–410.

Kho YO, Den Nijs APM, Franken J (1980) Interspecific hybridization in *Cucumis* L. II. The crossability of species, an investigation of in vivo pollen tube growth and seed set. *Euphytica* 29:661–672.

Kirkbride JH (1993) *Biosystematic Monograph of the Genus* Cucumis *(Cucurbitaceae): Botanical Identification of Cucumbers and Melons.* Parkway Publishers, Boone, North Carolina.

Koboldt DC, Steinberg KM, Larson DE, Wilson RK, Mardis E (2013) The next-generation sequencing revolution and its impact on genomics. *Cell* 155(1):27–38.

Kooistra E (1969) The inheritance of resistance to *Cucumis virus 1* in cucumber (*Cucumis sativus* L.). *Euphytica* 18(3):326–332.

Lecoq H, Desbiez C (2008) Watermelon mosaic virus and Zucchini yellow mosaic virus. *Encyclopedia of Virology* 5:433–440.

Lecoq H, Desbiez C (2012) Viruses of cucurbit crops in the Mediterranean region: an ever-changing picture. *Adv Virus Res* 84:67–126.

Lecoq H, Desbiez C, Delecolle B, Cohen S, Mansour A (2000) Cytological and molecular evidence that the whitefly-transmitted cucumber vein yellowing virus is a tentative member of the family Potyviridae. *J Gen Virol* 81(9):2289–2293.

Lei L, Goltsman E, Goodstein D, Wu GA, Rokhsar DS, Vogel JP (2021) Plant pan-genomics comes of age *Annu Rev Plant Biol* 72(1):411–435.

Leibman D, Kravchik M, Wolf D, Haviv S, Weissberg M, Ophir R, Paris HS, Palukaitis P, Ding SW, Gaba V, Gal-On A (2018) Differential Expression of Cucumber RNA-Dependent RNA polymerase 1 Genes during Antiviral Defence and Resistance. *Mol Plant Pathol* 19(2):300–312.

Leibman D, Wolf D, Saharan V, Zelcer A, Arazi T, Yoel S, Gaba V, Gal-On A (2011) A High Level of Transgenic Viral Small RNA Is Associated with Broad Potyvirus Resistance in Cucurbits. *Mol Plant-Microbe Interact* 24(10):1220–38.

Lewandowski DJ (2008) Tobammovirus. In: Mahy BWJ, van Regenmortel MHV (eds) *Encyclopedia of Virology* (Third Edition). Academic Press, San Diego, CA, pp 68–72.

Li H, Wang S, Chai S, Yang Z, Zhang Q, Xin H, Xu Y, Lin S, Chen X, Yao Z, Yang Q (2022) Graph-based pan-genome reveals structural and sequence variations related to agronomic traits and domestication in cucumber. *Nat Commun* 13(1):1–4.

Li Q, Li H, Huang W, Xu Y, Zhou Q, Wang S, Ruan J, Huang S, Zhang Z (2019) A chromosome-scale genome assembly of cucumber (*Cucumis sativus* L.), *GigaScience* 8(6):1–10.

Liu HW, Luo LX, Li JQ, Liu PF, Chen XY, Hao JJ (2014) Pollen and seed transmission of *Cucumber green mottle mosaic virus* in cucumber. *Plant Pathol* 63(1):72–77.

Lv J, Qi J, Shi Q, Shen D, Zhang S, Shao G, Li H, Sun Z, Weng Y, Shang Y, Gu X (2012) Genetic Diversity and Population Structure of Cucumber (*Cucumis sativus L.*). *PLoS ONE* 7(10):e46919.

Ma L, Wang Q, Mu Fu A, Wen C, Zhao X, Gao L, Li J, Sha K, Wang Y, Zhang X, Zhang X, Fei Z, Grierson D, Zuo J (2020) The genome and transcriptome analysis of snake gourd provide insights into its evolution and fruit development and ripening. *Hortic Res* 7:199.

Ma L, Wang Q, Zheng Y, Guo J, Yuan S, Fu A, Bai C, Zhao X, Zheng S, Wen C, Guo S (2022) Cucurbitaceae genome evolution, gene function, and molecular breeding. Hortic Res 9:uhab057.

Marathe R, Guan Z, Anandalakshmi R, Zhao H, Dinesh-Kumar SP (2004) Study of Arabidopsis thaliana resistome in response to Cucumber mosaic virus infection using whole genome microarray. *Plant Mol Biol* 55:501–520.

Martelli GP, Agranovsky AA, Bar-Joseph M, Boscia D, Candresse T, Coutts RHA, Dolja VV, Duffus JE, Falk BW, Gon-salves D, Jelkmann W, Karasev AV, Mi-nafra A, Murant A, Namba S, Niblett CL, Vetten HJ, Yoshikawa N (2000) Family Closteroviridae. In: Van-Regenmortel MHV, Fouquet CM, Bishop DHL, Carstens E, Estes MK, Lemon SM, Maniloff J, Mayo MA, McGeoch DJ, Pringle CR, Wickner RB (eds) *Virus Taxonomy. Seventh Report of the International Committee on Taxonomy of Viruses.* Academic Press, San Diego, CA, pp 943–952.

Martelli GP, Agranovsky AA, Bar-Joseph M, Boscia D, Candresse T, Coutts RH, Dolja VV, Falk BW, Gonsalves D, Jelkmann W, Karasev AV, Minafra A, Namba S, Vetten HJ, Wisler GC, Yoshikawa N (2002) ICTV Study Group on closteroviruses and allied viruses. The family Closteroviridae revised. Arch Virol 147(10):2039–44.

Martin-Hernandez AM, Pico B (2021) Natural resistances to viruses in cucurbits. *Agron* 11(1):23.

Monnot S, Cantet M, Mary-Huard T, Moreau L, Lowdon R, Van Haesendonck M, Ricard A, Boissot N (2022) Unravelling cucumber resistance to several viruses via genome-wide association studies highlighted resistance hotspots and new QTLs. Hortic Res 9:uhac184.

Montero-Pau J, Blanca J, Bombarely A, Ziarsolo P, Esteras C, Martin-Gomez C, Ferriol M, Gomez P, Jamilena M, Mueller L (2018) De novo assembly of the zucchini genome reveals a whole-genome duplication associated with the origin of the Cucurbita genus. *Plant Biotechnol J* 16(6):1161–1171.

Moriones E, Praveen S, Chakraborty S (2017) Tomato leaf curl New Delhi virus: an emerging virus complex threatening vegetable and fiber crops. *Viruses* 9(10):264.

Munger HM, Newhall AG (1953) Breeding for disease resistance in celery and cucurbits. *Phytopathology* 43:254–259.

Munshi AD, Panda B, Behera TK, Kumar R (2007) Genetic variability in *Cucumis sativus* var. *hardwickii* R. *(Alef.) germplasm. Cucurbit Genet Coop Rep* 30:5–10.

Munshi AD, Panda B, Mandal B, Bisht IS, Rao ES, Kumar R (2008) Genetics of resistance to *Cucumber mosaic virus* in *Cucumis sativus* var. *hardwickii* R. Alef. Euphytica 164(2):501–507.

Naegele RP, Wehner TC (2016) Genetic Resources of Cucumber. In: Grumet R, Katzir N, Garcia-Mas J (eds) *Genetics and Genomics of Cucurbitaceae. Plant Genetics and Genomics: Crops and Models, vol 20.* Springer, Cham, Switzerland, pp 61–86.

Navas-Castillo J, Lopez-Moya JJ, Aranda MA (2014) Whitefly-transmitted RNA viruses that affect intensive vegetable production. *Ann Appl Biol* 165(2):155–171.

Nerkar G, Devarumath S, Purankar M, Kumar A, Valarmathi R, Devarumath R, Appunu C (2022) Advances in crop breeding through precision genome editing. *Front Genet* 13:880195.

Nitzany FE (1970) *Virus Diseases of Annual Crops in Israel.* Volcani Institute of Agr. Res., Virus Res. Unit Bet Dagan (Suppl. To 18th Int. Hort. Congr.), pp 1–71.

Nitzany FE (1975) Review of vegetable virus diseases in Israel and their control. *FAO Plant Prot Bull* 23:148–154.

Oerke EC (1994) Estimated crop losses due to pathogens, animal pests and weeds. In: Oerke EC, Dehne HW, Schonbeck F, Weber A (eds) *Crop Production and Crop Protection Estimated Losses in Major Food and Cash Crops.* Elsevier, Amsterdam, The Netherlands, pp 72–741.

Oerke EC, Dehne HW (2004) Safeguarding production-losses in major crops and the role of crop protection. *J Crop Prot* 23(4):275–278.

Oren E, Dafna A, Tzuri G, Halperin I, Isaacson T, Elkabetz M, Meir A, Saar U, Ohali S, Thuy La T, Romay C, Tadmor Y, Schaffer AA, Buckler ES, Cohen R, Burger J, Gur A (2022) Melon pan-genome and multi-parental framework for high-resolution trait dissection. BioRxiv 2022:503186.

Osipowski P, Pawełkowicz M, Wojcieszek M, Skarzynska A, Przybecki Z, Pląder W (2020) A high-quality cucumber genome assembly enhances computational comparative genomics. *Mol Genet Genomics* 295(1):177–193.

Pan RS, More TA (1996) Screening of melon (Cucumis melo L.) germplasm for multiple disease resistance. *Euphytica* 88, 125–128.

Pandey S, Ansari WA, Mishra VK, Singh AK, Singh M (2013) Genetic diversity in Indian cucumber based on microsatellite and morphological markers. *Biochem Syst Ecol* 51:19–27.

Pandey S, Kujur SN (2022) Importance, distribution, botany, and genetics. In: Pandey S, Weng Y, Behera TK, Bo K (eds) *The Cucumber Genome. Compendium of Plant Genomes.* Springer, Cham, Switzerland, pp 1–13.

Paris HS (2008) Summer squash. In: Prohens J, Nuez F (eds) *Vegetables I. Handbook of Plant Breeding.* Volume 3. Springer, New York, NY, pp 351–375.

Paris HS, Daunay MC, Janick J (2012) Occidental diffusion of cucumber (*Cucumis sativus*) 500–1300 CE, two routes to Europe. *Ann Bot* 109:117–126.

Park YH, Sensoy S, Wye C, Antonise R, Peleman J, Havey MJ (2000) A genetic map of cucumber composed of RAPDs, RFLPs, AFLPs, and loci conditioning resistance to *papaya ringspot* and *zucchini yellow mosaic viruses. Genome* 43(6):1003–10.

Pawelkowicz M, Zeilinski K, Zeilinski D, Plader W, Yagi K, Wojcieszek M (2016) Next generation sequencing and omics in cucumber (*Cucumis sativus* L.) breeding directed research. *Plant Sci* 242:77–88.

Pico B, Sifres A, Martinez-Perez E, Leiva-Brondo M, Nuez F (2008) Genetics of the resistance to *CVYV* in cucumber. In *Modem Variety Breeding for Present and Future Needs.* Proceedings of the 18th EUCARPIA General Congress, Valencia, Spain, pp 452–456.

Pico B, Villar C, Nuez F, Weber WE (2003) Screening *Cucumis sativus* landraces for resistance to *cucumber vein yellowing virus. Plant Breed* 122:426–430.

Pitrat M (2016) Disease resistance in melon and its modification by molecular breeding techniques. In: Ezura H, Ariizumi T, Garcia-Mas J, Rose J (eds) *Functional Genomics and Biotechnology in Solanaceae and Cucurbitaceae Crops.* Springer, Berlin/Heidelberg, Germany, pp 175–197.

Porter RH (1929) Reaction of Chinese cucumbers to mosaic. *Phytopathology* 19:85

Porter, RH (1932) The reaction of cucumbers to types of mosaic. *Iowa State Coll Jour Sci* 6:95–129.

Provvidenti R (1985) Sources of resistance to viruses in two accessions of *Cucumis sativus. Rep Cucurbit Genet Coop USA* 8:9–12.

Provvidenti R (1987) Inheritance of resistance to a strain of *Zucchini yellow mosaic virus* in cucumber. *Hort Science* 22(1):102–103.

Pujol M, Alexiou KG, Fontaine AS, Mayor P, Miras M, Jahrmann T, Garcia-Mas J, Aranda MA (2019) Mapping *cucumber vein yellowing virus* resistance in cucumber (*Cucumis sativus* L.) by Using BSA-seq analysis. *Front Plant Sci* 10:1583.

Qi J, Liu X, Shen DI, Miao H, Xie B, Li X, Zeng P, Wang S, Shang Y, Gu X, Du Y (2013) A genomic variation map provides insights into the genetic basis of cucumber domestication and diversity. *Nat genet* 45(12):1510–1515.

Qin X, Zhang Z, Lou Q, Xia L, Li J, Li M, Zhou J, Zhao X, Xu Y, Li Q, Yang S (2021) Chromosome-scale genome assembly of *Cucumis hystrix*-a wild species interspecifically cross-compatible with cultivated cucumber. *Hortic Res* 8(1):1–4.

Raamsdonk LWD, den-Nijs APM, Jongerius MC (1989) Meiotic analyses of *Cucumis* hybrids and an evolutionary evaluation of the genus *Cucumis* (Cucurbitaceae). *Plant Syst Evol* 163:133–146.

Radouane N, Ezrari S, Belabess Z, Tahiri A, Tahzima R, Massart S, Jijakli H, Benjelloun M, Lahlali R (2021) Virus od cucurbit crops: current status in the Mediterranean Region. *Phytopathol Mediterr* 60(3):493–519.

Rajamony L, More TA, Seshadri VS (1987) Resistance to *Cucumber green mottle mosaic virus* (CGMMV) in muskmelon. *Cucurbit Gen Coop Rep* 10:58–59.

Rajamony L, More TA, Seshadri VS (1990b) Inheritance of resistance to *Cucumber green mottle mosaic virus* in muskmelon (*Cucumis melo* L.). *Euphytica* 47(2):93–97.

Rajamony L, More TA, Seshadri VS, Varma A (1990a) Reaction of muskmelon collections to *Cucumber green mottle mosaic virus*. *J Phytopathol* 129:237–244.

Ramirez-Madera AO, Havey MJ (2017) Different Haplotypes Encode the Same Protein for Independent Sources of *Zucchini Yellow Mosaic Virus* Resistance in Cucumber. *HortScience* 52(8):1040.

Ranjan KN, Rai AB, Rai M (2008) Export of cucumber and gherkin from India: performance, destinations, competitiveness, and determinants. *Agric Econ Res Rev* 21:130–8.

Ren Y, Zhang Z, Liu J, Staub JE, Han Y, Cheng Z, Li X, Lu J, Miao H, Kang H, Xie B (2009) An integrated genetic and cytogenetic map of the cucumber genome. *PloS one* 4(6):e5795.

Risser G, Pitrat M, Rode J C (1977) Etude de la résistance du melon (*Cucumis melo* L.) au virus de la mosaïque du concombre. *Ann Amél Plant* 27:509–522.

Rolnik A, Olas B (2020) Vegetables from the Cucurbitaceae family and their products: Positive effect on human health. *Nutrition* 78:110788.

Romay G, Pitrat M, Lecoq H, Wipf-Scheibel C, Millot P, Girardot G, Desbiez C (2019) Resistance against *melon chlorotic mosaic virus* and *tomato leaf curl New Delhi virus* in melon. *Plant Dis* 103(11): 2913–9.

Ruiz L, Lopez C, Pico B, Janssen D (2021) Resistance to *Cucumber Green Mottle Mosaic Virus* in *Cucumis melo. Plants (Basel)* 10(6):1077.

Sastry KS, Zitter TA (2014) Management of Virus and Viroid Diseases of Crops in the Tropics. In: Sastry KS, Zitter TA (eds) *Plant Virus and Viroid Diseases in the Tropics*. Springer, Dordrecht, The Netherlands, pp 149–480.

Schaible L (1954) *Some Aspects of Resistance to Cucumber Virus 1 in Cucumber and Squash*. Thesis. Cornell University, Ithaca, New York.

Seminis (2015) *Cucurbit Disease Guide-Seminis. A Disease Reference Guide for Cucumber, Melon, Squash, and Watermelon*. Agri Expo, South Africa, pp. 66–92. https://pdf.agriexpo.online/

Sharma BD, Hore DK (1996) Indian cucumber germplasm and challenges ahead. *Genet Resour Crop Evol* 43(1):7–12.

Shi L, Yang Y, Xie Q, Miao H, Bo K, Song Z, Wang Y, Xie B, Zhang S, Gu X (2018) Inheritance and QTL mapping of *cucumber mosaic virus resistance* in cucumber (*Cucumis Sativus* L.). *PLoS ONE* 13(7):e0200571.

Shifriss O, Myers CH, Chupp C (1942) Resistance to *mosaic virus* in the cucumber. *Phytopathology* 32:773–784.

Singh SJ, Verma R, Ahlawat YS, Singh RK, Prakash S, Pant RP (2003) Natural occurrence of a yellow mosaic' disease on zucchini in India caused by a potyvirus. *Indian Phytopathol* 56:174–179.

Staub JE, Bacher J (1997) Cucumber as a processed vegetable. In: Smith DS, Cash JN, Nip WK, Hui YH (eds) *Processing Vegetables: Science and Technology IV*. Tech-nomic Publishing Co., Inc., Lancaster, PA, pp. 129–193.

Staub JE, Knerr LD, Holder DJ, May B (1992) Phylogenetic relationships among several African *Cucumis* species. *Canad J Bot* 70:509–517.

Staub JE, Palmer MJ (1987) Resistance to Downy Mildew [*Pseudoperonospora cubensis* (Berk. & Curt.) Rostow.] and Scab (Spot Rot) [*Cladosporium cucumerinum* Ellis & Arthur] in *Cucumis* spp. Cucurbit Genet Coop Rept 10:21–23.

Staub JE, Robbins MD, Wehner TC (2008) Cucumber. In: Prohens J, Nuez F (eds) *Handbook of Plant Breeding; Vegetables I: Asteraceae, Brassicaceae, Chenopodiaceae, and Cucurbitaceae*. Springer, New York, NY, pp 241–282.

Staub JE, Serquen FC, Horejsi T, Chen J (1999) Genetic diversity in cucumber (*Cucumis sativus* L.): IV. An evaluation of Chinese germplasm. *Genet Resour Crop Evol* 46(3):297–310.

Staub JE, Serquen FC, McCreight JD (1997) Genetic diversity in cucumber (*Cucumis sativus* L.): III. An evaluation of Indian germplasm. *Genet Resour Crop Evol* 44(4):315–26.

Sugiyama M (2013) The present status of breeding and germplasm collection for resistance to viral diseases of cucurbits in Japan. *J Jpn Soc Hortic Sci* 82(3):193–202.

Sugiyama M, Kawazu Y, Fukino N, Yoshioka Y, Shimomura K, Sakata Y, Okuda M (2015) Mapping of quantitative trait loci for *Melon yellow spot virus* resistance in cucumber (*Cucumis sativus* L.). *Euphytica* 205(2):615–625.

Sugiyama M, Okuda M, Sakata Y (2006) A new source of resistance to Cucumber green mottle mosaic virus in melon. *J Jpn Soc Hortic Sci* 75:469–475.

Sugiyama M, Okuda M, Sakata Y (2007) Inheritance of resistance to Cucumber green mottle mosaic virus in Cucumis melo L. *"Chang Bougi". J Jpn Soc Hortic Sci* 76:316–318.

Sugiyama M, Okuda M, Sakata Y (2009) Evaluation of resistance to *melon yellow spot virus* in a cucumber germplasm collection. *Plant Breed* 128(6):696–700.

Sun H, Wu S, Zhang G, Jiao C, Guo S, Ren Y, Zhang J, Zhang H, Gong G, Jia Z, Zhang F, Tian J, Lucas WJ, Doyle JJ, Li H, Fei Z, Xu Y (2017). Karyotype stability and unbiased fractionation in the paleo-allotetraploid Cucurbita genomes. *Mol Plant* 10(10):1293–1306.

Sun J, Lv D, Chen Y, Pan J, Cai R, Pan J (2022a) QTL Mapping for Disease Resistance in Cucumber. In: Pandey S, Weng Y, Behera TK, Bo K (eds) *The Cucumber Genome. Compendium of Plant Genomes*. Springer, Cham, Switzerland, pp 81–92.

Sun J, Zhang Z, Zong X, Huang S, Li Z, Han Y (2013b) A high-resolution cucumber cytogenetic map integrated with the genome assembly. *BMC Genomics* 14(1):1–8.

Sun X, Liu D, Zhang X, Li W, Liu H, Hong W, Jiang C, Guan N, Ma C, Zeng H, Xu C (2013a) SLAF-seq: an efficient method of large-scale de novo SNP discovery and genotyping using high-throughput sequencing. *PLoS ONE* 8 (3):e58700.

Sun Y, Wang J, Li Y, Jiang B, Wang X, Xu W, Wang Y, Zhang P, Zhang Y, Kong X (2022b) Pan-genome analysis reveals the abundant gene presence/absence variations among different varieties of melon and their influence on traits. *Front Plant Sci* 13:835496.

Svoboda J, Leisova-Svobodova L, Amano M (2013) Evaluation of Selected Cucurbitaceous Vegetables for Resistance to *Zucchini yellow mosaic virus*. *Plant Dis* 97(10):1316–1321.

Takeuchi S, Okuda M, Hanada K, Kawada Y, Kameya-Iwaki M (2001) Spotted disease of cucumber (*Cucumis sativus*) caused by *Melon yellow spot virus*. *Jpn J Phytopathol* 67(1):46–51.

Tao Y, Zhao X, Mace E, Henry R, David Jordan D (2019) Exploring and Exploiting Pan-genomics for Crop Improvement. *Mo Plant* 12(2):156–169.

Tian G, Miao H, Yang Y, Zhou J, Lu H, Wang Y, Xie B, Zhang S, Gu X (2016) Genetic analysis and fine mapping of *Watermelon mosaic virus* resistance gene in cucumber. *Mol Breed* 36(9):131.

Tian G, Yang Y, Zhang S, Miao H, Lu H, Wang Y, Xie B, Gu X (2015) Genetic analysis and gene mapping of *papaya ring spot virus* resistance in cucumber. *Mol Breed* 35(4):110.

Van der Meer FW, Garnett HM (1987) Purification and identification of a South African isolate of watermelon mosaic virus–Morocco. Journal of *Phytopathol* 120(3):255–270.

Van Koot Y, Van Dorst HJM (1959) Virusziekten van komkommer in Nederland (with a summary: virus diseases of cucumber in the Netherlands). *Tijdschrift Over Plantenziekten* 65(6):257–71.

Velasco L, Ruiz L, Galipienso L, Rubio L, Janssen DA (2020) Historical Account of Viruses in Intensive Horticultural Crops in the Spanish Mediterranean Arc: New Challenges for a Sustainable Agriculture. *Agronomy* 10(6):860.

Vovlas C, Hiebert E, Russo M (1981) *Zucchini yellow fleck virus*, a new potyvirus of Zucchini squash. *Phytopath Mediterr* 20:123–128.

Wal T, Grumet R (1995) Inheritance of Resistance to *Watermelon Mosaic Virus* in the Cucumber Line TMG-1: Tissue-Specific Expression and Relationship to *Zucchini Yellow Mosaic Virus Resistance. Theor Appl Genet* 91(4):699–706.

Wal T, Staub JE, Kabelka E, Grumet R (1997) Linkage analysis of potyvirus resistance alleles in cucumber. *J Heredity* 88(6):454–458.

Wan H, Yuan W, Bo K (2013) Genome-wide analysis of NBS-encoding disease resistance genes in *Cucumis* sativus and phylogenetic study of NBS-encoding genes in Cucurbitaceae crops. *BMC Genomics* 14(1):109.

Wang C, Han J, Wang T, Chen C, Liu J, Xu Z, Zhang Q, Wang L, Ren Z (2022) Pan-Genome-wide identification and transcriptome-wide analysis of *DREB* genes that respond to biotic and abiotic stresses in cucumber. *Agriculture* 12(11):1879.

Wang HJ, Wu Y, Gu W, Sun XD, Qin ZW (2006) Extraction of DNA from cucumber by improved CTAB method. *Hei-longjiang Agric Sci* 5:124–125.

Wang JH (2010) Construction of a genetic linkage map and QTL analysis of the anti-*CMV* trait in cucumber (*Cucumis sativus* L.). Thesis. Northwest A&F University, Xianyang, Shaanxi, China.

Wang X, Bao K, Reddy UK, Bai Y, Hammar SA, Jiao C, Wehner TC, Ramirez-Madera AO, Weng Y, Grumet R (2018) The USDA cucumber (*Cucumis sativus* L.) collection: Genetic diversity, population structure, genome-wide association studies, and core collection development. *Hortic Res* 5:64.

Wang Y, Bo K, Gu X, Pan J, Li Y, Chen J, Wen C, Ren Z, Ren H, Chen X (2020) Molecularly tagged genes and quantitative trait loci in cucumber with recommendations for QTL nomenclature. *Hortic Res* 7:3.

Wang YJ, Provvidenti R, Robinson RW (1984) Inheritance of resistance to *watermelon mosaic 1 virus* in cucumber. *HortScience* 19(4):587–588.

Wasuwat SL, Walker J (1961) Inheritance of resistance in cucumber to *Cucumber mosaic virus. Phytopathology* 51:423–428.

Wei Q, Wang Y, Qin X, Zhang Y, Zhang Z, Wang J, Li J, Lou Q, Chen J (2014) An SNP-based saturated genetic map and QTL analysis of fruit-related traits in cucumber using specific-length amplified fragment (SLAF) sequencing. *BMC Genomics* 15:1158.

Weng Y (2021) *Cucumis sativus* Chromosome Evolution, Domestication, and Genetic Diversity. In: Goldman I (ed) *Plant Breed. Review* (Vol. 44). Wiley, NJ, pp 79–111.

Whitaker TW, Davis GN (1962) Cucurbits: Botany, cultivation, and utilization. *J Assoc Off Anal Chem* 45(4):1052.

Wu S, Shamimuzzaman S, Sun H, Salse J, Sui X, Wilder A, Wu ZJ, Levi A, Xu Y, Ling KS, Fei Z (2017) The bottle gourd genome provides insights into Cucurbitaceae evolution and facilitates mapping of a Papaya ring-spot virus resistance locus. *Plant J* 92(5):963–975.

Wu S, Wang X, Reddy U, Sun H, Bao K, Gao L, Mao L, Patel T, Ortiz C, Abburi VL (2019) Genome of "Charleston Gray", the principal American watermelon cultivar, and genetic characterization of 1365 accessions in the U.S. National Plant Germplasm System watermelon collection. *Plant Biotechnol J* 17(12):2246–2258.

Xanthopoulou A, Montero-Pau J, Mellidou I, Kissoudis C, Blanca J, Pico B, Tsaballa A, Tsaliki E, Dalakouras A, Paris HS (2019) Whole-genome re-sequencing of *Cucurbita pepo* morphotypes to discover genomic variants associated with morphology and horticulturally valuable traits. *Hortic Res* 6:94.

Xie D, Xu Y, Wang J, Liu W, Zhou Q, Luo S, Huang W et al. (2019) The wax gourd genomes offer insights into the genetic diversity and ancestral cucurbit karyotype. *Nat Commun* 10:5158.

Xu RF, Li H, Qin RY, Li J, Qiu CH, Yang YC, Ma H, Li L, Wei PC, Yang JB (2015a) Generation of inheritable and "transgene clean" targeted genome-modified rice in later generations using the CRISPR/Cas9 system. *Sci Rep* 5:11491.

Xu X, Lu L, Zhu B, Xu Q, Qi X, Chen X (2015b) QTL mapping of cucumber fruit flesh thickness by SLAF-seq. *Sci Rep* 5:15829.

Yang L, Koo DH, Li Y, Zhang X, Luan F, Havey MJ, Jiang J, Weng Y (2012) Chromosome rearrangements during domestication of cucumber as revealed by high-density genetic mapping and draft genome assembly. *Plant J* 71(6):895–906.

Yeh SD, Gonsalves D (1984) Evaluation of induced mutants of Papaya ring spot virus for control by cross protection. *Phytopathology* 74:1086–1091.

Zadoks JC, Schein RD (1979) *Epidemiology and Plant Disease Management*. Oxford University, Oxford.

Zhang C, Pratap AS, Natarajan S, Pugalenhdhi L, Kikuchi S, Sassa H, Senthil N, Koba T (2012) Evaluation of morphological and molecular diversity among South Asian germplasms of Cucumis sativus and C. melo. *ISRN Agron* 2012:1–11.

Zhang G, Ren Y, Sun H, Guo S, Zhang F, Zhang (2015a) A high-density genetic map for anchoring genome sequences and identifying QTLs associated with dwarf vine in pumpkin (*Cucurbita maxima* Duch.). *BMC Genomics* 16:1101.

Zhang Z, Mao L, Chen H, Bu F, Li G, Sun J, Li S, Sun H, Jiao C, Blakely R, Pan ZY, Cai H, Luo Y, Van de Peer Y, Jacobsen L, Fei Z, Huang S (2015b) Genome-wide mapping of structural variations reveals a copy number variant that determines reproductive morphology in cucumber. *Plant Cell* 27(6):1595–604.

Zhu WY, Huang L, Chen L, Yang JT, Wu JN, Qu ML, Yao DQ, Guo CL, Lian, HL, He HL, Pan JS, Cai R (2016) A high-density genetic linkage map for cucumber (*Cucumis sativus* L.): based on Specific Length Amplified Fragment (SLAF) sequencing and QTL analysis of fruit traits in cucumber. *Front Plant Sci* 7:437.

7 Allele Mining in Cucumber

Gayatri B. K. and Shyam Sundar Dey
Division of Vegetable Science, ICAR-Indian Agricultural Research Institute, New Delhi, India

7.1 INTRODUCTION

A diploid species ($2n = 14$), cucumber (*Cucumis sativus* L.) is one of the 2295 species and 129 genera that make up the *Cucurbitaceae* family of the order Cucurbitales (Berry et al. 2019). Cucumber originated in India, especially in the region between the Bay of Bengal and the Himalayan Mountain ranges. The wild relative *C. sativus* var. *hardwickii* is thought to be the cucumber's ancestor and its highly diverse wild and cultivated forms are found in the Indian subcontinent (Sebastian et al. 2010). Cucumber was the first horticultural crop to have a fully accessible draft genome (Huang et al. 2009; Wóycicki et al. 2011; Yang et al. 2012; Qi et al. 2013; Osipowski et al. 2020), greatly accelerating research into breeding methods, genome editing, and crop improvement.

Due to its short duration (between 4 and 5 months) and relatively small genomic size, cucumber genetic and genomic resources are well characterized. Building bioinformatics resources and facilitating the mapping and cloning of genes and quantitative trait loci (QTLs) of horticulturally significant traits of cucumber have undergone significant development over the past several decades (Weng 2020; Ma et al. 2022). To date, the genomes of six cucumber accessions have been sequenced: Chinese Long 9930, Gy-14, B10, PI183967, JEF, and KWS. B10 covers around 94% of the cucumber genome and is considered the most informative draft version to date (Osipowski et al. 2020). Turek et al. (2023) recently provided more information about the genome of the B10v3 accession by comparing it to the genomes of cucumber lines Gy-14 and 9930.

Recent advances in genomic sequencing have created exciting potential to accelerate research on cucumber breeding. Similar to other living species, plants have built up an impressive amount of advantageous alleles for numerous agronomic characteristics. A method for finding superior and/or distinctive alleles based on gene information is called allele mining. Allele mining enables both the identification of the precise genes underlying a given trait and the extraction of desirable alleles from the cucumber gene pool.

New alleles' evolution is significantly influenced by natural mutation. Mutations such as indels, transitions, and transversions are what are driving this evolution. In the post-genomics era, allele mining in various accessions may help with genomics research for crop development (Varshney et al. 2005). Superior alleles of genes from a variety of species are isolated using polymerase chain reaction (PCR) techniques.

As a result, allele mining requires the PCR-based amplification and sequencing of various gene variants found in inbred lines, varieties, landraces, and wild relatives. Variation in gene sequences that correlates with the phenotypic quality or performance of the accession allows for the identification of useful alleles for later evaluation. The entire procedure requires a combination of genetic searches and analyses of a carefully chosen list of genotypes as well as potential candidate genes

DOI: 10.1201/9781003376439-7

that may have particular functional importance (Kumari et al. 2013). The primary objective of this chapter is to provide a general overview of the allele mining strategy and the evolution of its application in cucumber breeding programs. The research work reported for important agronomic traits in cucumber are discussed as follows.

7.2 SHOOT ARCHITECTURE

Manipulation of plant architecture and especially reduced plant height (dwarf or compact growth habit) plays an important role in plant breeding. In many crops, plant height is an essential yield-related characteristic. Plants with dwarf, semi-dwarf, or bushy growth habits typically have short internodes and robust stems, allowing planting density to be manipulated to maximize productivity per unit area (Hedden 2003; Teng et al. 2013). The most well-known examples are the "Green Revolution" genes, which are involved in the signaling and biosynthesis of the plant hormone gibberellic acid (GA), such as the wheat reduced height (*Rht*) genes (Peng et al. 1999) and rice semi-dwarf (*sd1*) genes (Monna et al. 2002; Sasaki et al. 2002; Spielmeyer et al. 2002).

Cucumber, *Cucumis sativus* L., is a popular vegetable. Most cucumber cultivars have an indeterminate growth habit with tall vines, many lateral branches, and long internodes that adapt to trellis-supported production systems. Cucumbers with compact growth could benefit from mechanical harvesting to boost output and save labor costs (Cramer and Wehner 2000). A number of cucumber vine length or plant architecture mutants, including compact, extremely compact, dwarf, and short internode, have been described and characterized. The compact (*cp*) gene was discovered in two plant introduction (PI) lines, PI308915 and PI308916 (Kauffman and Lower 1976), and it was mapped to a 220-kb genomic area on cucumber chromosome Chr 4 (Li et al. 2011). A chemically generated compact mutant, *cp-2* (Kubicki et al. 1986), has shorter internodes and smaller seeds. Crienen et al. (2009) described compact cucumber mutant that differs from *cp* and *cp2*, which shows semi-dominant inheritance. The recessive gene (*scp*) that controls the ultra-compact mutant is responsible for its significantly reduced stem length, lack of branches, and wrinkled, dark-green leaves (Niemirowicz-Szczytt et al. 1996). A mutation in the gene for an F-box protein caused the EMS-induced cucumber mutant known as short internode (*si*) (Lin et al. 2016). Dwarf and determinate growth are caused by a loss in the cyclin-dependent protein kinase inhibitor gene (*CsSMR1*) (Li et al. 2022). A number of compact or super compact cucumber mutants, including super compact-1 (*scp-1*) encoding the cytochrome P450 protein *CsCYP85A1* (Wang et al. 2017a) and super compact-2 (*scp-2*) encoding the steroid 5-alpha-reductase (*CsDET2*) (Hou et al. 2017), and compact plant architecture (*cpa*) for the 7-dehydrocholesterol reductase (Zhang et al. 2021), have been reported. These investigations revealed that an array of processes govern plant height, with the BR biosynthesis/signaling system in cucumbers playing a key role in controlling plant architecture. Tables 7.1 and 7.2 present the details on genes and markers associated with cucumber shoot architecture traits.

7.3 FRUIT AND FLOWER-RELATED TRAITS

When the fruit is at its best flavor/brining quality, 10–15 days after anthesis, cucumber is harvested and consumed young. This is related to its exponential fruit growth, and the environment has a big impact on growth and development of fruits. Any fruit that is not within the commercially acceptable size range has to be eliminated during harvest in some market segments, which could significantly reduce yield. This is especially true for the once-over machine harvesting method used frequently in the US to harvest cucumbers (Pan et al. 2022). Fruit quality depends on many factors besides fruit size, including fruit shape, fruit skin color, fruit surface spines, fruit flesh color, fruit elongation, number of fruits, earliness, and female flowers per plant. These factors affect both consumer preference and commercial value. They impact directly the yield of cucumbers, the bottom line, and profitability. According to Simon and Navazio (1997) and Cuevas et al. (2010), the yellow color of the fruit mesocarp of some cucumber varieties depends on the level of b-carotene, which is the precursor for the biosynthesis of

TABLE 7.1
Details on Genes and Markers Associated with Cucumber Shoot Architecture Traits

Sl no.	Genes	Location	Donor	Source	Inheritance	Linked Marker/ Candidate Gene	Marker Type	Reference
1.	Little leaf (*ll*)	Chr6 (4.2cM) (3.6cM)	Gy7	Gy7 ×H-19 RIL	Recessive	SSR02355 SSR03940	SSR	Weng et al. (2010)
2.	Determinate (*de*)	Chr6 (1.4 cM) (4.2 cM)		Gy7 ×H-19 RIL	Recessive	CSWCTT14 SSR13251	SSR	Weng et al. (2010)
3.	Compact plant (*Cp*)	Chr.4 (2.1cM)	PI308915	USDA North Central Regional Plant Introduction Station.	Recessive	CKX-indel	Indels	Li et al. (2011)
4.		Chr.4 (2.1cM)				UW084979	SSR	
5.		Chr.4 (2.0cM)				UW084680	SSR	

vitamin A, and crucial for human health. Cucumbers with yellow mesocarp would therefore have a greater nutritional value than white or green colored varieties. Table 7.3 lists the markers and genes that have been identified in relation to fruit and flower-related traits.

7.4 SEX DETERMINATION

To increase crop yield and quality, cucumber breeders have been using a variety of techniques. These techniques include improving genetic diversity, using gynoecious sex types, and their connection to parthenocarpy, altering plant architecture, and improving disease resistance. Although the genetics of sex expression in cucumbers is a complex trait, it is crucial for the genetic advancement of the cucumber crop (Dhall et al. 2023). On cucumber plants, there are three different flower types that are possible: staminate (male), pistillate (female), and hermaphrodite/bisexual/perfect. Early research on cucumber's sex determination identified three genes: *F* (femaleness), *m* (andromonoecy), and *a* (androecy), all of which are members of the ACS gene family (*CsACS1G*, *CsACS2*, *CsACS11*, and *CsACO2*, which stand for *F*, *m*, *a*, and *a-1*, respectively) (Kubicki 1969; Pierce and Wehner 1990; Trebitsh et al. 1997; Mibus and Tatlioglu 2004; Li et al. 2009). Thus, the different sex expressions in a cucumber plant may be represented as monoecious (*MMffAA*), gynoecious (*MMFFA/a*), subgynoecious (*MMFfA/a*), andromonoecious (*mmffAA*), hermaphroditic (*mmFFA/a*), and androecious (*MMffaa/mmffaa*) (Li et al. 2020). Information about the genes in cucumbers that are related to fruit and flower traits are presented in Table 7.4.

7.5 DOWNY MILDEW (DM)

One of the diseases that poses the greatest threat to the production of cucumbers is downy mildew (DM), which is brought on by the obligate biotrophic oomycete *Pseudoperonospora cubensis* [Berkeley and Curtis] (Lebeda et al. 2004; Colucci et al. 2006). Cucumber DM symptoms first develop as chlorotic (yellow) lesions that are constrained by major leaf veins on the upper side of affected leaves. The pathogen requires moisture for germination and the initiation of a new infection; therefore symptom development is sensitive to environmental variables and favors damp and humid circumstances. Under the right environmental conditions, chlorotic lesions become necrotic and sporulation (with a brownish gray downy look) can be visible on the lower side of diseased leaves within days after infection. Infected leaves will gradually fall off as the disease spreads,

TABLE 7.2
Details on Genes and Markers Associated with Cucumber Shoot Architecture Traits

Sl. No.	No. Mutant or Transgenic Lines	Phenotype Description	Gene Name/ Locus	Location	Candidate Gene ID	Gene Annotation	References
1.	Short internode (*si*)	A dwarf phenotype with a reduced internode length	*CsVFB1*	Chr4: 21396485–21399329 (-)	*Csa4G641640*	Putative F-box/LRR-repeat protein 8	Lin et al. (2016)
2.	Compact (*cp*)	An extreme-dwarf-type plant with reduced internode length	*CsCKX*	Chr4: 22175499–22179648 (+)	*Csa4G647490*	Cytokinin oxidase/ dehydrogenase	Lin et al. (2016)
3.	tendril-less1 (*td-1*)	A dwarf shoot without tendrils and with less trichome	*sGCN5*	CChr6: 28883361. 28894330 (-)	*Csa6G 527060*	Histone acetyltransferase gcn5	Chen et al. (2017)
4.	super compact-2 (*scp-2*)	Super compact shoot architecture with dark green, wrinkle leaves	*CsDET2*	Chr3: 27762276. 27763359 (+)	*Csa3G732550*	3-Oxo-5-alpha-steroid 4-dehydrogenase	Hou et al. (2017)
5.	super compact-1 (*scp-1*)	Little hypocotyls elongation and drastically reduced internode length	*CsCYP85A1*	Chr5: 26718684. 26722145 (+)	*Csa5G641590*	Cytochrome P450; BR-C6-oxidase	Wang et al. (2017b)
6.	Cucumber dwarf (*Csdw*)	A dwarf phenotype with a reduced internode length	*CsCLV1*	Chr3: 36489540. 36494263 (-)	*Csa3G872760*	Receptor protein kinase-like protein	Xu et al. (2018a)
7.	determinate (*de*)	Determinate growth habit	*CsTFL1*	Chr6: 21554312. 21555486 (-)	*Csa6G452100*	Terminal flower 1a; Phosphatidylethanolamine-binding protein PEBP	Wen et al. (2019)
8.	tendril-less (*ten*) / ten-1	Modified tendrils like a branch-like structure	*CsTEN*	Chr5: 27044242. 27045612 (+)	*Csa5G644520*	PROLIFERATING CELL FACTORS (TCP) family of transcription factor	Yang et al. (2020)
9.	*ten-3*	Some tendrils are most of the normal morphology, while others display slight morphological defect with axillary meristems; both of the tendrils produce a reduced number of helical	*CsTEN*	Chr5: 27044242. 27045612 (+)	*Csa5G644520*	PROLIFERATING CELL FACTORS (TCP) family of transcription factor	Yang et al. (2020)
10.	*ten-4*	Some tendrils are most of the normal morphology, while others display slight morphological defect; both of the tendrils produce a reduced number of helical	*CsTEN*	Chr5: 27044242. 27045612 (+)	*Csa5G644520*	PROLIFERATING CELL FACTORS (TCP) family of transcription factor	Yang et al. (2020)

(*continued*)

TABLE 7.2 (Continued)
Details on Genes and Markers Associated with Cucumber Shoot Architecture Traits

Sl. No.	No. Mutant or Transgenic Lines	Phenotype Description	Gene Name/ Locus	Location	Candidate Gene ID	Gene Annotation	References
11.	*aco-2*	The tendril coiling capacity was substantially altered	*CsACO1*	Chr6: 11163849. 11165376 (+)	*Csa6G160180*	1-Aminocyclopropane-1-carboxylate oxidase	Yang et al. (2020)
12.	*aco-1*	The tendril coiling capacity was substantially altered	*CsACO1*	Chr6: 11163849-11165376 (+)	*Csa6G160180*	1-Aminocyclopropane-1-carboxylate oxidase	Yang et al. (2020)
13.	*cp*		*CsERECTA (CsER)*	chromosome 4	*CsGy4G024070*	Encodes a leucine-rich repeat (LRR) receptor-like serine/threonine-protein kinase.	Chen et. (2023)
14.	(Stem girth) *gSD*	Controlling stem development	*gSD1.1*	Chr 1: 15,337,811bp	*CsaV3_1G028310*	associated with antibiotic biosynthesis	Yang et al. (2022)
15.		Pentatricopeptide repeat	*gSD2.1*	Chr 2: 3,445,551	*CsaV3_2G006960*	proteins associated with TPR related to mitochondrial RNA processing	Yang et al. (2022)
16.		Role in maintaining the structure and function of stem-end meristem tissue	*gSD3.1*	Chr 3: 7,785,079	*CsaV3_3G009560*	LITTLE ZIPPER protein	Yang et al. (2022)
17.			*gSD5.2*	Chr5: 25,600,905	*CsaV3_5G031320*	encoded a kinase function in chloroplast	Yang et al. (2022)
18.			*gSD6.1*	Chr 6: 17,507,945	*CsaV3_6G031260*	associated with gibberellin	Yang et al. (2022)
19.	little leaf *(ll)*	Little leaf	*CsSAP*	Chr6: 7662821-7665390 (-)	*Csa6G111910*	Pentatricopeptide repeat-containing protein.	Yang et al. (2018)
20.	small and cordate leaf 1 (scl1) Small leaf	Small leaf size and cordate leaf with a blunt leaf base	*CsSL1*	Chr7: 3595111. 3599150 (+)	*Csa7G062760*	PHP domain-containing protein	Gao et al. (2017)
21.	CsHAN1-OE lines	Highly lobed leaves	*CsHAN1*	Chr4: 3624324. 3626040 (+)	*Csa4G046650*	GATA transcription factor	Ding et al. (2015)
22.	CsHAN1-RNAi lines	Highly lobed leaves	*CsHAN1*	Chr4: 3624324. 3626040 (+)	*Csa4G046650*	GATA transcription factor	Ding et al. (2015)

TABLE 7.3
List of Markers and Genes That Have Been Identified in Relation to Fruit and Flower-Related Traits

Sl. No.	Gene	Location	Donar	Source	Mapping Population	Identified or Candidate Gene	Linked Marker/ QTL	Protein /Gene Function	Reference
1.	Peduncle direction upward-pedicel *(up)*	Chr.1 22,736,083	*up* (spontaneous mutant)	China	C-8-6 (WT) x *up* (spontaneous mutant) F_1, F_2 and BC_1 population	*CsUP (CsGy1G024010.1)*	Indel	Auxilin-like protein with a conserved DnaJ-domain	Sun, J. et al. (2019)
2.	Size and shape Carpel number *(cn)*	Chr.1 10,816,556	WI2757	China	WI2757 (CN = 3) × True Lemon (CN = 5) F_2, F_3, and RILs	*CsCLV3 (CsGy1G014910.1)*	dCAPs, Bccl	*CLAVATA3*	Li et al. (2016)
3.	Size and shape Fruit size1.2 *(qfs1.2)*	Chr.1 25,331,715	`	China	WI7238 (long fruit) x WI7239 (round fruit)	*CsSUN2 (CsGy1G026840.1)*	Indel	Tomato fruit shape gene SUN homolog	Pan et al. (2017a)
4.	Size and shape Fruit size2.1 *(qfs2.1)*	Chr.2 11,224,424		China	WI7238 and WI7239	*CsTRM4(CsGy2G011350.1)*	dCAPs, Taql	Ortholog of *AtTRM5/SlTRM5*	Wu et al. (2018)
5.	Size and shape Fruitfull *(ful)*	Chr.1 3,947,246	6457 and ZND649		M85 and M573 are ethyl methane sulfonate mutants	*CsFUL1(CsGy1G006040.1)*	n.a.	MADS-box	Zhao et al. (2019)
6.	Size and shape Mango fruit *(mf)*	Chr.1 4,488,841	Extra Early Majestic	-	AM218W wild-type (WT) x AM218M (mutant)	*CsWOX1 (CsGy1G007020.1)*	SSR	WUSCHEL-related homeobox1 (WOX1	Niu et al. (2018)
7.	Size and shape Short fruit-1 *(sf-1)*	Chr.6 11,696,118			*sf1* (mutant) x Chinese long' 9930 (North China type)F_2 population		STS1 (sequence-tagged site 1)	n.a	Wang et al. (2017b)

(continued)

TABLE 7.3 (Continued)
List of Markers and Genes That Have Been Identified in Relation to Fruit and Flower-Related Traits

Sl. No.	Gene	Location	Donar	Source	Mapping Population	Identified or Candidate Gene	Linked Marker/ QTL	Protein /Gene Function	Reference
8.	Size and shape Short fruit–2 *sf-2*	Chr.2 10,113,710	Inbred line 406 (North China type)	China	*sf1* x wild-type 406, and F_2 population	*CsSF2 (CsGy2G010390.1)*	dCAPs	RING-type E3 ligase	Xin et al. (2019)
9.	Fruit flesh thickness 2.1 qfth2.1 *(fth2.1)*	Chr.2 4,434,893	-	China	D8 (thick fruit flesh parent) and XUE1 (thin fruit flesh parent) F_2 population	*fth2.1* n.a	-	Protein-lysine methyltransferase (PKMT)	Xu et al. (2015)
10.	Fruit diameter	Chr 1	PI618931 and PI249561	USA	WI7200 x WI7167 p F_2 and $F_{2:3}$ population	-	*fd1.1* (QTL)		Qu et al. (2014) Pan et al. (2017b)
11.	Fruit length	Chr 3,5, 6	PI618931 and PI249561	USA	WI7200 x WI7167 p F_2 and $F_{2:3}$ F_2 population	-	*fl3.1, fl5.1, fl6.1* (QTL)		Bo et al. (2012) Lu et al. (2017) Pan et al. (2017b)
12.	Length to diameter ratio	Chr 1	PI618931 and PI249561	USA	F_2	-	*ldl1.1* (QTL)		Pan et al. (2017b)
13.	Fruit shape (longer and slimmer fruits in cucumber)	Chr. 5 95.5 kb	PI 618931 is a semi-wild XIS cucumber	China	WI7167 × WI7200 F_2 and $F_{2:3}$ F_2 population	*CsCRC^G CsGy5G023910*	*FS5.2* (QTL)	encode a P450 protein	Pan et al. (2020) Pan et al. (2022)
14.	Fruit shape (shorter and wider fruit)	Chr. 5	PI 618931 is a semi-wild XIS cucumber	China	WI7167 x WI723 Near isogenic lines	*CsCRC^A* mutation *CsGy5G023910*	*fs5.2*	plant-specific YABBY transcription factor protein family	Pan et al. (2022)

15.	Fruit weight	Chr6 Chr.4 Chr.2	Beijing Jietou	China	CC3 x SWCC8 RIL-F_9	-	*fw6.1, fw4.1, fw2. 1* (QTL)		Bo et al. (2012)
16.	Fruit size/shape	Chr.5	PI 618931 (semi-wild XIS cucumber)	China	Near-isogenic lines (NILs)	*CsGy5G023910 CRABS CLAW (CsCRC)*	*f fs5.2* QTL	plant-specific YABBY transcription factor protein family	Pan et al. (2022)
17.	Fruit apex shape	Chr.4: 11 388 783 bp–17 666 521 bp	Delta star and Thaminbeit alpha	China	Round fruit apex (EC1 x IL52) and Sharp Fruit Apex (8419s-1 x CCMC)	*Csa4G303680 Csa4G363990 Csa4G385800 Csa4G418560*	*fa4.1*	Glutathione S-transferase JMS09K11.8 protein Receptor protein kinase KRI1-like protein	Zhu et al. (2022)
18.	Numerous spines (*ns*)	0.7 and 2.4 cM from *ns*	NCG-122	China	NCG-122 and NCG-121 were used as the parents for the F1, reciprocal F_1 (F_1'), and F_2 Populations.	*Csa2G285390*	*SSRns-127 and SSR04219*	Helix-loop-helix DNA-binding domain	Zhang et al. (2016)
19.	Spine color (*SPc*)	Chr.2: 17,761,336 17,763,934 +		37 countries	264 accessions, which included the local wild varieties and the breeding lines,	*Csa2G368270*	s2_17762942, s2_ 17762964 (SNP)	Pentatricopeptide repeat-containing protein	Lee et al. (2020)
20.	Spine color (*SPc*)	Chr.3: 15,235,551 15,240,040 –		37 countries	264 accessions, which included the local wild varieties and the breeding lines,	*Csa3G236570*	s3_15236835 (SNP)	Zinc-containing alcohol dehydrogenase quinone oxidoreductase	Lee et al. (2020)

(*continued*)

TABLE 7.3 (Continued)
List of Markers and Genes That Have Been Identified in Relation to Fruit and Flower-Related Traits

Sl. No.	Gene	Location	Donar	Source	Mapping Population	Identified or Candidate Gene	Linked Marker/ QTL	Protein /Gene Function	Reference
21.	Spine color (*SPc*)	Chr.5: 7,309,092 7,311,316 +		37 countries	264 accessions, which included the local wild varieties and the breeding lines,	*Csa5G175680*	s5_7311032 (SNP)	Pentatricopeptide repeat-containing protein	Lee et al. (2020)
22.	Spine color (*SPc*)	Chr.6: 21,137,853 21,139,047 +,		37 countries	264 accessions, which included the local wild varieties and the breeding lines,	*Csa6G448170*	s6_21138663 s6_ 21138684(SNP)	Thaumatin-like protein	Lee et al. (2020)
23.	Yellow flesh (*Yf*)	Chr.7:0.3cM	PI200815	North Carolina State University.	F_1, reciprocal F_1 (F_1 ') and F_2 populations.	-	yfSSR108 (SSR)		Lu et al. (2015)
1.		Chr.7: 0.6cM	PI200815		PI200815×931		yfIndel29 (indels)		
24.	Orange flesh β-carotene (*Ore*)	Chr.6 Chr. 3			F_2		*mc6.1/ec6.* *mc3.1/ec3.1* (QTL)		Staub et al. 2009
25.	β-carotene (*Ore*)	Chr3:1.9 cM Chr3:4.1 cM Chr3: 4.1 cM	XIS	China	F_7 RIL CC3 ×SWCC8		SSR07706 SSR23231 SSR11633		Bo et al. (2011)
26.	β-carotene (*Ore* gene) Endocarp	Chr. 3		China	CC3 x SWCC8 RIL-F_9	*CsBCH1* (*CsGy3G017310.1*)	*ec3.1* QTL (SSR markers)		Bo et al. (2012) Qi et al. (2013)
27.	Green flesh	718.225 Kb to 738,975 Kb	9110Gt	European	9110Gt (P_1) × PI183967 (P_2) and F_2 one BC_1P_1, BC_1P_2	*Csa5G021320*	*qgf5.1* (QTL)		Bo et al. (2019)

28.	Green flesh		9110Gt	European	9110Gt (P_1) × PI183967 (P_2) and F_2 one BC_1P_1, BC_1P_2	*Csa3G177380*	*qgf3.1* (QTL)		Bo et al. (2019)
29.	Fruit stalk-end color (*SDc)*	Chr.1: 1,180,448 1,182,094		37 countries	264 accessions, which included the local wild varieties and the breeding lines,	*Csa1G006300*	SNP+ (s1_1181919)	Response regulator 6	Lee et al. (2020)
30.		Chr.6: 21,137,853 21,139,047 +,		37 countries	264 accessions, which included the local wild varieties and the breeding lines,	*Csa6G448170*	s6_21138663 s6_ 21138684(SNP)	Thaumatin-like protein	Lee et al. (2020)
31.	Photoperiod flowering time	Chr 6			F_2 $F_{2:3}$	-	*FT6.1* (QTL)		Pan et al. (2017b)
32.	Male flowering time	Chr 1 Chr 6			F_2	-	*mft1.1* *mft6.* (QTL)		Pan et al. (2017b)
33.	Female flowering time	Chr 1,5,6 Chr 1			F_2, RIL- F_9	*CsaNFYA1* (*Csa1G613580*)	*fft1.1*, *fft5.1*,*fft6.2* *dff1.1* QTL, and Transcriptomic		Qu et al. (2014) Bo et al., (2015), Pan et al. (2017b) Tian et al. (2021)

TABLE 7.4
Details of Markers and Identified Genes Related to Fruit and Flower-Related Traits in Cucumber

Sl. No.	Gene	Location	Mapping Population	Linked Marker/QTL	Reference
1.	*F* locus	Chr. 1 5 cM and 0.8 cM from *F* locus	G421 (gynoecious) x H19 (monoecious) RILs and F_2	*sex1.1* (CSWCT28) and *sex1.2* (CSWCTT14) (QTLs with linked SSR)	Fazio et al. (2003)
2.	Gynoecious loci	Chr. 6 were at 7.7 and 6.8 cM, respectively	240-1-2-2-3-1 (gynoecious) x 3- 5-1-3-2-1-1-1-1-2 monoecious) F2	CSWCT25 and SSR18956 (SSR)	Zhou et al. (2013)
3.	F locus	1.2 and 1.7 cM,	9110Gt (gynoecious) x 9930 (monoecious) RILs	SSR13251 and CSWCT28	Miao et al. (2011)
4.	Gynoecious loci	Chr.6:7.7 and 6.8 cM,	240-1-2-2-3-1 (gynoecious) x 3- 5-1-3-2-1-1-1-1-2 (monoecious) F2	CSWCT25 and SSR18956	Shengjun et al. (2013)
5.	Sub-gynoecious	Chr.3: chr 6	S-2-98 (Sub-gynoecious) x M95 (monoecious) F_2 and BC_1	*sg3.1* linked to SSR13466 *sg6.1* and *sg6.2* SSR01308 and SSR02123 respectively (SSR)	Bu et al. (2016)
6.	Gynoecy trait	chr 6: 1.0 cM 4.5 cM	PPC-2 (gynoecious) x Pusa Uday (monoecious) F_2	SSR 13251 UW020605 SSR	Jat et al. (2019)
7.	Sub-gynoecious	Chr 3 Chr 1 274,377 and 267,386 polymorphic	LOSUAS (sub-gynoecious) x BMB (monoecious) F_2 and BC_1F_3	*sg1.1* and *sg1.2 sg3.1* (QTLs)	Win et al. (2019)
8.	Gynoecy trait	chr 6: at ≤1.5Cm	Gy-14 (gynoecious) x Pusa Uday (monoecious) BC_1F_2 and BC_1F_3	Three markers (SSR00233, SSR15516, and SSR13251) were tightly linked to QTLs (*qGyn6.5* and *qGyn 6.6*)	Boopalakrishnan et al. (2020)
9.	F-locus	1.5 and 4.5 cM	G421 (gynoecious) x Pusa Uday (monoecious) F_2	Markers SSR13251 and SSR15516	Behera et al. (2022)

which will reduce cucumber output because there will not be any leaves available for photosynthesis. Fanourakis and Simon (1987) found that *dm* is loosely linked to two *cla* (anthracnose resistance, 22.0 cm) and *cp* (compact habit, 38.6 cm). Recessive genes govern downy mildew resistance. Overview of mapped and cloned genes/quantitative trait loci (QTLs) for downy mildew resistance in cucumber is presented in Table 7.5.

TABLE 7.5
Overview of Mapped and Cloned Genes/Quantitative Trait Loci (QTLs) for Downy Mildew Resistance in Cucumber

Sl No.	Genes	Location	Donor	Source	Mapping Population	Identified or Candidate Gene	Linked Marker	Protein Associated	Reference
1.	Downy mildew resistance *(dm)*	9.5 and 17.7 cM, respectively	-	-	-	*CsC230/EcoR* and *CsC593/Dra,*	RAPD		Kennard et al. (1994)
2.	Downy mildew resistance *(dm)*	14 cM	-	-	-	*Pgm*	An isoenzyme marker		Meglic and Staub (1996)
3.	Downy mildew resistance *(dm)*	G14800 at 16.5 cM BC5191100 at 9.9 cM	PI 197087	Indian accession	F_3 families from two populations (55 from the WI 1983G × Straight 8 population)	-	G14800 BC5191100 RAPD	-	Horejsi et al. (2000)
4.	Downy mildew resistance *(dm)*	Chr. 1 SSR marker to *dm1.1* was 9.7 cM.	Chinese long cucumber hybrid, Yuanfeng	-	F_2 and $F_{2:3}$ families of Inbred line K8 (R) × K18 (S)	-	*dm1.1,* (QTL)	-	Bai et al. (2008)
5.	Downy mildew resistance *(dm)*	Chr.5: 1,371.8 kb	K8	China	F_2 and $F_{2:3}$ families derived from the cross K8 × K18.	*Csa008307* *Csa012579* *Csa012551*	*dm5.2* SSR	-	Zhang et al. (2013)
6.		Chr5:1,367.5 kb	K8			*Csa009587* *Csa009602* *Csa 009605* *Csa 009775*	*dm5.3* SSR	-	
7.		Chr.6: 195 kb	K8			none	*dm6.1* SSR	-	
8.	Downy mildew resistance *(dm)*	Chr.5	*Cucumis hystrix*	-	F_2 population of IL52 (R) × Changcunmici (S)	-	*DM_5.1,* *DM_5.2*	-	Pang et al. (2013) Zhang et al. (2018)

(continued)

TABLE 7.5 (Continued)
Overview of Mapped and Cloned Genes/Quantitative Trait Loci (QTLs) for Downy Mildew Resistance in Cucumber

Sl No.	Genes	Location	Donor	Source	Mapping Population	Identified or Candidate Gene	Linked Marker	Protein Associated	Reference
9.	Downy mildew resistance *(dm)*	Chr.1, 3, 5 1, 6, 7	Indian accession PI 197088 & Japanese cultivar Santou	India	RILs of CS-PMR1 (derived from PI 197088, R) × Santou (intermediate S)	-	7 QTLs from CS-PMR1; 3QTLs from Santousti	-	Yoshioka et al. (2014)
10.	Downy mildew resistance *(dm)*	Chr.2, 4, 5, 6	Hybrid Malini	India	$F_{2:3}$families of TH118FLM (derived from Malini (R) × WMEJ (S)	-	*dm2.2, dm4.1, dm5.1, dm5.2, dm6.1*	-	Win et al. (2017)
11.	Downy mildew resistance *(dm)*	All chromosomes except 7	Indian accession PI 197088	India	$F_{2:3}$ families of PI 197088 (R) × Coolgree (S)	-	11 QTLs of which *dm5.1, dm5.2,* and *dm5.3* are major QTLs	-	Wang et al. (2018)
12.	Downy mildew resistance *(dm)*	Chr.1, 3, 4, 5	Indian accession PI 197088	India	F_2 and $F_{2:3}$ families of PI 197088 × Changcunmici (S)	-	*dm1.1, dm3.1, dm4.1, dm5.1, dm5.2*	-	Li et al. (2018)
13.	Downy mildew resistance *(dm-1)*	Chr.5	PI 197087	Indian accession	RILs of Gy14 (R, derived from PI 197087) × 9930 and $F_{2:3}$ families of WTF23 (R, derived from PI 197087) × True Lemon	-	Mutant of *Sgr*	-	Wang et al. (2019)

14.	Downy mildew resistance *(dm)*	Chr.4	PI 197088	Indian accession	NILs and $F_{2:3}$ families of PI 197088 (R) × HS279 (S)	-	*DM4.1* and its derived subQTLs, *DM4.1.2* and *DM4.1.3* (SNP)	-	Berg J. (2020)
15.	Downy mildew resistance *(dm)*	Chr4:15766975	PI 197088	Indian accession	NILs and $F_{2:3}$ families of PI 197088 (R) × HS279 (S)	*CsLRK10L2* *Csa4M410850*	*DM4.1.2*	Receptor-like kinase; contains IPR011009 (Protein kinase-like domain)	Berg J. (2020)
16.	Downy mildew resistance *(dm)*	Chr.1: 21,753,524 to 21,774,330 bp	-	China	Core germplasm	*Csa1G575030*	*dmG1.4*	-	Liu et al. (2020)
17.	Downy mildew resistance *(dm)*	Chr.2: 45680765-4668765	-	China	Core germplasm	*Csa2G060360*	*dmG2.1*	-	Liu et al. (2020)
18.	Downy mildew resistance *(dm)*	Chr.4:5250-5350kb	-	China	Core germplasm	*Csa4G064680*	*dmG4.1*	an eukaryotic translation initiation factor 5A	Liu et al. (2020)
19.	Downy mildew resistance *(dm)*	Chr.5:22883182-22313023	-	China	Core germplasm	*Csa5G606470*	*dmG5.2*	a WRKY transcription factor 2–4	Liu et al. (2020)
20.	Downy mildew resistance *(dm)*	Chr.7:34900-134900	-	China	Core germplasm	*Csa7G004020*	*dmG7.1*		Liu et al. (2020)

7.6 POWDERY MILDEW

Powdery mildew (PM) is another damaging foliar disease that has a wide distribution and spreads quickly, resulting in significant production losses. Two distinct pathogenic species *Golovinomyces cichoracearum* (*Erysiphe cichoracearum*) and *Podosphaera xanthii (Sphaerotheca fuliginea)* can induce PM in cucumbers. (Block 2005). Although there are many fungicides available to reduce PM, their use puts growers under more financial and environmental strain (McGrath 2001). Meanwhile, long-term planting and varied pathogen adaptability have resulted in a progressive drop in plant resistance. Overall, temperature and area have an impact on the inheritance of cucumber PM resistance, which is complicated. Although an increasing number of PM-related genes have been found, providing insights into the genetic control of PM resistance in several cucumber cultivars, the QTLs revealed in these research are inconsistent, leaving the picture ambiguous. Recently a loss-of-susceptibility mutation of the highly conserved gene *STAYGREEN* (*CsSGR*) resulted in long-lasting, broad-spectrum disease resistance, as confirmed by QTL mapping (Wang et al. 2019). Information of the QTLs or genes for powdery mildew resistance identified by QTL or GWAS analysis in cucumbers given in Table 7.6.

7.7 VIRUS RESISTANCE

According to Anderson et al. (2004), plant viruses are thought to be the cause of 50 percent of newly emerging infectious diseases in plants. Additionally, when they affect crops, they can result in a complete harvest loss and significant economic losses (Nicaise 2014). One of the best defences against viral infections is the use of naturally resistant cultivars. While introgression, selection, and breeding have long been used to create resistant cultivars, the identification of the host factors governing resistance is now offering potential targets for strengthening the management of viral diseases. Recessive inheritance accounts for roughly half of the approximately 200 known virus resistance genes in plants (Kang et al. 2005). Given that recessive resistance appears to be more durable than dominant resistance (Fraser and Ronald 1992), focusing research on the genes involved provides an important tool for controlling plant diseases. The list of genes and markers for viral disease resistance is shown in the Table 7.7.

7.8 ABIOTIC STRESS TOLERANCE

Abiotic stresses like heat, cold, and salinity have a negative impact on the growth and development of plants. Heat stress (temperatures above 35–40 ℃) during the production of cucumbers frequently causes sunburn on the leaves, stunted growth on the stems and roots, malformed fruits, and even plant death, all of which have a significant impact on the quality and yield of the fruit. Cucumber seedling heat tolerance is a quantitative trait that is regulated by several genes. According to Staub and Wehneer (1996), the ideal temperature range for germination is 24–28 ℃. Cucumber (*Cucumis sativus* L.), a thermophilic crop native to the foothills of the Himalayas, is sensitive to low temperatures, likely as a result of its geographic origin (Cabrera et al. 1992). Temperature lower than optimal range leads to slow growth and chilling injury.

Depending on the salt concentration, salt stress generally limits plant physiological activity through osmotic and toxic effects. Cucumber seedlings are severely harmed by salt stress, which is most evident in the obvious symptoms of chlorosis and the decline of various plant and root system growth indicators (Shen et al. 2016; Liu et al. 2021a). The damage is also significantly exacerbated by increasing salt concentration and treatment time. Improved varieties can be bred by searching for genes that are tolerant to these abiotic stresses. Table 7.8 lists the genes and markers reported so far for tolerance to abiotic stress in cucumber.

TABLE 7.6
Information of the QTLs or Genes for Powdery Mildew Resistance Identified by QTL or GWAS Analysis in Cucumbers

Sl No.	Genes	Location	Donor	Source	Mapping Population	Identified or Candidate Gene	Linked Marker/ QTL	Protein	Reference
1.	*Pm* Cotyledon resistance	Chr.1 Chr.5 Chr.5 Chr.5: 27,072,020–31,877,374	PI197088-1	India, a sib-line of PI 197088-5	RILs of PI197088-1 × Santou		*pm1.1 (LGII, 26 ° C, 20 ° C), pm5.1 (LGIV, 20 ° C),pm5.2 (LGIV, 26 ° C, 20 ° C), pm5.3 (LGIV, 26 ° C), pm6.1 (LGI, 26 ° C)*		Sakata et al. (2006) Fukino et al. (2013)
2.	Leaf resistance	Chr.1 Chr.2 Chr.5 Chr.4 Chr.6	S06	European greenhouse-type	$F_{2:3}$ population and RILs of S94×SO6		*pm1.1, pm1.2, pm2.1,pm2.2, pm5.1, pm4.1, pm6.1*		(Liu et al. 2008a; 2008b)
3.	Leaf resistance; Hypocotyl resistance	Chr.	NPI		Natsufushnari x PI200815		*pm-l, pm-h*		de Ruiter et al. (2008), Hofstede et al. (2008)
4.	Leaf resistance	Chr.5 Chr.6	K8	China	F_2 and $F_{2:3}$ population of K8 × K18		*pm5.1, pm5.3, pm6.1*		Zhang et al. (2011)
5.	Leaf resistance	Chr.5 Chr.6	K8	China	F_2 and $F_{2:3}$ population of K8 × K18	*Csa009587, Csa009602, Csa009605 and BGICucGB009775*	*pm5.2*		Zhang et al. (2011)

(continued)

TABLE 7.6 (Continued)
Information of the QTLs or Genes for Powdery Mildew Resistance Identified by QTL or GWAS Analysis in Cucumbers

Sl No.	Genes	Location	Donor	Source	Mapping Population	Identified or Candidate Gene	Linked Marker/ QTL	Protein	Reference
6.	Leaf resistance; Cotyledon resistance; Hypocotyl resistance	Chr.1:18,577,265–21,339,881 Chr.3 Chr.4 Chr.5	WI2757		$F_{2:3}$ population of WI2757 × True lemon		*pm1.2, pm1.1, pm3.1, pm4.1, pm5.1, pm5.2*		He et al. (2013)
7.	Leaf resistance	Chr.1, Chr.5, Chr.5, Chr.6	CS-PMR1, Santou	CS-PMR1, derived from PI19708; Santou, Japanese	RILs of CSPMR1 ×Santou		*pm1.2, pm5.2, pm5.3 pm6.1*		Fukino et al. (2013)
8.	Leaf resistance	Chr.5	S1003	North China type	F_2 population; BC_3F_1, BC_2F_2 populations of S1003 ×S05	*CsMLO1 (Csa5G623470)*	*pm5.1*		Nie et al. (2015a), (2015b)
9.	Leaf resistance	Chr.1	Jin5-508	North China type	F_2 population of SSSL0.7×D8	*Csa1G064780, Csa1G064790*	*pm1.1*		Xu et al. (2016)
10.	Stem resistance	Chr.5	NCG122	European greenhouse-type	F_2 population of NCG122 × NCG121	*Csa5G623470*	*pm-s*		Liu et al. (2017)
11.	DMR and PMR	24,738,421	IL52		RILs changchunmici (CCMC) × IL52	*Csa5M622830.1*	*pm/dm5.2*	GATA transcriptional factor gene	Zhang et al. (2018)
12.	Leaf resistance	Chr.1, Chr.2, Chr.5, Chr.6	PI197088	Indian accession	RILs of PI197088 × Coolgreen	*Csa5G623470*	*pm1.1, pm2.1, pm5.1, pm6.1*		Wang et al. (2018)
13.	Leaf resistance		IL52	Introgression line derived from cucumber and *Cucumis hystrix* Chakr.	RILs of IL52 × changchunmici	*Csa5G622800, Csa5G622830, Csa5G623490*	*pm*		Zhang et al. (2018)

14.	Powdery mildew resistance	Chr.5: 16,045,687-16,048,001 +		37 countries	264 accessions, which included the local wild varieties and the breeding lines,	*Csa5G453160*	SNP (s5_16047445)	Short-chain dehydrogenase/ reductase	Lee et al. (2020)
15.	Powdery mildew resistance	Chr.5: 16,621,524–16,631,782		37 countries	264 accessions, which included the local wild varieties and the breeding lines,	*Csa5G471070*	SNP (s5_16623037)	Anaphase-promoting complex subunit	Lee et al. (2020)
16.	Leaf resistance	Chr.5 Chr.6	PM-R	Korea	F_2 population of PM-R × PM-S	*CsGy5G015660*	*pm5.2, pm6.1*		Zhang et al. (2021)
17.	Leaf resistance	Chr.1 Chr.2 Chr.3 Chr.4 Chr.5 Chr.6	Natural population	Selected from more than 3,000 germplasms worldwid	A cucumber core germplasm consisting of 94 resequencing line	*Csa2G030010, Csa2G030020, Csa2G030030* for *pmG2.1*; *Csa5G488800* for *pmG5.2*; *Csa5G603950, Csa5G603960* for *pmG5.3*	*pmG1.1, pmG1.2, pmG2.1, pmG2.2, pmG3.1, pmG4.1, pmG4.2, pmG5.1, pmG5.2, pmG5.3, pmG5.4, pmG6.1, pmG6.2*		Lui et al. (2021c)
18.	Leaf resistance	Chr.1 Chr.5	Natural population	Selected from cucumber core germplasms	231 cucumber core germplasms collected from the world	68 significantly associated SNPs. 15 genes for *pm1.1*; 2 genes for *pm5.7*	*pm1.1* *pm5.7*		Jia et al. (2021)
19.	Leaf resistance	Chr.1 Chr.5 Chr.3 Chr.5	K5	Progeny of CS- PMR1	$F_{2:3}$ population of K5 × KA		*pm1.1 pm5.3 pm3.2 pm5.1*		Shimomura et al. (2021)

Note: Chr.5, chromosome 5; QTL, quantitative trait locus; and GWAS, genome-wide association study. a The same name does not mean the same QTL. The first number in the QTL/gene name represents the chromosome number. b The major QTL or gene indicated in reference is shown in bold. c The causal gene identified.

TABLE 7.7
The List of Genes and Markers for Viral Disease Resistance

Sl no.	Genes	Location/Position	Donor	Source	Mapping Population	Identified or Candidate Gene	Linked Marker/ QTL	Protein	Reference
1.	*Cucumber Mosaic Virus* resistance	Chr.6 8,004,868 to 9,628,895 bp	02245	China	65Gx 02245 RIL	*Csa6M133680*	*cmv6.1* (SSR9-56–SSR11-177)	a bZIP domain-containing transcriptional activator	Shi et al. (2018)
2.	*Cucumber Vein Yellowing Virus* (CVYV) resistance	Chr05:5,088,092-8,208,448 bp	CE0749 CVYV-resistant Long Dutch type cucumber	China	CE0754×CE0749 F_1, F_2, and $F_{2:3}$ plants were inoculated mechanically with CVYV-AILM	*CsCvy-1*	CVYV_181 and CVYV_122 (SNP markers)	n.a.	Martínez-García et al. (2004) Pujol et al. (2019)
3.	Resistance to *Cucumber Mosaic Virus* (CMV)	Chr6:7688887	02245	China	65G x 02245 RILs	*cmv6.1*	SSR	SSR9-56 and SSR11-177	Shi et al. (2018)
4.	Resistance to *Papaya Ringspot Virus* (PRSV)	Chr6:9726336	02245	China	65Gx 02245 $F_{2:3}$ population	*PRSV*	SSR	SSR11-177 and SSR11-1	Tian et al. (2015)
5.	Resistance to *Watermelon Mosaic Virus* (WMV)	Chr6:22530869	02245	China	65Gx 02245 $F_{2:3}$ population	*wmv*	SSR	SSR11-177 and SSR11-1	Tian et al. (2015)
6.	Resistance to *Zucchini Yellow Mosaic Virus* (ZYMV)	Chr6:10,962,805	A192-18	Japan	'CS-PMR1' x 'A192-18' $F_{2:3}$ population	*zym*	CAPs, MnlI	n.a.	Amano, M. et al. (2013) Ramírez-Madera and Havey et al. (2017)

7.	Resistance to *Melon Yellow Spot Virus* (MYSV)	Chr.1:1,315,863 6,041,674	Tokiwa	Japan	accession 27028930 x the susceptible cultivar "Tokiwa"	n.a.	*mysv1.1* SSR	n.a.	Sugiyama et al. (2015)
8.	Resistance to *Melon Yellow Spot Virus* (MYSV)	Chr.3:5,926,467 12,326,572	Tokiwa	Japan	accession 27028930 x the susceptible cultivar "Tokiwa"	n.a.	*mysv3.1* SSR	n.a.	Sugiyama et al. (2015)
9.	Resistance to *Melon Yellow Spot Virus* (MYSV)	Chr.4:10,211,019 26,166,864	Tokiwa	Japan	accession 27028930 x the susceptible cultivar "Tokiwa"	n.a.	*mysv4.1* SSR	n.a.	Sugiyama et al. (2015)
10.	Resistance to *Melon Yellow Spot Virus* (MYSV)	Chr.7 :12,693,676 1,723,853	Tokiwa	Japan	accession 27028930 x the susceptible cultivar "Tokiwa"	n.a.	*mysv7.1* SSR	n.a.	Sugiyama et al. (2015)

TABLE 7.8
List of Genes and Markers for Tolerance to Abiotic Stress

Sl No.	Trait	Location	Donor	Source	Mapping Population	Identified or Candidate Gene	Linked Marker/ QTL	Protein	Reference
1.	Heat tolerance	Chr.3: 31,699,402–33,960,084 bp	02245	China	Recombinant inbred lines (RILs) from 65G × 02245	*Csa3G822410* *Csa3G824920* *Csa3G824940* *Csa3G823060* *Csa3G825010* *Csa3G824990* *Csa3G822440* *Csa3G825020* *Csa3G824850*	*qHT3.2*	HSP20-likechaperone NBS-containing resistance like protein TIR-NBS-LRR resistance protein Calmodulin-like protein 1 Calmodulin-likeProtein NAC domain protein AP2-like ethylene-responsive transcription factor Dof zinc finger protein MYB transcription factor	Dong et al. (2020)
2.	Heat tolerance	Chr.4: 17,744,278–18,744.278bp		China	86 cucumber germplasms		*gHII4.2*		Wei et al. (2019)
3.	Heat tolerance	Chr.4: 17,294,615–18,294,615bp		China	86 cucumber germplasms		*gHII4.1*		Wei et al. (2019)
4.	Heat tolerance	Chr1: 987,005–990,966bp	99281	China	931 x 99281 F_2 and RIL population	*Csa1G004990*	*qHT1.1*	RING finger protein	Liu et al. 2021b
5.	Low-temperature tolerance	Chr1: 14,855,251–15,424,716 bp	65G	Europe	65G × 02245 F_9 RILs	*Csa1M408720*	*qLTG1.1* (SSR11340)	encodes the DELLA protein	Song et al. (2018)
6.	Low-temperature tolerance	Chr1: 29,185,530-29,524,806bp			151 accessions from different geographical origins	*CsaV3_1G044080*	*gLTG1.2*	encoded a pentatricopeptide repeat-containing protein (CsPPR)	Li et al. 2023

7.	Salt tolerance	Chr 2: 23,409,202~23,509, 202 bp		China	220 accessions	*CsaV3_2G035120*	*gST2.1*	encoding an isocitrate lyase	Liu et al. (2022b)
8.	Salt tolerance	Chr.3: 21,496,729~21,596, 729 bp		China	220 accessions	*CsaV3_3G023710*	*gST3.1*, *gST3.2*, ,	encoding an ankyrin repeat family protein	Liu et al. (2022b)
9.	Salt tolerance	Chr.4: 23,423,770~23,523, 770 bp		China	220 accessions	*CsaV3_4G033150*	*gST4.2*	encoding a phosphoenolpyruvate carboxylase protein	Liu et al. (2022b)
10.	Salt tolerance	Chr.6: 7,916,377~8,016, 377 bp		China	220 accessions	*CsaV3_6G009810*	*gST6.1*	encoding a pesticidal crystal cry8Ba protein	Liu et al. (2022b)
11.	Waterlogging	Chr6: 28825007	Zaoer-N	China	Zaoer-N (landrace) x Pepino, F_1 and F_2	*CsARN6.1* (*CsGy6G030800.1*)	*qARN6.1* (mutant)	AAA ATPase domain-containing protein	Xu et al. (2017) Xu et al. (2018b)

Significant advancement has been made in cucumber in terms of identification of genomic regions, development of molecular markers, and identification of possible candidate genes associated with number of economically important traits. However, cloning of the candidate genes and their functional validation through transformation and/or targeted mutation are required to taken up in near future for getting greater insight about the underlying molecular mechanisms associated with these traits. Besides identification of new genetic resources, understanding the genetic basis and molecular characterization for different emerging problems needs to be taken in war footing to make cucumber cultivation more eco-friendly.

REFERENCES

Amano M, Mochizuki A, Kawagoe Y, Iwahori K, Niwa K, et al. (2013) High-resolution mapping of *zym*, a recessive gene for Zucchini yellow mosaic virus resistance in cucumber. *Theor Appl Genet* 126(12):2983–2993. doi:10.1007/s00122-013-2187-5

Anderson PK, Cunningham AA, Patel NG Morales FJ, Epstein PR, et al. (2004) Emerging infectious diseases of plants: pathogen pollution, climate change and agrotechnology drivers. *Trends Ecol Evol* 19(10):535–544. doi:10.1016/j.tree.2004.07.021

Bai Z, Yuan X, Cai R, Liu L, He H et al. (2008) QTL analysis of downy mildew resistance in cucumber. *Prog Nat Sci* 18:706–710.

Behera TK, Boopalakrishnan G, Jat GS, Munshi AD, Choudhary H, et al. (2022) Deriving stable tropical gynoecious inbred lines of slicing cucumber from American pickling cucumber using MABB. *Hortic Environ Biotechnol* 63(2):263–274

Berg JA, Hermans FW, Beenders F, Lou L, Vriezen WH, et al. (2020) Analysis of QTL *DM4.1* for downy mildew resistance in cucumber reveals multiple subQTL: a novel RLK as candidate gene for the most important subQTL. *Front Plant Sci* 11:1601.

Berry PE (2019) Cucurbitales. In: *Encyclopsedia Britannica*. Chicago, IL, USA. www.britannica.com/plant/Cucurbitalesn

Block CC. (2005) Powdery mildew resistance in the U.S. national plant germplasm system cucumber collection. *Hortic Sci* 40:416–420.

Bo K, Ma Z, Chen J, Weng Y (2015) Molecular mapping reveals structural rearrangements and quantitative trait loci underlying traits with local adaptation in semi-wild Xishuangbanna cucumber (Cucumis sativus L. var. xishuangbannanesis Qi et Yuan). *Theor Appl Genet* 128(1):25–39. doi: 10.1007/s00122-014-2410-z

Bo K, Song H, Shen J, Qian C, Staub JE et al (2011) Inheritance mapping of the Ore gene controlling the quantity of β-carotene in cucumber (Cucumis sativus L.) endocarp. *Mol Breed* 30(1):335–344.

Bo K, Song H, Shen J, Qian Q, Lou J, et al. (2012) Inheritance mapping of the *Ore* gene controlling the quantity of β-carotene in cucumber (*Cucumis sativus* L.) endocarp. *Mol Breed* 30(1):335–344

Bo, K, Wei, S, Wang, W, Miao H, Dong S et al. (2019) QTL mapping and genome-wide association study reveal two novel loci associated with green flesh color in cucumber. *BMC Plant Biol* 19,(243)1–13. https://doi.org/10.1186/s12870-019-1835-6

Boopalakrishnan G, Jayavel S, Behera TK, Munshi, AD, Choudhary H, et al. (2020) Mapping and linkage analysis of epistatic QTLs for gynoecious trait (*F*) in cucumber (*Cucumis sativus* L.). *J Hortic Sci Biotechnol* 96:3, 344–355

Bu F, Chen H, Shi Q, Zhou, Q, Gao, D et al. (2016) A major quantitative trait locus conferring subgynoecy in cucumber. *Theor Appl Genet.* 129(1):97–104

Cabrera RM, Saltveit ME Owens K (1992) Cucumber cultivars differ in their response to chilling temperatures. *J Am Soc Hortic Sci* 117(5):802–807

Chen F, Fu B, Pan Y, Zhang C, Wen H, et al. (2017) Fine mapping identifies CsGCN5 encoding a histone acetyltransferase as putative candidate gene for tendril-less1 mutation (td-1) in cucumber. *Theor Appl Genet* 130(8):1549–1558. doi:10.1007/s00122-017-2909-1

Chen F, Yong J, Zhang G, Liu M, Wang Q, et al. (2023) An LTR retrotransposon insertion inside *CsERECTA* for an LRR receptor-like serine/threonine-protein kinase results in compact (*cp*) plant architecture in cucumber. *Theor Appl Genet* 136(3):31.

Colucci SJ, Wehner TC, Holmes GJ. (2006) The downy mildew epidemic of 2004 and 2005 in the eastern United States. In: *Proceedings of the Cucurbitaceae*. Asheville, North Carolina, pp 403–410.

Cramer CS and Wehner TC (2000) Path analysis of the correlation between fruit number and plant traits of cucumber populations. *Hortscience*. 35(4):708–711

Crienen J, Reuling G, Segers B, van de Wal M (2009) New cucumber plants with a compact growth habit [patent], *International publication number WO*, vol 2009(059777):A1.

Cuevas HE, Song H, Staub JE, Simon PW (2010) Inheritance of beta-carotene-associated flesh color in cucumber (Cucumis sativus L.) fruit. *Euphytica* 171:301–311

de Ruiter W, Hofstede R, de Vries J, van den Heuvel, H, et al. (2008) Combining QTL for resistance to CYSDV and powdery mildew in a single cucumber line. In: *Proceedings of the IXth EUCARPIA Meeting on Genetics and Breeding of Cucurbitaceae*, p 181.e188.

Dhall RK, Kaur H, Manchanda P, Eshanee (2023) Genetics and marker assisted breeding for sex expression in cucumber. *Front Genet* 14:1180083

Ding L, Yan S, Jiang L, Liu M, Zhang J et al. (2015) HANABA TARANU regulates the shoot apical meristem and leaf development in cucumber (*Cucumis sativus* L.). *J Exp Bot* 66(22):7075–7087.

Dong, Song Z, Shuang W, Liu Y, Li C (2020) Identification of quantitative trait loci controlling high-temperature tolerance in cucumber (*Cucumis sativus* L.) Seedlings. *Plants* 9(9):1155 doi:10.3390/plants909115

Fanourakis NE, Simon PW (1987) Analysis of genetic linkage in the cucumber. *J Hered* 78(4):238–242. doi:10.1093/oxfordjournals.jhered.a110374

Fazio G., Staub J. E. and Stevens M. R. (2003) Genetic mapping and QTL analysis of horticultural traits in cucumber (Cucumis sativus L.) using recombinant inbred lines. *Theor Appl Genet*. 107:864–874 DOI 10.1007/s00122-003-1277-1

Fraser, Ronald SS (1992) The genetics of plant-virus interactions: implications for plant breeding. *Euphytica* 63(1–2):175–185. doi:10.1007/BF00023922

Fukino N, Yoshioka Y, Sugiyama M, Sakata Y, Matsumoto S (2013) Matsumoto S. Identification and validation of powdery mildew (*Podosphaera xanthii*)-resistant loci in recombinant inbred lines of cucumber (*Cucumis sativus* L.). *Mol Breed* 32(2):267–277. doi:10.1007/s11032-013-9867-3

Gao D, Zhang C, Zhang S, Hu B, Wang S et al. (2017) Mutation in a novel gene SMALL AND CORDATE LEAF 1 affects leaf morphology in cucumber. *J Integr Plant Biol* 59(10):736–741. doi:10.1111/jipb.12558

He X, Li Y, Pandey S, Yandell BS, Pathak M et al. (2013) QTL mapping of powdery mildew resistance in WI 2757 cucumber (*Cucumis sativus* L.). *Theor Appl Genet* 126(8):2149–2161. doi:10.1007/s00122-013-2125-6

Hedden P (2003) The genes of the Green Revolution. *Trends Genet* 19(1):5–9. doi:10.1016/s0168-9525(02)00009-4

Horejsi T, Staub JE, Thomas C. (2000) Linkage of random amplified polymorphic DNA markers to downy mildew resistance in cucumber (*Cucumis sativus* L.). *Euphytica* 115(2):105–113. doi:10.1023/A:1003942228323

Hou S, Niu H, Tao Q, Wang S, Gong Z et al. (2017) A mutant in the *CsDET2* gene leads to a systemic brassinosteriod deficiency and super compact phenotype in cucumber (*Cucumis sativus* L.). *Theor Appl Genet* 130(8):1693–1703. doi:10.1007/s00122-017-2919-z

Huang S, Li RQ, Zhang ZH, Li L, Gu X, et al. (2009) The genome of the cucumber, *Cucumis sativus* L. *Nat Genet* 41(12):1275–1281. doi:10.1038/ng.475

Jat GS, Munshi AD, Behera TK, Choudhary H, Dash P et al. (2019) Genetics and molecular mapping of gynoecious (F) locus in cucumber (*Cucumis sativus* L.). *J Hortic Sci Biotechnol*. 94(1):24–32. doi:10.1080/14620316.2018.1449671

Jia HX, Li XX, Song JP, Lin YE, Zhang XJ et al. (2021) Genome-wide association study of powdery mildew resistance of cucumber core germplasms. *Acta Hortic Sin* 48:1371e1385 (in Chinese).

Kang BC, Yeam I, Jahn MM (2005) Genetics of plant virus resistance. *Annu Rev Phytopathol* 43:581–621. doi:10.1146/annurev.phyto.43.011205.141140

Kauffman C, Lower R (1976) Inheritance of an extreme dwarf plant type in the cucumber. *J Am Sci Hortic Sci* 101:150–151

Kennard WC, Poetter K, Dijkhuizen A, Meglic V, Staub JE et al. (1994) Linkage among RFLP, RAPD, isozyme, disease resistance and morphological markers in narrow and wide crosses of cucumber. *Theor Appl Genet* 89(1):42–48

Kubicki B, Sołtysiak U, Korzeniewska A (1986) Induced mutation in cucumber (*Cucumis sativus* L.). *V: Compact type of growth. Genet Pol* 27:289–298.

Kubicki B. (1969) Investigation of sex determination in cucumber (*Cucumis sativus* L). *Genet Pol* 10:123–143.

Kumari A, Das A, Devanna BN, Thakur S, Singh P et al. (2013) Mining of rice blast resistance gene *Pi54* shows effect of single nucleotide polymorphisms on phenotypic expression of the alleles. *Eur J Plant Pathol* 137(1):55–65

Lebeda A, Urban J (2004) Distribution, harmfulness and pathogenic variability of cucurbit downy mildew in the Czech Republic. *Acta Fytotech Zootech* 7:170–173.

Lee HY, Kim JG, Kang B (2020) Assessment of the genetic diversity of the breeding lines and a genome wide association study of three horticultural traits using worldwide cucumber (*Cucumis spp.*) *Germplasm Collection. Agronomy* 1736;10(11): 1–15.

Li C, Dong S, Beckles DM, Liu X, Guan J et al. (2023) GWAS reveals novel loci and identifies a pentatricopeptide repeat-containing protein (*CsPPR*) that improves low temperature germination in cucumber. *Front Plant Sci* 14:1116214.

Li L, He H, Zou Z, Li Y (2018) QTL analysis for downy mildew resistance in cucumber inbred line PI 197088. *Plant Dis* 102(7):1240–1245.

Li S, Pan Y, Wen C, Li Y, Liu X, et al. (2016) Integrated analysis in bi-parental and natural populations reveals *CsCLAVATA3 (CsCLV3)* underlying carpel number variations in cucumber. *Theor Appl Genet* 129(5):1007–1022

Li S, Zhang Q, Zhang H, Wang J, Sun J et al. (2022) Deletion of a cyclindependent protein kinase inhibitor, *CsSMR1*, leads to dwarf and determinate growth in cucumber (*Cucumis sativus* L.). *Theor Appl Genet* 135(3):915–927

Li Y, Yang L, Pathak M, He X, Weng Y (2011) Fine genetic mapping of *cp*: a recessive gene for compact (dwarf) plant architecture in cucumber, *Cucumis sativus* L. *Theor Appl Genet* 123(6):973–983.

Li Z, Han Y, Niu H, Wang Y, Jiang B, et al. (2020) Gynoecy instability in cucumber (*Cucumis sativus* L.) is due to unequal crossover at the copy number variation-dependent Femaleness (F) locus. *Hortic Res* 7:32

Li Z, Huang S, Liu S, Pan J, Zhang Z, et al. (2009) Molecular isolation of the M gene suggests that a conserved-residue conversion induces the formation of bisexual flowers in cucumber plants. *Genetics* 182(4):1381–138

Lin T, Wang S, Zhong Y, Gao D, Cui Q, et al. (2016) A truncated F-Box protein confers the dwarfism in cucumber. *J Genet Genom* 43(4):223–6

Liu D, Dong S, Bo K, Miao H, Li C, et al. (2021a) Identification of QTLs controlling salt tolerance in cucumber (*Cucumis sativus* L.) seedlings. *Plants (Basel)* 10(1):85.

Liu D, Dong S, Miao H, et al. (2022b) A large-scale genomic association analysis identifies the candidate genes regulating salt tolerance in cucumber (*Cucumis sativus* L.) seedlings. *Int J Mol Sci* 23(15):8260

Liu L, Cai R, Yuan X, He H, Pan J et al. (2008a) QTL molecular marker location of powdery mildew resistance in cucumber (*Cucumis sativus* L.). *Sci China C Life Sci* 51(11):1003–1008. doi:10.1007/s11427-008-0110-0

Liu L, Yuan X, Cai R, Pan J, Zhu L. (2008b) Quantitative trait loci for resistance to powdery mildew in cucumber under seedling spray inoculation and leaf disc infection. *J Phytopathol* 156:691.e697.

Liu PN, Miao H, Lu HW, Cui JY, Tian GL et al. (2017) Molecular mapping and candidate gene analysis for resistance to powdery mildew in *Cucumis sativus* stem. *Genet Mol Res.* 16(3):gmr16039680. doi:10.4238/gmr16039680

Liu X, Gu X, Lu H, Liu P, Miao H, et al. (2021c) Identification of novel loci and candidate genes for resistance to powdery mildew in a resequenced cucumber germplasm. Genes 12(4):584. doi:10.3390/genes12040584

Liu X, Lu H, Liu P, Miao H, Bai Y, et al. (2020) Identification of novel loci and candidate genes for cucumber downy mildew resistance using GWAS. Plants (Basel) 9(12):1659. doi:10.3390/plants9121659

Liu Y, Dong S, Wei S, Han M, Bo K et al. (2021b) QTL mapping of heat tolerance in cucumber (*Cucumis sativus* L.) at adult stage. *Plants (Basel)* 10(2):324. doi:10.3390/plants10020324

Lu HW, Miao H, Tian GL, et al. (2015) Molecular mapping candidate gene analysis for yellow fruit flesh in cucumber. Mol Breed 35(2):64. doi:10.1007/s11032-015-0263-z

Lu P (2016) *Analysis of β-carotene biosynthesis and expression of genes related to fruit development in cucumber of Xishuangbanna [D].* Chinese Academy of Agricultural Sciences.

Ma L, Wang Q, Zheng Y, Guo J, Yuan S et al. (2022) Cucurbitaceae genome evolution, gene function, and molecular breeding. *Hortic Res* 9:uhab057. doi:10.1093/hr/uhab057

Martínez-García B, Marco CF, Goytia E, López-abella D, María S, et al. (2004) Development and use of detection methods specific for *Cucumber vein yellowing virus* (CVYV). *Eur J Plant Pathol* 110(8):811–821

McGrath MT. (2001) Fungicide resistance in cucurbit powdery mildew: experiences and challenges. *Plant Dis* 85(3):236–245. doi:10.1094/PDIS.2001.85.3.236

Meglic V, Staub JE. (1996) Inheritance and linkage relationships of isozyme and morphological loci in cucumber (*Cucumis sativus* L.). *Theor Appl Genet.* 92(7):865–872. doi:10.1007/BF00221899

Mibus H, Tatlioglu T. (2004) Molecular characterization and isolation of the *F/f* gene for femaleness in cucumber (*Cucumis sativus* L.). *Theor Appl Genet* 109(8):1669–1676. doi:10.1007/s00122-004-1793-7

Monna L, Kitazawa N, Yoshino R, Suzuki J, Masuda H, et al. (2002) Positional cloning of rice semi dwarfing gene, *sd-1*: rice "Green Revolution Gene" Encodes a Mutant Enzyme Involved in gibberellin Synthesis. *DNA Res* 9(1):11–17

Nicaise V. (2014) Crop immunity against viruses: outcomes and future challenges. *Front Plant Sci.* 5:660

Nie J, He H, Peng J, Yang X, Bie B et al. (2015a) Identification and fine mapping of *pm5. 1:* a recessive gene for powdery mildew resistance in cucumber (*Cucumis sativus* L.). *Mol Breed* 35(1):7. doi:10.1007/s11032-015-0206-8

Nie J, Wang Y, He H, Guo C, Zhu W et al. (2015b) Loss-of-function mutations in CsMLO1 confer durable powdery mildew resistance in cucumber (*Cucumis sativus* L.). *Front Plant Sci* 6: 1155.

Niemirowicz-Szczytt K, Rucinska M, Korzeniewska (1996) A. An induced mutation in cucumber (*Cucumis sativus* L.): super compact. *Cucurbit Genet Coop Rep* 19:1–3:(article 1).

Niu H, Liu X, Tong C, Wang H, Li S, et al. (2018) The WUSCHEL-related homeobox1 gene of cucumber regulates reproductive organ development. *J Exp Bot* 69(22):5373–5387. doi:10.1093/jxb/ery329

Osipowski P, Pawełkowicz M, Wojcieszek M, Skarzyńska A, Przybecki Z et al. (2020) A high-quality cucumber genome assembly enhances computational comparative genomics. *Mol Genet Genom* 295:177–193. doi:10.1007/s00438-019- 01614-3

Pan Y, Chen B, Qiao L, Chen F, Zhao J et al. (2022) Phenotypic characterization and fine mapping of a major-effect fruit shape QTL *FS5.2* in Cucumber, *Cucumis sativus* L., with Near-Isogenic Line-Derived Segregating Populations. *Int J Mol* Sci 23(21):13384. doi:10.3390/ijms232113384

Pan Y, Liang X, Gao M, Liu H, Meng H, et al. (2017a) Round fruit shape in WI7239 cucumber is controlled by two interacting quantitative trait loci with one putatively encoding a tomato SUN homolog. *Theor Appl Genet* 130(3):573–586. doi:10.1007/s00122-016-2836-6

Pan Y, Qu S, Bo K, Gao M, Haider KR et al. (2017b) QTL mapping of domestication diversifying selection related traits in round-fruited semi-wild Xishuangbanna cucumber (*Cucumis sativus* L. var *xishuangbannanesis*). *Theor Appl Genet* 130(7):1531–1548. doi:10.1007/s00122-017-2908-2

Pan YP, Wang YH, McGregor C, Liu S, Luan F et al. (2020) Genetic architecture of fruit size and shape variation in cucurbits: a comparative perspective. *Theor Appl Genet* 133(1):1–21. doi:10.1007/s00122-019-03481-3

Pang X, Zhou X, Wan H, Chen J (2013) QTL mapping of downy mildew resistance in an introgression line derived from interspecific hybridization between cucumber and *Cucumis hystrix*. *J Phytopathol* 161(7–8):536–543. doi:10.1111/jph.12103

Peng J, Richards DE, Hartley NM, Murphy GP, Devos KM et al. (1999) "Green revolution" genes encode mutant gibberellin response modulators. *Nature* 400(6741):256–261. doi:10.1038/22307

Pierce LK, Wehner TC. (1990) Review of genes and linkage groups in cucumber. *HortScience* 25(6):605–615. doi:10.21273/HORTSCI.25.6.605

Pujol M, Alexiou KG, Fontaine AS, Mayor P, Miras M, et al.(2019) Mapping cucumber vein yellowing virus resistance in cucumber (*Cucumis sativus* L.) by using BSA-seq analysis. *Front Plant Sci* 10:1583. doi:10.3389/fpls.(2019).01583

Qi J, Liu X, Shen D, Miao H, Xie B et al. (2013) A genomic variation map provides insights into the genetic basis of cucumber domestication and diversity. *Nat Genet* 45(12):1510–1515. doi:10.1038/ng.2801

Qu SP, Pan YP, Weng Y (2014) QTL mapping of flowering time and fruit shape in Xishuangbana cucumber (*Cucumis sativus* L. var *xishuangbannanesis* Qi et Yuan). In: *Proceedings of the Cucurbitaceae*. East Lansing, Michigan, pp 54–56.

Ramírez-Madera AO, Havey MJ (2017) Different haplotypes encode the same protein for independent sources of *Zucchini yellow mosaic virus* resistance in cucumber. *HortScience* 52(8):1040–1042.

Sakata Y, Kubo N, Morishita M, itadani E, Sugiyama M et al. (2006) QTL analysis of powdery mildew resistance in cucumber (*Cucumis sativus* L.). *Theor Appl Genet* 112(2):243–250.

Sasaki A, Ashikari M, Ueguchi-Tanaka M, Itoh H, Nishimura A et al. (2002) Green revolution: A mutant gibberellin-synthesis gene in rice. *Nature* 416(6882):701–702. doi:10.1038/416701a

Sebastian P., Schaefer H., Telford I. R. H. and Renner S. S. (2010). Cucumber (Cucumis sativus) and melon (C. melo) have numerous wild relatives in Asia and Australia, and the sister species of melon is from Australia. *PNAS*. 107 (32) 14269–14273. https://doi.org/10.1073/pnas.1005338107

Shen J, Jiang J, Tian S et al. (2016) Effects of NaCl stress on growth and cell membrane permeability of cucumber seedlings. *Guizhou Agric Sci* 8:19.

Shengjun Z., Peng Z., Yuquing Z., Xinjuan C., Liping C. (2013). Identification of SSR markers linked to gynoecious loci in cucumber (Cucumis sativus L). *J. Zhejiang Univer.* 39, 291–298. 10.3785/j.issn.1008-9209.2012.11.141

Shi L, Yang Y, Xie Q, Miao H, Bo K, et al. (2018) Inheritance and QTL mapping of cucumber mosaic virus resistance in cucumber (*Cucumis sativus* L.). *PLoS One* 13(7):e0200571. doi:10.1371/journal.pone.0200571

Shimomura K, Sugiyama M, Kawazu Y Yoshioka Y (2021) Identification of quantitative trait loci for powdery mildew resistance in highly resistant cucumber (*Cucumis sativus* L.) using ddRAD-seq analysis. *Breed* Sci 71(3):326–333. doi:10.1270/jsbbs.20141

Song Z, Wang W, Shi L, Zhang S, Xie Q, et al. (2018) Identification of QTLs controlling low-temperature tolerance during the germination stage in cucumber (*Cucumis sativus* L.). *Plant Breed* 137(4):629–637. doi:10.1111/pbr.12601

Spielmeyer W, Ellis MH, Chandler PM (2002) Semidwarf (*sd-1*), "green revolution" rice, contains a defective gibberellin 20-oxidase gene. *Proc Natl Acad Sci USA* 99(13):9043–9048. doi:10.1073/pnas.132266399

Staub J. (2009) QTL analyses of orange color carotenoid content mapping of carotenoid biosynthesis genes in cucumber (*Cucumis sativus* L.). In: *Proceedings of the IV International Symposium on Cucurbits, Changsha*, China 871, pp 607–614.

Staub JE (1996) Noninfectious disorders: temperature stress. In: Zitter TA, Hopkins DL, Thomas CE (eds) *Compendium of Cucurbit Diseases, Part II.* APS Publishing, St. Paul, MN, pp 66–67.

Sugiyama M, Kawazu Y, Fukino N Shimomura K, Sakata Y et al. (2015) Mapping of quantitative trait loci for *Melon yellow spot virus* resistance in cucumber (*Cucumis sativus* L.). *Euphytica* 205(2):615–625.

Sun J, Xiao T, Nie J, Chen Y, Lv D et al. (2019) Mapping and identification of *CsUp*, a gene encoding an auxilin-like protein, as a putative candidate gene for the upward-pedicel mutation (*up*) in cucumber. *BMC Plant Biol* 19(1):157. doi:10.1186/s12870-019-1772-4

Teng F, Zhai L, Liu R, Bai W, Wang L, et al. (2013) ZmGA3ox2, a candidate gene for a major QTL, *qPH3.1*, for plant height in maize. *Plant J* 73(3):405–416. doi:10.1111/tpj.12038

Tian G, Yang Y, Zhang S, et al. (2015) Genetic analysis and gene mapping of papaya ring spot virus resistance in cucumber. *Mol Breed* 35(4):110. doi:10.1007/s11032-015-0279-4

Tian Z, Jahn M, Qin X, Obel HO, Yang F et al. (2021) Genetic transcriptomic analysis reveal the molecular basis of photoperiod-regulated flowering in Xishuangbanna cucumber (*Cucumis sativus L. var Xishuangbannesis* Qi et Yuan). Genes 12(7):1064. doi:10.3390/genes12071064

Trebitsh T, Staub JE, O'Neill SD. (1997) Identification of a 1- aminocyclopropane-1-carboxylic acid synthase gene linked to the Female (F) locus that enhances female sex expression in cucumber. *Plant Physiol* 113(3):987–995. doi:10.1104/pp.113.3.987

Turek S, Pląder W, Hoshi Y, Skarzyńska A, Pawełkowicz M (2023) Insight into the organization of the B10v3 cucumber genome by integration of biological and bioinformatic data. *Int J Mol Sci* 24(4):4011. doi:10.3390/ijms24044011

Varshney RK, Graner A, Sorrells ME. (2005) Genomics-assisted breeding for crop improvement. *Trends Plant Sci* 10(12):621–630. doi:10.1016/j.tplants.2005.10.004

Wang H, Li W, Qin Y, Pan Y, Wang X et al. (2017a) The cytochrome P450 gene CsCYP85A1 is a putative candidate for Super Compact-1 (Scp-1) plant architecture mutation in cucumber (Cucumis sativus L.). *Front Plant Sci* 8:266. doi:10.3389/fpls.2017.00266

Wang L, Cao C, Zheng S, Zhang H, Liu P et al. (2017b) Transcriptomic analysis of short-fruit 1 (*sf1*) reveals new insights into the variation of fruit-related traits in *Cucumis sativus*. *Sci Rep* 7(1):2950. doi:10.1038/s41598-017-02932-5

Wang Y, Tan J, Wu Z, VandenLangenberg K, Wehner TC et al. (2019) STAYGREEN, STAY HEALTHY: A loss-of-susceptibility mutation in the *STAYGREEN* gene provides durable, broad-spectrum disease resistances for over 50 years of US cucumber production. *New Phytol* 221(1):415–430. doi:10.1111/nph.15353

Wang Y, Vanden LK, Wen C Wehner TC, Weng Y et al. (2018) QTL mapping of downy and powdery mildew resistances in PI 197088 cucumber with genotyping-by-sequencing in RIL population. *Theor Appl Genet* 131(3):597–611. doi:10.1007/s00122-017-3022-1

Wei S, Zhang S, Bo KL, et al. (2019) Evaluation and genome-wide Association Study (GWAS) of seedling thermotolerance in cucumber core germplasm. *J Plant Genet Resour* 20:1223–1231.

Wen C, Zhao W, Liu W, Yang L, Wang Y et al. (2019) *CsTFL1* inhibits determinate growth and terminal flower formation through interaction with CsNOT2a in cucumber. *Development* 146(14):dev180166. doi:10.1242/dev.180166

Weng Y (2020) *Cucumis sativus* chromosome evolution, domestication, and genetic diversity. *Plant Breed Rev* 44:79–111. doi:10.1002/9781119717003.ch4

Weng Y, Johnson S, Staub JE, Huang S (2010) An extended intervarietal microsatellite linkage map of cucumber, *Cucumis sativus* L. *Hortic Sci* 45(6):882–886.

Win KT, Vegas J, Zhang C, Song K, Lee S. (2017) QTL mapping for downy mildew resistance in cucumber via bulked segregant analysis using next-generation sequencing and conventional methods. *Theor Appl Genet* 130(1):199–211.

Win KT, Zhang C, Silva RR Lee JH, Kim YC, et al. (2019) Identification of quantitative trait loci governing subgynoecy in cucumber. *Theor Appl Genet* 132(5):1505–1521. doi:10.1007/s00122-019-03295-3

Wóycicki R, Witkowicz J, Gawroński P, Dąbrowska J, Lomsadze A et al. (2011) The genome sequence of the north-European cucumber (*Cucumis sativus* L.) unravels evolutionary adaptation mechanisms in plants. *PLoS One* 6(7):22728. doi:10.1371/journal.pone.0022728

Wu S, Zhang B, Keyhaninejad N, Rodríguez GR, Kim HJ, et al. (2018) A common genetic mechanism underlies morphological diversity in fruits and other plant organs. *Nat Commun* 9(1):4734. doi:10.1038/s41467-018-07216-8

Xin T, Zhang Z, Li S, Zhang S, Li Q, et al. (2019) Genetic regulation of ethylene dosage for cucumber fruit elongation. *Plant Cell* 31(5):1063–1076. doi:10.1105/tpc.18.00957

Xu L, Wang C, Cao W, Zhou S, Wu T (2018a). CLAVATA1-type receptor-like kinase *CsCLAVATA1* is a putative candidate gene for dwarf mutation in cucumber. *Mol Genet Genom* 293(6):1393–1405.

Xu X, Ji J, Xu Q Qi X, Weng Y, et al. (2018b) The major-effect quantitative trait locus *CsARN6.1* encodes an AAA ATPase domain-containing protein that is associated with waterlogging stress tolerance by promoting adventitious root formation. *Plant J* 93(5):917–930. doi:10.1111/tpj.13819

Xu X, Ji J, Xu Q, Qi X, Chen X et al. (2017) Inheritance and quantitative trail loci mapping of adventitious root numbers in cucumber seedlings under waterlogging conditions. *Mol Genet Genom* 292(2):353–364. doi:10.1007/s00438-016-1280-2

Xu X, Lu L, Zhu B, Xu Q, Qi X et al. (2015) QTL mapping of cucumber fruit flesh thickness by SLAF-seq. *Sci Rep* 5:15829. doi:10.1038/srep15829

Xu X, Yu T, Xu R, Shi Y, Lin X et al. (2016) Fine mapping of a dominantly inherited powdery mildew resistance major-effect QTL, *Pm1.1. Theor Appl Genet* 129(3):507–516. doi:10.1007/s00122-015-2644-4

Yang L, Koo DH, Li Y, Zhang X, Luan F, et al. (2012) Chromosome rearrangements during domestication of cucumber as revealed by high-density genetic mapping and draft genome assembly. *Plant J* 71(6):895–906. doi:10.1111/j.1365-313X.2012.05017.x

Yang L, Liu H, Zhao J, Pan Y, Cheng S et al. (2018) LITTLELEAF (LL) encodes a WD40 repeat domain-containing protein associated with organ size variation in cucumber. *Plant J* 95:834–847. doi:10.1111/tpj.13991

Yang X, Yan J, Zhang Z, Lin T, Xin T et al. (2020) Regulation of plant architecture by a new histone acetyltransferase targeting gene bodies. *Nat Plants* 6(7):809–822. doi:10.1038/s41477-020-0715-2

Yang Y, Dong S, Miao H, Liu X, Dai Z et al. (2022) Genome-wide association studies reveal candidate genes related to stem diameter in cucumber (*Cucumis sativus* L.). *Genes* 13(6):1095. doi:10.3390/genes13061095

Yoshioka Y, Sakata Y, Sugiyama M et al. (2014) Identification of quantitative trait loci for downy mildew resistance in cucumber (*Cucumis sativus* L.). *Euphytica* 198(2):265–276. doi:10.1007/s10681-014-1102-8

Zhang C, Badri Anarjan M, Win KT, Begum S, Lee S (2021) QTL seq analysis of powdery mildew resistance in a Korean cucumber inbred line. *Theor Appl Genet* 134(2):435–451. doi:10.1007/s00122-020-03705-x

Zhang K, Wang X, Zhu W, Qin X, Xu J, et al. (2018) Complete resistance to powdery mildew and partial resistance to downy mildew in a *Cucumis hystrix* introgression line of cucumber were controlled by a co-localized locus. *Theor Appl Genet* 131(10):2229–2243. doi:10.1007/s00122-018-3150-2

Zhang M, Song M, Cheng F, Yang Z, Davoudi M, et al. (2021) Identification of a putative candidate gene encoding 7-dehydrocholesterol reductase involved in brassinosteroids biosynthesis for compact plant architecture in Cucumber (*Cucumis sativus* L.). *Theor Appl Genet.* 134(7): 2023–2034. doi:10.1007/s00122-021-03802-5

Zhang SP, Liu MM, Miao H, Zhang SQ, Yang YH, et al. (2013) Chromosomal mapping and QTL analysis of resistance to downy mildew in *Cucumis sativus*. *Plant Dis* 97(2):245–251. doi:10.1094/PDIS-11-11-0941-RE

Zhang SP, Liu MM, Miao H, Zhang SQ, Yang YH, Xie BY, et al. (2011) QTL mapping of resistance genes to powdery mildew in cucumber. *Sci Agric Sinica*. 44:3584–3593.

Zhao J, Jiang L, Che G, Pan Y, Li Y et al. (2019) A functional allele of *CsFUL1* regulates fruit length through repressing *CsSUP* and inhibiting auxin transport in cucumber. *Plant Cell* 31(6):1289–1307. doi:10.1105/tpc.18.00905

Zhou S., Zhang P., Zhu Y., Chen X., Chen L. (2013). Identification of SSR marker linked to gynoecious loci in cucumber (Cucumis sativus L). *J. Zhejiang Univ. Agric. Life Sci.* 39, 291–298. 10.3785/J.ISSN.1008-9209.2012.11.141

Zhu P, Meng Y, Zhang K, Wang X, Liang K et al. (2022) Mapping of fruit apex shape related QTLs across multi-genetic backgrounds in cucumber (Cucumis sativus L.). *Hortic Plant J* 8(3) 328–340. https://doi.org/10.1016/j.hpj.2021.12.001

8 Allele Mining in *Brassica oleracea* Crops

Simone Treccarichi[1], Marwen Amari[1], Hajer Ben Ammar[1], and Ferdinando Branca[1]*
[1]Department of Agriculture, Food and Environment (Di3A), University of Catania, 95131 Catania, Italy
*Corresponding author

8.1 INTRODUCTION

The selection of new genetic or inheritable traits and alleles has become the main objective of modern agriculture mostly in the framework of addressing ever increasing global climate change, requiring more resilient crop cultivars, which, through their allelic combinations, can induce a stress response (Tao et al. 2011; Kohorn et al. 2012). In the face of increasing world population, all the major crops must be enhanced as a response for increasing food, feed, fiber, and bioenergy demands (Quian et al. 2017; Bhat et al. 2021). Genome wide association studies (GWAS) can provide an essential route to discover the genetic basis for the complex traits in plants and animals, establishing that several genetic regions are inherited as a "genetic block" and their mapping and variation can represent a new way to hasten the breeding process to enhance superior crop traits (Buntjer et al. 2005; Wu et al. 2021). Haplotypes selection has occurred in the evolutionary history of several cultivated plants generating bottlenecks, which allowed for the human selection of the most interesting haplotypes related to specific agronomic or biochemical traits, such as sucrose synthase in wheat, the encoding of ZmCCT in maize, which increases photoperiodic sensitivity and the Gn1a related to the number of grains in a head of rice (Hou et al. 2014; Jordan et al. 2015; Li et al. 2017; Gouda et al. 2021). The development of several typologies of molecular markers for harnessing DNA polymorphism in the selection of superior plant traits has been carried out also thanks to new molecular biology and biotechnology approaches. Linkage disequilibrium (LD) is a statistical analysis tool used to associate different alleles in different *loci,* which tend to be inherited and which belong to the same ancestral haplotype. LD-based approaches aim to identify and map quantitative trait loci (QTLs) and detect the polymorphisms responsible for the expression of different phenotypes (Palaisa et al. 2004; Gupta et al. 2005; Sorkheh et al. 2008).

The *Brassica* genera belong to the asymmetrical evolution of polyploid genomes, which are linked to the U's triangle that includes the three primary diploids *Brassica* gene pools (*B. rapa*, *B. nigra* and *B. oleracea*: AA, BB and CC, respectively). From the diploid species the amphiploids *B. napus*, *B. carinata* and *B. juncea* having the AACC, BBCC and AABB, were respectively generated (Liu et al. 2014; Koh et al. 2017; Bayer et al. 2019; Li et al. 2021).

Plastid genome analysis of the *Brassicaceae* family allows for a deep study of the origins, diversity, and genetic relationships among *Brassica napus*, *B. rapa*, and *B. oleracea,* focusing on the different distribution of their haplotypes, which shows that the cpDNA of the three *Brassica* species is well differentiated and conserved (Flannery et al. 2006; Allender and King 2010; Qiao et al. 2016). *Brassica* genome-wide analysis of haplotypes can represent a useful tool to predict traits based on a single nucleotide polymorphism (SNP) *loci* calculating the general combining ability (GCA) among the population studied (Werner et al. 2018; Jan et al. 2019). Brassica vegetables are rich in bioactive compounds like polyphenols, vitamins, and glucosinolates

DOI: 10.1201/9781003376439-8

developed by their secondary metabolism which confers high sensorial properties (Di Bella et al. 2020, ; Ben Ammar et al. 2022). Several authors were involved in the detection of alleles of the *Brassica* oilseed *crops* (*B. napus* and *B. juncea*) for the selection of the most interesting traits related to plant morphology and to oilseed production such as plant height, the flower's maturity and morphology, and the yellow seedcoat (Jambhulkar et al. 2015; Kumar et al. 2015; Gill et al. 2021). In *B. oleracea* and *B. rapa,* genomes occurred 13–17 million years ago, the whole genome triplication (WGT) event in the *Brassica* ancestor, followed by an extensive gene fractionation and chromosome reduction due to the rediploidization process produced a stable diploid species. For this reason, *B. oleracea* and *B. rapa* genomes present in each feature three paleosubgenomes and multiple copies of paralogous genes and several of them, with their presence/absence and copy number variation determine different gene expression and morphotype diversification (Cheng et al. 2016; Golicz et al. 2016).

The following genome organization allowed for the development of a BeadChip array with about 50.000 SNPs markers for population structure (Pelc et al. 2015; Mei et al. 2017).

The origin of *B. oleracea* in the Mediterranean region is supported by several authors and was substantiated based on the new genotyping techniques such as the genotyping-by-sequencing (GBS) (Maggioni 2015; Maggioni et al. 2018; Stansell et al. 2018; Mabry et al. 2021; Stansell and Björkman 2020). Cabbage is the most exploited *B. oleracea* vegetable, with the largest harvested area and production worldwide of about 2.414.000 ha and about 70.900.000 t of production, respectively, followed by cauliflower and broccoli cultivated on about 1.357.000 ha with about 25,000,000 t of production, respectively. *B. oleracea* vegetables include different morphotypes showing enormous phenotypic variation, which can be exploited by different crops like broccoli (*B. oleracea* var. *italica*), cauliflower (*B. oleracea* var. *botrytis*), cabbage (*B. oleracea* var. *capitata*), kale (*B. oleracea* var. *acephala*), savoy cabbage (*B. oleracea* var. *sabauda*), and Brussel sprouts (*B. oleracea* var. *gemmifera*) (Maggioni et al. 2010; Ciancaleoni et al. 2014).

Several landraces of broccoli, which are highly valued by growers and producers for their organoleptic qualities, are now grown in both large fields and household gardens (Ciancaleoni et al. 2014; Cartea et al. 2020). *Brassica oleracea* L. ($2n = 2 = 18$) is an insect-pollinated crop with substantial outcrossing (72–95%) due to a sporophytic self-incompatibility system regulated by a polymorphic S-locus (Takayama et al. 2001). The species is one of the most striking examples of structural evolution among domesticated plants (Babula et al. 2007). Regarding significant agronomic properties, most landraces and synthetics clearly separate themselves from hybrid types. Many morpho-physiological features (such as harvesting time, head size, color, and number of leaves) separate landraces and synthetics, as previously observed in other investigations (Branca 2008). Their names, on the other hand, already allude to identifiable and/or specific characteristics (Iannarisi and Marzolo, that flower in January and March, respectively; Nero, which has a very dark color, etc.). This is most likely owing to a long history of adaptation to local pedo-climatic circumstances and farmers' selection in favor of crucial characteristics like maturity date, corymb form, or flavor. It is also worth noting that detailed characterization aids in the promotion of products derived from local variations, as is now being done for many cultures (Negri 2012). Furthermore, it is necessary to use the diversity of local races in the selection. The development of new varieties from local varieties may also help to keep their diversity alive in the fields.

Domestication and breeding are typically linked with large changes in genetic diversity for a subset of target genes or genomic areas as a result of linkage drag (Suzuki 2010). This has resulted in the development of novel genome-wide techniques that expose the pattern of genetic alterations and allow for the discovery and dissection of areas that contribute to complicated trait variation without the requirement for previous trait assessment. These methods include genome-wide scans of nucleotide diversity, LD decay patterns, and the identification of outliers in a genomic scan to select the chromosomic areas based on genetic divergence across populations (Motsinger et al. 2006). Organic plant breeding requires several conditions related to the integrity of the crop cultivations and several authors have focused on the relationship between the ethical aspect related to genetic engineering approaches and plant breeding (Van Bueren et al. 2003; Nuijten et al. 2016).

The evolution and the selection techniques of the new superior alleles in *B. oleracea* crops were examined and reviewed highlighting the new biotechnological methods allowing for the identification of candidate genes for the elite traits. The allele identification in *B. oleracea* is usually developed based on the candidate genes previously identified in the *Arabidopsis* or *B. rapa* genome and transcriptomic analysis mediate the process detecting the differential expressed genes in plants grown in contrasting conditions. In this review, the methods performed on *B. oleracea* for superior alleles identification were detailed, specifying the validated QTLs and candidate genes, useful for breeding programs by the exploitation of *B. oleracea* genetic diversity.

8.2 ALLELE MINING METHODS FOR *BRASSICA OLERACEA* VEGETABLES

8.2.1 Approaches and Application

Detection of major genes and QTLs in *B. oleracea* vegetables is ascertained by several methods, which will be discussed in the following section, focusing on the molecular methodology. The aim of the crop breeding programs is the improvement of the most important agronomic traits and several of which show complex quantitative inheritance, and their analysis and mapping allow for the identification of the QTLs of the candidate genes. Advances in molecular biology and in the knowledge of the genetics of complex traits allowed for the development of several methodologies to associate traits with their proper genomic regions through the molecular markers, which can be exploited in the marker-assisted selection (MAS) (Jaganathan et al. 2020). The phenotypic variation related to a quantitative trait belongs to its segregation and is closely related to the internal and external environments, and combining the QTL mapping and cloning to mutagenesis approaches which can accelerate the discovery of the underlying genes QTLs (Paran and Zamir 2003; Salvi and Tuberosa 2007).

Polymerase chain reaction (PCR) strategies have been developed based on gene and genome sequences to isolate useful alleles of genes from a diverse range of species (Latha et al. 2004). This capability allows for direct access to key alleles that confer resistance to biotic and abiotic stresses, increased water and nutrient use efficiency, increased yield, and improved quality of the products. Similar alleles responsible for a given trait and their variants in other genotypes can be identified using novel genomic tools. This is known as dissection of naturally occurring variation at candidate genes/*loci* or allele mining. The discovery of allelic variants in germplasm collections not only provides new germplasm for delivering novel alleles for targeted trait improvement, but also categorizes germplasm entries for conservation (Kumar et al. 2010).

The conservation and the exploitation of plant genetic resources are critical for future agricultural progress. The development of genetic markers that not only distinguishes individuals and accessions but also reflects the inherent variation; and genetic relationships among collection holdings is critical to this conservation effort. Kresovich et al. (1992) characterized genetic identities and relationships of *B. oleracea* via a random amplified polymorphic DNA (RAPD) assay. They investigated the suitability of the RAPD assay for quick, cost-effective, and dependable use in addressing the organization of these collections and management needs. Twenty-five decamer oligonucleotide primers were individually tested in *B. oleracea* using a test array made up of individuals representing a variety of genetic relationships. Over 140 reproducible polymorphic fragments were created for research purposes. Each member of the test array had a distinct molecular genotype, and composites specific to accessions and botanical varieties could be identified. The known genetic relationships of the selected entries were reflected in a similarity analysis based on amplified DNA fragments. These findings demonstrated that RAPD markers can be extremely useful in gene bank management for purposes of identification, variation measurement, and establishing genetic similarity at the intraspecific level.

The distribution of genetic diversity in a *B. oleracea* gene bank collection was linked to the effects of diversity regeneration, as measured by amplified fragment length polymorphisms (AFLPs) (van Hintum et al. 2007). These issues were investigated in an existing gene bank collection of a cross-pollinating crop, using a selection of very similar Dutch white cabbage accessions, as well as additional groups of reference material representing the Dutch and the global white cabbage gene pool. Six different accessions were sampled before and after a standard regeneration. AFLPs were used to characterize 30 plants from each of the 50 accessions and six regeneration populations were included in the study, with scores for 103 polymorphic bands. The genetic changes caused by standard gene bank regenerations, as measured by AFLPs, were found to be of comparable magnitude to the differences between some of the more similar accessions. The observed changes are primarily the result of highly significant changes in allele frequencies for a few fragments, whereas most fragments have allele frequencies that are similar before and after regeneration. It is argued that because accessions change over generations, accessions with similar levels of differentiation can be safely combined.

New molecular tools can improve the superior allelic variants' individuation detecting oligonucleotide variation that can express important phenotypic changes. Alleles from different genotypes are amplified by PCR, and then DNA sequencing is used to identify nucleotide variation. To study haplotypes' diversity and structure and infer genetic association studies in plants, sequencing-based allele mining might be helpful. Contrary to Eco-TILLING (Targeting Induced Local Lesions In Genomes), sequencing-based allele mining entails significant sequencing costs but does not call for specialized machinery or time-consuming procedures. Despite the assertion that Eco-TILLING is less expensive than SNP/haplotyping approaches, the intricate equipment, specialized knowledge, and requirement for confirmation sequencing in the Eco-TILLING procedure make its broad applicability questionable (Banik et al. 2007). Each technology does, however, have advantages and disadvantages that should be carefully weighed in relation to the desired application.

Sequencing-based allele mining is highly efficient since it involves one step to perform the identification of sequence variations, and the detection of any type of nucleotide polymorphism. On the other hand, Eco-TILLING requires high technical expertise starting from DNA pooling, to detection of cleavage of heteroduplexes. It is less efficient due to the high chances of false positives, non-specific cleavage and chances of non-detection in DNA pools, proposed as effective in the detection of SNPs rather than InDels. With regard to Eco-TILLING application costs per data point are comparatively high due to the need for the intervention of a confirmatory sequencing step, also requiring more time, especially for sample preparation. In addition, associated complexity reduces the throughput and only fewer samples can be processed (Kumar et al. 2010).

Although the DNA sequence offers fundamental insights into how the genome is organized, early sequencers that relied on either chain termination sequencing or cycle sequencing had significant drawbacks in terms of cost, complexity, and time. Faster sequencing platforms have made the sequencing process easier and less expensive. The emergence of massively parallel techniques in recent years has also resulted in the creation of next-generation sequencing systems with higher throughput and accuracy. The primary purposes of these techniques were resequencing, alignment of the sequence data, and comparison of those results with reference genomes. The first of its kind, which relied on pyrosequencing and did not require cloning, was commercialized by 454 Life Sciences (Margulies et al. 2005). It was possible to generate 100 Mb of sequence with 99.5% accuracy and increased read length average over 250 bases using this 454-sequencing process. Illumina/Solexa genome analyzer, a different massively parallel sequencer, has been created (Bennett 2004). It is capable of sequencing one billion bases (Gb) of 30- to 40-base sequence reads in a single run in a short amount of time (3 to 4 days). A massively parallel sequencer using supported oligonucleotide ligation and detection system (SOLiD) has also been created by Applied Biosystems Inc. (USA) in addition to the aforementioned (Hutchison 2007). This system can produce 2–3 Gb of sequences every run despite its shorter read length of 25–35. Recently, a brand-new massively parallel system built on single molecule sequencing (SMS) technology was commercialized by HelicosHeliscope (www.helicosbio.com). SMS technology has been referred to as third-generation or next to next-generation sequencing technology since it is significantly faster and less expensive than Applied Biosystems and Illumina second-generation sequencing technologies (Gupta 2008). Early in 2010, Pacific Biosciences SMRT (www.pacificbiosciences.com) anticipated beginning commercializing yet another such third-generation sequencing technology.

As the DNA sequencing technologies advance toward the long-awaited $1000 genome (or sequencing a human genome for $1,000), and sequencing costs fall by a factor of two or three each year (Pettersson et al. 2009), sequencing at a lower cost will become a reality. Given the availability of low-cost, longer-read-length, high-throughput sequencing platforms, sequencing-based allele mining would result in a faster generation of allelic data at a lower cost in the future. Furthermore, sequencing-based allele mining of specific genes in identified accessions and their association with phenotypic variations will provide a significant boost to crop plant precision breeding programs.

The availability of genome and gene sequence information for a specific crop species would be critical for successful and efficient allele mining activity. Availability of efficient and reliable phenotyping techniques; genomic resources; availability of high-throughput techniques for quick and easy generation of allelic data points; cost-effective sequencing platforms; efficient bioinformatic tools for identifying nucleotide variation and molecular marker construction technologies for marker-assisted selection (MAS) are a few other important factors. Allele mining can be started for any crop with sequence information and even for a related member of its family based on these criteria.

Allele mining can be used effectively to discover superior alleles by mining the gene of interest from various genetic resources. It can also shed light on the molecular basis of novel trait variations and for identifying nucleotide sequence changes linked to superior alleles. Furthermore, the rate of evolution of alleles, allelic similarity/dissimilarity of a candidate gene, and allelic synteny with other members of the family can be investigated. Allele mining may also pave the way for molecular discrimination among related species, the development of allele-specific molecular markers, the facilitation of novel allele introgression via MAS, or deployment via genetic engineering (Kumar et al. 2010).

Transcriptome sequencing has been shown to be a valuable tool for gene discovery, allele mining, and marker development. Because it reflects transcribed portions of the genome, markers based on expressed sequence tags (ESTs) are preferred. Izzah et al. (2014) describes the creation of simple sequence repeat (SSR) and single nucleotide polymorphism (SNP) markers derived from cabbage transcriptome sequences, as well as their application in map construction. This genetic map supplements the existing *B. oleracea* map with genetic information. Furthermore, the newly developed set of EST-based SSR and dCAPS markers is a valuable resource for genetic studies and will aid in cabbage breeding. Additionally, this study shows that next-generation sequencing (NGS) transcriptomes can be used to develop genetic maps even when the mapping population has little genetic diversity.

QTL research and MAS are rapidly evolving and widely used in plant breeding, including wheat (Singh et al. 2004), potato (Gebhardt et al. 2006), and cabbage (Chen et al. 2013). Though some QTLs have been identified in cabbage, research is still in its early stages because QTL cloning has yet to be reported and functional analysis studies are uncommon. To date, cabbage QTLs have primarily been studied for their effects on fertility (Wang et al. 2000), plant size (Lan and Paterson 2001), plant regeneration (Holme et al. 2004), flowering time (Bohuon et al. 1998; Okazaki et al. 2007; Uptmoor et al. 2008), clubroot resistance (Nagaoka et al. 2010), and black rot resistance (Kifuji et al. 2013).

Previous research has identified QTLs for flowering time in *Brassica oleracea* (Kennard et al. 1994), discovering significant QTLs using single-factor ANOVA on an F_2 from a cabbage × broccoli cross, whereas Camargo and Osborn (1996) used F_3 families from a different cabbage × broccoli cross. Indeed, Rae *et al.* (1999) over the course of 5 years developed, 79 recombinant backcross substitution lines from a cross between *B. oleracea* var. *italic* and *B. oleracea* var. *alboglabra* were grown in field trials alongside the *alboglabra* recurrent parent. Plants were scored based on the number of days it took from sowing to the first flower opening, and lines significantly showing reduction or increase of the flowering time from that of the recurrent parent, were identified. Evidence for 11 QTLs on chromosomes O1, O2, O3, O5, and O9 was found based on the size of the substitutions, five of them were mapped in similar regions to the ones which were previously analyzed in doubled haploid (DH) lines from the same cross. Several QTLs were found to be closely related in terms of repulsion.

The isolation of genomic *BoFLC* genes and identification of QTLs controlling flowering time using a linkage map and flowering data from an F_2 population, derived from a broccoli DH line by a cabbage DH line to better understand the genetic control of flowering time in *B. oleracea*, and also the role of *BoFLC2* in flowering through vernalization is discussed by Okazaki et al. (2007). In fact, four *FLC* homologs (*BoFLCs*) from *B. oleracea* were cloned. Three of these, *BoFLC1*, *BoFLC3*, and *BoFLC5*, have previously been identified. The fourth novel sequence shared 98% sequence homology with the previously identified gene *BoFLC4*, but also 91% homology with *B. rapa*'s *BrFLC2*. Phylogenetic analysis revealed that this clone is a member of the *FLC2* clade. As a result, they named this gene *BoFLC2*. A detailed linkage map of *B. oleracea* was constructed in the F_2 progeny obtained from a cross of *B. oleracea* non-vernalization and vernalization types which covered 540 cM and nine major linkage groups. Six QTLs that control flowering time were discovered. The QTLs controlling flowering time were not linked to *BoFLC1*, *BoFLC3*, or *BoFLC5*. The largest QTL effect, however, found in the region where *BoFLC2* was mapped. Due to a frameshift in exon 4, the *BoFLC2* homologs found in non-vernalization plants were non-functional. Furthermore, *BoFLC2* duplications and deletions were found in broccoli and a rapid cycling line, respectively. These findings suggest that *BoFLC2* plays a role in the regulation of flowering time in *B. oleracea*.

Finch-Savage et al. (2005) presented a study supported by quantitative genetic analysis to determine the sensitivity of *B. oleracea* seed germination to oxygen deprivation, as well as to investigate whether variation in seed sensitivity to hypoxia has a genetic basis, and to identify QTLs involved in the determination of this trait in a *B. oleracea* DH population. The DH lines are useful for QTL mapping because they have homozygous alleles at all loci and, like recombinant inbred lines, can be propagated and distributed while retaining genetic fidelity. In fact, significant QTL was found on linkage group O2, and the presence of others in linkage groups O8 and O9 was suggested. The data show that this trait is genetically controlled and the QTL on O8 may be a general stress QTL influencing the germination rate, whereas the QTL on O2 may be oxygen stress-specific and influencing a threshold that determines the number of seeds capable of germination.

Sebastian et al. (2002) discovered some QTLs that control the development of *B. oleracea* L. In fact, a segregating population of *B. oleracea* F_1-derived DH lines was used to detect and locate QTLs controlling 27 morphological and developmental traits such as leaf, flowering, axillary bud, and stem characteristics. The population was created by crossing two distinct *B. oleracea* crop types, an annual cauliflower, and a biennial Brussels sprout. A PCA based on the linear means classified all the traits into distinct categories based on the first five principal components. They were as follows: leaf traits (PC1), flowering traits (PC2), axillary bud traits (PC3 and 5), and stem traits (PC6) (PC4). Using the multiple-marker regression approach to QTL mapping, researchers found between zero and four putative QTL(s) per trait that explained between 6% and 43% of the additive genetic variation. Two QTLs with opposing effects were found on the same linkage groups for lamina width, bare petiole length, and stem length. Intra- and inter-specific comparative mapping with restriction fragment length polymorphism (RFLP) markers revealed a QTL on linkage group O8 that accounts for variation in vernalization and is likely synonymous with a QTL found on *B. napus* linkage group N19. Furthermore, a QTL for petiole length discovered on O3 of this study appeared to be homologous to a QTL discovered on another *B. oleracea* genetic map (Camargo et al. 1996).

Wang et al. (2012) conducted research on the creation and comprehensive analysis of a high-density linkage map of *B. oleracea.* The map spans 1197.9 cM and is divided into nine linkage groups, one for each of the *B. oleracea* chromosomes, with an average distance of 0.98 cM between adjacent markers. Significantly, 1.063 (86.63%) of the 1,227 mapped markers were sequence-based markers for *B. oleracea.* SSRs are effective anchor markers with high polymorphism and single sites. They can be used to connect various linkage maps and chromosomes. SSRs are also useful probes for fluorescent *in situ* hybridization (FISH), which is used to integrate genetic and cytogenetic maps. The main reason for the uneven distribution of molecular markers is that some chromosomal regions in the parental lines are devoid of polymorphisms. As a result, developing mapping populations from different crosses and employing new molecular markers such as SNPs and SSRs are effective ways to fill in gaps between makers in order to obtain saturated genetic maps.

Lv et al. (2014) created a linkage map using a DH population of 196 lines and SSR markers based on the reference genome sequence and InDel markers resulting from whole-genome re-sequencing data of the two parental lines. They also evaluated the heading characteristics in three field experiments. They hoped to map novel genomic regions and relatively significant QTL regions influencing heading traits using the map and data from three seasons in order to gain a better understanding of the genetic control of these characters. In fact, on five chromosomes, six active regions with multiple QTLs were identified, one of which contained major QTLs associated with five traits. This study's QTLs should be useful for marker-assisted selection in cabbage breeding as well as understanding the genetic control of these traits concerning the heading traits including head mature period, head weight, core length (Cl), head vertical diameter (Hvd), and the ratio of Cl to Hvd (Cl/Hvd).

In contrast to defining the coding region for allele mining, the demarcation of putative promoter regions is one of the challenges in promoter mining analysis. Promoters and regulatory elements are dispersed throughout the upstream region, and their locations vary from gene to gene. The available literature clearly indicates a lack of agreement on the size of the region for mining promoter and regulatory elements. The size of this window ranges from 100 to 3000 bp (Brazma et al. 1998, Park et al. 2002; Molina and Grotewold 2005; Veerla and Hoglund 2006). However, high regulatory element densities have only been reported between 100 and 600 bp in the upstream region.

Mining for sequence variation, particularly in regulatory regions of genes, has grown in importance in the context of gene expression in recent years. Since it is well understood that promoter elements play an important role in gene regulation, any changes in their sequences will have a significant impact on gene expression, resulting in variable trait expression. Such differential gene expression is typically caused by sequence polymorphism in upstream cis-acting/trans-acting regulatory element binding sites.

In contrast to identifying variation in coding regions of the genome, characterizing the extent of cis-acting regulatory variation is far more difficult, as it is not discernible even in fully sequenced genomes. Screening for regulatory variants based on transcript level differences between individuals is hampered by potential trans-acting factors or environmental differences (Dai et al. 2007). As the promoter regions can function in a bi-directional pathway, prediction of the orientation will be difficult even by through software. Furthermore, the complexity of analyzing these promoters is increased because transcription factors typically control more than one gene (Hehl and Wingender 2001).

This type of promoter sequence analysis also provides information on the conservation of regulatory motifs specific to specific differentiated cell/tissue types. Finally, promoter mining will provide insights into the molecular mechanisms of gene expression as well as the isolation of novel and efficient promoters for use in genetic engineering applications. As the number of global gene expression experiments to identify sets of genes with similar expression patterns grows, it is critical to incorporate approaches such as *in vitro* and *in silico* promoter mining into regular allele mining programs to identify and categorize cis-regulatory elements and understand their role in gene regulation.

Using hypocotyl and cotyledon explants, *Agrobacterium rhizogenes* A4M70GUS-mediated transformation of Savoy cabbage (*B. oleracea* L. var. *sabauda*) and two local lines of cabbage (*B. oleracea* L. var. *capitata*) was achieved. In all genotypes, the percentage of explants that formed roots was very high: 92.3% in Savoy Gg-1, 64.4% in cabbage P22I5, and 87.2% in P34I5. In all three genotypes, spontaneous shoot regeneration of excised root cultures grown on a hormone-free medium occurred. Shoot regeneration was higher in cabbage lines P22I5 and P34I5 (9.3 and 2.6%, respectively) than in Savoy cabbage Gg-1 (1.3%). The glucuronidase (GUS) histological test and PCR analysis were used to determine the transgenic nature of hairy root-derived plants. GUS expression was detected in all the tested cabbage shoots, whereas GUS expression was detected in only 55% of the tested Savoy cabbage clones. The GUS gene was found in regenerated shoot clones and T1 progeny via PCR analysis (Sretenović-Rajičić et al. 2006).

Rhizobium radiobacter-mediated transformation transferring the *Bacillus thuringiensis* (*Bt*) *cry* genes, provided the transgenic cabbage (*B. oleracea* var. *capitata*) resistance to diamondback moth (*Plutellaxy lostella*) *larvae*. Cabbage cultivar; preculture and coculture of explants on a callus initiation medium; use of appropriate amount; and delay in the initial application of selective agents were all important transformation factors. Several transformation experiments yielded a total of 15 independent transformed lines with over 100 plants, representing an overall transformation efficiency of 1%. Cabbage plants transgenic with a synthetic *Bt* gene, *cry1Ab3*, were all resistant to diamondback moth larvae, whereas plants transgenic with a wild-type *Bt* gene, *cryIIa3*, were susceptible (Jin et al. 2000).

Allele mining has the potential to be used to identify nucleotide variation at a genomic region (candidate gene) that is associated with phenotypic variation for a trait. This allows one to assess the frequency, type, and extent of occurrence of new haplotypes, as well as the resulting phenotypic changes. There are numerous cases of studies available for analyzing haplotype diversity and identifying new haplotypes incandidate genes. The most common haplotype changes and their prevalence in populations would serve as the foundation for association mapping studies (Li et al., 2021). Many crops have reported the precise establishment of mini-core collections using molecular markers (Upadhyaya et al. 2006b; Upadhyaya et al. 2008). To further reveal the genetic use of germplasm accessions in allele mining, molecular characterization of mini core and trait-specific subsets was proposed (Upadhyaya et al. 2008a). The data generated by morphological and molecular characterization studies could be used to define the genetic structure of the global composite collection and, finally, to select a "reference sample" of approximately 300 accessions (depending on the size and genetic diversity of that crop's core collection) representing the maximum diversity for the purpose of identifying and isolating allelic variants of candidate genes associated with beneficial traits (Upadhyaya et al. 2006a). It has been proposed that mini core collections or reference sets (subsets of mini core) could be used to begin dissecting naturally occurring allelic variation at candidate genes via allele mining efforts. The methods used for the identification of the QTLs or major genes were resumed in Table 8.1, focusing also on the regions in which the candidate *loci* were identified.

8.2.2 Data Analysis and Database

Allele mining's main goal is to identify a small number of genomes and accessions from a large collection using either sequence or phenotypic information. The genome data serves as a digital filter for identifying a small number of accessions for targeted phenotyping. The dataset of different genomes makes it easy to map traits and validate specific target sequences. Accessions with unique sequences and haplotypes can be identified using known sequences to search the genome database (Leung et al. 2015).

CropSNPdb, a database of SNP array data for Brassica crops and hexaploid wheat, is a user-friendly resource for exploring the genetic diversity of a large collection of germplasm accessions. This database allows users to search for SNP haplotypes, germplasm accession names, passport data,

TABLE 8.1
Allele Mining Application and Approaches in *B. oleracea* Vegetables with Their Reference

Method	Target Function	Genomic Region	Reference
RAPDs markers	Genetic diversity		Kresovich et al. 1992
AFLPs markers	Genetic diversity		van Hintum, 2007
QTLs analysis by RFLP markers	Stem length; leaf lamina length; lamina width and petiole length, respectively; stem weight and number of nodules.	C01, C04 and C07; C07; C01 and C04; C07 and C09.	Lan and Paterson 2001
QTLs analysis by ERPAR markers	Fertility		Wang et al. 2000
QTLs analysis	Vernalization; petiole length	C08; C03	Sebastian et al. 2002
	Sensitivity of seed germination to oxygen deprivation	C02 (66.9 cM), C08 (49.7 cM) and C09 (25.3 cM).	Finch-Savage et al. 2005
	Flowering time	Whole genome, but the most consistent regions were in O2, O3, O5 and O9.	Okazaki et al. 2007
	Club rot resistance (CR *loci*)	Major QTLs on C02 and C01.	Nagaoka *et al.* 2010
Transcriptomic analysis	Development of EST-SSRs, SSRs and SNPs for black rot resistance and genetic map construction.	C03 (six loci within 6.3 cM); C07 (three loci within 2.2 cM); C08 (3 loci within 1.1 cM).	Izzah et al. 2014

and basic agronomic information. As a result, the SNP-Seek database provides a simple and easy-to-use platform for searching for new alleles. The SNP data will evolve over time as the sequences are aligned to additional high-quality reference genomes. Previously, the *Brassicaceae* Database (BRAD) was a searchable database of *Brassicaceae* genomic data hosted on the website "http://brassicadb.org." They relocated it to the new website "http://brassicadb.cn" for policy and security reasons. It is now a new database that includes not only newly released *Brassicaceae* genome sequences, but also published genomic data from the majority of other *Brassicaceae* species. They provided services for 41 genomes or genome versions from 26 species in total. Genomic data primarily consists of genome assemblies, predicted gene models, and gene annotations (Scheben et al. 2019).

Other public databases for *Brassica oleracea* genome sequence data such as the *Brassica* Genome Gateway (http://brassica.bbsrc.ac.uk/), Brassica.info (www.brassica.info/resource/databases.php), and AAFC Comparative Genome Viewer (http://brassica.agr.gc.ca/navigation/viewer e.shtml) provide access to *B. oleracea* genome sequence and annotations. These databases only provide partial genomic data for *B. oleracea*, including QTLs, ESTs, and cloned genes, including data on genome sequences and comparative genomics. These user-friendly databases are exploited by students and researchers to perform *in silico* analysis investigating the molecular function of genes, comparative genomics, and phylogenetic analysis related to the *Brassicaceae* family, as well as to promote advancements in molecular breeding within *Brassicaceae* (Yu et al. 2019). The Swiss-Prot and

TrEMBL databases can be used for gene characterization and were used by Zou (2019) to individuate candidate genes involved in *B. oleracea* head length and weight.

In 2019, Yu and his colleagues established Bolbase, a comprehensive database that allows users to access *Brassica oleracea* genome data and gene annotations. *Brassica oleracea*'s whole genome includes nine fully sequenced chromosomes and 1.848 scaffolds, as well as 45.758 predicted genes, 13.382 transposable elements, and 3.581 non-coding RNAs. There is information available on comparative genomics, such as syntenic regions between *Brassica oleracea*, *Brassica rapa*, and *Arabidopsis thaliana*, synonymous (Ks) and non-synonymous (Ka) substitution rates between orthologous gene pairs, gene families, or clusters, and differences in the quantity, category, and distribution of transposable elements on chromosomes (Yu et al. 2019).

Bolbase represent the first database developed for *B. oleracea* genome and genomic comparisons with relatives, and it will help researchers better understand the function and evolution of Brassica genomes, as well as improve molecular breeding research. This database is regularly updated with new features, genome annotation improvements, and new genomic sequences, it is available for free at http://ocri-genomics.org/bolbase. The data from various genomes could be used as a community-wide resource, allowing the global community to explore diversity. Another challenge is to link phenotypes to sequenced accessions, as well as to use disease resistance traits to demonstrate the identification of new diversity from traditional varieties or landraces.

The genome-wide association study (GWAS) is a molecular technique for identifying specific genomic regions or loci that govern simple to complex traits. GWAS can be used to determine whether a genomic variant is associated with a trait of interest by using germplasm, a segregation population, or a diverse genotype collection (Challa and Neelapu 2018; Liu and Yan 2019). Using the GWAS approach, various candidate genes or quantitative trait loci responsible for traits of interest have been identified.

8.3 EVOLUTION AND SELECTION OF SUPERIOR *B. OLERACEA* ALLELES RELATED TO THE AGRONOMIC TRAITS

8.3.1 Genes Related to the Flowering Traits

Brassica oleracea has several classes of genes involved in floral development and several of them belong to mutation events, which determined the induction of hypertrophic inflorescence in the *Brassica* crop. Most of these genes are related to the MADS-box family and were studied by several authors (Bowman et al. 1993; Smith and King 2000; Duclos and Björkman 2008; Irish 2010; Wils and Kaufmann 2017; Sheng et al. 2019). The MADS-box family includes key regulator genes involved in reproductive development and plays an important role in the floral organ's differentiation, flower development, flowering time, inflorescence architecture, and seed development (Castelán-Muñoz et al. 2019). This large family of genes is widely distributed in eukaryotes, and it encodes several known transcription factors having 58–60 conserved amino acids determining the MADS domain which was discovered by founding its first four genes encoded in *Saccharomyces cerevisiae*, *Arabidopsis thaliana*, *Antirrhinum majus* and *Homo sapiens*, respectively (Passmore et al. 1988; Sommer et al. 1990; Honma et al. 2001). Flower development genes were studied by Bowman et al. who found several genes involved such as *apetala* 1 and *cauliflower* in *Arabidopsis*. These genes are closely related to members of the MADH-box genes family and a mutant copy of them is present in *B. oleracea* genome. Irish and Sussex (1990) characterized several floral phenotypes and identified an *Arabidopsis* mutant expressing the recessive homeotic *apetala* 1 (*ap1*) mutation, and the homozygote for this mutation showed weak inflorescence affecting floral primordia formation. Several genes such as *apetala* 1 (**AP1**) have been demonstrated to be involved into floral structures that control meristematic inflorescence development (Wils and Kaufmann, 2017). CAULIFLOWER gene (**CAL**) was identified by Bowman et al. (1993) showing the same *ap1or CAL* phenotype in *Arabidopsis* exhibiting curd resembling cauliflower. The orthologue of this gene was found in cauliflower by Kempin et al. (1995) and was called *BoCAL*. The following genes, in different genotypes, showed different alleles and one of them named *Bocala* was responsible of the premature stop codon in exon 5, arresting the inflorescence primordia proliferation (Schilling et al. 2018). The variation and selection at the *CAULIFLOWER* floral homeotic gene accompanying the evolution of domesticated *Brassica oleracea*. Sicilian Purple type is considered an intermediate of the Calabrese Broccoli and cauliflower genotype in fact, it shows heterozygosity at the *BoAP1* and *BoCALloci* exhibiting the following genes at the status AAcc or aaCC with the alleles *BoAP1*-a and *Bocal-a*, or *Boap*1-a and *BoCAL*-a, respectively allowing the study of the following alleles distribution (Smith and King 2000; Purugganan et al. 2000). The allelic variation in MADS-box genes show various developmental processes that generate several different morphotypes in *Brassica* species (Schilling et al. 2018). These genes are related to the development of reproductive organs and belong to **MADS-box genes family** (Wils and Kaufmann 2017). Another gene named *BoTFL1* was identified as a strong repressor of the inflorescence development in Arabidopsis but did not arrest its induction in *B. oleracea* confirming that the floral development in broccoli and cauliflower is not managed exclusively by the homologues of the genes of Arabidopsis (Duclos and Björkman 2008).

The development of the floral primordia in *Arabidopsis* is supported by the suppression of *BoCAL*, *BoAP1-a* or *BoLFY* genes or a failure in the suppression of *BoTFL1* gene, which is a strong repressor of flowering in Arabidopsis, but which does not suppress the development of floral primordia in *B. oleracea* (An et al. 2007). AP1 paralog gene like *LEAFY* (LFY) and *CAL*, interact among their transcription factors to encode proteins to control the onset of flower development and in this regulatory model key regulators like *TERMINALFLOWER1* (*TFL1*) are also present which play against the *AP1* and *CAL* genes (An et al. 2007; Goslin et al. 2017). To resume, *B. oleracea* differs from the *Arabidopsis thaliana* floral genetic model because of the different genetic pathways involved.

Flower development in cauliflower is also linked to the expression of the *FLC* gene (flowering *locus*) and its paralogues like *FLC2*, which in a segregating F2 population of late flowering cauliflower with the allele *BoFLC2,* to an early flowering cauliflower with the allele *Boflc2* accounted for about 65% of flowering time variation (Ridge et al. 2015). FLC gene is linked to the flowering vernalization governing flowering time regulation and it was mapped by Okazaki et al. detecting several gene's homologs (2007), and its activity in flowering regulation was also confirmed for *B. rapa* (Okazaki et al. 2007; Ridge et al. 2015).

Single nucleotide polymorphism can affect the expression of the allelic patterns involved in the flower induction and this can occur in *BoFLC4* for *B. oleracea* showing a different response in the vernalization process (Irwin et al. 2016). The functional gene *FRIGIDA* (*FRI*) in *Arabidopsis thaliana* was studied and it encodes the requirements useful for the vernalization process through the overexpression of the MADS box transcriptional repressor *FLOWERING LOCUS C* (*FLC*). Functional alleles of the FRIGIDA genes act as flowering time regulators in the *Brassicaoleracea* genotypes (Irwin et al. 2012). FRI orthologs were mapped in *B. oleracea* detecting *BolC.FRI.a* and *BolC.FRI.b* alleles showing an abundant amino acid concentration in C-terminal and divergences from *A. thaliana* for the amino acid in N-terminal (Sun et al. 2013). FRI allelic orthologs were also mapped in *B. rapa* and they differ from *B. oleracea* in the chromosome region, which is A3 and A4 and C3 and C9, in A and C genomes, respectively (Fadina et al. 2013; Sun et al. 2013; Takada et al. 2019). *FRI* paralogs are highly conserved in the subgenomes A and C of tetraploid *B. napus* suggesting the *loci* duplication in the speciation of the *Brassica* genus (Takada et al. 2019).

MADS-box domain includes two genetic groups named type I and type II which comprise the *STRUBBELIGRECEPTOR* family (SRF) and MEF2-like, respectively (De Bodt et al. 2003). Sheng et al. (2019) characterized and mapped 91 MADS-box transcription factors distinguishing from type I (Mα, Mβ, Mγ) and type II (MIKC^{C}, MIKC*) groups as a consequence of phylogeny and gene structure analysis: 59 genes randomly distributed on 9 chromosomes and 23 were in 19 scaffolds and nine of them were not located due to lack of information on the NCBI database (Sheng et al. 2019). SSRs developed by the MADS box genes can represent a useful tool to detect polymorphisms inducing a hypertrophic curd induction (Treccarichi et al. 2021, 2023).

Basic helix-loop-helix genes represent the second largest plants' family genes, which are involved in pistil development and in the abiotic and biotic stresses response and belong to a family of transcriptional regulators present in three eukaryotic kingdoms (Heim et al. 2003; Skinner et al. 2010). The study of Miao et al. (2020) allowed for the identification of 268 *bHLH* genes in *B. oleracea*, 440 genes in *B. napus*, and 251 genes in *B. rapa*, including 21 new *bHLH* members and the gene structures, Phylogeny threes and the conservative motif analysis were carried out showing similar expression patterns between *B. rapa* and *B. napus* in roots and contrasting ones in stems, leaves and in reproductive organs (Miao et al. 2020). Genome wide analysis of the *bHLH* transcription factor was studied for Chinese-cabbage showing the interaction network associated to the *ArabidopsisbHLH* genes and in cabbage, which revealed several chilling response patterns (Song et al. 2019; Shan et al. 2019). The divergence of the S-Haplotypes between *B. oleracea* and *B. napus* was also studied to investigate the recognition specificity of self-incompatibility which is determined by the SRK alleles encoded in the stigma and by the SP11 ones in the pollen (Kusaba et al. 1999; Kimura et al. 2002; Sato et al. 2003).

Brassica oleracea's S-allele collection has been improved over the last 30 years by detecting and eliminating duplicate alleles, as well as cleaning up lines where the desired allele was contaminated by a second S-allele. These issues are frequently caused by recessive S-alleles, which can cause a weak self-incompatibility reaction when homozygous, or for no reaction at all when heterozygous with a dominant S-allele. The original 47 numbered lines were eventually reduced to 36. A total of 50 S-alleles have been added to the collection as a result of extensive S-allele surveys of most of the species' cultivated and wild forms. A complete list of all S-allele numbers is provided, including their current status, details on duplication and contamination, and distribution. Initially, seed sets were used in the collection, but pollen tube growth counts were later used because they provided faster and more accurate results. Recently, molecular methods based on DNA analysis have been developed, and now that DNA sequence data for most S-alleles is available, this may become the preferred method of identification. The S-allele collection has become an international standard and is widely used for breeding and for the study of self-incompatibility. The Wellesbourne Genetic Resources Unit is critical in preserving authenticated samples of the collection and providing material for distribution (Ockendon 2000).

The *Brassica oleracea* S-allele reference collection has been analyzed using a molecular technique for identifying S-alleles involved in self-incompatibility by Brace et al. (1994). There are nearly 50 different lines in the reference collection, each with a different S-allele present in the homozygous state. The technique entails polymerase chain reaction (PCR) amplification of sequences from the S multigene sequence family using a single pair of conserved primers. Following digestion with six restriction enzymes and gel electrophoresis of the digestion products, the PCR products are further examined. A straightforward method for estimating the band sizes of digestion products is described. The restriction patterns distinguish S-locus-related sequences from S-locus glycoprotein and S-receptor kinase genes. Furthermore, several alleles showed the same restriction pattern with any one restriction enzyme. As a result, alleles could be grouped together. Except for two individuals, each member of the S-allele reference collection displayed a distinct set of restriction patterns. The use of pollen tube growth tests to investigate the exceptions revealed that these accessions were duplicated within the collection. As a result, this technique provides a simple and useful method for identifying different S-alleles.

In 1993, Brace et al. have developed a method for identifying S alleles in *Brassica oleracea* based on restriction endonuclease digestion of PCR-amplified DNA, when the "consensus" primers were used in the PCR, and the preferential amplification of one of the alleles of the heterozygotes was observed in the F2 population. It was able to amplify both alleles of the heterozygotes equally by using different primers that were homologous to another region of the S sequence. The genotypes of the plants were determined using restriction analysis of PCR products and concurreded with pollen-tube growth test results.

Ockendon (1982) investigated the S-allele of cabbage (*B. oleracea* var. *capitata*), indeed, a total of 31 S-alleles were discovered analyzing a population of 197 cabbage plants divided into 11 different cultivars. Several variants of the S-alleles were also found in kale or Brussels sprouts, but five of them were not previously discovered appearing to be unique to cabbage vegetables. A deeper examination of five spring cabbage cultivars revealed only 12 S-alleles in total, with 6–10 S-alleles in four older cultivars and only three S-alleles in the newer, more highly selected cultivar. S2 was by far the most common S-allele, as it is in *B. oleracea* in general. The recessive alleles S5 and S15 were not prevalent in cabbage, which may explain why the sib problem in F1 hybrids appears to be less frequent in cabbage than in Brussels sprouts. In three cases, an S-allele was completely recessive in both the stigma and the pollen. The following unusual conditions can determine issues for the breeder, which are discussed. The highly recessive alleles S5 and S15 were not particularly prevalent in cabbage, which may explain why the sib problem in F1 hybrids appears to be less prevalent in cabbage than in Brussels sprouts. In three cases, an S-allele was totally recessive in stigma and pollen tissues. This unusual situation causes problems for the breeder, which are discussed. S5, S14, S15, and S45 appear to be the most recessive S-alleles in the stigma, while S23, S29, S35, S36, and S39 appear to be the most dominant. The importance of dominance between S-alleles in the breeding system of Brussels sprouts is discussed, and an explanation for the high frequency of S2 based on its unusual dominance behavior is proposed (Ockendon 1975). The *loci* related to MADH box domain and for the flowering traits are listed in Table 8.2.

TABLE 8.2
Identification of Candidate Genes or QTLs Related to Agronomic Traits in *B. oleracea*

Target Function	Genes, QTLs or Domains	Reference
Inflorescence size, Inflorescence period.	MADH box domain (*BoAP1, BoCAL, BoFLC, BoTFL1, BoLFY, BoFRI*)	Irish and Sussex 1990; Bowman et al. 1993; Kempin et al. 1995; Smith and King 2000; Purugganan et al., 2000; An et al. 2007; Irwin et al. 2012 Fadina et al. 2013; Sun et al. 2013; Ridge et al. 2015; Irwin et al. 2016; Wils and Kaufmann, 2017; Sheng et al. 2019; Takada et al. 2019
Sexual compatibility and flower *primordia* induction.	Basic helix loop helix domain (S-allele, *SRK, SP1*)	Ockendon 1982; Brace et al. 1994; Kusaba et al. 1999; Kimura et al. 2002; Sato et al. 2003; Shan et al. 2019; Miao et al. 2020
Leaf traits	*BoLl* (lobed leaf) *MF1* (curled leaf); *KANADI* and *ARFs* (leaf adaxial identity); *YABBY APUM23* (leaf blade development); *BoFL* (feathered leaves)	Kalve et al. 2014; Cheng et al. 2016; Feng et al. 2020
Hollow stem resistance.	*QHS.C09-2*	Boersma et al. 2009; Boersma et al. 2013; Yu et al. 2019
Head splitting resistance.	*SPL-2-1* locus and *Bol016058* gene	Pang et al. 2015; Su et al. 2015
Dwarfmutant.	Hormonal pathway (SA, JA, ET, ABA, BR, IAA, SI) genes	Xing et al. 2020
Bloomed leaves.	*BoWax1, LACS1, KCS1, KCR1, ECR, CER3* and *MAH1*	Lee et al. 2015; Liu et al. 2018
Hybridlethality.	*BoHL; Bo1g153320; Bo1g153380; Bo4g173780; Bo4g173810.*	Xiao et al. 2017

8.3.2 Genes Related to Leaf Traits

Different candidate genes have been discovered in the scientific literature regarding variation in leaf development in species belonging to the genus *Brassica*, and are thus thought to be more important regarding variation in *Brassica oleracea* leaf shape and size (Table 8.2). *LATEMERISTEMIDENTITY* 1 (*LMI1*) is required for simple serrated leaves (Ni et al. 2017). There are three types of LMI1-like genes in *Brassicaceae*, and overexpression could change the leaf shape by producing deep lobes (Vlad et al. 2014). The gene *BnLL1* is essential for obtaining a lobed-leaf phenotype in rapeseed, *B. napus* (Ni et al. 2015). On *B. napus* chromosome A10, the *BNLL1* gene is located between BnaA10g26280D and BnaA10g26340D.The two genes *BoATHB15.2* (MF1) and *BoKAN2.2* (*MF1*) are orthologues for the gene families HD-ZIPIII and KAN in *B. oleracea*, and both are involved in the abaxial/adaxial pathway in *Arabidopsis thaliana* (Cheng et al. 2016). When comparing heading cabbages to non-heading morphotypes, these genes are both found in selective sweeps (Cheng et al. 2016). The gene *REVISRADIX* (*BRX*) in *Arabidopsis thaliana* has an orthologue in *B. oleracea BoBRX.2* (*MF1*), which is involved in the pathway for the curled leaves (Cheng et al. 2016). In 2015 and 2016, Wageningen University conducted association studies on 471 and 465 *B. oleracea* accessions, respectively. The accessions were grown in a field experiment both years, and both leaf and leafy head traits were scored. The accessions were also subjected to genomic studies. The Swiss-Prot and TrEMBL databases were used to characterize genes. The following candidate genes were discovered; in fact, *SG1* and *NAC098* are thought to be involved in head length. Blistering appears to involve *CyCu2*-1, *EXPB4&6*, *NAC054*, *WSD1*, *AVT1*, *MYB81/104*, *CDC48A*, and *ARF6*. Head weight is thought to be influenced by *TMK1*&4, *IAA9*, *TFL1*, *APUM5*, *MKK5*, *IRX9*, *GTE4*, *CLO*, *CHC1*, *LOB21*, *SPL10*, and *PIP5K3* (Zou, 2019).

The leaf has two distinct functional surfaces, with photosynthesizing cells concentrated on the upper side and gas-exchanging cells concentrated on the lower. Specific transcription factors determine the polarities that form around young leaf primordia. Adaxial identity is promoted by HD-ZIP IIIs and the ASs complex, whereas abaxial identity is promoted by the KANADI gene family and auxin-dependent ARFs. Cell polarity fates become mutually exclusive and mechanistically linked with the development of adaxial and abaxial leaf blades under the control of YABBY gene family transcription factors, *APUM23*, and transacting short interfering RNAs (Kalve et al. 2014). Following cytoplasmic growth, cell division, endoreduplication, and cell expansion, the primordial leaf grows. *ANGUSTIFOLIA3* and *GROWTH-REGULATINGFACTOR5* are primarily responsible for that until the leaf reaches maturity (Dkhar and Pareek 2014). Cells then differentiate into different types of cells, such as guard cells, vascular tissue cells, and trichomes. Gibberellin and cytokinin are phytohormones that regulate cell proliferation and expansion rate, compensating for cell number and volume in the integrated system.

Several *B. oleracea* crops like kale (*B. oleracea* var. *acephala*), are exploited for their attractive and colorful leaves which are characterized by several shapes showing different anthocyanin patterns (Schmidt et al. 2010; Liu et al. 2021). In *B. oleracea* the gene *lobedleaf* (BoLl), responsible of the lobed leaf trait, respectively, was identified by its orthologue on *Arabidopsis* genome, named *LATE MERISTEM IDENTITY 1* (*LMI1*) (Ren et al. 2019). *BoLl1*, was subsequently named *BoLMI1-a* and its expression was significantly higher in lobed plants than unlobed ones (Zhang et al. 2021). Another gene, named *BoFL* was identified as a candidate to confer the feathered-leaved trait of ornamental kale and is located on chromosome nine, in 374.532-kb region (Feng et al. 2020). The conserved domain BoALG10 responsible for the encoding of an α-1,2 glycosyltransferase, also plays an important role in the maintenance of the leaf margin shape (Feng et al. 2022).

8.3.3 Identification and Selection of Superior Alleles Related to Plant Physiology

The investigation of the traits affecting important economic losses allowed the localization of the essential target genes involved in the physiological stresses individuating the superior alleles regulating them. The development of high-quality genetic maps can improve the knowledge of the economic traits related to broccoli (*B. oleracea* var. *italica*) quality such as the hollow stem, which is a physiological disorder causing the internal broccoli stem tissue to crack and it is multi-genically regulated (Boersma et al. 2009, 2013). Yu et al. (2019) constructed a high-quality genetic map of 4.787 specific length amplified segment (SLAF) markers in a DH population identifying different loci responsible for the hollow stem phenotype. The locus *QHS.C09-2* controls the hollow stem trait and explained 14.1% of the total phenotypic variation (Yu et al. 2019). Head splitting in cabbage is an important physiological disorder, which causes huge economic losses for farmers, affecting both the quality and the yield, and has been observed in several fruits and vegetables including Brassica Chinese cabbage (Niu et al. 2010; Pang et al. 2015). Several authors investigated the genetic region involved in head splitting resistance and they identified several QTLs conferring resistance and which were detected on chromosomes 2, 4, 6, 7 and 9 (Pang et al. 2015; Su et al. 2015). The markers BRPGM0676 and BRMS137 were tightly linked to the head splitting resistance and were associated with the QTL *SPL-2-1* (Pang et al. 2015). SNP genotyping identified *Bol016058* as the candidate genomic region flanking 5.6 Mb on chromosome 3 responsible for head splitting resistance which is homologous to the *proline-richprotein4* (*PRP4*) of *Arabidopsis* which encode structural cell wall protein. Brassica plant architecture and growing cycle are linked to the plant height and the selection of dwarf hybrids can be useful for several vegetables like cabbage and Chinese cabbage (*B. rapa* ssp. *pekinensis*) and several authors identified genes related to the dwarf phenotype which are in turn, related to several phytohormones like indole-3-acetic acid (IAA), abscisic acid (ABA), salicylic acid (SA) and jasmonic acid (JA) (Wang et al. 2018; Wei et al. 2018). Xing et al. (2020) associated the transcription factors of the dwarf cabbage mutant showing up-regulation for the genes controlling the hormonal pathways of salicylic acids (SA), jasmonic acid (JA), abscisic acid (ABA), ethylene (ET) and the down-regulation of the genes involved in the hormones related to the plant height, like gibberellin (GA), brassinolide (BR), indole-3-acetic acid (IAA), and strigolactone (SL).

Brassica crops may be subjected to several biotic and abiotic stresses and the cuticular wax cover layer provides protection against the various above-cited stresses. The following waxy coverage consists of long-chain hydrocarbon compounds, which include alkanes, alcohols, aldehydes, ketones, esters, and other derived compounds (Shepherd and Griffiths 2006; Xue et al. 2017). Genetic analysis of the F_2 segregant population generated by the crossing between the bloomless line of *B. oleracea* and the bloomed cultivar identified Bol013612 as a candidate for the *BoWax1* gene, responsible of the glossy green trait in *B. oleracea*. The analyzed locus was mapped in the region 158.5 kb on the C01 (Liu et al. 2018). In broccoli cultivars, the cuticular wax biosynthetic genes *LACS1*, *KCS1*, *KCR1*, *ECR*, *CER3,* and *MAH1* were differentially expressed among the bloomed and bloomless, respectively, and they were significantly up-regulated as a consequence of drought stress (Lee et al. 2015).

The production of F_1 hybrids, a key point for vegetable production for obtaining higher yields through the effect of heterosis in *Brassica* species, represents an important goal for improving the productivity of the crops (Shen et al. 2005; Teklewold and Becker 2005; Yu et al. 2019). Fifty-one heterosis related genes selected from Arabidopsis were identified, as orthologs, on the *Brassica oleracea* genome and the expression analysis revealed that correlation among the detected genes and the cotyledons and young leaf size (Jeong et al. 2017). Hybrid lethality is a post zygotic reproductive isolation process pertaining to the genetic incompatibility model that occurs in seedling or adult plants causing wilting, chlorosis, and lethality (Maheshwari et al. 2011). The above-mentioned process was studied in a cabbage performing linkage map for the detected genes, which were *BoHL1*, located on the first chromosome between the markers BoHLTO124 and BoHLTO130, and *BoHL1* on chromosome 4 between the InDel markers HL234 and HL235 (Xiao et al. 2017). RNA-seq analysis identified and detected several differentially expressed genes like *Bo1g153320*, *Bo1g153380* on chromosome one, and *Bo4g173780* and *Bo4g173810* on chromosome four (Xiao et al. 2017). All the resistance genes involved in physiological disorder response are mentioned in Table 8.2.

8.4 IDENTIFICATION AND SELECTION OF SUPERIOR *B. OLERACEA* ALLELES RELATED TO ACCUMULATION OF BIOACTIVE COMPOUNDS

8.4.1 Genes Related in Glucosinolate Biosynthesis and Accumulation

Brassica genotypes are full of bioactive compounds and glucosinolates (GSL) representing the most important class of secondary metabolites, which form isothiocyanates (ITCs) as a consequence of their hydrolysis catalyzed by the enzymatic complex of myrosinases that hydrolyze the thioglucosidic bond (Blažević et al. 2020). It is interesting to see how much the glucosinolate profile varies between species and varieties. The distribution of GLS commonly varies among plant organs (roots, stem, leaves, sprouts, seedlings), both qualitatively and quantitatively. Unfortunately, both qualitative and quantitative information concerning the expression of genes involved in glucosinolate biosynthesis in the edible organs of *B. oleracea* cultivars is insufficient. GSL amount differs for each organ and their expression is mediated by the *MYB transcriptional factors* (*MYBTFs*), which regulate the GSL biosynthesis. Several authors have focused on the identification of the QTLs involved in glucosinolates synthetic pathways (Kliebenstein 2009; Sotelo et al. 2014). The molecular markers used for identification of the QTLs underlying the glucosinolates were developed by Gao et al. (2007). Several studies focused on the identification of the QTLs involved in the glucosinolate biosynthesis pathways (Kliebenstein 2009; Sotelo et al. 2014; He et al. 2018). Thus, candidate genes underlying QTLs that affect GLS content can be found using the strong co-linearity between *Arabidopsis thaliana* and *Brassica* species. So far, *A. thaliana* has been used as a model plant to identify and to characterize the main genes and transcription factors involved in the three phases of glucosinolate biosynthesis. Glucosinolate biosynthesis occurs in three independent stages: chain elongation of the precursor amino acid, which is supported by methylthioalkylmalate synthase enzymes (MAM). Then the formation of the core structure in which several phases are clearly distinguished: aldoxime formation catalyzed by the CYP79 family of cytochromes P450 and the aldoxime oxidation by the CYP83 family. The thiohydroximic acid formation by conjugation to an S donor and after C-S bond cleavage is carried out by the enzyme S-glucosyltransferase (S-GT) inducing the desulfoGSL formation followed by the GSL formation by the enzyme sulfotransferase. The last step is the side-chain modifications consisting of oxidation, hydroxylation, methoxylation, desaturation, sulfation and glycosylation. The GSL pathway has also been well known as a "model" for secondary metabolites. In the study of Yi et al. (2015), a total of 84 *B. oleracea* genes orthologous to *B. rapa* genes related to glucosinolate biosynthesis, transcriptional regulation and breakdown were identified. According to the genome of *A. thaliana* three *B. oleracea* loci were found and cloned using homology. Though undoubtedly similar, the glucosinolate production and catabolism in *A. thaliana*, *B. rapa*, and *B. oleracea* differ in the proportion of tandem genes, the number of *MAM* and 2-oxoglutarate-dependent dioxygenase genes, and the roles (*AOP*) and functions of the *MAM* gene family members for condensation, and by the side chain elongation and the chain length production during GLSs biosynthesis. The biosynthesis differs in *Arabidopsis* compared to *B. rapa* and *B. oleracea*. Multiple genes, including *BCAT4* (branched-chain aminotransferase 4), *MAM1* (methylthioalkylmalate synthase 1), *CYP79F1* (dihomomethionine N-hydroxylase), and *AOP2* (2-oxoglutarate-dependent dioxygenases) have been shown to regulate and affect the amount of glucoraphanin in cells. Nowadays, several MYB transcription factors have been verified, and they are able to regulate GSL biosynthesis in different plants. In *B. oleracea* var. *italica*, the overexpression of BoMYB28 allows an increase in the amount of Met-derived 4-(methylsulfinyl) butylGSL and the post transcriptional gene silencing of its homolog in *B. juncea* determining a significant reduction of GLSs derived from Met in seed tissues; overexpressing *BoMYB29* homologs could upregulate GSL biosynthesis from Met in *B. oleracea* (Shang et al. 2022). MAM3 is assumed to contribute to the production of all GSL chain length. In addition to 4C glucosinolate, biosynthesis of sinigrin, a 3C glucosinolate, in *B. oleracea* is assumed to be related to high expression of the *Bol017070* gene, while its ortholog, *Bra013007*, remains silenced in *B. rapa*. By contrast, *B. rapa* biosynthesizes more of the 5C glucosinolate glucobrassicanapin due to higher expression of *MAM3* in comparison to *B. oleracea*. *MYB76* transcription factor is present in *Arabidopsis* but *B. oleracea* and *B. rapa* lack this factor. *B. oleracea* has only one functional *AOP* gene (*AOP2*), whereas *Arabidopsis* and *B. rapa* have four and three functional AOP genes, respectively. Concerning theindolic GSL, QTL mapping has been performed analyzing the transcriptome profile followed by the cis-expression QTL to prompt the regulation gene of 4-methoxy indolyl-3-methyl GSL. In addition, the variations of the indole GLSs biochemical pathway can determine decompensations in terms of plant growth due to several of the following pathways often overlap the biosynthesis of phytohormones. Under different environmental conditions a complex network of transcription factors from the *R2R3-MYB* family regulate the glucosinolate biosynthetic pathways. *MYB28* and *MYB29* are related to biosynthesis of aliphatic glucosinolates while *MYB51*, *MYB122* and *MYB34* are related to indolicglucosinolate biosynthesis (Kim et al. 2013). The identified genes related to the GLSs accumulation are listed in Table 8.3.

8.4.2 Genes Related to the Accumulation of Polyphenols

Among several secondary metabolites produced by the plants, phenolic compounds represent the second largest family, and their biosynthesis begins with the aromatic amino acids produced by the shikimate pathway (Santos-Sánchez et al. 2019). Gardner et al. (2016) have identified 60 QTLs involved in the accumulation of phenolic compounds which were found in all the nine *B. oleracea* chromosomes. Twenty-three of them were identified analyzing three phenolic and antioxidant assays carried out over 4 years (from 2009 to 2014). The QTLs found between the markers Bn-C1-p40578360 and Bn-C1-p41497525 represented the most consistent ones showing the highest whole phenotypic variation, and the putative candidate gene MYB65 localized in this QTL. The second most representative QTL identified was localized between the markers Bn-C8- p29068278 and Bn-C8-p32125534 and it explained 9.9–17.6% of the phenotypic variation of the total polyphenolic content, 2,2'-azino-bis 3-ethylbenzothiazoline-6-sulfonic acid (ABTS) and 2,2-diphenyl-1-picrylhydrazyl (DPPH). Its candidate gene is the chalcone-flavanone isomerase 1 [CHI, aka *TRANSPARENTTESTA* 5 (TT5)] which is located on chromosome 8 (Gardner et al. 2016). The candidate genes involved in the phenylpropanoid pathway were also mapped by Francisco et al.

TABLE 8.3
Identification of Candidate Genes or QTLs Related to the Biosynthesis and Accumulation of Bioactive Compounds in *B. oleracea*

Target Function	Genes, QTLs or Domains	Reference
Glucosinolates	Biosynthesis: *MYBTFs* (Aliphatic: *MYB28* and *MYB29*; Indolic: *MYB51*, *MYB122* and *MYB34*); Regulation: *BCAT4*, *MAM1*, *MAM3*, *CYP79F1* and *AOP2*.	Gao et al. 2007; Kim et al. 2013; Sotelo et al. 2014; He et al. 2018; Shang et al. 2022.
Polyphenols	*MYB65*, *TT5*.	Gardner et al. 2016; Francisco et al. 2016.
Anthocyanins	*F3'H*; *Purple* gene encoding *R2R3MYBTF; BoMYB2; Bo9g058630* (anthocyanin leaves).	Chiu et al. 2010; Liu et al. 2017; Liu et al. 2022; Gu et al. 2022.
Carotenoids	*Orange* (*Or*)	Li et al. 2001

(2016) detecting the chromosome position of 33 independent consensus QTLs involved in polyphenol metabolism in different organs like leaf, flower buds and seeds. The detected QTLs shared all of the nine chromosomic linkage groups and were: six on C3; five on C2 and C9; four on C8; three on C4, C5, C6, and C7; and two on C1. Twenty-three of the detected QTLs showed plant organ specificity and 13 were found exclusively in flower buds, six in leaves, and four in seeds (Francisco et al. 2016). The above-mentioned genes and QTLs are listed in Table 8.3.

8.4.3 Genes Related to Anthocyanin Accumulation

Anthocyanins are secondary metabolites biosynthesized from the phenylalanine (PAL) which also plays an important role to contrast plants abiotic and biotic environmental stresses and they can be also applied to induce the desirable colour (Guo et al. 2008; Ahmed et al. 2015). The most common anthocyanins detected in plants were cyanidin, delphinidin, pelargonidin, peonidin, petunidin, and malvidin (Houghton et al. 2021). The study carried out by Liu et al. (2022) developed new green non anthocyanin broccoli and anthocyanin rich varieties using the molecular markers F30H and Pur7.1-K1 among two parental line and their crossing populations. The F1 genotypes were confirmed by the molecular markers and back crossed (BC) to the recurrent parental line. For this case the green broccoli and subsequently the recurrent line was crossed to the BC1F1 and BC2F1 obtaining the BC3F2 population in which, after the selection using the molecular markers F3'H and Pur7.1-K1. Several self-crossing cycles were performed to stabilize the traits obtaining finally three purple broccoli and two green broccoli lines. The development of purple cauliflower could be supported by the Purple (Pr) gene mutation which encodes the R2R3 MYB transcription factor, responsible of an abnormal anthocyanin accumulation pattern in the mutant populations (Chiu et al. 2010). Gu et al. (2022) identified a 43bp deletion in F3'H gene which plays against the anthocyanin coloration it is responsible to keeping green buds under cold conditions. The sequencing of the BoMYB2 gene in purple cabbage and kohlrabi showed the insertion of 7606 bp CACTA-like transposon in its promoter region upregulating the expression of the following gene (Yan et al. 2019). Furthermore, also the BoMYB2 gene activation in purple cabbage is linked to the 1 bp point mutations in the promoter region (Yan et al. 2019). The genetic analysis performed by Liu et al. (2017) showed that the purple leaf trait is controlled by the single dominant gene Bo9g058630 which was named Pr which was mapped in chromosome 9, between the markers BRID485 and BRID490 flanking the gene at the genetic distance of 0,1 and 0,2 cM, respectively. The high throughput sequencing performed, showed by Blast analysis high homology to the gene AT5G442800 of Arabidopsis which encodes the dihydroflavonol reductase (DFR) (Liu et al. 2017).

8.4.4 Genes Related to Accumulation of Carotenoids

Carotenoids are ubiquitous pigments, which belong to the isoprenoid pathway and their synthesis is supported by all the photosynthetic organisms like cyanobacteria, algae, plants, and some fungi (Young and Lowe 2018). The above-mentioned pigments play an important role in photosynthesis absorbing the solar spectrum in the blue-green region and transferring the absorbed energy into the chlorophylls by the singlet–singlet energy transfer mechanism (Hashimoto et al. 2016). They are also essential in plants for the photoprotection carried out by the triplet–triplet energy transfer from chlorophylls to carotenoids contrasting the harmful effects of excess light exposure (Domonkos et al. 2013). Accumulation of carotenoids can enhance the response to multiple abiotic stresses (Wang et al. 2018; Ke et al. 2019). The target gene involved in their accumulation is *Orange* (*Or*), which is responsible for the carotenoids' homeostasis and enhancing plant tolerance to the abiotic stress (Kim et al. 2018; Osorio et al. 2019). The *Or* gene encodes for the Orange protein located in the nucleus suggesting its interaction with nuclear elements, like transcription factors, necessary to start the plastid transition and carotenoids' accumulation (Sun et al. 2016). In *B. oleracea* var.

botrytis, *Or* gene can alter the color of the cauliflower curd from the cream colored wild type and the *Or* homozygous generated by the cross between the orange curd parental line and the wild type showing an intense orange colored curd phenotype, exhibiting an important size reduction from the parental lines (Li et al. 2001). Furthermore, the heterozygous of the above cited cross showed less intense orange pigments but normal size curd (Li et al. 2001). In *B. oleracea,* Brown et al. (2014) identified QTLs related to the carotenoid biosynthesis developing a high-resolution genetic linkage map of *B. oleracea* starting from a *B. napus* SNP array. Genes and references related to carotenoids accumulation are listed in Table 8.3.

Brassica vegetables, as is well known, are full of bioactive compounds and some of them show high antioxidant activity (AA) which can be related to the presence of several metabolites in Brassica vegetables, like phenolic compounds, vitamin C, carotenoids, and anthocyanin (Podsedek et al. 2006; Soengas et al. 2012). The mapping of QTLs related to the AA was performed by Sotelo et al. (2014), which identified and mapped seven QTLs in leaves and six in flower buds by the analysis of the AA, while, based on the metabolites content, two and six QTLs were founded, in leaves and flower buds, respectively.

8.5 EVOLUTION AND SELECTION OF SUPERIOR *B. OLERACEA* ALLELES RELATED TO BIOTIC AND ABIOTIC STRESS RESISTANCE

Plant disease can cause important yield losses hindering the whole food supply chain and to overcome this issue, the detection, and the improvement of the pathogen resistance genes in crops, is required (Gururani et al. 2012). The development of breeding strategies to enhance the resistance crop traits is a key point to reduce crop losses caused by disease and most of these resistant traits are linked to the production of secondary metabolites, generating a quantitative resistance (Boyd et al. 2013; Nelson et al. 2018). Genes related to plant disease resistance are listed in Table 8.4.

Cabbage Fusarium wilt is an important soil-borne fungal disease causing severe economic losses all over the world and the new discoveries about the Brassica genomes allowed an investigation into new resistant genes and varieties (Tong et al. 2022). It is caused by *Fusarium oxysporum* f. sp. *conglutinans* and several authors have developed linkage maps through molecular tools to detect a single dominant gene (Pu et al. 2012; Lv et al. 2014; Shimizu et al. 2015; Xing et al. 2016; Tong et al. 2022). Lv et al. (2014) have identified and mapped the candidate resistance gene in cabbage showing high identification with the region Bol037156 in the resistant lines and there were two InDels of 1 and 10 bp, respectively, as a difference to the susceptible lines, suggesting that

TABLE 8.4
Identification of Candidate Genes or QTLs for Disease Resistance

Disease	Genes, QTLs, or Domains	Reference
Fusarium wilt	*FocBo1.*	Pu et al. 2012; Lv et al. 2014; Shimizu et al. 2015
Clubrot	CR *loci* (*CRk, CRc, CRa, CRre,Bol037115* and *Bol042270.*	Hirai et al. 2004; Saito et al. 2006; Nagaoka et al. 2010; Kato et al. 2013; Yu et al. 2019 Peng et al. 2018; Ce et al. 2021.
Blackrot	*Xcalbo*; *Bol003711*; *Bol010135*; *Bol010559*; *Bol022784*; *Bol029866*; *Bol042121*; *Bol031422*; *Bol040045*; *Bol042095.*	Izzah et al. 2014; Kalia et al. 2017; Afrin et al. 2018; Iglesias-Bernabé et al. 2019.
Black spot	*LRR-RLK7*;*WRKY33*;*FAH1*;*CCoAMT.*	Macioszek et al. 2020

resistance is driven by a single mutation. The single dominant resistance gene *FocBo1*, is located on chromosome 7 between the region BSA8 and KBrS003O10N1 and it has a high homology to the *Arabidopsis thaliana* region in chromosome 4 (Pu et al. 2012; Shimizu et al. 2015).

Clubroot is a typical soil-borne Brassica disease caused by the obligate fungal parasite *Plasmodiophora brassicae*, which survives inside the soil through three different life cycles, which include, survival into the soil, root hair infection, and cortical infection (Dixon et al. 2009; Donald and Porter 2009; Kageyama et al. 2009). Different authors have investigated the role of QTLs involved in clubroot resistance (CR) in *B. rapa* and have identified several genes like *CRk*, *CRc*, *CRa*, *CRre* (Hirai et al. 2004; Saito et al. 2006; Okazaki et al. 2007; Kato et al. 2013; Yu et al. 2019). Nagaoka et al. (2010) generated a fine density genetic linkage map based on the specific markers linked to *B. rapa CR* genes and seven of them were mapped into the *B. oleracea* genome. Peng et al. (2018) developed a 60K SNPs based array, which was performed to associate the specific SNP markers to different pathogenetic traits like disease incidence, number of fibrous roots and the microorganism content in roots. The comparative analysis among *CR loci* and among the syntenic regions between *B. oleracea* and *B. rapa* genome by the Bolbase database show a significant QTL on the third *B. oleracea* chromosome (6.28–6.32 Mb), which overlaps the number of fibrous roots and the *P. brassicae* content in roots traits, respectively, and it was partially syntenic to the 0.59–6.22 Mb on the second *B. rapa* chromosome (Yu et al. 2019; Peng et al. 2018). Among the several QTLs detected, the one on the *B. oleracea* chromosome 8 corresponding to the 32.7–33.3 Mb, showed 10.7–17.8% of the total phenotypic variation while two overlapping QTL regions were found on the sixth chromosome (Peng et al. 2018). Ce et al. (2021) identified by genomic resequencing three QTLs on the seventh chromosome and RNAseq, and qRT-PCR analysis was also performed identifying two potential candidate genes (*Bol037115* and *Bol042270*) which were upregulated and downregulated in the resistant and susceptible cabbage line, respectively, after the pathogen inoculation. Based on the previously mentioned work, a functional marker was also constructed from the QTL *qCRc7-4* on the chromosome 7 (Ce et al. 2021).

Black rot is the most destructive and widespread disease which infects all cultivated varieties of worldwide Brassicas and several authors identified different genes against the race 1 and 4 (Singh et al. 2011; Doullah et al. 2011; Vicente et al. 2013; Singh et al. 2018; Bella et al. 2019). The causal agent of the above mentioned disease is *Xanthomonas campestris* pathovar. *Campestris* causes severe losses in agricultural yield worldwide (Qian et al. 2005; Gupta et al. 2013). Genotyping by different molecular markers followed by transcriptome analysis was carried out in cabbage resistant lines building a genetic map comprising 265 loci, mapping the position of 98 EST-SSRs, 21 SNP-based dCAPS and 55 IBP markers derived from *B. rapa* genome sequence and 91 publicly available SSRs (Izzah et al. 2014). Sharma et al. (2016) identified the resistance locus *Xca1bc* on chromosome 7 of Ethiopian mustard (*B. carinata*), which was introgressed in cauliflowers generating an interspecific hybrid and developing backcross progeny by *in vitro* embryo rescue techniques (Sharma et al. 2017). Useful RAPD and ISSR derived SCAR markers of the cauliflower *Xca1bo* locus were developed and validated in 17 cauliflower lines (Kalia et al. 2017). Nucleotide binding site (NBS) genes were identified as candidate resistant genes in cabbage against black rot disease and several of them like *Bol003711*, *Bol010135*, *Bol010559*, *Bol022784*, *Bol029866*, *Bol042121*, *Bol031422*, *Bol040045*, and *Bol042095*, were significantly up regulated in the resistant cabbage lines, in comparison to the sensitive ones (Afrin et al. 2018). Iglesias-Bernabé et al. (2019) confirmed the previous candidate resistant QTLs highlighting the close linkage between the genes involved in anthocyanin accumulation and in the leaf area and shape traits.

Another important, destructive disease for *B. oleracea* crop is represented by black spot disease, caused by the pathogen *Alternaria brassicicola,* which belongs to the fungal division of *Ascomycetes*. The above-mentioned pathogen is a seed-transmitted nectrotrophic fungus infecting young seedlings or mature plants (Dillard et al. 1998; Pedras et al. 2009). In adult plants the pathogen is wind or insect spread and the infection evolves in black spots on leaves, which can also cause the whole plant decay. The above-mentioned pathogen also represents an important problem during the post-harvest storage of the *B. oleracea* vegetable (Troncoso-Rojas et al. 2014). Macioszek et al. (2020) identified several genes involved in black spot resistance performing microarray analysis of the whole transcriptome based on their differential expression discovering 167 genes, with 87 of them, mainly involved in cell response, signaling, degradation of the cell wall, protein degradation, auxin-induced metabolism, being downregulated. Some of the stress related genes were LRR-RLK7 encoding leucine-rich repeat transmembrane protein receptor-like kinase 7, WRKY33 codifying for its own transcription factor, PDF1.2 encoding jasmonate-responsive plant defensin, FAH1, which encode for ferulate-5-hydroxylase 1 and CCoAMT encoding caffeoyl-CoA 3-O-methyltransferase involved in lignin biosynthesis.

8.6 CONCLUSION

Allele mining is a promising technique that can be useful for enhancing and exploiting *Brassica oleracea* genetic diversity in breeding programs having the aim of generating productive and resilient genotypes to obtain more sustainable vegetable production. Breeders that develop new varieties, in fact, must consider the reality of an ever-increasing world population and the challenging issue of malnutrition, which affects many millions of people. *B. oleracea* crops provide vegetal products rich in bioactive compounds available for the worldwide population. A key point for breeders is represented by the identification of the allelic patterns which can enhance the above-mentioned biochemical compounds guaranteeing a high yield.

The identification of resistant genes against Brassica pests and diseases is another point of interest for breeders having identified several candidate resistance genes based on linkage maps developed for other species or based on their differential expression. The demand for high-nutraceutical produce, i.e., superfood, increases the requirement for new resilient cultivars addressing the main biotic and abiotic stresses encountered by the growers and reducing and or avoiding the use of pesticides and chemical compounds during the growing cycles. In this way, we can satisfy the increasing demand of healthy food with the overall aim of improving human wellbeing.

REFERENCES

Afrin KS, Rahim MA, Park JI, Natarajan S, Kim HT, Nou IS (2018) Identification of NBS-encoding genes linked to black rot resistance in cabbage (*Brassica oleracea* var. *capitata*). *Mol Biol Rep,* 45(5), 773–785.

Ahmed NU, Park JI, Jung HJ, Hur Y, Nou IS (2015) Anthocyanin biosynthesis for cold and freezing stress tolerance and desirable color in Brassica rapa. *Funct Integr Genom, 15*(4), 383–394.

Allender CJ, and King GJ (2010) Origins of the amphiploid species Brassica napusL. investigated by chloroplast and nuclear molecular markers. *BMC Plant Biol, 10*(1), 1–9.

An YH, Li XF, Zhu JX, Shao XH, Sun Y, Xiong LJ (2007) The screening of interaction factors with BoCAL and BoAP1 related to curd formation. *J Mol Cell Biol, 40*(2), 130–136.

Babula D, Kaczmarek M, Ziółkowski PA, Sadowski J (2007) *Brassica oleracea*. In: Kole C (ed) *Genome Mapping and Molecular Breding in Plants*. Vol. Vegetables. Springer, Berlin, Heidelberg, pp 227–285.

Banik M, Liu S, Yu K, Poysa V, Park SJ (2007) Molecular TILLING and EcoTILLING: effective tools for mutant gene detection in plants. *Genes Genomes Genom, 1*(2). 123–132.

Bayer PE, Golicz AA, Tirnaz S, Chan CK, Edwards D, Batley J (2019) Variation in abundance of predicted resistance genes in the Brassica oleracea pangenome. *J Plant Biotechnol, 17*(4), 789–800.

Bella P, Moretti C, Licciardello G, Strano CP, Pulvirenti A, Alaimo S, Zaccardelli M, Branca F, Buonaurio R, Vicente JG, Catara V (2019) Multilocus sequence typing analysis of Italian Xanthomonas campestris pv. campestris strains suggests the evolution of local endemic populations of the pathogen and does not correlate with race distribution. *Plant Pathol, 68*(2), 278–87.

Bennett S (2004) Solexa Ltd. *J Pharmacogenom, 5*(4), 433–438.

Bhat JA, Yu D, Bohra A, Ganie SA, Varshney RK (2021) Features and applications of haplotypes in crop breeding. *Commun Biol, 4*(1), 1–2.

Blažević I, Montaut S, Burčul F, Olsen CE, Burow M, Rollin P, Agerbirk N (2020) Glucosinolate structural diversity, identification, chemical synthesis and metabolism in plants. *Phytochemistry, 169*, 112100.

Boersma M, Gracie AJ, Brown PH (2009) Relationship between growth rate and the development of hollow stem in broccoli. *Crop Pasture Sci, 60*(10), 995–1001.

Boersma M, Gracie AJ, Brown PH (2013) Evidence of mechanical tissue strain in the development of hollow stem in broccoli. *Sci Hortic, 164*, 353–358.

Bohuon EJ, Ramsay LD, Craft JA, Arthur AE, Marshall DF, Lydiate DJ, Kearsey M (1998) The association of flowering time quantitative trait loci with duplicated regions and candidate loci in *Brassica oleracea. Genetics, 150*(1), 393–401.

Bowman JL, Alvarez J, Weigel D, Meyerowitz EM, Smyth DR (1993) Control of flower development in *Arabidopsis thaliana* by APETALA1 and interacting genes. *Development, 119*(3), 721–43.

Boyd LA, Ridout C, O'Sullivan DM, Leach JE, Leung H (2013) Plant–pathogen interactions: disease resistance in modern agriculture. *Trends Genet, 29*(4), 233–240.

Brace J, King GJ, Ockendon DJ (1994) A molecular approach to the identification of S-alleles in *Brassica oleracea. Sexual Plant Reprod, 7*(4), 203–208.

Brace J, Ockendon DJ, King GJ (1993) Development of a method for the identification of S alleles in *Brassica oleracea* based on digestion of PCR-amplified DNA with restriction endonucleases. *Sexual Plant Reprod, 6*(2), 133–138.

Branca F (2008) Cauliflower and broccoli. In: Prohens J, Nuez F (eds) *Vegetables I. Handbook of Plant Breeding*, vol 1. Springer, New York, NY, pp 151–186.

Brāzma A, Jonassen I, Vilo J, Ukkonen E (1998) Predicting gene regulatory elements in silico on a genomic scale. *Genome Res, 8*(11), 1202–1215.

Brown AF et al. (2014) High-density single nucleotide polymorphism (SNP) array mapping in Brassica oleracea: identification of QTL associated with carotenoid variation in broccoli florets. *Theor Appl Genet*, 127, 2051–2064. DOI 10.1007/s00122-014-2360-5

Buntjer JB, Sørensen AP, Peleman JD (2005) Haplotype diversity: the link between statistical and biological association. *Trends Plant Sci, 10*(10), 466–471.

Camargo LE, Osborn TC (1996) Mapping loci controlling flowering time in *Brassica oleracea. Theor Appl Genet, 92*(5), 610–616.

Cartea ME, Di Bella MC, Velasco P, Soengas P, Toscano S , Branca F (2020) Evaluation of Italian and Spanish Accessions of *Brassica rapa* L.: Effect of flowering earliness on fresh yield and biological value. *Agronomy, 11*(1), 29.

Castelán-Muñoz N, Herrera J, Cajero-Sánchez W, Arrizubieta M, Trejo C, García-Ponce B, Sánchez MD, Álvarez-Buylla ER, Garay-Arroyo A (2019) MADS-box genes are key components of genetic regulatory networks involved in abiotic stress and plastic developmental responses in plants. *Front Plant Sci, 10*, 853.

Ce F, Mei J, He H, Zhao Y, Hu W, Yu F, Li Q, Ren X, Si J, Song H, Qian W (2021) Identification of candidate genes for clubroot-resistance in Brassica oleracea using quantitative trait loci-sequencing. *Front Plant Sci, 12*, 703520–703520.

Challa S, Neelapu NR (2018) Genome-wide association studies (GWAS) for abiotic stress tolerance in plants. In: Hussain Wan S (ed) *Biochemical, Physiological and Molecular Avenues for Combating Abiotic Stress Tolerance in Plants*. Academic Press, San Diego, pp 135–150.

Chen CH, Zhuang M, Fang ZY, Wang QB, Zhang YY, Liu YM, Yang LM, Cheng F (2013) A co-dominant marker BoE332 applied to marker-assisted selection of homozygous male-sterile plants in cabbage (*Brassica oleracea* var. capitata L.). *J Integr Agric, 12*(4), 596–602.

Cheng F, Sun R, Hou X, Zheng H, Zhang F, Zhang Y, Liu B, Liang J, Zhuang M, Liu Y, Liu D (2016) Subgenome parallel selection is associated with morphotype diversification and convergent crop domestication in *Brassica rapa* and *Brassica oleracea. Nat Genet, 48*(10), 1218–1224.

Cheng F, Wu J, Cai C, Fu L, Liang J, Borm T, Zhuang M, Zhang Y, Zhang F, Bonnema G, Wang X (2016) Genome resequencing and comparative variome analysis in a *Brassica rapa* and *Brassica oleracea* collection. *Sci Data, 3*(1), 1–9.

Chiu LW, Zhou X, Burke S, Wu X, Prior RL, Li L (2010) The purple cauliflower arises from activation of a MYB transcription factor. *Plant Physiol, 154*(3), 1470–1480.

Ciancaleoni S, Chiarenza GL, Raggi L, Branca F, Negri V (2014) Diversity characterisation of broccoli (*Brassica oleracea* L. var. *italic* Plenck) landraces for their on-farm (in situ) safeguard and use in breeding programs. *Genet Resour Crop Evol, 61*(2), 451–464.

Dai LY, Liu XL, Xiao YH, Wang GL (2007) Recent advances in cloning and characterization of disease resistance genes in rice. *J Integr Plant Biol*, *49*(1), 112–119.

Dkhar J, Pareek A (2014). What determines a leaf's shape?. *EvoDevo, 5*(1), 1–19.

De Bodt S, Raes J, Van de Peer Y, Theißen G (2003) And then there were many: MADS goes genomic. *Trends Plant Sci, 8*, 475–83.

Di Bella MC, Niklas A, Toscano S, Picchi V, Romano D, Lo Scalzo R, Branca F (2020) Morphometric characteristics, polyphenols and ascorbic acid variation in *Brassica oleracea* L. novel foods: Sprouts, microgreens and baby leaves. *Agronomy*, *31*, 10(6), 782.

Di Bella MC, Toscano S, Arena D, Moreno DA, Romano D, Branca, F (2021) Effects of growing cycle and genotype on the morphometric properties and glucosinolates amount and profile of sprouts, microgreens and baby leaves of Broccoli (*Brassica oleracea* L. var. *italica* Plenck) and Kale (*B. oleracea* L. var. *acephala* DC.). *Agronomy*, *11*(9), 1685.

Duclos DV, Björkman T (2008) Meristem identity gene expression during curd proliferation and flower initiation in *Brassica oleracea. J Exp Bot*, *59*(2), 421–433.

Dillard HR, Cobb AC, Lamboy JS (1998) Transmission of Alternaria brassicicola to cabbage by flea beetles (Phyllotretacruciferae). *Plant Dis*, *82*(2), 153–157.

Dixon GR (2009) The occurrence and economic impact of Plasmodiophorabrassicae and clubroot disease. *J Plant Growth Regul*, *28*(3), 194–202.

Domonkos I, Kis M, Gombos Z, Ughy B (2013) Carotenoids, versatile components of oxygenic photosynthesis. *Progr Lipid Res*, *52*(4), 539–561.

Donald C, Porter I (2009) Integrated control of clubroot. *J Plant Growth Reg*, *28*(3), 289–303.

Doullah M, Mohsin GM, Ishikawa K, Hori H, Okazaki K (2011) Construction of a linkage map and QTL analysis for black rot resistance in *Brassica oleracea* L. *Int J Nat Sci,* 1(1), 1–6.

Fadina OA, Pankin AA, Khavkin EE (2013) Molecular characterization of the flowering time gene FRIGIDA in Brassica genomes A and C. *Russ J Plant Physiol*, *60*(2), 279–289.

Feng X, Li X, Yang X, Zhu P (2020) Fine mapping and identification of the leaf shape gene *BoFL* in ornamental kale. *Theor Appl Genet*, *133*(4), 1303–1312.

Feng X, Yang X, Zhong M, Li X, Zhu P (2022) BoALG10, an α-1, 2 glycosyltransferase, plays an essential role in maintaining leaf margin shape in ornamental kale. *Hort Res*, *9*, p.uhac137.

Finch-Savage WE, Côme D, Lynn JR Corbineau F (2005) Sensitivity of *Brassica oleracea* seed germination to hypoxia: a QTL analysis. *Plant Sci*, *169*(4), 753–759.

Flannery ML, Mitchell FJ, Coyne S, Kavanagh TA, Burke JI, Salamin N, Dowding P, Hodkinson TR (2006) Plastid genome characterisation in Brassica and Brassicaceae using a new set of nine SSRs. *Theor Appl Genet*, *113*(7), 1221–1231.

Francisco M, Ali M, Ferreres F, Moreno DA, Velasco P, Soengas P (2016) Organ-specific quantitative genetics and candidate genes of phenylpropanoid metabolism in *Brassica oleracea*. Front Plant Sci, *6*, 1240.

Gao M, Li G, Yang B, Qiu D, Farnham M, Quiros C (2007) High-density *Brassica oleracea* linkage map: identification of useful new linkages. *Theor Appl Genet, 115*, 277–287.

Gardner AM, Brown AF, Juvik JA (2016) QTL analysis for the identification of candidate genes controlling phenolic compound accumulation in broccoli (*Brassica oleracea* L. var. italica). *Mol Breed, 36*(6), 1–12.

Gebhardt C, Bellin D, Henselewski H, Lehmann W, Schwarzfischer J, Valkonen JPT (2006) Marker-assisted combination of major genes for pathogen resistance in potato. *Theor Appl Genet, 112*(8), 1458–1464.

Gill KS, Kaur G, Kaur G, Kaur J, Sra SK, Kaur K, Gurpreet K, Sharma M, Bansal M, Chhuneja P, Banga SS (2021) Development and validation of kompetitive allele-specific pcr assays for erucic acid content in Indian mustard [*Brassica juncea* (L.) Czern and Coss.]. *Front Plant Sci*, *12*, 738805.

Golicz AA, Bayer PE, Barker GC, Edger PP, Kim H, Martinez PA, Chan CKK, Severn-Ellis A, McCombie WR, Parkin IA, Paterson AH (2016) The pangenome of an agronomically important crop plant *Brassica oleracea. Nat Commun*, *7*(1), 1–8.

Goslin K, Zheng B, Serrano-Mislata A, Rae L, Ryan PT, Kwaśniewska K, Thomson B, Ó'Maoiléidigh DS, Madueño F, Wellmer F, Graciet E (2017) Transcription factor interplay between *LEAFY* and *APETALA1/CAULIFLOWER* during floral initiation. *Plant Physiol*, *174*(2), 1097–1109.

Gouda G, Gupta MK, Donde R, Kumar J, Parida M, Mohapatra T, Dash SK, Pradhan SK, Behera L (2020) Characterization of haplotypes and single nucleotide polymorphisms associated with *Gn1a* for high grain number formation in rice plant. *Genomics*, *112*(3), 2647–2657.

Gu H, Yu H, Wang J, Shen Y, Sheng X, Shaw RK, Branca F (2022) A 43 bp-deletion in the F3'H gene reducing anthocyanins is responsible for buds-keeping-green at low temperature in broccoli. *Authorea Preprints.*

Guo J, Han W, Wang M (2008) Ultraviolet and environmental stresses involved in the induction and regulation of anthocyanin biosynthesis: a review. *Afr J Biotechnol 7*(25).

Gupta PK (2008) Single-molecule DNA sequencing technologies for future genomics research. *Trends Biotechnol, 26*(11), 602–611.

Gupta PK, Rustgi S, Kulwal PL (2005) Linkage disequilibrium and association studies in higher plants: present status and future prospects. *Plant Mol Biol, 57*(4), 461–485.

Gupta M, Vikram A, Bharat N (2013) Black rot-A devastating disease of crucifers: a review. *Agric Rev, 34*(4), 269–278.

Gururani MA, Venkatesh J, Upadhyaya CP, Nookaraju A, Pandey SK, Park SW (2012) Plant disease resistance genes: current status and future directions. *Physiol Mol Plant Pathol, 78*, 51–65.

Hashimoto H, Uragami C, Cogdell RJ (2016) Carotenoids and photosynthesis. Carotenoids in nature. *Subcellular Biochem, 79*, 111–139.

He Y, Fu Y, Hu D, Wei D, Qian W (2018) QTL mapping of seed glucosinolate content responsible for environment in *Brassica napus. Front Plant Sci, 9*, 891.

Hehl R, Wingender E (2001) Database-assisted promoter analysis. *Trends Plant Sci, 6*(6), 251–255.

Heim MA, Jakoby M, Werber M, Martin C, Weisshaar B, Bailey PC (2003) The basic helix–loop–helix transcription factor family in plants: a genome-wide study of protein structure and functional diversity. *Mol Biol Evol, 20*(5), 735–747.

Hirai M, Harada T, Kubo N, Tsukada M, Suwabe K, Matsumoto S (2004) A novel locus for clubroot resistance in Brassica rapa and its linkage markers. *Theor Appl Genet, 108*(4), 639–643.

Holme IB, Torp AM, Hansen LN, Andersen SB (2004) Quantitative trait loci affecting plant regeneration from protoplasts of *Brassica oleracea. Theor Appl Genet*, 108, 1513–1520.

Honma T, Goto K (2001) Complexes of MADS-box proteins are sufficient to convert leaves into floral organs. *Nature, 409*(6819), 525–529.

Hou J, Jiang Q, Hao C, Wang Y, Zhang H, Zhang X (2014) Global selection on sucrose synthase haplotypes during a century of wheat breeding. *Plant Physiol, 164*(4), 1918–1929.

Houghton A, Appelhagen I, Martin C (2021) Natural blues: Structure meets function in anthocyanins. *Plants, 10*(4), 726.

Hutchison III CA (2007) DNA sequencing: bench to bedside and beyond. *Nucleic Acids Res, 35*(18), 6227–6237.

Iglesias-Bernabé L, Madloo P, Rodríguez VM, Francisco M, Soengas P (2019) Dissecting quantitative resistance to *Xanthomonas campestris* pv. *campestris* in leaves of *Brassica oleracea* by QTL analysis. *Sci Rep, 9*(1), 1–11.

Irish VF (2010) The flowering of *Arabidopsis* flower development. *Plant J, 61*(6), 1014–1028.

Irish VF, Sussex IM (1990) Function of the apetala1 gene during Arabidopsis floral development. *Plant Cell,* 2, 741–753.

Irwin JA, Lister C, Soumpourou E, Zhang Y, Howell EC, Teakle G, Dean C (2012) Functional alleles of the flowering time regulator FRIGIDA in the *Brassica oleracea* genome. *BMC Plant Biol, 12*(1), 1–15.

Irwin JA, Soumpourou E, Lister C, Ligthart JD, Kennedy S, Dean C (2016) Nucleotide polymorphism affecting FLC expression underpins heading date variation in horticultural brassicas. *Plant J, 87*(6), 597–605.

Izzah NK, Lee J, Jayakodi M, Perumal S, Jin M, Park BS, Ahn K, Yang TJ (2014) Transcriptome sequencing of two parental lines of cabbage (*Brassica oleracea* L. var. *capitata* L.) and construction of an EST-based genetic map. *BMC Genom, 15*(1), 1–13.

Jaganathan D, Bohra A, Thudi M, Varshney RK (2020) Fine mapping and gene cloning in the post-NGS era: advances and prospects. *Theor Appl Genet, 133*(5), 1791–1810.

Jambhulkar SJ (2015) Induced mutagenesis and allele mining. In: Kumar A, Banga SS, Meena PD, Kumar PR (eds) *Brassica Oilseeds: Breeding and Management. CABI*, pp 53–67.

Jan HU, Guan M, Yao M, Liu W, Wei D, Abbadi A, Zheng M, He X, Chen H, Guan C, Nichols RA (2019) Genome-wide haplotype analysis improves trait predictions in *Brassica napus* hybrids. *Plant Sci, 283*, 157–164.

Jeong SY, Ahmed NU, Jung HJ, Kim HT, Park JI, Nou IS (2017) Discovery of candidate genes for heterosis breeding in *Brassica oleracea* L. *Acta Physiol Planta, 39*(8), 1–12.

Jin RG, Liu YB, Tabashnik BE, Borthakur D (2000) Development of transgenic cabbage (*Brassica oleracea* var. *capitata*) for insect resistance by Agrobacterium tumefaciens-mediated transformation. In *Vitro Cell Dev Biol Plant, 36*(4), 231–237.

Jordan KW, Wang S, Lun Y, Gardiner LJ, MacLachlan R, Hucl P, Wiebe K, Wong D, Forrest KL, Sharpe AG, Sidebottom CH (2015) A haplotype map of allohexaploid wheat reveals distinct patterns of selection on homoeologous genomes. *Genome Biol*, *16*(1), 1–18.

Kageyama K, Asano T (2009) Life cycle of *Plasmodiophora brassicae*. *J Plant Growth Regul*, *28*(3), 203–211.

Kalia P, Saha P, Ray S (2017) Development of RAPD and ISSR derived SCAR markers linked to Xca1Bo gene conferring resistance to black rot disease in cauliflower (Brassica oleracea var. botrytis L.). Euphytica, *213*, 1–12.

Kalve S, De Vos D, Beemster GT (2014) Leaf development: a cellular perspective. *Front Plant Sci, 5*, 362.

Kato T, Hatakeyama K, Fukino N, Matsumoto S (2013) Fine mapping of the clubroot resistance gene CRb and development of a useful selectable marker in *Brassica rapa*. *Breed Sci*, *63*(1), 116–124.

Ke Q, Kang L, Kim HS, Xie T, Liu C, Ji CY, Kim SH, Park WS, Ahn MJ, Wang S, Li H (2019) Down-regulation of lycopene ε-cyclase expression in transgenic sweetpotato plants increases the carotenoid content and tolerance to abiotic stress. *Plant Sci, 281*, 52–60.

Kempin SA, Savidge B, Yanofsky MF (1995) Molecular basis of the cauliflower phenotype in Arabidopsis. *Science, 267*, 522–525.

Kennard WC, Slocum MK, Figdore SS, Osborn TC (1994) Genetic analysis of morphological variation in Brassica oleracea using molecular markers. *Theor Appl Genet*, *87*(6), 721–732.

Kliebenstein DJ (2009) A quantitative genetics and ecological model system: understanding the aliphatic glucosinolate biosynthetic network via QTLs. *Phytochem Rev*, *8*(1), 243–254.

Kifuji Y, Hanzawa H, Terasawa Y, Nishio T (2013) QTL analysis of black rot resistance in cabbage using newly developed EST-SNP markers. *Euphytica*, *190*(2), 289–295.

Kim HS, Ji CY, Lee CJ, Kim SE, Park SC, Kwak SS (2018) Orange: a target gene for regulating carotenoid homeostasis and increasing plant tolerance to environmental stress in marginal lands. *J Exp Bot*, *69*(14), 3393–3400.

Kim Y, Li X, Kim S-J, Kim H, Lee J, Kim H, Park S (2013) MYB transcription factors regulate glucosinolate biosynthesis in different organs of Chinese cabbage (Brassica rapa ssp. pekinensis). *Molecules*, 18(7), 8682–8695.

Kimura R, Sato K, Fujimoto R, Nishio T (2002) Recognition specificity of self-incompatibility maintained after the divergence of *Brassica oleracea* and *Brassica rapa*. *Plant J*, *29*(2), 215–223.

Koh JC, Barbulescu DM, Norton S, Redden B, Salisbury PA, Kaur S, Cogan N, Slater AT (2017) A multiplex PCR for rapid identification of *Brassica* species in the triangle of U. *Plant Methods*, *13*(1), 1–8.

Kohorn BD, Kohorn SL, Todorova T, Baptiste G, Stansky K, McCullough M (2012) A dominant allele of Arabidopsis pectin-binding wall-associated kinase induces a stress response suppressed by MPK6 but not MPK3 mutations. *Mol Plant*, *5*(4), 841–851.

Kresovich SJGK, Williams JGK, McFerson JR, Routman EJ, Schaal BA (1992) Characterization of genetic identities and relationships of *Brassica oleracea* L. via a random amplified polymorphic DNA assay. *Theor Appl Genet*, *85*(2), 190–196.

Kumar A, Banga SS, Meena PD, Kumar PR (eds) (2015) *Brassica Oilseeds: Breeding and Management*. CABI.

Kumar GR, Sakthivel K, Sundaram RM, Neeraja CN, Balachandran SM, Rani NS, Viraktamath BC, Madhav MS (2010) Allele mining in crops: prospects and potentials. *Biotechnol Adv* *28*(4), 451–461.

Kusaba M, Nishio T (1999) Comparative analysis of S haplotypes with very similar SLG alleles in *Brassica rapa* and *Brassica oleracea*. *Plant J*, *17*(1), 83–91.

Lan TH, Paterson AH (2001) Comparative mapping of QTLs determining the plant size of *Brassica oleracea*. *Theor Appl Genet*, *103*(2), 383–397.

Latha R, Rubia L, Bennett J, Swaminathan MS (2004) Allele mining for stress tolerance genes in *Oryza* species and related germplasm. *Mol Biotechnol*, *27*(2), 101–108.

Lee J, Yang K, Lee M, Kim S, Kim J, Lim S, Kang GH, Min SR, Kim SJ, Park SU, Jang YS (2015) Differentiated cuticular wax content and expression patterns of cuticular wax biosynthetic genes in bloomed and bloomless broccoli (*Brassica oleracea* var. italica). *Process Biochem, 50*(3), 456–462.

Leung H, Raghavan C, Zhou B, Oliva R, Choi IR, Lacorte V, Jubay ML, Cruz CV, Gregorio G, Singh RK, Ulat VJ (2015) Allele mining and enhanced genetic recombination for rice breeding. *Rice*, *8*(1), 1–11.

Li L, Paolillo DJ, Parthasarathy MV, DiMuzio EM, Garvin DF (2001) A novel gene mutation that confers abnormal patterns of β-carotene accumulation in cauliflower (*Brassica oleracea* var. *botrytis*). *Plant J*, *26*(1), 59–67.

Li Y, Tong L, Deng L, Liu Q, Xing Y, Wang C, Liu B, Yang X, Xu M (2017) Evaluation of ZmCCT haplotypes for genetic improvement of maize hybrids. *Theor Appl Genet*, *130*(12), 2587–2600.

Li B, Zhang X, Liu Z, Wang L, Song L, Liang X, Dou S, Tu J, Shen J, Yi B, Wen J (2021) Genetic and molecular characterization of a self-compatible *Brassica rapa* line possessing a new class II S haplotype. *Plants*, *10*(12), 2815.

Li P, Zhang S, Li F, Zhang S, Zhang H, Wang X, Sun R, Bonnema G, Borm TJ (2017) A phylogenetic analysis of chloroplast genomes elucidates the relationships of the six economically important Brassica species comprising the triangle of U. *Front Plant Sci, 8*, 111.

Liu D, Dong X, Liu Z, Tang J, Zhuang M, Zhang Y, Lv H, Liu Y, Li Z, Fang Z, Yang L (2018) Fine mapping and candidate gene identification for wax biosynthesis locus, BoWax1 in *Brassica oleracea* L. var. *capitata*. *Front Plant Sci*, *9*, 309.

Liu Y, Feng X, Zhang Y, Zhou F, Zhu P (2021) Simultaneous changes in anthocyanin, chlorophyll, and carotenoid contents produce green variegation in pink–leaved ornamental kale. *BMC Genom*, *22*(1), 1–16.

Liu XP, Gao BZ, Han FQ, Fang ZY, Yang LM, Zhuang M, Lv HH, Liu YM, Li ZS, Cai CC, Yu HL (2017) Genetics and fine mapping of a purple leaf gene, BoPr, in ornamental kale (*Brassica oleracea* L. var. *acephala*). *BMC Genom*, *18*(1), 1–9.

Liu S, Liu Y, Yang X, Tong C, Edwards D, Parkin IA, Zhao M, Ma J, Yu J, Huang S, Wang X (2014) The Brassica oleracea genome reveals the asymmetrical evolution of polyploid genomes. *Nat Commun*, *5*(1), 1–11.

Liu HJ, Yan J (2019) Crop genome-wide association study: a harvest of biological relevance. *Plant J*, *97*(1), 8–18.

Liu C, Yao X, Li G, Huang L, Liu C, Xie Z (2022) Development of novel markers and creation of non-anthocyanin and anthocyanin-rich broccoli (*Brassica oleracea* var. italica) Cultivars. *Appl Sci*, *12*(12), 6267.

Lv H, Fang Z, Yang L, Zhang Y, Wang Q, Liu Y, Zhuang M, Yang Y, Xie B, Liu B, Liu J (2014) Mapping and analysis of a novel candidate Fusarium wilt resistance gene FOC1 in *Brassica oleracea*. *BMC Genom*, *15*(1), 1–10.

Mabry M.E, Turner-Hissong SD, Gallagher EY, McAlvay AC, An H, Edger PP, Moore JD, Pink DA, Teakle GR, Stevens CJ, Barker G (2021) The evolutionary history of wild, domesticated, and feral *Brassica oleracea* (*Brassicaceae*). *Mol Biol Evol*, *38*(10), 4419–4434.

Macioszek VK, Wielanek M, Morkunas I, Ciereszko I, Kononowicz AK, (2020) Leaf position-dependent effect of *Alternaria brassicicola* development on host cell death, photosynthesis and secondary metabolites in *Brassica juncea*. *Physiol Planta*, *168*(3), 601–616.

Maggioni L (2015) Domestication of *Brassica oleracea* L. *Acta Univ Agric Suec*, 2015, 74.

Maggioni L, Bothmer R, Poulsen G, Branca F (2010) Origin and domestication of cole crops (*Brassica oleracea* L.): linguistic and literary considerations1. *Econ Bot*, *64*(2), 109–123.

Maggioni L, von Bothmer R, Poulsen G, Lipman E (2018) Domestication, diversity and use of *Brassica oleracea* L., based on ancient Greek and Latin texts. *Genetic Res Crop Evol,* 65(1), 137–159.

Maheshwari S, Barbash DA (2011) The genetics of hybrid incompatibilities. *Annu Rev Genet*, *45*, 331–355.

Margulies M, Egholm M, Altman WE, Attiya S, Bader JS, Bemben LA, Berka J, Braverman MS, Chen YJ, Chen Z, Dewell SB (2005) Genome sequencing in microfabricated high-density picolitre reactors. *Nature*, *437*(7057), 376–380.

Mei J, Wang J, Li Y, Tian S, Wei D, Shao C, Si J, Xiong Q, Li J, Qian W (2017) Mapping of genetic locus for leaf trichome in *Brassica oleracea*. *Theor Appl Genet*, *130*(9), 1953–1959.

Miao L, Gao Y, Zhao K, Kong L, Yu S, Li R, Liu K Yu X (2020) Comparative analysis of basic helix–loop–helix gene family among *Brassica oleracea*, *Brassica rapa*, and *Brassica napus*. *BMC Genom*, *21*(1), 1–18.

Molina C, Grotewold E (2005) Genome wide analysis of Arabidopsis core promoters. *BMC Genom*, *6*(1), 1–12.

Motsinger AA, Reif DM, Dudek SM, Ritchie MD (2006) Understanding the evolutionary process of grammatical evolution neural networks for feature selection in genetic epidemiology. In: *2006 IEEE Symposium on Computational Intelligence and Bioinformatics and Computational Biology* 2006 Sep 28, 1–8. IEEE.

Nagaoka T, Doullah MAU, Matsumoto S, Kawasaki S, Ishikawa T, Hori H, Okazaki K (2010) Identification of QTLs that control clubroot resistance in Brassica oleracea and comparative analysis of clubroot resistance genes between *B*. rapa and B. oleracea. *Theor Appl Genet*, *120*(7), 1335–1346.

Negri V (2012) Policies supportive of on-farm conservation and their impact on custodian farmers in Italy. In: *On Farm Conservation of Neglected and Underutilized Species: Status, Trends and Novel Approaches to Cope with Climate Change*. Proceedings of an International Conference, Frankfurt, 14–16 June, 2011, p 211.

Nelson R, Wiesner-Hanks T, Wisser R, Balint-Kurti P (2018) Navigating complexity to breed disease-resistant crops. *Nat Rev Genet, 19*(1), 21–33.

Ni X, Huang J, Ali B, Zhou W, Zhao J (2015) Genetic analysis and fine mapping of the LOBED-LEAF 1 (BnLL1) gene in rapeseed (*Brassica napus* L.). *Euphytica, 204*(1), 29–38.

Ni X, Liu H, Huang J, Zhao J (2017) LMI1-like genes involved in leaf margin development of *Brassica napus. Genetica, 145*(3), 269–274.

Niu N, Zheng C, Zhang L, Xu Y, Zhang L, Fu W (2010) Inheritance and RAPD marker of head-splitting trait in Chinese cabbage. *China Vegetables,* (14), 44–48.

Nuijten E, Messmer MM, Lammerts van Bueren ET (2016) Concepts and strategies of organic plant breeding in light of novel breeding techniques. *Sustainability, 9*(1), 18.

Ockendon DJ (1975) Dominance relationships between S-alleles in the stigmas of Brussels sprouts (*Brassica oleracea* var. *gemmifera*). *Euphytica, 24*(1), 165–172.

Ockendon DJ (1982) An S-allele survey of cabbage (*Brassica oleracea* var. *capitata*). *Euphytica, 31*(2), 325–331.

Ockendon DJ (2000) The S-allele collection of *Brassica* oleracea. In: *III International Symposium on Brassicas and XII Crucifer Genetics Workshop 539*, September 5, pp 25–30.

Okazaki K, Sakamoto K, Kikuchi R, Saito A, Togashi E, Kuginuki Y, Matsumoto S, Hirai M (2007) Mapping and characterization of FLC homologs and QTL analysis of flowering time in *Brassica oleracea. Theor Appl Genet, 114*(4), 595–608.

Osorio CE (2019) The role of orange gene in carotenoid accumulation: manipulating chromoplasts toward a colored future. *Front Plant Sci, 10*: 1235.

Palaisa K, Morgante M, Tingey S, Rafalski A (2004) Long-range patterns of diversity and linkage disequilibrium surrounding the maize Y1 gene are indicative of an asymmetric selective sweep. *Proc Natl Acad Sci USA 101*, 9885–9890.

Pang W, Kim YY, Li X, Choi SR, Wang Y, Sung CK, Im S, Ramchiary N, Zhou G, Lim YP (2015) Anatomic characteristics associated with head splitting in cabbage (*Brassica oleracea* var. capitata L.). *PLoS One, 10*(11), e0142202.

Paran I, Zamir D (2003) Quantitative traits in plants: beyond the QTL. *Trends Genet, 19*(6), 303–306.

Park PJ, Butte AJ, Kohane IS (2002) Comparing expression profiles of genes with similar promoter regions. *Bioinformatics, 18*(12), 1576–1584.

Passmore S, Maine GT, Elble R, Christ C, Tye BK (1988) *Saccharomyces cerevisiae* protein involved in plasmid maintenance is necessary for mating of MATα cells. *J Mol Biol*, 204, 593–606.

Pedras MSC, Chumala PB, Jin W, Islam MS, Hauck DW (2009) The phytopathogenic fungus *Alternaria brassicicola*: phytotoxin production and phytoalexin elicitation. *Phytochemistry, 70*(3), 394–402.

Pelc SE, Couillard DM, Stansell ZJ, Farnham MW (2015) Genetic diversity and population structure of collard landraces and their relationship to other Brassica oleracea crops. *Plant Genome, 8*(3), 2015–04.

Peng L, Zhou L, Li Q, Wei D, Ren X, Song H, Mei J, Si J, Qian W (2018) Identification of quantitative trait loci for clubroot resistance in *Brassica oleracea* with the use of Brassica SNP microarray. *Front Plant Sci, 9*, 822.

Pettersson E, Lundeberg J, Ahmadian A (2009) Generations of sequencing technologies. *Genomics, 93*(2), 105–111.

Podsedek A, Sosnowska D, Redzynia M, Anders B (2006) Antioxidant capacity and content of *Brassica oleracea* dietary antioxidants. *Int J Food Sci Technol,* 41, 49–58.

Pu ZJ, Shimizu M, Zhang YJ, Nagaoka T, Hayashi T, Hori H, Matsumoto S, Fujimoto R, Okazaki K (2012) Genetic mapping of a fusarium wilt resistance gene in *Brassica oleracea. Mol Breed, 30*(2), 809–818.

Purugganan MD, Boyles AL, Suddith JI (2000) Variation and selection at the CAULIFLOWER floral homeotic gene accompanying the evolution of domesticated *Brassica oleracea. Genetics, 155*(2), 855–862.

Qian L, Hickey LT, Stahl A, Werner CR, Hayes B, Snowdon RJ, Voss-Fels KP (2017) Exploring and harnessing haplotype diversity to improve yield stability in crops. *Front Plant Sci, 8*, 1534.

Qian W, Jia Y, Ren SX, He YQ, Feng JX, Lu LF, Sun Q, Ying G, Tang DJ, Tang H, Wu W (2005) Comparative and functional genomic analyses of the pathogenicity of phytopathogen *Xanthomonas campestris* pv. *campestris. Genome Res, 15*(6), 757–767.

Qiao J, Cai M, Yan G, Wang N, Li F, Chen B, Gao G, Xu, K, Li J, Wu X (2016) High-throughput multiplex cp DNA resequencing clarifies the genetic diversity and genetic relationships among *Brassica napus, Brassica rapa* and *Brassica oleracea. Plant Biotechnol J, 14*(1), 409–418.

Rae AM, Howell EC, Kearsey MJ (1999) More QTL for flowering time revealed by substitution lines in *Brassica oleracea. Heredity*, *83*(5), 586–596.

Ren J, Liu Z, Du J, Fu W, Hou A, Feng H (2019) Fine-mapping of a gene for the lobed leaf, BoLl, in ornamental kale (*Brassica oleracea* L. var. acephala). *Mol Breed*, *39*(3), 1–14. IntechOpen.

Ridge S, Brown PH, Hecht V, Driessen RG, Weller JL (2015) The role of BoFLC2 in cauliflower (*Brassica oleracea* var. botrytis L.) reproductive development. *J Exp Bot*, *66*(1), 125–135.

Saito M, Kubo N, Matsumoto S, Suwabe K, Tsukada M, Hirai M (2006) Fine mapping of the clubroot resistance gene, Crr3, in *Brassica rapa. Theor ApplGenet*, *114*(1), 81–91.

Salvi S, Tuberosa R (2007) Cloning QTLs in plants. In: Varshney RK, Tuberosa R (eds) *Genomics-Assisted Crop Improvement.* Springer, Dordrecht, pp 207–225

Santos-Sánchez NF, Salas-Coronado R, Hernández-Carlos B, Villanueva-Cañongo C (2019) Shikimic acid pathway in biosynthesis of phenolic compounds. In: Soto-Hernández M, García-Mateos R, Palma-Tenango M (eds) *Plant Physiological Aspects of Phenolic Compounds 1*, pp 1–15.

Sato Y, Fujimoto R, Toriyama K, Nishio T (2003) Commonality of self-recognition specificity of S haplotypes between *Brassica oleracea* and *Brassica rapa. Plant Mol Biol, 52*(3), 617–626.

Scheben A, Verpaalen B, Lawley CT, Chan CKK, Bayer PE, Batley J, Edwards D (2019) CropSNPdb: a database of SNP array data for Brassica crops and hexaploid bread wheat. *Plant J*, *98*(1), 142–152.

Schilling S, Pan S, Kennedy A, Melzer R (2018) MADS-box genes and crop domestication: the jack of all traits. *J Exp Bot*, *69*(7), 1447–1469.

Schmidt S, Zietz M, Schreiner M, Rohn S, Kroh LW, Krumbein A (2010) Genotypic and climatic influences on the concentration and composition of flavonoids in kale (*Brassica oleracea* var. *sabellica*). *Food Chem, 119*(4), 1293–1299.

Sebastian RL, Kearsey MJ, King GJ (2002) Identification of quantitative trait loci controlling developmental characteristics of *Brassica oleracea* L. *Theor Appl Genet*, *104*(4), 601–609.

Shan X, Zhang W, Yu F, Wang S, Li J, Tang J, Dai Z (2019) Genome-wide analysis of basic helix–loop–helix superfamily members reveals organization and chilling-responsive patterns in cabbage (*Brassica oleracea* var. capitata L.). *Genes, 10*(11), 914.

Shang G, Zhao H, Tong L, Yin N, Hu R, Jiang H, Kamal F, Zhao Z, Xu L, Lu K, Li J (2022) Genome-wide association study of phenylalanine derived glucosinolates in *Brassica rapa*. *Plants,* 11, 1274.

Sharma BB, Kalia P, Yadava DK, Singh D, Sharma TR (2016). Genetics and molecular mapping of black rot resistance locus Xca1bc on chromosome B-7 in Ethiopian mustard (*Brassica carinata* A. *Braun*). *PLoS One, 11*(3), e0152290.

Sharma BB, Kalia P, Singh D, Sharma TR (2017) Introgression of black rot resistance from *Brassica carinata* to cauliflower (*Brassica oleracea botrytis* group) through embryo rescue. *Front Plant Sci*, *8*, 1255.

Shen JX, Fu TD, Yang GS, Ma CZ, Tu JX (2005). Genetic analysis of rapeseed self-incompatibility lines reveals significant heterosis of different patterns for yield and oil content traits. *Plant Breed*, *124*(2), 111–116.

Sheng XG, Zhao ZQ, Wang JS, Yu HF, Shen YS, Zeng XY, Gu HH (2019) Genome wide analysis of MADS-box gene family in *Brassica oleracea* reveals conservation and variation in flower development. *BMC Plant Biol*, *19*(1), 1–15.

Shepherd T, Wynne Griffiths D (2006) The effects of stress on plant cuticular waxes. *New Phytol*, *171*(3), 469–499.

Shimizu M, Pu ZJ, Kawanabe T, Kitashiba H, Matsumoto S, Ebe Y, Sano M, Funaki T, Fukai E, Fujimoto R, Okazaki K (2015) Map-based cloning of a candidate gene conferring Fusarium yellows resistance in *Brassica oleracea. Theor Appl Genet*, *128*(1), 119–130.

Skinner MK, Rawls A, Wilson-Rawls J, Roalson EH (2010) Basic helix-loop-helix transcription factor gene family phylogenetics and nomenclature. *Differentiation*, *80*(1), 1–8.

Smith LB, King GJ (2000) The distribution of BoCAL-a alleles in Brassica oleracea is consistent with a genetic model for curd development and domestication of the cauliflower. *Mol Breed*, *6*(6), 603–613.

Soengas P, Cartea ME, Francisco M, Sotelo T, Velasco P (2012) New insights into antioxidant activity of Brassica crops. *Food Chem*, *134*(2), 725–733.

Sommer H, Beltran JP, Huijser P, Pape H, Lönnig W, Saedler H, Schwarz-Sommer Z (1990) *Deficiens*, a homeotic gene involved in the control of flower morphogenesis in *Antirrhinum majus*: the protein shows homology to transcription factors. *EMBO J*, 9, 605.

Song, XM, Huang ZN, Duan WK, Ren J, Liu TK, Li Y, Hou X.L (2014) Genome-wide analysis of the bHLH transcription factor family in Chinese cabbage (*Brassica rapa* ssp. *pekinensis). Mol Genet Genom*, *289*(1), 77–91.

Sorkheh K, Malysheva-Otto LV, Wirthensohn MG, Tarkesh-Esfahani S, Martínez-Gómez P (2008) Linkage disequilibrium, genetic association mapping and gene localization in crop plants. *Genet Mol Biol, 31*, 805–814.

Sotelo T, Cartea ME, Velasco P, Soengas P (2014) Identification of antioxidant capacity-related QTLs in *Brassica oleracea. PLoS One*, *9*(9), 107290.

Sotelo T, Soengas P, Velasco P, Rodríguez VM, Cartea ME (2014) Identification of metabolic QTLs and candidate genes for glucosinolate synthesis in *Brassica oleracea* leaves, seeds and flower buds. *PLoS One*, *9*(3), 91428.

Singh AL (2004) Mineral nutrient requirement, their disorders and remedies in groundnut. In: Basu MS, Singh NB (eds) *Groundnut Research in India*, Edition: 1, Publisher: National Research Center for Groundnut (ICAR). Junagadh, India, pp 137–159.

Singh S, Dey SS, Bhatia R, Batley J, Kumar R (2018) Molecular breeding for resistance to black rot [*Xanthomonas campestris* pv. *campestris* (Pammel) Dowson] in Brassicas: recent advances. *Euphytica*, *214*(10), 1–17.

Singh D, Dhar S, Yadava DK (2011) Genetic and pathogenic variability of Indian strains of *Xanthomonas campestris* pv. *campestris* causing black rot disease in crucifers. *Curr Microbiol*, *63*(6), 551–560.

Sretenović-Rajičić T, Ninković S, Miljuš-Đukić J, Vinterhalter B, Vinterhalter D (2006) Agrobacterium rhizogenes-mediated transformation of Brassica oleracea var. sabauda and B. oleracea var. capitata. *Biol Planta, 50*(4), 525–530.

Stansell Z, Hyma K, Fresnedo-Ramírez J, Sun Q, Mitchell S, Björkman T, Hua J (2018) Genotyping-by-sequencing of *Brassica oleracea* vegetables reveals unique phylogenetic patterns, population structure and domestication footprints. *Hortic Res, 5*(2018), 38.

Stansell Z, Björkman T (2020) From landrace to modern hybrid broccoli: the genomic and morphological domestication syndrome within a diverse *B. oleracea* collection. *Hortic Res, 7*(2020), 159.

Su Y, Liu Y, Li Z, Fang Z, Yang L, Zhuang M, Zhang Y (2015) QTL analysis of head splitting resistance in cabbage (*Brassica oleracea* L. var. *capitata*) using SSR and InDel makers based on whole-genome re-sequencing. *PLoS One, 10*(9), 0138073.

Sun Z, Li N, Huang G, Xu J, Pan Y, Wang Z, Tang Q, Song M, Wang X (2013) Site-specific gene targeting using transcription activator-like effector (TALE)-based nuclease in *Brassica oleracea. J Integr Plant Biol, 55*(11), 1092–1103.

Sun TH, Zhou F, Liu C-J, Zhuang Z, Lu S (2016) The DnaJ-like zinc finger domain protein Orange localizes to the nucleus in etiolated cotyledons of *Arabidopsis thaliana. Protoplasma,* 253 (6), 1599–1604.

Suzuki, Y. 2010. Statistical methods for detecting natural selection from genomic data. *Genes Genet Syst*, *85*(6), 359–376.

Takada S, Akter A, Itabashi E, Nishida N, Shea DJ, Miyaji N, Mehraj H, Osabe K, Shimizu M, Takasaki-Yasuda T, Kakizaki T (2019) The role of FRIGIDA and FLOWERING LOCUS C genes in flowering time of *Brassica rapa* leafy vegetables. *Sci Rep*, *9*(1), 1–11.

Takayama S, Shimosato H, Shiba H, Funato M, Che FS, Watanabe M, Iwano M, Isogai A (2001) Direct ligand–receptor complex interaction controls Brassica self-incompatibility. *Nature*, *413*(6855), 534–538.

Tao Z, Kou Y, Liu H, Li X, Xiao J, Wang S (2011) OsWRKY45 alleles play different roles in abscisic acid signalling and salt stress tolerance but similar roles in drought and cold tolerance in rice. *J Exp Bot*, *62*(14), 4863–4874.

Teklewold A, Becker HC (2005) Heterosis and combining ability in a diallel cross of Ethiopian mustard inbred lines. *Crop Sci*, *45*(6), 2629–2635.

Timpanaro G, Branca F, Cammarata M, Di Bella MC, Foti VT, Scuderi A (2022). Biodiversity enhancement for improving the sustainability of broccoli (*Brassica oleracea* vr. *italica* Plenk) organic seed production. *Sustainability*, 14(11), 6417.

Tong L, Zhao C, Liu J, Yang L, Zhuang M, Zhang Y, Wang Y, Ji J, Kuang B, Tang K, Fang Z (2022) Resource screening and inheritance analysis of *Fusarium oxysporum* sp. *conglutinans* Race 2 resistance in cabbage (*Brassica oleracea* var. *capitata*). *Genes*, *13*(9), 1590.

Treccarichi S, Di Gaetano C, Di Stefano F, Gasparini M, Branca F (2021). Using simple sequence repeats in 9 brassica complex species to assess hypertrophic curd induction. *Agriculture, 11*(7), 622.

Treccarichi S, BenAmmar H, Amari M, Cali R; Tribulato A, Branca F (2023) Molecular markers for detecting inflorescence size of *Brassica oleracea* L. Crops and *B. oleracea* complex species (n = 9) useful for breeding of Broccoli (B. oleracea var. italica) and Cauliflower (B. oleracea var. botrytis). *Plants*, (*12*), 407.

Troncoso-Rojas R, Tiznado-Hernández ME (2014) *Alternaria alternata* (black rot, black spot). In: Bautista-Baños S (ed) *Postharvest Decay*. Academic Press, pp 147–187.

Upadhyaya HD, Gowda CLL, Buhariwalla HK, Crouch JH (2006a) Efficient use of crop germplasm resources: identifying useful germplasm for crop improvement through core and mini-core collections and molecular marker approaches. *Plant Genet Resour*, *4*(1), 25–35.

Upadhyaya HD, Gowda CLL, Pundir RPS, Reddy VG, Singh S (2006b) Development of core subset of finger millet germplasm using geographical origin and data on 14 quantitative traits. *Genet Resour Crop Evol*, *53*(4), 679–685.

Upadhyaya HD, Gowda CLL, Sastry DVSSR (2008) Plant genetic resources management: collection, characterization, conservation and utilization. *J SAT Agric Res*, *6*, 16.

Uptmoor R, Schrag T, Stützel H, Esch, E (2008) Crop model based QTL analysis across environments and QTL based estimation of time to floral induction and flowering in *Brassica oleracea*. *Mol Breed*, *21*(2), 205–216.

Van Bueren EL, Struik PC, Tiemens-Hulscher M, Jacobsen E (2003) Concepts of intrinsic value and integrity of plants in organic plant breeding and propagation. *Crop Sci*, *43*(6), 1922–1929.

van Hintum TJ, van De Wiel CCM, Visser DL, Van Treuren R, Vosman B (2007) The distribution of genetic diversity in a *Brassica oleracea* gene bank collection related to the effects on diversity of regeneration, as measured with AFLPs. *Theor Appl Genet,* 114(5), 777–786.

Veerla S, Höglund M (2006) Analysis of promoter regions of co-expressed genes identified by microarray analysis. *BMC Bioinform*, *7*(1), 1–15.

Vicente JG, Holub EB (2013) *Xanthomonas campestris* pv. *campestris* (cause of black rot of crucifers) in the genomic era is still a worldwide threat to brassica crops. *Mol Plant Pathol*, *14*(1), 2–18.

Vlad D, Kierzkowski D, Rast MI, Vuolo F, DelloIoio R, Galinha C, Gan X, Hajheidari M, Hay A, Smith RS, Huijser P (2014) Leaf shape evolution through duplication, regulatory diversification, and loss of a homeobox gene. *Science*, *343*(6172), 780–783.

Wang X, Fang Z, Huang S, Sun P, Liu Y, Yang L, Zhuang M, Qu D (2000) An extended random primer amplified region (ERPAR) marker linked to a dominant male sterility gene in cabbage (*Brassica oleracea* var. *capitata*). *Euphytica*, *112*(3), 267–273.

Wang W, Huang S, Liu Y, Fang Z, Yang L, Hua W, Yuan S, Liu S, Sun J, Zhuang M, Zhang, Y (2012) Construction and analysis of a high-density genetic linkage map in cabbage (*Brassica oleracea* L. var. *capitata*). *BMC Genom*, *13*(1), 1–12.

Wang Y, Huang S, Liu Z, Tang X, Feng H (2018) Changes in endogenous phytohormones regulated by microRNA-target mRNAs contribute to the development of Dwarf Autotetraploid Chinese Cabbage (*Brassica rapa* L. *ssp. pekinensis*). *Mol Genet Genom*, *293*(6), 1535–1546.

Wang Z, Xu W, Kang J, Li M, Huang J, Ke Q, Kim HS, Xu B, Kwak SS (2018) Overexpression of alfalfa Orange gene in tobacco enhances carotenoid accumulation and tolerance to multiple abiotic stresses. *Plant Physiol Biochem*, *130*, 613–622.

Wei C, Zhu L, Wen J, Yi B, Ma C, Tu J, Shen J, Fu T (2018) Morphological, transcriptomics and biochemical characterization of new dwarf mutant of Brassica napus. *Plant Sci, 270*, 97–113.

Wils CR, Kaufmann K (2017) Gene-regulatory networks controlling inflorescence and flower development in *Arabidopsis thaliana*. *Biochim Biophys Acta (BBA)-Gene Regulatory Mechanisms*, 1860(1), 95–105.

Wu X, Jiang W, Fragoso C, Huang J, Zhou G, Zhao H, Dellaporta S (2021) Causal haplotype block identification in plant genome-wide association studies. bioRxiv, 2021–10.

Werner CR, Qian L, Voss-Fels KP, Abbadi A, Leckband G, Frisch M, Snowdon RJ (2018) Genome-wide regression models considering general and specific combining ability predict hybrid performance in oilseed rape with similar accuracy regardless of trait architecture. *Theor Appl Genet*, *131*(2), 299–317.

Xiao Z, Hu Y, Zhang X, Xue Y, Fang Z, Yang L, Zhang Y, Liu Y, Li Z, Liu X, Liu Z (2017) Fine mapping and transcriptome analysis reveal candidate genes associated with hybrid lethality in cabbage (*Brassica oleracea*). *Genes*, *8*(6), 147.

Xing M, Lv H, Ma J, Xu D, Li H, Yang L, Kang J, Wang X, Fang Z (2016) Transcriptome profiling of resistance to *Fusarium oxysporum* f. sp. *conglutinans* in cabbage (*Brassica oleracea*) roots. *PLoS One*, *11*(2), 0148048.

Xing M, Su H, Liu X, Yang L, Zhang Y, Wang Y, Fang Z, Lv H (2020) Morphological, transcriptomics and phytohormone analysis shed light on the development of a novel dwarf mutant of cabbage (Brassica oleracea). *Plant Sci, 290,* 110283.

Xue D, Zhang X, Lu X, Chen G, Chen ZH (2017) Molecular and evolutionary mechanisms of cuticular wax for plant drought tolerance. *Front Plant Sci, 8,* 621.

Yi GE, Robin AH, Yang K, Park JI, Kang JG, Yang TJ, Nou IS (2015) Identification and expression analysis of glucosinolate biosynthetic genes and estimation of glucosinolate contents in edible organs of *Brassica oleracea* subspecies. *Molecules*, 20, 13089–13111.

Young AJ, Lowe GL (2018) Carotenoids—antioxidant properties. *Antioxidants*, 7(2), 28.

Yu H, Wang J, Zhao Z, Sheng X, Shen Y, Branca F, Gu H (2019) Construction of a high-density genetic map and identification of loci related to hollow stem trait in broccoli (*Brassic oleracea* L. italica). *Front Plant Sci*, 10, 45.

Zhang B, Chen W, Li X, Ren W, Chen L, Han F, Zhang Y (2021) Promoter variations strongly enhance the transcription level of the BoLMI1 Gene, causing lobed leaves in Ornamental Kale (*Brassica Oleracea* L. Var. *Acephala*).

Zou C (2019) Genetic analysis of leaf related phenotypic data of a very diverse *Brassica oleracea* collection. Netherlands.

9 Allele Mining for Genome Designing in Carrot (*Daucus carota* L.)

Raman Selvakumar[1] *and Pritam Kalia*[2]
[1]Center for Protected Cultivation, ICAR-Indian Agricultural Research Institute, New Delhi, India
[2]Division of Vegetable Science, ICAR- Indian Agricultural Research Institute, New Delhi, India

9.1 INTRODUCTION

The conventional farming methods, plant landraces, and vegetable types have all been supplanted by more modern cultivars that have been produced over the course of the previous century. The gradual erosion of genetic variety and financial resources within the vegetable business is hastened because of this. On the other hand, several gene banks specializing in vegetable crops have been successful in rescuing a considerable amount of the essential genetic capital lost from elite content (Till et al. 2003a). These protected accessions from gene banks are used as the foundation for the genetic development of vegetable crops, which is accomplished by breeding for certain desirable characteristics. In addition, they offer a large supply of unusual allelic diversity that may be used in the cultivation of agricultural plants, which is a significant benefit. Because of the progress that has been achieved in genomics over the course of the last quarter century, it is now more likely that vegetable crops will be genetically modified. This will help in the resolution of concerns connected to the preservation of food and nutrition for the growing population of the world. A comprehensive understanding of the genomic sequence of a vegetable crop is required for the development of novel molecular markers for commercially beneficial vegetable crops as well as the transgenic modification of gene expression. In recent years, several allele mining strategies, such as targeted induced local lesions in genomes (TILLING) and ecotype-TILLING (Eco-TILLING), have been developed to make use of the DNA sequence data that is presently accessible to investigate specific gene functions. These strategies aim to make use of the fact that there is such data. Investigation of gene function may now go in a variety of different ways, thanks to the expansion of the library of gene sequences containing information on vegetable crop genes. To discover the underlying genetic basis of a feature, one can apply the tried-and-true method of forward genetic analysis of a phenotype. This approach has been around for quite some time. This method of inverse genetics involves changing the gene or the degree at which it is expressed to investigate the phenotype that is produced consequently (Portis et al. 2016). A forward genetic method is recommended rather than a reverse genetic approach since the latter makes it possible to determine how each gene contributes to the overall function of the organism. Reverse genetic approaches have been found to be highly beneficial for analyzing fundamental gene function. Because of this, utilizing approaches that are based on forward genetics is the best option. Expressed sequence tags (ESTs) and complementary DNA (cDNA) sequences are now being created, and additional vegetable crop genome sequencing projects are either already ongoing or have been recommended (www.ncbi.nlm.nih.gov/sites/entrez?db=genomeplj). Genome sequencing paves the way for comparative genomic

 DOI: 10.1201/9781003376439-9

research, which has the potential to eventually provide proof of genomics (Prochnik et al. 2012). TILLING is a method that was developed to fill in the blanks that were left behind by other, more well-known methods such as gene attribute projections and genome sequencing. This indicates that it may be beneficial for vegetable species that have not yet been subjected to intensive research. Because it does not include any genetically modified organisms, it is permissible to use in research that aims to improve vegetable crops without running the risk of getting into trouble with the law (McCallum et al. 2000a). This makes it possible to utilize it without the risk of breaking any laws. The discoveries of functional genomics that are connected to essential characteristics of one crop might potentially influence another crop. TILLING is, in general, the greatest strategy for enhancing vegetable crops that does not include the use of transgenic organisms since it gives a mechanism to access practically every gene for functional study. The expansion of the genetic variety of domesticated wild vegetable crop relatives is the means through which this goal can be accomplished. The use of transgenics is not required to successfully cultivate vegetable crops using TILLING as an alternative strategy. In this review, we discussed not only the myriad of techniques that have proven successful in the past when used to the genetic improvement of carrot but also the unique challenges that these crops provide. The positive impacts that conventional TILLING and Eco-TILLING have on vegetable crops are seen (Selvakumar et al. 2023).

9.2 ALLELE MINING

An allele is an alternative form of a gene that may be found on a certain chromosome. Mining is just the search for new alleles in natural germplasm. Thus, "allele mining" refers to the search for useful alleles of genes from a broad range of cultivars, similar species, and even unrelated species (Kumari et al. 2018). To put it another way, it is a field of research that seeks to find allelic variation in key aspects of genetic resource collections. Because these modifications were thought to affect the encoded protein structure and/or function, early studies of allele mining concentrated solely on the detection of single nucleotide polymorphisms (SNPs) and insertion/deletion (InDels) polymorphisms at the gene's coding regions or exons. There are various examples available to show the influence of such genic sequence changes on phenotypes. Many studies have shown that nucleotide alterations in noncoding regions (5′ UTR including promoter, introns, and 3′ UTR) have a major influence on transcript production and accumulation, changing trait expression.

In recent years, mining for sequence variation, especially in regulatory regions of genes, has gained relevance in the context of gene expression. Because promoter elements are widely established to play a key role in gene regulation, alterations in their sequences will have a considerable influence on gene expression, resulting in varied trait expression. The cause of such variable gene expression is generally variation in upstream cisacting/transacting regulatory element binding locations. Thus, to capture the most functional variants of importance in the genes, "real" allele mining should involve examination of noncoding and regulatory areas of candidate genes, in addition to sequence differences in the coding sections (Latha et al. 2004). Recognizing the importance of "mining" in the context of regulatory domains, researchers identified gene promoters as a possible target for allele mining, also known as "promoter mining". The goal of promoter mining and *in silico* analysis was to find nucleotide variation within transcription factor binding motifs (TFBMs) and to count and distribute regulatory element binding sites within promoter regions. Furthermore, the goal of this research was to discover transcription factors linked with unique expression patterns in response to diverse kinds of regulation, such as constitutive, developmental, tissue-specific, hormone, and environmental variables (Veerla and Hoglund 2006). This sort of promoter sequence analysis also provides information on the preservation of regulatory motifs that are specific to distinct types of differentiated cells or tissues. Finally, promoter mining will provide insights into the molecular mechanisms of gene expression as well as the extraction of innovative and effective promoters for use in genetic engineering applications. As the number of global gene expression experiments to identify sets of genes with similar expression patterns grows, it is critical to incorporate approaches such as *in vitro* and *in silico*promoter mining into regular allele mining programs to identify and categorize *cis* regulatory elements and understand their role in gene regulation.

9.2.1 Mining Approaches

Several strategies for allele mining may be utilized, depending on the objective but also on the genetic, genomic, and financial resources available, ranging from a quick and affordable polymerase chain reaction (PCR) to next-generation sequencing (NGS) and everything in between. For certain applications, just partial sequence information or allele molecular polymorphism is required. The complete allele must be cloned for additional uses. In general, all DNA-based technologies need precise targeting of genes. Following validation of the target gene model, a thorough design of selective primers will enable the detection of new alleles at potential loci in the complete or core germplasm collection. PCR amplification of one or more alleles is used in all molecular marker procedures. Polymorphisms between amplified alleles might be identified using single-strand specific nucleases. Nicking of heteroduplexes of PCR products may be readily recognized with this method, which is often employed in TILLING techniques. The findings may be apparent as a DNA fragment of a specified size on an agarose gel. However, when using gene-specific markers in distinct germplasm material, specific annealing of the primers might occur, necessitating the need to sequence the resultant PCR fragments to ensure their identity and homogeneity.

9.2.1.1 TILLING

TILLING is widely utilized as a reverse genetic method for discovering mutations. This approach combines high-throughput mutation detection approaches with repetitive mutagenesis by DNA defective nuclease cleavage (Tadele 2016; Till et al. 2003b; Till et al. 2007a, b). The first organizations to use it on *Arabidopsis thaliana* were the Fred Hutchinson Cancer Research Centre and the Howard Hughes Medical Institute (McCallum et al. 2000a; Tadele 2016). Seeds and genomic DNA derived from mutagenic materials are two common types of germplasm stored in the TILLING library. It is possible that DNA and seed material, among other things, can be safely stored for decades. Screening for multiple characteristics is simplified by the availability of a wide range of germplasms, and the library's storage materials may be freely exchanged among the numerous breeders working on the genetic improvement of vegetables (Till et al. 2003b). This exemplifies a significant improvement above standard forward genetics or mutation breeding techniques. In these techniques, just a subset of mutant alleles is preserved, while the rest of the modifications that were introduced are weeded out. About 30 potentially dangerous mutations are expected to be found in a screening of a diploid generation of 6,000 chemically mutagenic *Arabidopsis* plants (Till et al. 2003b; Greene et al. 2003). Potentially beneficial alleles that do not induce phenotypic changes on their own but do so when paired with other alleles can be amplified *via* TILLING. Increased agricultural yields can be achieved using targeted mutation breeding in conjunction with genomics, although this approach necessitates the phenotypic analysis of many plants. *Drosophila* has been subjected to chemical mutagenesis since the 1940s, and plant species have been mined for mutants for reverse-genesis strategies for decades (Natarajan 2005). Many new varieties of tomato, chilli pepper, okra, sweet potato, garden pea, dolichos bean, French bean, and other plants have been generated thanks to mutants (Ahloowalia et al. 2004). Using these artificial mutations, a desired phenotype may be combined with a desired genotype. Ethylmethane sulfonate (EMS), N-ethyl-N-nitrosoure (ENU), N-nitroso-N-methylurea (NMU), and ionizing radiation are some of the chemical agents utilized in the production of mutants. Since EMS generates nuclear adducts, these mutagens can interact with complementary bases and integrate modifications to the baseline during replication (Haughn and Somerville 1987; Ashburner et al. 2005). As a result of the alkylation, guanine bases are mispaired with thymine instead of cytosine. Therefore, the next mutations will be those that change G/C to A/T (Henikoff and Comai 2003; Nagy et al. 2003). TILLING can determine the allelic sequence of missense and truncation mutations by using chemical mutagens that primarily generate random point mutations at high density (Greene et al. 2003). SNPs and/or INDELS in genes of interest in a mutagenic population are categorized by TILLING. To begin TILLING, a mutagenic species must be created and then treated with a chemical mutagen, such as EMS. Randomly causing single nucleotide changes or tiny insertions/deletions at a high frequency (30 nucleotides) (Till et al. 2007a) is what makes TILLING potent. TILLING is a chemical muagenesis-based reverse genetics technology used for phenotype-altering point mutation detection *via* heterogeneity and adaptive molecular screening.

Methods for detecting mediated polymorphism or mutations that rely on the structural features of single-stranded DNA are widely used (Myers et al. 1987; Nataraj et al. 1999; Rozycka et al. 2000). Since the detection sensitivity is limited to amplicons of 450 bp dHPLC, which identify mismatches in heteroduplexes created by melting and annealing of heteroallele DNA, this method is optimal for target genes with small exons that are separated by large introns. Among the defects identified (Jeong et al. 2012), there were important alterations that led to deception and irrational behavior. Phenotypic analysis might benefit from these alterations. All TILLING screens have revealed the presence of a basic nuclease cleavage, which is what people mean when they talk about TILLING methods (Figure 9.1).

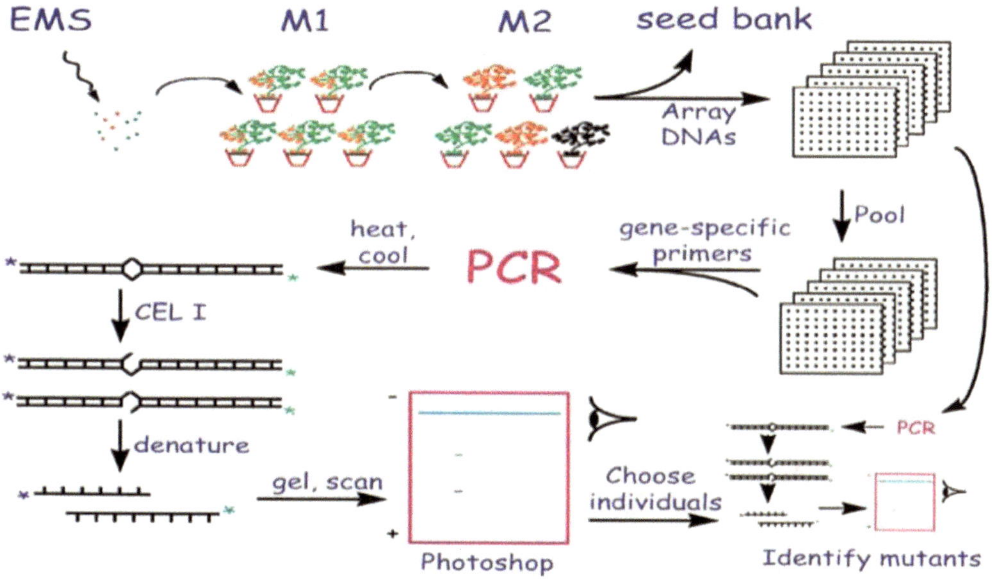

FIGURE 9.1 Steps involved in TILLING: seeds are mutagenized by treatment with alkylating agents, such as EMS, which primarily introduces G/C to A/T transitions; M_1 plants are self-fertilized, and M_2 individuals are used to prepare DNA samples for mutational screening, whilst an inventory of their seeds is established for future and downstream research. For mutation screening, DNAs are pooled to maximize the efficiency of mutation detection. PCR is performed using 5′-end-labeled gene-specific primers to target the desired locus, and hetero-duplexes are formed by heating and cooling the PCR products. CEL I nuclease is used to cleave at base mismatches, and the products representing induced mutations are visualized with denaturing polyacrylamide gel electrophoresis.

9.2.1.1.1 Conventional TILLING

Self-fertilization is a trait shared by most plant species, making this method particularly well suited to them. It is also possible to store the seeds these plants produce for years to come (Borevitz et al. 2003; McCallum et al. 2000a, b; Colbert et al. 2001). The most recent version of TILLING combines standard mutagenesis techniques with state-of-the-art mutation detection methods. The most crucial parts of TILLING are mutagenesis, non-chemical culture, germplasm collection, DNA isolation, sample pooling, mutation identification (by sampling a population for induced mutations), and mutant assessment steps. The TILLING method is commonly accepted because the steps involved may be used to a wide variety of vegetable crops. By using TILLING-revealed mutants, we may study the structure and function of genes that could be included into crop breeding systems through genetic enhancement. Plants that create a lot of chemical mutagens during TILLING can cause one-of-a-kind single base pair changes in a culture if their seeds or pollens are handled. Concurrently, these societies are entering a phase in which mutations are passed on very effectively from one generation to the next. The mutagen fed population seeds are developed into M_1 seeds and distributed to the population to foster the development of the M_2 populace. Mutation screening was carried out using DNA taken from M_2 plants (Colbert et al. 2001). The goal of spreading the seeds across the population is to create a renewable source of supply. DNA is only collected from one M_2 plant per M_1 to avoid reintroducing a previously removed mutation (Till et al. 2007a, b). M_3 seeds, which originate from M_2 offspring, may be stored for a very long time. A DNA pool consisting of eight human samples can be useful for detecting mutations in diploid animals (Henikoff and Comai 2003). Based on the ploidy stage of the organisms, the heterozygosity, and the extent of the beneficial effects of the SNP, a remarkable pooling of the plant species that are of concern should be identified in nature. These things need to be considered. Microtitre plates can be used to culture this DNA mixture for further study. Polymerase chain reaction (PCR) primers are tailored to amplify target genes in preparation for TILLING assays. Leveraging relevant genomics data across a wide range of organisms using methods from the field of bioinformatics can also help in the investigation of possible targets in novel plant species. Dye-labeled primers are used to enhance the DNA PCR findings from a group of people. Primers developed for the detection of mutations must be used for amplification of the desired genes. Primers labeled with an IR dye allow for mechanized amplification by capillary electrophoresis, and PCR products may be identified using tools like LI-COR. In both cases, labeling the primers is essential. Denaturing and reannealing PCR products helps produce heteroduplexes, the building blocks of mediated natural PCRs, which are used to detect inequalities? Endonucleases such as the celery-isolated CEL-1 endonuclease, the S1 nucleases, and the T4 endonuclease have all been used in TILLING technology. Pools that have been exposed to induced mutation create the combination of homo and heteroduplexes. When using cleaved heteroduplexes, the size of the full-length product was calculated by summing the sizes of two products (Till et al. 2007a). IRDye 700 and IRDye 800 were the designations given to the two products that had lower molecular weights. The series will be able to detect and confirm the mutation's location roughly based on the size of the split pieces relative to the size norm. People who visit the website www.proweb.org/parsesnp and utilize the PARSESNP software (which stands for Project Compatible Linked Sequences and SNPs Evaluation) will be given guidance when they enter a specific number of nucleotide gene variants. The program tracks how each variation affects the expression of a gene variable by comparing it to a reference DNA sequence and an exon/intron location model. Polymorphisms within the gene(s) can be categorized and their functions seen using the software PARSESNP (Taylor et al. 2003). Researchers working on the *Arabidopsis* TILLING project (ATP) have created two web-based software programs, CODDLE (Codons Programmed to Identify Deleterious Lesions) (www.proweb.org/input) and SIFT (Sorting Intolerant from Tolerant), to evaluate the effect of mediated or normal polymorphisms on gene expression (Gilchrist et al. 2005, Ng and Henikoff et al. 2003). Throughput can be maximized when anywhere from four to eight separate entities exchange DNA. Genomic sequences analyzed varied in size from 0.3 to 1.6 kilobytes. By increasing pooling by 8%, cutting the fragments into 1.4 kb units (excluding the ends of the fragments because those are the only places where the detection of SNPs is noiseless), and performing the experiment with 96 lanes, it is possible to scan up to one million base pairs of genomic DNAs in a single experiment. The bulk of existing TILLING techniques involve two distinct and sequential approaches: (1) pool scanning, and (2) individual marking.

9.2.1.1.2 TILLING by Sequencing

As part of this strategy, we can rule out the possibility of mistakes in the sequence being caused by non-random noise (Druley et al. 2009; Koboldt et al. 2009; Bansal et al. 2010). Broad parallel sequencing, which is made possible by high throughput sequencing, is being used to build mutation finding approaches (Gilchrist and Haughn 2005). In recent years, researchers have resequenced the whole *Arabidopsis* genome to learn more about the frequency and breadth of spontaneous mutations in this plant (Ossowski et al. 2010). One-nucleotide polymorphisms, on the other hand, need extensive sequencing coverage for identification. The cost of resequencing such a sizeable population is also very high. Relying on a smaller selection of genes helps keep the cost of resequencing down. It is possible to achieve this by sequencing PCR products (Porreca et al. 2007, Rigola et al. 2009) or by releasing an oligonucleotide-induced sequence, which can then be sequenced (Nijman et al. 2010). In addition, standards have been set to address the challenges of detecting rare polymorphisms in living things. Tomato (*Solanum lycopersicum*) was used by Rigola et al. (2009) to prove the use of 454 sequences from overlapping pools in conjunction with a machine workflow for mutation analysis. Non-overlapping pools of next-generation technologies for SNP identification include CRISP (Bansal et al. 2010), SNP Seeker (Druley et al. 2009), SAM tools, and MAQ (Li et al. 2009). For a simultaneous multidimensional Illumina-GA-based approach, however, these methodologies and computational procedures have yet to be created. TILLING by sequencing is an adaptable method that allows a thorough search for mutations at chosen regions using next-generation sequencing and a mix of molecular and bioinformatics tools. Effective TILLING populations across various crops can begin using the technology immediately. Rare SNP and mutation discovery can be hinted at even in a complicated sample. This approach provides scalable features because of the use of multidimensional pooling and a random threshold, which finds a mutation that has a matching change in bases. It is not dependent on variable endonuclease digestion or unstable fluorescent primers. Its versatile, letting you mix and match a wide range of pooling techniques and species.

Rather than sequencing the whole fragment, the precise position of the induced mutation may be determined using a TILLING technique (Colbert et al. 2001, Henikoff et al. 2004). This allows for selective sequencing of the caused mutation. It is flexible enough to pick up both naturally occurring SNPs and those that have been artificially introduced (Dahm and Geisler 2006) as well as the ability to identify heterozygotes. Mining for uncommon SNPs is more efficient and cost-effective than having to resequence an entire population. To categorize homozygous and heterozygous mutations in an eight-fold pool of one of the 16 genomes in the diploid species pool, this method is adaptable enough (Colbert et al. 2001; Slade and Knauf 2005). To concentrate the desired genotype sequence with the chemically produced variant alleles is the primary goal of TIILLING. This TILLING method calls for neither intricate nor pricey modifications. It eliminates the need for the time-consuming and baffling procedures of gene isolation and tissue culture and speeds up the screening of mutant pools to determine the activities of specific genes. Additionally, it may categorize alleles of interest according to biotic and abiotic gene voltage. Backcross breeding and mutation selection utilizing SNP-based markers for each generation make it easy to erase the mutant's past if it is to be employed in the crop breeding program (Allen et al. 2011). This TILLING method does not need a plant growing facility since it may be used on extremely low-density populations of any plant species. The major downside of TILLING is that most mutations are quiet since most positive pools do not contain target gene knockouts. The costly equipment is required to detect mutations.

9.2.1.1.3 TILLING by Next-Generation Sequencing

NGS technologies are gradually replacing Sanger sequencing methods in the majority of genome and transcriptome sequencing studies. Previously, Sanger sequencing was the predominant sequencing method. These next-generation sequencing methods can create data at a pace that is quicker by many orders of magnitude than older technologies, and at a cost that is far lower. According to Voelkerding et al. (2009), there are now several next-generation sequencers available for purchase. Although the fundamental biology of most of these systems is different, all these methods sequence populations of DNA molecules have been amplified by PCR. The Oxford Nanopore and Pacific Biosciences systems are two notable outliers since they can sequence individual molecules. Sequencing-based allele mining would result in the quicker collection of allelic data at a cheaper cost in the future because of the availability of low-cost, longer-read-length, and high-throughput sequencing technologies. In addition, the mining of alleles for specific genes in recognized accessions based on sequencing, as well as the correlation of these alleles with phenotypic differences, would offer a considerable boost to precision breeding efforts in agricultural plants. Characterizing genomic variation and locating SNPs requires the use of a technology that is quickly gaining in significance, and that tool is sequencing-based genotyping.

9.2.1.2 Eco-TILLING

Like TILLING process, but without the use of chemicals to induce mutations, this molecular strategy has been used to categorize natural polymorphisms in target alleles in the primary and secondary gene pools of melons, beans, and peppers (Comai et al. 2004; Nieto et al. 2007; Comai and Henikoff 2006; Gilchrist et al. 2006a, b). This method was initially used in *Arabidopsis thaliana* to extract natural polymorphisms (Comai et al. 2004). Rather than using genetically modified plants, this method relies on the DNA of naturally occurring vegetable species. TILLING is used to detect polymorphisms by comparing these DNA samples to a reference DNA set. Ecotypes were pooled using the method with the usual 1: 1 ratio (Comai et al. 2004). If fewer haplotypes were detected in the population than individuals, Eco-TILLING offers significant advantages over resequencing. Effective analyses were conducted on crop variety trials (Gilchrist et al. 2006a), potato polyphenol oxidase (Jin et al. 2010), and mining of various types of *Brassica* vegetables (Wang et al. 2010) to determine their effects on viral resistance in melons and capsicums. It is useful for screening genetic variation in species where conventional genetic analysis presents challenges (Gilchrist et al. 2006b). This technique may be utilized to discover natural variations and their possible gene functions in certain wild crop species that are resistant to induced mutations (Fery 2002). To identify naturally occurring SNPs and small InDels, this may be done with the SNP/haplotyping approaches at a minimal cost. This approach may be used to examine many samples for the presence of a target gene (Comai et al. 2004). It enables speedy identification of individual differences. It is economical too, as only one person needs to be sequenced to determine a haplotype. This method works great with cross-pollinated vegetable species as well as heterozygous and polyploid individuals. DNA polymorphisms, such as variations in the number of satellite repeats, can be detected and utilized as a proxy for population genetic diversity (Gilchrist et al. 2006b; Zerr and Henikoff 2005). Eco-TILLING's premise to improve economic properties like fruit size, fruit structure, fruit color, fruit consistency, nutritional and nutraceutical capacity, shelf life, insect pest and disease resistance, drought tolerance, and so on is obvious given its focus on detecting natural alleles in the genomic resources of vegetable crops. Furthermore, this approach is applicable to all heterozygous and polyploid organisms. The amount of data collected is the primary differentiator between TILLING and Eco-TILLING. In a sufficiently mutagenized diploid population, the finding of four mediated mutations in a 1.5 kb region by 768 people is expected. Screening 96 samples for a 1.5 kb target gene revealed 100 polymorphisms in *Arabidopsis* and humans using Eco-TILLING. GelBuddy was developed to aid in the analysis of Li-Cor gel images for use in the investigation of Eco-TILLING gels (Zerr and Henikoff 2005). Automatic lane and molecular weight identification, as well as manual

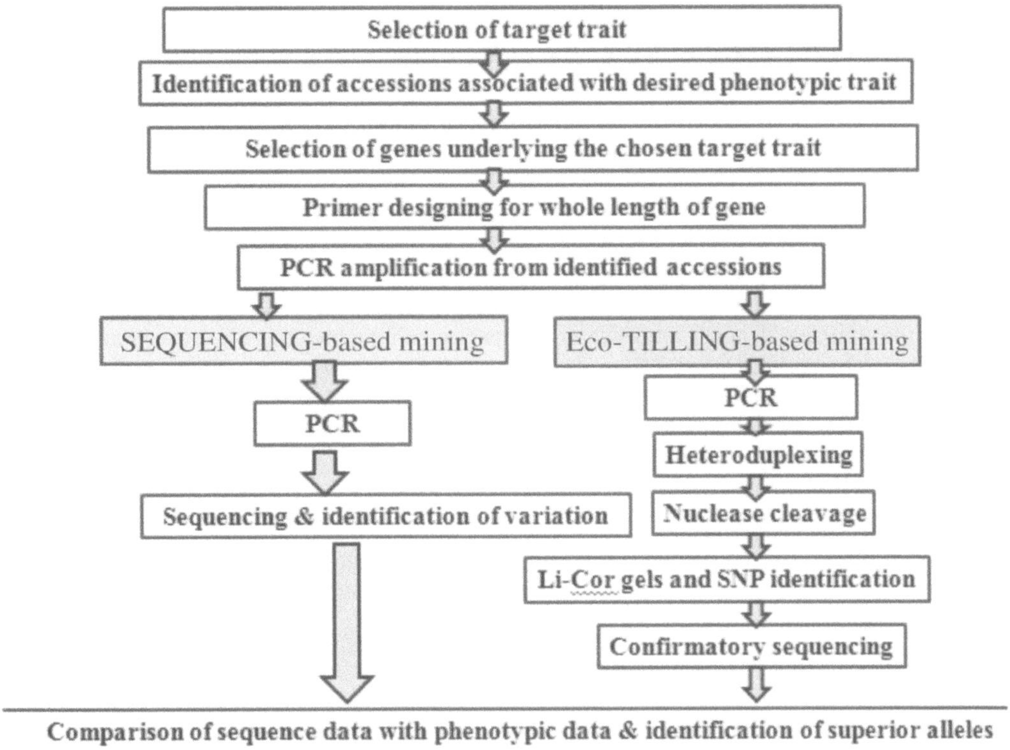

FIGURE 9.2 Steps involved in Eco-TILLING.

and automatic polymorphism findings are all features of the open-source program gelbuddy.org (Till et al. 2006). Different allelic variants were created in two resistance genes using Eco-TILLING technology. Several vegetable crops might benefit from using these genes to increase their tolerance to biotic and abiotic stressors. Natural allelic variation in *Cucumis* species for Melon Necrotic Spot Virus (MNSV) resistance was described using Eco-TILLING. This speeds up the process of identifying useful alleles for growing high-quality enhanced vegetable and hybrid varieties (Nieto et al. 2007). Although it is not often utilized now, SNP mining in several vegetable germplasms, measuring heterozygosity, exposing disease resistance variations, and defining the significance of genes by discovering natural variants are all areas in which this technique might be useful. Success in characterizing morphological data is possible. The steps involved in Eco-TILLING are given in Figure 9.2.

9.2.2 Adoption of Next-Generation Sequencing (NGS) in TILLING Experiments

Both traditional breeding approaches and more advanced transgenic technologies have been proved to make important contributions to agricultural production growth. However, regulatory barriers in several countries have hindered the process of widespread commercialization of genetically modified (GM) crops. As a result, it is critical to create a strategy that excludes genetic alteration of agricultural plants in any form. Next-generation sequencing, often known as NGS, is increasingly replacing classical TILLING as the technique of choice for finding mutations. This is because next-generation sequencing allows for the analysis of a high number of amplicons in a short period of time. When using next-generation sequencing (NGS) techniques, the sequencing of specific amplified sections allows for the accurate and rapid detection of rare mutations in target genes (Guo et al.

2015; Gupta et al. 2017). Using NGS has various advantages, and this is one of them. The cost of DNA sequencing has decreased dramatically because of the advancement of NGS technology, and as a result, the adoption of these technologies into TILLING research initiatives has increased rapidly. Only 5 years ago, the cost of sequencing a million base pairs was roughly $1,000. This figure has now more than doubled. According to the study of Irshad et al. (2020), the identical sequence may now be sequenced for as little as 10 cents. Whole genome sequencing (WGS) is expected to cost roughly $5 for a single genotype of a three gigabase genome sequenced 30 times. When compared to the cost of WGS, the cost of doing restriction site associated DNA sequencing, or RAD-seq, is reduced by a factor of 35. Furthermore, RAD-seq can simulate 200,000 SNPs in those populations while providing same coverage. According to Rothberg et al. (2011), depending on whether the complete genome sequence of the species being targeted is already available, the cost of genotyping by sequencing (GBS) or multiplexed shotgun genotyping (MSG) might be lowered by 10–14 times. If a genome has not yet been entirely decoded, targeted sequencing will be the most effective way for identifying large-scale markers. Because the cost of next-generation sequencing (NGS) and whole-genome sequencing (WGS) is likely to fall by a factor of several in the not-too-distant future, next-generation sequencing is expected to replace partial genome sequencing as the preferred approach. As a result, plant researchers are benefiting from the evaluation of numerous high-throughput sequencing technologies, including as PacBio and Ion Torrent, which are supporting them in lowering the cost of sequencing, which will coexist with TILLING research.

9.2.3 Approaches of Genome-Editing Towards TILLING Platforms

Genome editing methods, sometimes known as GE, are gaining favor in crop improvement for several reasons, including the fact that they are extremely effective, simple, and long-lasting. *CRISPR/Cas9* and clustered regularly interspaced short palindromic repeats (*CRISPR/Cas9*) are two examples of precise genome editing methods that have recently gained popularity due to their ease of use, scalability, and low cost (Jaganathan et al. 2018). Other GE techniques that might be used to change gene sequences include ZFNs (zinc finger nucleases), MNs (maganucleases), and TALENs (transcription activator-like effector nucleases); however, their use is restricted due to the need for technical expertise. To allow innovative genome editing technologies such as *CRISPR/Cas9* and others to be commercialized, the regulatory framework must first be updated. Plant breeders may be able to increase crop variety by combining TILLING with *CRISPR/Cas9,* and nuclease is led to its target region in the genome by a single guide RNA (sgRNA), resulting in the formation of double strand breaks (DSBs) and subsequent repair of these breaks, resulting in a deletion (InDel) or insertion mutation (Palan et al. 2021). These mutations result in the generation of alleles that no longer function. The utilization of mutant screening approaches that make use of TILLING DNA libraries and *CRISPR/Cas9* allows for the investigation of developmental effects in relation to gene sequencing. TILLING, a kind of sequencing, might be used to assist speed up the mutation screening process. Because of the advancement of next-generation sequencing (NGS) technology, improved software, and artificial intelligence (AI), the number of wrongly diagnosed mutations has greatly reduced. When genetic engineering (GE) is used, mutations occur only in the genes of interest; however, when TILLING is used, mutations may occur everywhere in the genome. To employ TILLING effectively, researchers must be acquainted with the gene sequence as well as the copy number associated with the desired trait. TILLING the population is required to search for mutations in the gene of interest, and after these mutations are detected, the mutant population must be evaluated to look for phenotypic changes linked with the mutation. When compared to TILLING, genome editing has a reduced incidence of off-target mutations. As a result, using GE technologies like *CRISPR/Cas9,* it is conceivable to produce deletions or insertions at locations in the genomes of diploid or polyploid animals (Palan et al. 2021). CRISPR/Cas9-induced adjustments to improve blast resistance have helped rice and wheat (Shan et al. 2014), while maize protoplast has profited from the same changes (Xu et al. 2017). Point mutation analysis has recently been shown to be successful in detecting single locus heterosis using both GE and TILLING technologies (Kashtwari et al. 2019). The purpose of this investigation was to find single locus heterosis. The CRISPR/Cas9 technology has greatly aided plant research and breeding. Using 20-bp short guide RNAs that are conserved among prolific genes might be a future technique for changing many members of a gene family on a big scale throughout the whole genome. Given this knowledge, it is not implausible that combining the two technologies may increase the efficacy and accuracy of identifying mutations that could be exploited to boost agricultural production.

9.2.4 Bioinformatic Tools for Allele Mining

No matter what methods are used, it is important for gene mining to find variation by comparing strings from different genotypes. There are many software tools, such as PLACE, plantCARE, TRANSFAC, JASPAR, MEME, Plantprom DB, DCPD, SCPD, BioEdit, ClustalW, etc. These tools can be used to handle complex nucleotide data, predict the putative function or structural components of complex macromolecules, predict TFBM, find sequence polymorphisms, and predict the amino acid changes that cause changes in the structure and/or function of encoded proteins. These tools make it easy to access and examine DNA sequences. They also help find sequence polymorphism and check if the difference in TFBM matches the difference in sequence polymorphism. These tools are also good for finding overrepresented motifs in upstream regions while defining the main promoter region. This makes it easier to find new variations in regulatory elements by narrowing the focus. Many of these kinds of computer methods will usually give you a list of patterns that are most likely to match the real binding sites, along with many chance changes. It is important to note that computer-based guesses are based on algorithms that have already been made. These algorithms are mostly made based on patterns that have already been figured out and are available in databases. So, these kinds of predictions should always be tested in an organized way in a wet lab.

Carrot DB (Carrot Data Base) was built using the genomic and transcriptome database of *Daucus carota* sp. *sativus*, which can be accessible at http://apiaceae.njau.edu.cn/carrotdb (Table 9.1). It provides the Basic Local Alignment Search Tool (BLAST) and Genome Map tools for finding and searching scaffold and target gene sequences, assembled transcriptomic sequences, expressed gene in the transcriptome, *de novo* assembled whole-genome sequences, putative gene sequences or gene fragments, and putative protein sequences of carrot (DC-27), as well as gene annotation checking and submission to the genome map group (Xu et al. 2014). Furthermore, 45 carrot genotype photos, accession numbers, names, countries of origin, and colors for the cortex, phloem, and xylem parts of each genotype are extracted from the Carrot DB's germplasm section. *De novo* assembly and analysis of DC-27 were carried out using whole genome sequencing. The Sequence Read Archive (SRA) database (www.ncbi.nlm.nih.gov/sra/) was used to successfully retrieve DC-27 transcriptomic sequences from the National Center for Biotechnology Information (NCBI). The complete genome sequence of 14 carrot genotypes' combined transcriptome sequences was mapped,

TABLE 9.1
Genomic Tool for Carrot

Database	Database Link	References
RoBuST	http://robust.genome.com	Bhasi et al. 2010
CarrotDB	http://apiaceae.njau.edu.cn/carrotdb	Xu et al. 2014
Carrot genome	https://phytozome.jgi.doe.gov/pz/portal.html#!info?alias=Org_Dcarota	Iorizzo et al. 2016
CarrotOmics	https://carrotomics.org/	Rolling et al. 2022

along with fragments per kilobase of transcript sequence per millions of base pairs of sequencing data (FPKM). The Carrot DB database now includes an interface for the 57 families and 2826 transcription factor (TF) genes discovered in the whole genome sequencing (Xu et al. 2014). Users may instantly begin exploring and analyzing the sequence annotations of 3663 genes, 5959 RNAs, 22,723 ESTs, and 11,438 regulatory sequence elements from the Apiaceae and Alliaceae plant families using the RoBuST database (http://robust.genome.com). Users may also search and download RBV crop features, biochemical pathways, genetic linkage maps, and molecular taxonomy information from the database. Furthermore, the RoBuST database allows for comparative genomic analysis of the 659,369 splice signals from 6015 plant species (Bhasi et al. 2010). CarrotOmics (https://carrotomics.org/) is a comprehensive, well-curated, open access database of carrot genomic resources recently developed by MainLab bioinformatics, Washington State University (Rolling et al. 2022) and is timely for carrot geneticists and breeders worldwide to exploit genomic information for trait-based breeding in climate change scenarios and newer niches.

9.2.5 Considerations for Allele Mining

The existing genetic base of a crop species, as well as the availability of genome and gene sequence information, are two of the most essential elements determining the feasibility and effectiveness of an allele mining operation. Availability of efficient and dependable phenotyping techniques, genomic resources, high-throughput techniques for the rapid and simple generation of allelic data points, cost-effective sequencing platforms, efficient bioinformatic tools for the identification of nucleotide variation, and molecular marker construction technologies for marker-assisted selection (MAS) are all important factors.

9.2.6 Utilization of Carrot Genetic Resources in Mining Approaches

The root is essentially an enlarged and swollen base of the tap root, and it also contains hypocotyls. There are four main parts that make up the carrot root: the peel (periderm), the cortex or store root (phloem, which is food conducting tissue), the cambium and the center core (xylem, which is water conducting tissue). The commercial product is the storage root, which is collected when the plant is still in its vegetative phase. There are various categories of cultivated carrots (Table 9.2), which belong to the *Daucus carota* subsp. *sativus* Schubl. & Martens species. These categories differ from one another in terms of the color, shape, size, and flavor of the roots. Therefore, broadly classified into two groups based on (a) presence of root pigments such as the carotenoid group (*D. carota* ssp. *sativus* var. *sativus*) and anthocyanin or purple group (*D. carota* ssp. *sativus* var. *atrorubens*), and (b) the need of vernalization to induce flowering such as the tropical/eastern/oriental carrot. Over 13,000 accessions, largely from Europe and the United States, but also from Asia and South America, are maintained in collections across the world to conserve carrot genetic variety (Allender 2019). Larger collections of wild carrot germplasm, however, were not achieved until recently (Castaneda-Alvarez et al. 2016) (Table 9.3). The introduction of new cultivars threatens carrot landraces in the Mediterranean Region and Central Asia, increasing the urgency with which they must be gathered. Because carrots are not clonally replicable, the only method to protect their genetic material is to preserve their seeds in the refrigerator. It has been recommended to preserve landraces in their original places (Allender2019), however this strategy has limitations since it depends on landraces being used on farms. Carrot landraces that are not kept in *ex situ* germplasm collections fear extinction as new cultivars replace them globally. Some wild subspecies of carrots, such as *D. carota* L. ssp. *gummifer* Hooker Fil., are only found in specific geographic locations, raising the chance that they may become endangered if they are lost (Spooner 2019). Wild carrots (*Daucus carota* ssp. *carota*) are native to Europe, Asia, North Africa, and the Americas. The perusal of Table 9.4 shows the status of *Daucus* collections worldwide.

TABLE 9.2
Status of *Daucus* Conservation in *Ex Situ* and *In Situ*

Ex situ in Genebanks		Conserved in Genebanks	
Daucus hispidifolius	1	D. carota subtaxon	
Daucus halophilus	1	rupestris	1
Daucus montevidensis	1	drepanensis	1
Daucus durieua	1	azoricus	1
Daucus mauritii	1	fontanesii	2
Daucus durieua	1	gadecaei	3
Daucus tenuisectus	2	hispanicus	4
Daucus glochidiatus	2	commutatus	10
Daucus littoralis	3	hispidus	12
Daucus involucratus	4	sativus	20
Daucus sahariensis	7	gummifer	21
Daucus glaber	7	maximus	51
Daucus syrticus	11	maritimus	59
Daucus capillifolius	12	carota	308
Daucus broteri	30	**Total**	**535**
Daucus bicolor	34		
Daucus guttatus	34		
Daucus sativus	34		
Daucus broteri	38		
Daucus crinitus	46		
Daucus pusillus	49		
Daucus aureus	52		
Daucus muricatus	70		
Daucus species	147		
Daucus carota	6099		
Total	**6687**		

Data originate from the Genesys database (www.genesys-pgr.org; accessed 17 June 2023).
Data originate from the Genesys database (www.genesys-pgr.org; accessed 18 June 2023).

9.2.6.1 Selection of Genotypes

The most difficult barrier to disclosing the hidden variation that already exists is the selection of germplasm to be "mined". If a reliable way for characterizing a gene is provided, screening the whole collection would surely be advantageous in discovering unusual alleles; nonetheless, this screening strategy is very time-consuming and unsuccessful. As a result, we must devise a strategy for transitioning from genotyping the whole composite collection to screening the accessions in an efficient way to discover new alleles and allele combinations while lowering the number of accessions that must be screened. Prioritizing genotypes for allele mining may be achieved in a variety of methods, some of which are covered in the following sections.

9.2.6.2 Development of Core/Mini Core Collections

Handling of whole germplasm, whether for conventional plant breeding or allele mining, is a Herculean task. Consequently, sampling processes are necessary to limit the population to a more manageable level while maintaining variety. It has been proposed that the whole collection be separated into a core collection and numerous smaller sub collections to simplify both the conservation of germplasm resources and the effective use of the diversity currently existing in gene banks.

TABLE 9.3
Wild Genetic Resources of Carrot

Taxon	2n	Fruit Type	Distribution
Daucus bicolor Sm.	-	Spiny	Greece, Turkey
Daucus biseriatus Murb.	-	Spiny	Algeria
Daucus conchitae Greuter	-	Spiny	Greece
Daucus gracilis Steinh.	-	Spiny	Algeria, Tunisia
Daucus hochstetteri A. Braun ex Drude	-	Spiny	Eritrea, Ethiopia
Daucus jordanicus Post	-	Spiny	Libya, Israel, Jordan
Daucus littoralis Sm.	-	Spiny	Egypt, Libya, Cyprus, Iran, Israel, Jordan, Lebanon, Syria, Turkey
Daucus mauritii Sennen	-	Spiny	Morocco
Daucus setulosus Guss. ex DC.	-	Spiny	Greece, Turkey
Daucus syrticus Murb.	-	Spiny	Egypt, Libya, Tunisia
Daucus tenuisectus Coss. ex Batt.	-	Spiny	Morocco
Daucus virgatus (Poir.) Maire	-	Winged	Algeria, Tunisia
Daucus carota subsp. *capillifolius* (Gilli) C. Arbizu	18	Spiny	Libya, Tunisia
Daucus carota subsp. *carota* L.	18	Spiny	Widely naturalized worldwide
Daucus carota subsp. *gummifer* (Syme) Hook. f.	18	Spiny	Coastal Mediterranean, coastal Atlantic in UK, France, Tunisia, Italy
Daucus carota subsp. *maximus* (Desf.) Ball	18	Spiny	Spain, Algeria, Morocco, Tunisia, Afghanistan, Cyprus, Iran, Israel, Jordan, Lebanon, Syria, Turkey, Pakistan, Greece, Italy (former) Yugoslavia, France, Portugal, Spain
Daucus carota subsp. *sativus* (Hoffm.) Arcang.	18	Spiny	Cultivated worldwide
Daucus sahariensis Murb.	18	Spiny	Algeria
Daucus arcanus García-Martín and Silvestre	22	Spiny	Spain
Daucus aureus Desf.	22	Spiny	Spain (Canary Islands), Algeria, Egypt, Libya, Morocco, Tunisia, Cyprus, Israel, Lebanon, Syria, Italy
Daucus crinitus Desf.	22	Spiny	Algeria, Morocco, Tunisia, Portugal, Spain
Daucus durieua Lange	22	Spiny	Spain, Algeria, Libya, Morocco, Tunisia, Cyprus, Israel, Lebanon, Syria, Portugal, Spain
Daucus muricatus (L.) L.	22	Spiny	Portugal, Algeria, Libya, Morocco, Tunisia, Italy, France, Portugal, Spain
Daucus pusillus Michx. (= D. montevidensis Link ex Spreng.*)*	22	Spiny	Canada, United States, Mexico, Brazil, Argentina, Chile, Uruguay
Daucus setifolius Desf.	22	Spiny	Algeria, Morocco, Tunisia, Portugal, Spain
Daucus guttatus Sm.	20, 22	Spiny	Egypt, Libya, Cyprus, Iran, Iraq, Israel, Lebanon, Syria, Turkey, Albania, Bulgaria, Greece, Italy, Romania (former), Yugoslavia
Daucus involucratus Sm.	20, 22	Spiny	Cyprus, Turkey, Greece
Daucus glochidiatus (Labill.) Fisch. and C. A. Mey.	44	Spiny	Australia, New Zealand
Daucus montanus Humb. and Bonpl. ex Schult.	66	Spiny	Mexico, Costa Rica, El Salvador, Guatemala, Honduras, Venezuela, Bolivia, Colombia, Ecuador, Peru, Argentina, Chile

TABLE 9.4
Country and Institute-Level of *Daucus* Germplasm Collections

Country	Code	Institute	Total Accessions
Croatia	HRV041	Faculty of Agriculture, University of Zagreb	24
Azerbaijan	AZE015	Genetic Resources Institute	25
Brazil	BRA020	Embrapa Clima Temperado	25
Taiwan	TWN001	World Vegetable Center	30
Switzerland	CHE001	Agroscope Changins	40
Ukraine	UKR008	Ustymivka Experimental Station of Plant Production	48
Israel	ISR002	Israel Gene Bank for Agricultural Crops, Agricultural Research Organisation, Volcani Center	73
Spain	ESP027	Gobierno de Aragón. Centro de Investigación y Tecnología Agroalimentaria. Banco de Germoplasma de Hortícolas	78
Romania	ROM007	Suceava Genebank	90
Spain	ESP004	Centro Nacional de Recursos Fitogenéticos	101
Bulgaria	BGR001	Institute for Plant Genetic Resources "K. Malkov"	105
USA	USA995	National Center for Genetic Resources Preservation	110
Portugal	PRT001	Portuguese Bank of Plant Germplasm	145
USA	USA005	National Seed Storage Laboratory, USDA-ARS	165
United Kingdom	GBR004	Millennium Seed Bank Project, Seed Conservation Department, Royal Botanic Gardens, Kew, Wakehurst Place	178
Sweden	SWE054	Nordic Genetic Resource Center	194
Hungary	HUN003	Institute for Agrobotany	203
USA	USA974	Seed Savers Exchange	239
Ukraine	UKR021	Institute of Vegetable and Melon Growing	389
Czech Republic	CZE122	Gene bank, Crop Research Institute	404
Germany	DEU146	Genebank, Leibniz Institute of Plant Genetics and Crop Plant Research	479
Poland	POL003	Plant Breeding and Acclimatization Institute	483
USA	USA020	North Central Regional Plant Introduction Station, USDA-ARS, NCRPIS	1326
United Kingdom	GBR006	Warwick Genetic Resources Unit—UK Vegetable Genebank	1481
		Total	6435

A core collection, according to Brown (1989), is a subset of the whole collection that comprises accessions that reflect the bulk of the genetic diversity that is presently available in a species. Mining for alleles demands the creation of a sufficient core germplasm collection, which generally contains fewer than a thousand accessions and is intended to be representative of the genetic variability currently present in the human population (Hofinger et al. 2009). According to Mackay et al. (2004) and Bhullar et al. (2009), the Focused Identification of Germplasm Strategy (FIGS) may be used to efficiently construct such collections.

9.2.7 Accurate Phenotyping Methods

Phenotyping will be critical to properly narrowing down the list of genotypes to be mined for alleles since phenotypes are considered as the most accurate markers of genotypes. The possibility of "mining" possible genotypes grows as phenotyping methods get more accurate. Furthermore, the number of genotypes that must be "mined" is considerably reduced because of the generation of trait-specific subsets obtained by phenotyping testing. Consequently, there is an urgent need to develop phenotyping approaches to make allele mining more successful.

9.2.7.1 Flexible Computational Tools

The ever-increasing amount of plant genetic resources held in gene banks has resulted in the problem of repetitiveness and genetic redundancy within today's agricultural gene pools. This creates a barrier to the effective management and use of these resources in the future. Therefore, efficient computational tools are necessary to ensure access to advantageous alleles. One of the most major difficulties that exist today is the development of flexible computational techniques for the selection of desirable alleles, which is seen as a road for the rapid rise and extension of GenBank DNA data. Park et al. (2007) developed Power Core as a computational tool to aid in the building of core and allele mining sets. It does this by increasing the number of viable alleles while decreasing their redundancy. When it comes to deploying favorable alleles from gene banks, having a variety of such tools at your disposal is critical to accelerate the future development of core subsets that encompasses all the observed traits or classes.

9.2.8 Handling of Genomic Resources

Because of the rapid collection of nucleotide and gene expression data, computer algorithms for assessing functional nucleotide diversity and predicting specific nucleotide modifications that are responsible for changing function are required. This is required to keep up with the rate of data collection. One solution that has been proposed is leverage on current achievements in allele mining, association genetics, and comparative genomics by combining the expertise of professionals from a variety of domains, including molecular genetics, statistics, and bioinformatics.

9.2.9 Traditional TILLING Platforms in Carrot

In the latter half of the 20th century, substantial research was conducted on the generation of mutations in plant tissue culture, and carrot was one of the crops that was evaluated (Sung1976). Although genetic characterization in regenerated plants was not always undertaken, putative mutants for numerous features, including temperature sensitivity and antibiotics, were developed (Lo Schiavo et al. 1988). These putative mutants were tested for their abilities. Few practical applications of mutation breeding have been found in the enhancement of carrots. Al-Safadi and Simon (1990) observed substantial affects of up to 40 krad of gamma-radiation on the growth and development of carrot callus. They also observed that exposures of up to 1 krad stimulated growth, while greater doses caused a higher incidence of albino plants and leaf abnormalities. Root weight and the content of soluble solids differed between M_1 and M_2 generations, and this variation was heritable. There are now several carrot mutagenesis projects ongoing that make use of both radiation-induced and chemically induced mutagenesis. Recent research conducted by Beejan (2018) found that the growth reduction dose 50 for carrot seeds was 1075 Gy when combined with 137 Cesium.

It is interesting to note that effective mutagen dose (EMD) of callus from purple carrots exposed to 60 Cobalt radiation found by Turan Buyukdinc (2019) was 8.36 Gy, while the EMD of callus from orange carrots exposed to 60 Cobalt radiation was 23.605 Gy. The recent resurgence of interest in radiation-induced mutation breeding of carrots may result in the production of fascinating mutations that can be compared to the outcomes of gene editing.

9.2.10 Carrot Genome and Sequence Information

Before attempting to sequence the whole carrot genome, researchers investigated its structure and content. The predicted carrot genome size is ~473 Mb (Arumuganathan and Earle 1991). Flow cytometry was used to achieve this estimate. The carrot is a diploid species with nine pairs of chromosomes, according to early cytogenetic research (Simon 1984). A chromosomal karyotype revealed five pairs of submetacentric chromosomes and four pairs of metacentric chromosomes (Iovene et al. 2011). They combined the genetic and cytogenetic maps before assigning each linkage group to a specific chromosome. Six of the linkage groups were found along the chromosome's short (north)/long (south) arm. Using a DNA association curve and thermal denaturation, the proportion of repetitive sequences in the carrot genome was determined to be about 46%, with a GC content of 37–38% (Simon 1984). According to the results of the first in-depth research of repetitive DNA in carrots, which used BAC-end sequences (BES), transposable elements (TE) accounted for most of the repetitive DNA (11.9%). Class I retrotransposons account for most transposable elements (TEs) discovered in genomic background sequences and transcriptome sequences (Cavagnaro et al. 2009; Iorizzo et al. 2011). The *Ty1/copia* and *Ty3/gypsy* superfamilies are two of the most common LTR families among Class I retrotransposons. Aside from Class I retrotransposons, many families of Class II DNA transposons have been discovered and identified. *Tdc1* was the name given to the first carrot TE that received substantial investigation. It was discovered to be an insertion produced in the 5' flanking region of phenylalanine ammonialyase (*gDcPAL1*) by a non-autonomous member of the CMC (CACTA) superfamily (Ozeki et al. 1997). According to an examination of conserved domains and amplification-based characterization of related elements, three subfamilies of Tdc elements were discovered in the carrot genome (Itoh et al. 2003). *TdcA, TdcB*, and *TdcC* were assigned to these subfamilies. *DcMaster*, a member of the PIF/Harbinger superfamily, similarly causes a knock-out mutation in the acid-soluble invertase isozyme II gene (Grzebelus et al. 2006; Yau and Simon 2003). *KrakL1, KrakL2, Midi*, and the most prevalent Tourist-like DcMaster-related MITE, Krak, were used to find multiple-related TE (Grzebelus and Simon 2009). Nine families of carrot Stowaway MITEs have recently been discovered, according to Macko-Podgorni et al. (2013). *DcSto1* through *DcSto9* were assigned to these families.

The carrot genome was also investigated to establish its gene makeup. Cavagnaro et al. (2009) discovered that roughly 10% of the carrot genome includes coding sequences by comparing translated BES with protein databases. They also discovered that 25,000 genes are expected, with one gene for every 19 kb of genomic DNA. These studies not only give a basic overview of the carrot genome's structure and repetitive DNA content, but they also provide the framework for future study, such as whole-genome sequencing. First Draft Carrot Genome Sequence Over the past 10 years, the rapid expansion of next-generation sequencing (NGS) technology has aided breeding efforts in a variety of crops, including carrot, by giving critical genomic resources for speedier molecular understanding of biological traits. These components have also resulted in the development of new carrot varietals. The findings of the first draft assembly of the carrot genome were reported by Xu et al. (2014). They sequenced the orange "DC-27" inbred line using Illumina and Roche 454 sequencing technologies. The genome assembly encompassed around 372 Mb and had approximately 185,000 contigs and scaffolds. The scaffold N50, or the smallest contig that makes up half of the overall genome length, is 4.8 kb long, and it was longer. According to specialists, there are around 75,000 genes, each with an average of 833 nucleotide bases. Even though this resource was useful in identifying transcription factors, potential SSR markers, and genes involved in the flavonoid pathway, the genome assembly was significantly fragmented and did not provide chromosomal information (Wang et al. 2018). As a result, its use was limited to in-depth comparative genomic and marker trait association research, as well as the identification of prospective genes controlling economically relevant elements.

9.2.11 Application of Allele Mining in Carrot

The *rp* mutation differs from other carrot genes that impact carotenoid color in that it affects plant vitality. Although this has not been verified, the *rp* mutant may block the carotenoid pathway (Goldman and Breitbach 1996). Given the specific features of *rp*, more study may reveal helpful insights regarding carrot carotenoid metabolism.

Vrn1 is the sole gene known to impact flower mutation, according to recent research by Alessandro et al. (2012). *Vrn1*'s position on chromosome 2 was discovered to be 0.36 cM. Because of the vast range of variation recorded for this feature, it is likely that many more genes affecting carrot vernalization will be identified when a wider variety of germplasm is studied (Alessandro et al. 2012). *CENH3* genes with different basic chromosome numbers were identified and functionally cloned in *Daucus carota* and three distantly related wild *Daucus* species (*D. glochidiatus* Labill., *D. pussillus* Michx., and *D. muricatus* L.) to lay the groundwork for a potential alternative haploidization strategy based on centromere-mediated genome elimination in cultivated carrots. Researchers searched the carrot transcriptome for homologous coding sequences for *CENH3*s in four *Daucus* species, and then cloned the sequences using PCR, utilizing a carrot as a model. Three of the four species'*CENH3* ORFs shared the same 146-aa amino acid sequence (Dunemann et al. 2014).

The enzyme alternative oxidase (*AOX*) is important in the alternative respiration pathway, which plants utilize to adapt to stress. Carrots'*DcAOX1* gene has been shown to be highly sensitive to environmental changes. The most polymorphic region was discovered to be intron 1; where a 400-bp insertion event generates large differences in sequence depending on the carrot genotype. The insertion was discovered in a heritable and genotype-specific area of single-stranded repeats (SSRs). The FMs obtained from this intronic insertion, if verified, have the potential to greatly contribute to marker-assisted selection for better carrot genotypes (Nogales et al. 2016). *In vitro* chromosome-anchored simple sequence repeat (SSR) marker building was done using the carrot genome assembly. In the 55,386 markers developed by Uncu and Uncu (2019), marker loci were found to relate to protein coding sequences. Researchers saturate carrot chromosomes with sequence-specific markers to produce a physical map of the carrot genome. The markers will be employed as useful molecular tools in this important agricultural species for characterizing germplasm, tagging genes, and undertaking molecular breeding studies.

To develop a mutant library, carrot "17005" inbred line seeds were treated with ethyl methyl sulfone (EMS). The "17005" inbred line was treated to optimum EMS mutagenesis conditions, including a 6-hour seed treatment with 0.5% EMS. They changed the morphology of the cotyledon, growth point, leaf, and taproot during the vegetative phase, and the height and color of the main stem, secondary branch, flower, and seed during the reproductive phase. Crossing the yellow taproot and purple-red epidermal taproot mutants with the wild type resulted in an F_2 segregating population, demonstrating that the phenotypes of both mutants were regulated by a single recessive gene. Carrots with inheritable mutations not only increase the species' genetic diversity, but they also give instruments for studying the genes'functions (Wu et al. 2020). A 180 kb area on carrot chromosome 1 contributed for 10% of the reported phenotypic variation in shoulder diameter, according to a GWAS. The *DcDCAF1 and DcBTAF1* genes are thought to have a function in controlling secondary development of the carrot storage root at that place (Macko-Podgórni et al. 2020). White-rooted carrots were shown to have greater levels of *DcCCD4* gene expression than orange-rooted carrots (Li et al. 2021). *DcCCD4* gene overexpression reduced beta- and γ-carotene levels in carrots with an orange taproot. Mutant carrots with a faulty *DcCCD4* gene are less orange and contain more of the pigment β-carotene. The functional investigation further revealed the relevance of the *DcCCD4* gene in the degradation of β- and γ-carotene content.

Carotenoid accumulation in underground carrot taproot may be aided by the *DcPAR1* (PHYTOCHROME RAPIDLY REGULATED1) gene, which is more highly expressed in underground carrot taproot than light-grown carrot taproot (Arias et al. 2022). Since the conclusion of the genomic era, carrot transcriptome sequencing databases have emerged. miRNA, or micro-RNA, research may one day lead to superior carrots in a variety of ways. With the integration of emerging methods and omics sciences (genomics, transcriptomics, proteomics, and metabolomics) in the post-genomic era, better carrot breeding programs for multiple characteristics, such as carotenoids, will be possible, thanks to gene mining (Singh et al. 2023) for carrot functional genes.

9.2.12 Challenges in TILLING and Eco-TILLING Approaches

The primary difficulty of this species is the need to gather germplasm to "mine" it. If the stock is too big, a micro core selection is made (consisting of 1% of the collection) (Upadhyaya and Ortiz 2001; Mazzucato et al. 2015). Ten percent of the core collection, or one percent of the whole collection, is a micro core selection. Another issue is the time it takes to raise a healthy population, which can take anywhere from a year and a half to two years (Slade and Knauf 2005; Caldwell et al. 2004). The optimal dosage produces several mutations with little embryo mortality (Henikoff and Comai 2003). There is a threshold dosage of chemical mutagens that results in a disproportionately large number of mutations across different crops. The results of the different chemical mutagenesis techniques will vary based on the plant or plant variant; therefore, this is not a major concern. Some perennial vegetables with a long-life cycle and those that may be propagated vegetatively may be able to inhibit the spread of mutants (Slade and Knauf 2005). Normal polymorphisms reduce the efficiency with which TILLING and Eco-TILLING can detect mutations in highly heterozygous species. Finding rare, induced mutations may become more difficult if several fragments are generated (Till et al. 2003a; b). These methods further complicate matters by requiring a wide range of activities, such as the creation of test genotypes and DNA pools for comparative purposes. This is only one possible way in which using these techniques adds complexity. Recently, a fast and cost-effective method called "recognition of novel allelic variations of existing agarose gel candidates and identification of their usage in candidate gene mapping" (Raghavan et al. 2007) has been discovered, which can drastically reduce the price of Eco-TILLING. This strategy can be useful.

First and foremost, among these tasks is the selection of mutagens, the ability to mutagenize the tissues, and the capacity to reach a sufficient density of dominant mutations or the creation of two haploid plants, both of which are necessary for the cultivation of a functional TILLING community of vegetatively propagated vegetables. Research on plants that can be produced from their own spores is just getting started, and experience is the only way to learn what works. TILLING allows for the quick generation and screening of small populations, allowing for the selection of just those plantlets with interesting mutations for future study (Irshad et al. 2020; Szurman-Zubrzycka et al. 2018). This is especially useful in settings with limited access to *in vitro* procedures.

TILLING might be difficult for some vegetable crops because of unforeseen circumstances. In some cases, polyploidy might be problematic. TILLING allows for the detection of extreme alleles in different homeologists in polyploids. Once these alleles are discovered, the desired phenotype may be produced by cross-fertilization. It is sufficient to amplify a single gene of interest (homeologist) by PCR before using the result in an enzymatic cleavage TILLING test for mutation detection. This is a necessary procedure. Therefore, multiplexing boils down to the issue of identifying several gene targets in a single PCR run. The inability to find mutations using this method has been shown, however (Cooper et al. 2008a;Cooper et al. 2008b). Target gene amplification (Cooper et al. 2008a; Prasanna and Jain 2017; Chen et al. 2014) and target gene deletion (Slade and Knauf 2005) are both possible with the right prime architecture. This is achieved by minimizing genomic DNA digestion before performing PCR. There have been serious worries voiced about the lack of genome sequencing data. Using the sequence information from cDNA or EST, inefficient TILLING primers may be generated. Mutational load, or the amount of background mutations present in a plant, is thought to be a major obstacle to the TILLING method being used in vegetable breeding. Previous mutations can lead to a phenotype with both desirable and undesirable traits, and genetic engineering is to blame for a drop in production. All plants, before being harvested, release chemicals that, if ingested by humans, would be harmful (Getahun et al. 2003; Vetter 2000). Erosion, soil acidity, salinity, rodents, diseases, and weeds are only some of the environmental factors that might reduce production and yield. As a result, the world needs an agricultural revolution to boost the food supply of under-researched crops and meet the growing demand for food in developing nations.

9.2.13 Future Perspectives

A great number of vegetable crops have profited from the utilization of vegetable mutants in genetic improvement initiatives, and currently, vegetable mutants are being directly exploited in improvement activities. However, the mutation screening technologies that are being covered in this chapter (TILLING and Eco-TILLING) will move swiftly forward by boosting advantageous gene changes of proven horticultural utility and stimulating the deployment of applicant genes in *Daucus* species. This would be improved by the eventual explosion of sequence data for *Daucus* species owing to the introduction of next-generation sequencing technologies (NGSs), which would allow for the sequencing of far larger datasets. In addition, in contrast to technologies such as RNAi, which may target several genes all at once, point mutagenesis can only impact a single copy of an identical gene in paralogists or homeologists. This is the case regardless of the kind of gene. It is possible that crossing is the sole way to steer the mutagenesis process towards a certain aim, but it is not guaranteed. Recent reverse genetic techniques, like TILLING and Eco-TILLING, have shown promise as valuable tools for carrying out realistic genomic research in carrot genomic era. It has not been established that TILLING is species-specific, unlike other techniques to reverse genetics (such as RNAi or homologous recombination), which makes the possibility of its application to many types of carrots more encouraging.

9.2.14 Conclusion

There will be a shift in consumer demand from traditional cuisine towards food that is more wellness and health conscious because of the world's rising population, climate anomalies, and the need to diversify food baskets and place a greater emphasis on nutritional diet. People are becoming more aware of the need to consume foods that are abundant in nutrients and are adjusting their dietary habits because of the COVID-19 epidemic, which has led to a new normal in agricultural practices throughout the world. This process has already begun in earnest. Therefore, it is crucial that we boost production of nutrient-dense vegetable crops if we are to meet the changing needs of the population. In this view, vegetable crops will be developed for improved nutrient content, including more calories, starch, mineral, and immune-boosting natural therapeutic component content. Carrot breeding techniques and programs also play an essential role in increasing tolerance to biotic and abiotic stress, decreasing pollution, and boosting industrial attributes like carotenoids and anthocyanin intensity. Furthermore, both induced and natural mutations offer a powerful tool for enhanced functioning. However, mutations may also be employed in reverse genetic techniques to categorize lesions in specific target genes, thanks to the rapid accumulation of plant genomic sequence information and genetic feature predictions. These gene-driven strategies aim to increase the rate at which novel phenotypes may be generated and may even make possible the development of phenotypes that would be impossible to achieve using more traditional, forward-looking approaches. TILLING was created as a low-cost, high yield reverse genetic method for detecting mediated mutations. The methodology of classifying natural nucleotide polymorphisms has been updated to include a method called Eco-TILLING. The need for modern carrot cultivation is rising, and with it comes the necessity of boosting carrot production in the 21st century. This is the case even while public acceptance of transgenic or genetically modified crops remains low. Furthermore, *Agrobacterium*-based insertional mutation approaches are limited in their applicability to a broad range of vegetable plant species. So, using allele-mining techniques like TILLING, which focuses on EMS mutagenesis, and Eco-TILLING will help to address this deficiency. Cutting-edge TILLING technology alleles might be used in carrot without being hindered by the strict regulatory frameworks that regulate genetically modified crops. This method can be rapidly used in carrot breeding initiatives by the simple act of TILLING. These methods may improve the efficiency with which large-scale findings of induced point mutations across necessary populations are processed. Arcadia employs this method to keep tomatoes fresher for longer, but so far, no vegetable crop produced from TILLING has been established and published. It has been proven without a shadow of a doubt that both induced mutations and natural polymorphism may be advantageous for the aim of vegetative proliferation. Carrot breeding relies heavily on the identification of alleles associated with significant qualitative and quantitative elements of phonotypical and economic traits; TILLING and/or Eco-TILLING provide a viable option for this purpose.

REFERENCES

Ahloowalia BS, Maluszynski M, Nichterlein K (2004) Global impact of mutation-derived varieties. *Euphytica* 135:187–204.

Alessandro MS, Galmarini CR, Iorizzo M, Simon PW (2012) Molecular mapping of vernalization requirement and fertility restoration genes in carrots. *Theor Appl Genet* 126(2):415–423. https://doi.org/10.1007/S00122-012-1989-1

Allen AM, Barker GLA, Berry ST, Coghill JA, Gwilliam R, Kirby S, Robinson P, Brenchley RC, D'Amore R, McKenzie N, Waite D, Hall A, Bevan M, Hall N, Edwards KJ (2011) Transcript specific, single nucleotide polymorphism discovery and linkage analysis in hexaploid bread wheat (*Triticum aestivum* L.). *Plant Biotechnol J* 9:1086–1099.

Allender C (2019) Genetic resources for carrot improvement. In: Simon P, Iorizzo M, Grzebelus D, Baranski R (eds) *The carrot genome, compendium of plant genomes*. Springer Nature Switzerland AG, Cham, Switzerland, pp 93–100.

Al-Safadi B, Simon PW (1990) The effects of gamma irradiation on the growth and cytology of carrot tissue culture. *Env Exp Bot* 30:361–371.

Arias D, Ortega A, González-Calquin C, Quiroz LF, Moreno-Romero J, Martínez-García JF, Stange C (2022). Development and carotenoid synthesis in dark-grown carrot taproots require PHYTOCHROME RAPIDLY REGULATED 1. *Plant Physiol* 189(3):1450–1465. https://doi.org/10.1093/PLPHYS/KIAC097

Arumuganathan K, Earle ED (1991) Nuclear DNA content of some important plant species. *Plant Mol Biol Rep* 9:208–218.

Ashburner M, Hawley RS, Golic KG (2005) *Drosophila*. A Laboratory Handbook. Cold Spring Harbor Lab.

Bansal V, Harismendy O, Tewhey R, Murray SS, Schork NJ, Topol EJ, Frazer KA (2010) Accurate detection and genotyping of SNPs utilizing population sequencing data. *Genome Res* 20:537–545.

Beejan PBF (2018) Study to determine the growth reduction dose 50 (GR50) for gamma rays induced mutagenesis in carrot (*Daucus carota* (L). FAO/IAEA Int Symp Plant Mutation Breeding and Biotech, Vienna.

Bhasi A, Senalik D, Simon PW, Kumar B, Manikandan V, Philip P, Senapathy P (2010) RoBuST: an integrated genomics resource for the root and bulb crop families Apiaceae and Alliaceae. *BMC Plant Biol* 10:161.

Bhullar NK, Street K, Mackay M, Yahiaoui N, Keller B (2009) Unlocking wheat genetic resources for the molecular identification of previously undescribed functional alleles at the Pm3 resistance locus. *Proc Natl Acad Sci USA* 106:9519–9524.

Borevitz JO, Liang D, Plouffe D, Chang HS, Zhu T, Weigel D, Berry CC, Winzeler E, Chory J (2003) Large-scale identification of single-feature polymorphisms in complex genomes. *Genome Res* 13:513–523.

Brown AHD (1989) The Case for Core Collections in The Use of Plant Genetic Resources. In: Brown AHD, Frankel OH, Marshall DR, Williams JT (eds), *The Use of Plant Genetic Resources*. Cambridge University Press, London pp. 136–156.

Caldwell DG, McCallum N, Shaw P, Muehlbauer GJ, Marshall DF, Waugh R. (2004) A structured mutant population for forward and reverse genetics in Barley (*Hordeum vulgare* L.). *Plant J* 40:143–150.

Castaneda-Alvarez NP, Khoury CK, Achicanoy HA, Bernau V, Dempewolf H, Eastwood RJ, Guarino L, Harker RH, Jarvis A, Maxted N, Muller JV, Ramirez-Villegas J, Sosa CC, Struik PC, Vincent H, Toll J (2016) Global conservation priorities for crop wild relatives. *Nat Plants* 2:16022.

Cavagnaro PF, Chung S-M, Szklarczyk M, Grzebelus D, Senalik D, Atkins AE, Simon PW (2009) Characterization of a deep-coverage carrot (*Daucus carota* L.) BAC library and initial analysis of BAC-end sequences. *Mol Genet Genom* 281:273–288.

Chen L, Hao L, Parry MAJ, Phillips AL, Hu YG (2014) Progress in TILLING as a tool for functional genomics and improvement of crops. *J Integr Plant Biol* 56:425–443.

Colbert T, Till BJ, Tompa R, Reynolds S, Steine MN, Yeung AT, McCallum CM, Comai L, Henikoff S (2001) High-throughput screening for induced point mutations. *Plant Physiol* 126:480–484.

Comai L, Henikoff S (2006) TILLING: practical single nucleotide mutation discovery. *Plant J* 45:684–694.

Comai L, Young K, Till BJ, Reynolds SH, Greene EA, Codomo CA, Enns LC, Johnson JE, Burtner C, Odden AR, Henikoff S (2004) Efficient discovery of DNA polymorphisms in natural populations by Ecotilling. *Plant J* 37:778–786.

Cooper JL, Greene EA, Till BJ, Codomo CA, Wakimoto BT, Henikoff S (2008a) Retention of induced mutations in a *Drosophila* reverse-genetic resource. *Genetics* 180:661–667.

Cooper JL, Till BJ, Henikoff S (2008b) Fly-TILL: reverse genetics using a living point mutation resource. *Fly (Austin)* 2:300–302.

Dahm R, Geisler R (2006) Learning from small fry: the zebrafish as a genetic model organism for aquaculture fish species. *Mar Biotechnol* 8:329–345.

Druley TE, Vallania FLM, Wegner DJ, Varley KE, Knowles OL, Bonds JA, Robison SW, Doniger SW, Hamvas A, Cole FS, Fay JC, Mitra RD (2009) Quantification of rare allelic variants from pooled genomic DNA. *Nat Methods* 6:263–265.

Dunemann F, Schrader O, Budahn H, Houben A (2014) Characterization of Centromeric Histone H3 (CENH3) variants in cultivated and wild carrots (*Daucus* sp.). *PLoS One* 9(6):e98504. https://doi.org/10.1371/JOURNAL.PONE.0098504

Fery RL (2002) New opportunities in *Vigna*. In: *Trends in New Crops and New Uses* (Eds. Janick J and Whipkey A). ASHS Press, Virginia, VA.

Getahun H, Lambein F, Vanhoorne M, Van derStuyft P (2003) Food-aid cereals to reduce neurolathyrism related to grass-pea preparations during famine. *Lancet* 362:1808–1810.

Gilchrist EJ, Haughn GW (2005) TILLING without a plough: a new method with applications for reverse genetics. *Curr Opin Plant Biol* 8:211–215.

Gilchrist EJ, Haughn GW, Ying CC, Otto SP, Zhuang J, Cheung D, Hamberger B, Aboutorabi F, Kalynyak T, Johnson L, Bohlmann J, Ellis BE, Douglas CJ, Cronk QC (2006a) Use of Ecotilling as an efficient SNP discovery tool to survey genetic variation in wild populations of *Populus trichocarpa. Mol Ecol* 15:1367–1378.

Gilchrist EJ, O'Neil NJ, Rose AM, Zetka MC, Haughn GW (2006b) TILLING is an effective reverse genetics technique for *Caenorhabditis elegans. BMC Genomics* 7:1–16.

Goldman IL, Breitbach DN (1996) Inheritance of a recessive character controlling reduced carotenoid pigmentation in carrot (*Daucus carota* L.). *J Hered* 87(5):380–382. https://doi.org/10.1093/OXFORDJOURNALS.JHERED.A023018

Greene EA, Codomo CA, Taylor NE, Henikoff JG, Till BJ, Reynolds SH, Enns LC, Burtner C, Johnson JE, Odden AR, Comai L, Henikoff S (2003) Spectrum of chemically induced mutations from a large-scale reverse-genetic screen in *Arabidopsis*. *Genetics* 164:731–740.

Grzebelus D, Simon PW (2009) Diversity of *DcMaster* like elements of the *PIF/Harbinger* superfamily in the carrot genome. *Genetica* 135:347–353.

Grzebelus D, Yau YY, Simon PW (2006) Master: a novel family of PIF/Harbinger-like transposable elements identified in carrot (*Daucus carota* L.). *Mol Genet Genom* 275:450.

Guo Y, Abernathy B, Zeng Y, Ozias-Akins P (2015) TILLING by sequencing to identify induced mutations in stress resistance genes of peanut (*Arachis hypogaea*). *BMC Genomics* 16:157.

Gupta P, Reddaiah B, Salava H, Upadhyaya P, Tyagi K, Sarma S, Datta S, Bharti Malhotra, Thomas S, Sunkum A, Devulapalli S, Till BJ, Sreelakshmi Y Sharma R (2017) Next-generation sequencing (NGS)-based identification of induced mutations in a doubly mutagenized tomato (Solanum lycopersicum) population. *Plant J* 92(3):495–508.

Haughn G, Somerville CR (1987) Selection for herbicide resistance at the whole-plant level. In: *Biotechnology in Agricultural Chemistry* (Eds: Homer M. LeBaron, Ralph O. Mumma, Richard C. Honeycutt, John H. Duesing, J. F. Phillips, Michael J. Haas). ACS Publication Chapter 7 pp 98–107, ISBN13: 9780841210196, eISBN: 9780841211773. DOI: 10.1021/bk-1987-0334

Henikoff S, Comai L (2003) Single-nucleotide mutations for plant functional genomics. *Annu Rev Plant Biol* 54:375–401/

Henikoff S, Till BJ, Comai L (2004) TILLING. Traditional mutagenesis meets functional genomics. *Plant Physiol* 135:630–636

Hofinger BJ, Jing HC, Hammond-Kosack KE et al. (2009) High-resolution melting analysis of cDNA-derived PCR amplicons for rapid and cost-effective identification of novel alleles in barley. *Theor Appl Genet* 119:851–865.

Iorizzo M, Ellison S, Senalik DA, et al. (2016) A high quality carrot genome assembly provides new insights into carotenoid accumulation and asterid genome evolution. *Nat Genet* 48:657–666.

Iorizzo M, Senalik DA, Grzebelus D, Bowman M, Cavagnaro PF, Matvienko M, Ashrafi H, Van Deynze A, Simon PW (2011) De novo assembly and characterization of the carrot transcriptome reveals novel genes, new markers, and genetic diversity. *BMC Genomics* 12:389–389.

Iovene M, Cavagnaro PF, Senalik D, Buell CR, Jiang J, Simon PW (2011) Comparative FISH mapping of *Daucus* species (Apiaceae family). *Chromosome Res* 19:493–506.

Irshad A, Guo H, Zhang S, Liu L (2020) TILLING in cereal crops for allele expansion and mutation detection by using modern sequencing technologies. *Agronomy* 10:405.

Itoh Y, Hasebe M, Davies E, Takeda J, Ozeki Y (2003) Survival of Tdc transposable elements of the *En/Spm* superfamily in the carrot genome. *Mol Genet Genom* 269:49–59.

Jaganathan D, Ramasamy K, Sellamuthu G, Jayabalan S and Venkataraman G (2018) CRISPR for crop improvement: An update review. *Front Plant Sci* 9:985. doi: 10.3389/fpls.2018.00985

Jeong H-J, Kwon J-K, Pandeya D, Hwang JN, Hoang NH, Bae JH, Kang BC (2012) A survey of natural and ethyl methane sulfonate-induced variations of *eIF4E* using high-resolution melting analysis in *Capsicum*. *Mol Breed* 29:349–360.

Jin J, Chen L, Yao M, et al (2010) Simplification of EcoTILLING technique for tea plant. *J Tea Sci* 30:19–26.

Kashtwari M, Wani AA, Rather RN (2019) TILLING: An alternative path for crop improvement. *J Crop Improv* 33:83–109.

Koboldt DC, Chen K, Wylie T, et al (2009) VarScan: variant detection in massively parallel sequencing of individual and pooled samples. *Bioinformatics* 25:2283–2285.

Kumari R, Kumar P, Sharma VK, Kumar H (2018) Allele mining for crop improvement. *Int J Pure Appl Biosci* 6(1):1456–1465.

Latha R, Rubia L, Bennett J, Swaminathan MS (2004) Allele mining for stress tolerance genes in *Oryza* species and related germplasm. *Mol Biotechnol* 27:101–108.

Li H, Handsaker B, Wysoker A, Fennell T, Ruan J, Homer N, Marth G, Abecasis G, Durbin R (2009) 1000 genome project data processing subgroup. The sequence alignment/map format and SAMtools. *Bioinformatics* 25:2078–2079.

Li T, Deng YJ, Liu JX, Duan AQ, Liu H, Xiong AS (2021) DcCCD4 catalyzes the degradation of α-carotene and β-carotene to affect carotenoid accumulation and taproot color in carrot. *Plant J* 108:1116–1130.

Lo Schiavo FL, Giuliano G, Sung ZR (1988) Characterization of a temperature-sensitive carrot cell mutant impaired in somatic embryogenesis. *Plant Sci* 54:157–164.

Mackay MC, et al. (2004) Focused identification of germplasm strategy—FIGS. Cereals 2004. Proceedings of the 54th Australian Cereal Chemistry Conference and the 11th Wheat Breeders' Assembly, 21–24 September 2004, Canberra, Australian Capital Territory. eds Black CK, Panozzo JF, Rebetzke GJ (Royal Australian Chemical Institute, Melbourne), pp 138–141.

Macko-Podgorni A, Nowicka A, Grzebelus E. et al. (2013) DcSto: carrot Stowaway-like elements are abundant, diverse, and polymorphic. *Genetica* 141; 255–267.

Macko-Podgórni A, Stelmach K, Kwolek K, Machaj G, Ellison S, Senalik DA, Simon PW, Grzebelus D (2020) Mining for candidate genes controlling secondary growth of the carrot storage root. *Int J Mol Sci* 21(12):4263.https://doi.org/10.3390/IJMS21124263

Mazzucato A, Cellini F, Bouzayen M, Zouine M, Mila I, Minoia S, Petrozza A, Picarella ME, Ruiu F, Carriero F (2015) A TILLING allele of the tomato Aux/IAA9 gene offers new insights into fruit set mechanisms and perspectives for breeding seedless tomatoes. *Mol Breed* 35:1–15.

McCallum CM, Comai L, Greene EA, Henikoff S (2000a) Targeted screening for induced mutations. *Nat Biotechnol* 18:455–457.

McCallum CM, Comai L, Greene EA, Henikoff S (2000b) Targeting induced local lesions in genomes (TILLING) for plant functional genomics. *Plant Physiol* 123:439–442.

Myers RM, Maniatis T, Lerman LS (1987) Detection and localization of single base changes by denaturing gradient gel electrophoresis. *Methods Enzymol* 155:501–27. doi: 10.1016/0076-6879(87)55033-9

Nagy A, Perrimon N, Sandmeyer S, Plasterk R (2003) Tailoring the genome: the power of genetic approaches. *Nat Genet* 33:276–284.

Nataraj AJ, Olivos Glander I, Kusukawa N, Highsmith Jr WE (1999) Single strand conformation polymorphism and heteroduplex analysis for gel based mutation detection.*Electrophoresis* 20:1177–1185.
Natarajan AT (2005) Chemical mutagenesis: From plants to human. *Curr Sci* 89:312–317.
Ng PC, Henikoff S (2003) SIFT: Predicting amino acid changes that affect protein function. *Nucleic Acids Res* 31:3812–3814.
Nieto C, Piron F, Dalmais M, Marco CF, Moriones E, Gómez-Guillamón ML, Truniger V, Gómez P, Garcia-Mas J, Aranda MA, Bendahmane A (2007) EcoTILLING for the identification of allelic variants of melon eIF4E, a factor that controls virus susceptibility. *BMC Plant Biol* 7:1–9.
Nijman IJ, Mokry M, Van Boxtel R, Toonen P, de Bruijn E, Cuppen E (2010) Mutation discovery by targeted genomic enrichment of multiplexed barcoded samples. *Nat Methods* 7:913–915.
Nogales A, Nobre T, Cardoso HG, Muñoz-Sanhueza L, Valadas V, Campos MD, Arnholdt-Schmitt B (2016) Allelic variation on *DcAOX1* gene in carrot (*Daucus carota* L.): An interesting simple sequence repeat in a highly variable intron. *Plant Gene* 5:49–55 https://doi.org/10.1016/J.PLGENE.2015.11.001
Ossowski S, Schneeberger K, Lucas-Lledó JI, Warthmann N, Clark RM, Shaw RG, Weigel D, Lynch M (2010) The rate and molecular spectrum of spontaneous mutations in *Arabidopsis thaliana*. *Science* 327:92–94.
Ozeki Y, Davies E, Takeda J (1997) Somatic variation during long term subculturing of plant cells caused by insertion of a transposable element in a phenylalanine ammonia-lyase (PAL) gene. *Mol Gen Genet* 254:407–416.
Palan B, Bhattacharya A, Char B (2021) TILLING in the era of precise genome editing. *Indian J Biotechnol* 20:9–16.
Park YJ, Kim KV, Chung HK, Cho GT, Ma KH, Chandrabalan D, Gwag JG, Kim TS, Cho EG (2007) PowerCore: a Program applying the advanced M strategy with a heuristic search for establishing core sets. *Bioinformatics* 23(16):2155–62.
Porreca GJ, Zhang K, Li JB, Xie B, Austin D, Vassallo SL, LeProust EM, Peck BJ, Emig CJ, Dahl F, Gao Y, Church GM, Shendure J (2007) Multiplex amplification of large sets of human exons. *Nat Methods* 4:931–936.
Portis E, Portis F, Valente L, Moglia A, Barchi L, Lanteri S, Acquadro A (2016) A genome-wide survey of the microsatellite content of the globe artichoke genome and the development of a web-based database. *PLoS One* 11:e0162841.
Prasanna S, Jain SM (2017) Mutant resources and mutagenomics in crop plants. *Emir J Food Agric* 29:651–657.
Prochnik S, Marri PR, Desany B, Rabinowicz PD, Kodira C, Mohiuddin M, Rodriguez F, Fauquet C, Tohme J, Harkins T, Rokhsar DS, Rounsley S (2012) The cassava genome: current progress, future directions. *Trop Plant Biol* 5:88–94.
Raghavan C, Naredo M, Elizabeth B, Wang H, Atienza G, Liu B, Qiu F, McNally KL, Leung H (2007) Rapid method for detecting SNPs on agarose gels and its application in candidate gene mapping. *Mol Breed* 19:87–101.
Rigola D, van Oeveren J, Janssen A, Bonné A, Schneiders H, van der Poel HJ, van Orsouw NJ, Hogers RC, de Both MT, van Eijk MJ (2009) High-throughput detection of induced mutations and natural variation using KeyPoint™ technology. *PLoS One* 4:e4761.
Rolling WR, Senalik D, Iorizzo M, Ellison S, Van Deynze A, Simon PW (2022) CarrotOmics: a genetics and comparative genomics database for carrot (Daucus carota). Database (Oxford) 2022:1–8.
Rothberg JM, Hinz W, Rearick TM, Schultz J, Mileski W, Davey M, Leamon JH, Johnson K, Milgrew MJ, Edwards M (2011) An integrated semiconductor device enabling non-optical genome sequencing. *Nature* 475:348.
Rozycka M, Collins N, Stratton MR, Wooster R (2000) Rapid detection of DNA sequence variants by conformation-sensitive capillary electrophoresis. *Genomics* 70:34–40.
Selvakumar R, Jat GS, Manjunathagowda DC (2023) Allele mining through TILLING and EcoTILLING approaches in vegetable crops. *Planta* 258:15. https://doi.org/10.1007/s00425-023-04176-2
Shan Q, Wang Y, Li J, Gao C (2014) Genome editing in rice and wheat using the CRISPR/Cas system. *Nat Protoc* 9:2395.
Simon P (1984) Carrot genetics. *Plant Mol Biol Rep* 2:54–63.
Singh S, Singh R, Sharma D, Sharma SK, Dey SS, Bhatia R, Ghemeray H, Kumar R (2023) Insight into carrot carotenoids in post-genomic world for higher nutrition. In: *Smart plant breeding for vegetable crops in post-genomics era* (Eds: Singh, S., Sharma, D., Sharma, S.K., Singh, R.). Springer, Singapore. pp 367–382. https://doi.org/10.1007/978-981-19-5367-5_15/COVER

Slade AJ, Knauf VC (2005) TILLING moves beyond functional genomics into crop improvement. *Transgen Res* 14:109–115.

Spooner DM (2019) *Daucus*: Taxonomy, Phylogeny, Distribution. In: *The carrot genome. Compendium of plant genomes* (Eds: Simon P, Iorizzo M, Grzebelus D, Baranski R). Springer, Cham pp 9–26. https://doi.org/10.1007/978-3-030-03389-7_2

Sung ZR (1976) Mutagenesis of cultured plant cells. *Genetics* 84:51–57.

Szurman-Zubrzycka ME, Zbieszczyk J, Marzec M, et al (2018) HorTILLUS—a rich and renewable source of induced mutations for forward/reverse genetics and pre-breeding programs in barley (*Hordeum vulgare* L.). *Front Plant Sci* 9:216.

Tadele Z (2016) Mutagenesis and TILLING to dissect gene function in plants. *Curr Genomics* 17:499–508.

Taylor NE, Greene EA (2003) PARSESNP: a tool for the analysis of nucleotide polymorphisms. *Nucleic Acids Res* 31:3808–3811.

Till BJ, Colbert T, Codomo C, Enns L, Johnson J, Reynolds SH, Henikoff JG, Greene EA, Steine MN, Comai L, Henikoff S (2006) High-throughput TILLING for *Arabidopsis*. *Methods Mol Biol* 323:127–35. doi: 10.1385/1-59745-003-0:127

Till BJ, Colbert T, Tompa R, Enns LC, Codomo CA, Johnson JE, Reynolds SH, Henikoff JG, Greene EA, Steine MN, Comai L, Henikoff S (2003a) High-throughput TILLING for functional genomics. *Methods Mol Biol* 236:205–20. doi: 10.1385/1-59259-413-1

Till BJ, Comai L, Henikoff S (2007a) TILLING and EcoTILLING for crop improvement. In: *Genomics-assisted crop improvement* (Eds: Varshney RK, Tuberosa R). Springer, Dordrecht, pp 333–349.https://doi.org/10.1007/978-1-4020-6295-7_15

Till BJ, Cooper J, Tai TH, Colowit P, Greene EA, Henikoff S, Comai L (2007b) Discovery of chemically induced mutations in rice by TILLING. *BMC Plant Biol* 7:1–12.

Till BJ, Reynolds SH, Greene EA, Codomo CA, Enns LC, Johnson JE, Burtner C, Odden AR, Young K, Taylor NE, Henikoff JG, Comai L, Henikoff S (2003b) Large-scale discovery of induced point mutations with high-throughput TILLING. *Genome Res* 13:524–530.

Turan Buyukdinc (2019) Determination of effective mutagen dose for carrot (Daucus carota ssp. sativus var. atrorubens alef and D. carota) callus cultures. *Intnl J Sci Technol Res* 5: 15–23.

Uncu AO, Uncu AT (2019) High-throughput simple sequence repeat (SSR) mining saturates the carrot (*Daucus carota* L.) genome with chromosome-anchored markers. *Biotechnol Biotechnol Equip* 34(1):1–9. https://doi.org/10.1080/13102818.2019.1701551

Upadhyaya HD, Ortiz R (2001) A mini core subset for capturing diversity and promoting utilization of chickpea genetic resources in crop improvement. *Theor Appl Genet* 102:1292–1298.

Veerla S, Hoglund M (2006) Analysis of promoter regions of co-expressed genes identified by microarray analysis. *BMC Bioinform* 7:384.

Vetter J (2000) Plant cyanogenic glycosides. *Toxicon* 38:11–36.

Voelkerding K, Dames S, Durtschi J (2009) Next-generation sequencing: From basic research to diagnostics. *Clin Chem* 55:641–58.

Wang F, Wang G-L, Hou X-L, Li M-Y, Xu Z-S, Xiong A-S (2018) The genome sequence of "Kurodagosun", a major carrot variety in Japan and China, reveals insights into biological research and carrot breeding. *Mol Genet Genom* 293:861–871.

Wang N, Shi L, Tian F, Ning H, Wu X, Long Y, Meng J (2010) Assessment of FAE1 polymorphisms in three Brassica species using EcoTILLING and their association with differences in seed erucic acid contents. *BMC Plant Biol* 10:1–11.

Wu Z, Liu Z, Chang S, Zhao Y (2020) An EMS mutant library for carrot and genetic analysis of some mutants. *Breed Sci* 70(5):540–546. https://doi.org/10.1270/JSBBS.20020

Xu Y, Li P, Zou C, Lu Y, Xie C, Zhang X, Prasanna BM, Olsen M (2017) Enhancing genetic gain in the era of molecular breeding. *J Exp Bot* 68:2641–2666.

Xu ZS, Tan HW, Wang F, Hou XL, Xiong AS (2014) CarrotDB: a genomic and transcriptomic database for carrot. *Database: J Biol Databases Curation* 2014:1–8.

Yau YY and Simon PW (2003) A 2.5-kb Insert eliminates acid soluble invertase isozyme II transcript in a carrot (Daucus carota L.) roots, causing high sucrose accumulation. *Plant Mol Biol* 53;151–162.

Zerr T, Henikoff S (2005) Automated band mapping in electrophoretic gel images using background information. *Nucleic Acids Res* 33:2806–2812.

10 Allele Mining and Genomic Tools to Improve Capsicums

Bala Rathinasabapathi,[1,] Dominick Padilla,[1] Jingwei Fu,[1] Corina Hayano-Kanashiro,[2] and Octavio Martínez[3]*
[1]Horticultural Sciences Department, 2247 Fifield Hall, University of Florida, Gainesville, FL, USA
[2]Departamento de Investigaciones Científicas y Tecnológicas de la Universidad de Sonora, Universidad de Sonora, Hermosillo, Sonora, Mexico
[3]Unidad de Genómica Avanzada (Langebio), Centro de Investigación y de Estudios Avanzados del Instituto Politécnico Nacional (Cinvestav), Irapuato, Guanajuato, Mexico
*Corresponding author: brath@ufl.edu

10.1 INTRODUCTION

Five species of the genus *Capsicum* form economically significant vegetable and spice crops of the world. Many aspects relating to improving sweet and pungent peppers for fruit quality traits, yield and resistance to diseases, pests, and harsh climatic conditions were reviewed recently (Jarret et al., 2019; Rathinasabapathi, 2020; Lozada et al., 2021c). Availability of genome sequences, transcriptomes, and DNA-based markers have provided powerful new tools for vegetable crop breeders. The objective of this review is to present recent developments in the field of *Capsicum* genomics with special emphasis on the discovery and deployment of DNA-based markers linked to multiple traits that are valuable for crop breeding.

We summarize the basic genomic information on the edible *Capsicum* species in Table 10.1. Whole genome sequences are available for several accessions or landraces of *C. annuum* (Kim et al. 2014; Qin et al. 2014) and more recently for *C. chinense* and *C. baccatum*. One notable feature of *Capsicum* genomes is that their genome sizes are about three time larger than the genomes of related crops tomato and eggplant (Table 10.1). During the evolution of peppers *Capsicum* genome went through a rapid amplification of retrotransposon elements resulting in a large amount of repetitive DNA (Qin et al. 2014). Relatively larger DNA content in *C. pubescens* (Table 10.1) suggests that this species could likely have more heterochromatin in its genome than *C. annuum*. Because of the availability of genetic and genomic studies in *C. annuum*, we will selectively use examples of studies in this pepper species and refer to the other edible Capsicums for comparative purposes only.

10.2 PHENOTYPIC VARIATION AND DNA MARKERS

To best utilize the genetic diversity available within a species, it is imperative to identify inheritance patterns of genes that underlie the phenotypic variations that could potentially be used for breeding.

 DOI: 10.1201/9781003376439-10

TABLE 10.1
Chromosome Number in Somatic Cells (2*n*), Nuclear DNA Content (pg/1C), and Genome Size Compared to Two Other Major Crops in the Solanaceae Tomato (*S. lycopersicum*) and Eggplant (*S. melongena*)

Species	2n	Nuclear DNA (pg/1C)	Genome Size (Gbp)	References
C. annuum	24	3.38	3.34 to 3.48	(Kim et al. 2014) (Qin et al. 2014) (Moscone et al. 2003)
C. frutescens	24	3.40	Not available	(Moscone et al. 2003)
C. chinense	24	3.42	3.09	NCBI PRJNA873099 (Moscone et al. 2003)
C. baccatum	24	3.71	3.22	NCBI PRJNA308879 (Ahn et al. 2018) (Moscone et al. 2003)
C. pubescens	24	4.47	Not available	(Moscone et al. 2003)
Solanum melongena	24	1.14	1.21	(Barchi et al. 2019) Arumuganathan and Earle 1991
Solanum lycopersicum	24	1.88	0.90	(Arumuganathan and Earle, 1991.; Sato et al. 2012)

Since the phenotype is the result of both the influence of gene(s) and the environment and their interactions, one needs to understand the heritability of the traits of interest so that the influence of selection can be inferred.

In pepper several morphological traits can be visually scored and studied using transmission genetics. Figure 10.1 shows phenotypic variations for fruit size, fruit shape, and color in *C. annuum.* A gene list by Wang and Bosland (2016) reviewing various studies documented the nomenclature rules for naming genes and described 292 genes relating to morphological, physiological traits, and resistance to pests and diseases. Up to that point in time, inheritance of many traits in *Capsicum* were reported, but linkage information was not available for most. Technological progress to identify DNA-based markers and computational methods to analyze the marker data to infer linkages have led to major breakthroughs in genomic studies of crops including *Capsicum.*

In genotyping-by-sequencing (GBS) methodology the genome complexity is reduced by restriction enzyme digestion and size selection of DNA prior to high-throughput sequencing. This method can be used independent of genome sequence and hence was applied to obtain marker data from many crops including peppers.

Mapping DNA-based markers and identification of loci associated with desired phenotypes (via QTL mapping and genome wide association mapping) will be useful for marker-assisted selection in breeding and map-based cloning of specific genes. To provide an overview of progress in this area, we will summarize examples of linkage maps using DNA markers and then describe examples for studies that have identified quantitative trait loci, and candidate genes for various traits. Table 10.2 lists some examples of studies where DNA markers were used to generate linkages maps in *Capsicum annuum.* While earlier studies used DNA markers such as RFLP, RAPD, and SSR, after the availability of the genome sequences for *Capsicum* and development of newer genotyping platforms, more recent studies have focused on high-density SNP markers for linkage mapping and for QTL detection (Table 10.2).

DNA markers can be used as fingerprints to identify germplasm and for classifying plants within a collection. Such analyses are important for germplasm conservation and for selecting specific lines

FIGURE 10.1 An illustration of phenotypic variations in fruit shape, fruit size, and fruit color in *Capsicum annuum*.

for breeding. Compared to RFLP and RAPD markers, simple sequence repeat (SSR) markers, also known as microsatellite markers, are more user friendly. SSRs are based on nucleotide variations of a repeating 1–6 bp units. DNA fragments amplified from the genomic DNA are fluorescently labeled and analyzed for variations in amplicon length using capillary electrophoresis. Kwon et al. (2013) used SSRs to group 170 commercial pepper cultivars grown in South Korea.

Others tested COSII primer sets (developed based on conserved sequences between tomato and Arabidopsis) to identify polymorphisms between pepper inbreds and showed that pungent and non-pungent peppers can be identified with high confidence (Jung et al. 2010).

A study on the diversity of the *Capsicum* collection at Zaragosa, Spain examined 51 accessions of *C. annuum* and 51 accessions of other *Capsicum* species using 39 SSR markers (González-Pérez et al. 2014). Lozada et al (2021a) used genotyping-by-sequencing (GBS) methodology to generate SNP markers to test the genetic diversity of 165 accessions from New Mexico pepper breeding program. (Lozada et al. 2021a). These studies showed that *C. annuum* lines separated from *C. frutescens* and *C. chinense*, the varieties in the latter two species grouped together.

Others have mined transcriptomic data to identify SNP markers. For example, Ahn et al (2014) compared transcripts between "Mandarin" and "Black cluster" to identify more than 2000 potential SSR motifs (Ahn et al. 2014).

TABLE 10.2

Examples for the Types of DNA-Based Markers Used to Generate Linkage Maps of *C. annuum*. Abbreviations: AFLP – Amplified Fragment Length Polymorphism, CAPS – Cleaved Amplified Polymorphisms, RFLP – Restriction Fragment Length Polymorphism, RIL – Recombinant Inbred Line, SSAP – Sequence-Specific Amplification Polymorphism, SPP – Single-Position Polymorphism, SSR – Simple Sequence Repeat, STS – Sequence-Tagged Sites

Population Used	Marker Type(s)	Genotyping Technology	Number of Markers Mapped	References
F_2 derived from 'NuMex RNaky' x 'PI 159234'	AFLP, Isozyme, RAPD, RFLP	Restriction digestion of DNA, Gel electrophoresis, Polymerase chain reaction	677	Livingstone et al. (1999)
$F_{2:5}$ inbreds derived from 'Yolo Wonder' X 'CM334'	AFLP, SSR, RFLP, SSAP, and STS	Restriction digestion of DNA, gel electrophoresis, Polymerase chain reaction.	489	Barchi et al. (2007)
'CM334' x 'Chilsungcho' F_2 and three other populations	AFLP, CAPS, RFLP, SSAP, SSR	Restriction digestion of DNA, gel electrophoresis, Polymerase chain reaction	2052	Lee et al. (2009)
Doubled haploid derived F_1 D4 families of cross 'California Wonder' X 'LS2341'	AFLP and SSR	Polymerase chain reaction, gel electrophoresis	72	Mimura et al. (2010)
Doubled haploid derived F_1 D4 families of cross 'California Wonder' X. 'LS2341'	SSR	Polymerase chain reaction, gel electrophoresis	253	Mimura et al. (2012)
RIL of 'BG 2814-6' x 'NuMex RNaky' RIL of 'Early Jalepeno' x 'CM334'	SPP	30K Genotyping array DNA-DNA hybridization	3878/16,167	Hill et al. (2015)
'Tabasco' X 'P4' F_2	SNP	Infinium Illumina array PepperSNP 16k array	13,800	Hulse-Kemp et al. (2016)
RIL of 'Perennial' x 'Dempsey'	Bins of SNPs	Genome sequencing	2578	Han et al. (2016)

10.3 IDENTIFICATION OF QUANTITATIVE TRAIT LOCI (QTL) FOR DESIRABLE TRAITS USING BIPARENTAL MAPPING POPULATIONS

Most of the traits of value in pepper breeding are quantitative traits controlled by many genes. Quantitative trait loci (QTLs) demarcate the chromosomal regions involved in controlling phenotypic variability in the expression of a quantitative trait. Multiple studies identifying QTLs for yield, fruit qualities, morphological traits, and resistance to diseases have used biparental mapping populations. These mapping populations are usually derived by crossing two inbreds of *C. annuum* or *C. annuum* and another *Capsicum* species. Some of the studies used F_2 populations and others recombinant inbred lines (RILs) of F_7 or later (Table 10.3). Figure 10.2 depicts the steps involved in studies aiming to do identification of QTLs using biparental mapping populations and Figure 10.3 shows the sequence for genome wide association studies. Table 10.3 shows only those QTLs that have major effects on the phenotype, but many other minor QTLs have also been identified in these

TABLE 10.3
Examples for Studies Using Biparental Mapping Populations. Traits Studied, Major QTLs Identified, the Germplasm Used, and DNA-Based Marker(s) Linked to Them or Chromosomal Regions. Abbreviations: BC – Backcross; LG – Linkage Group; RIL – Recombinant Inbred Lines

Traits and QTLs	Germplasm	Linked Markers or Region for Major QTL	References
Axillary shooting (axl), flowering date (fd), primary stem length, number of leaves (NL). Twelve QTLs.	Doubled haploid derived F_1 D4 families of cross 'California Wonder' X. 'LS2341'	QTLs for axl, fd and NL on P12 and P8.	Mimura et al. (2010)
Fruit and plant trait related QTLs 76. Fruit weight, shape, length, fruit diameter	RIL of 'Yolo Wonder' x 'CM334'	In all chromosomes except P5.	Palloix et al. (2009)
Fruit traits, Fruit weight, diameter, pericarp thickness, pedicel diameter	F_3 population of 'Maor' x 'Perennial'	Fruit shape QTL on chromosome 3.	Chaim et al. (2001)
Root-knot nematode resistance	F2 of 'Carolina Wonder' x 'AZN1'	An SSR marker 0.8 cM from *N* gene on P9	Celik et al. (2016)
Fruit length, Five QTLs	F_2 of 'F68' x 'Habanero'	Chromosomes 3, 5, 11 and P3. Genetic marker ATCC260 detected in QTL region (MW20N6-M87N8) in chromosome 3 explained 27%	Lee et al. (2011)
Fruit length, Two QTLs	'FL 201' x 'TC07245'	Two QTLs on LG 2.	Arjun et al. (2018)
Male sterility, *ms8* locus	F_2 population of 'Line 320' x 'Elf'	Lower arm of Chromosome P4. SCAR markers linked to *ms8*: P2, V17,N2, V01.	Bartoszewski et al. (2012)
Yield and physiological traits, leaf area, proportion of biomass that is fruit, total fruit number, internode length. Forty-seven regions with QTLs detected.	RIL of 'Yolo Wonder' x 'CM334'	QTLs on linkage groups 1b, 2, 3a, 4, 6 and 12 had pleiotropic effects.	Alimi et al. (2013b, 2013a)

TABLE 10.3 (Continued)
Examples for Studies Using Biparental Mapping Populations. Traits Studied, Major QTLs Identified, the Germplasm Used, and DNA-Based Marker(s) Linked to Them or Chromosomal Regions. Abbreviations: BC – Backcross; LG – Linkage Group; RIL – Recombinant Inbred Lines

Traits and QTLs	Germplasm	Linked Markers or Region for Major QTL	References
Twelve morphological and horticultural traits. *Phytophthora capsici* resistance	F_6 RIL of 'YCM334' x 'Yolo Wonder'	Phytophthora resistance QTLs on LGs 4, 5 and d.	Lu et al. (2012a)
Phytophthora capsici root rot resistance	RIL of 'CM334' x 'Early Jalepeno'	QTLs on LGs 5, 8 and 9.QTL on LG5 was major and QTL on 8 had additive effect.	Lozada et al. (2021b)
Root-knot nematode resistance	Doubled haploid line with *Me1*–DH330. F_2 and BC from 'DH330' x '0516'	Mapped to 132 kb region in P9.	Wang et al. (2018)
Microspore embryogenesis; Seven QTLs. Major *Me2.2* on linkage group 2.	F_{10} RILS of 'FS871' x 'PM702'	Chromosomal Segment 166,621,667–167,038,758 bp,	Bai et al. (2023)
Resistance to Bacterial Spot Disease	RILs of 'ECW50R' x 'ECW60R'	Fine mapped *bs5* and *bs6*	Sharma et al. (2023)
Fruit wall thickness	RILs of 'YCM334' x 'Taen'	Mapped *qFWT-2-1, qFWT-3-1*	Lu et al. (2012b)
Yellow leaf curl virus resistance	F_2 of 'LP97' x 'ECW30R'	Three QTL identified. Markers validated for two: Chr7-LCV7 and Chr12-LCV12.	Siddique et al. (2022)

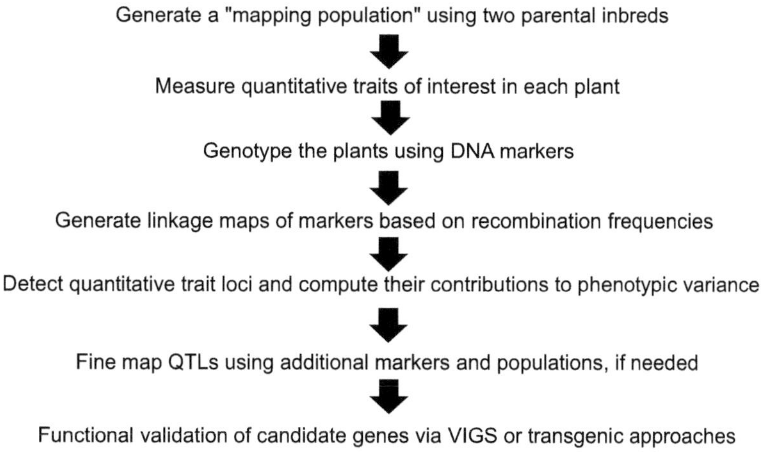

FIGURE 10.2 Steps involved in mapping studies where QTLs are detected using biparental mapping populations.

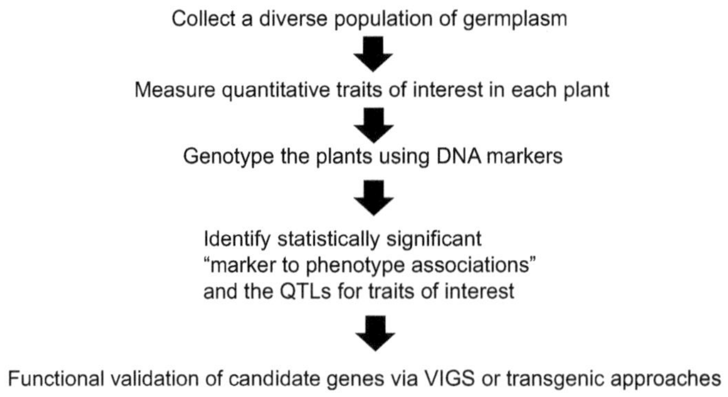

FIGURE 10.3 Steps involved in mapping studies where QTLs are detected using genome-wide association.

studies but not listed in the table. In Figure 10.4, we have illustrated the location of some of the major QTLs with reference to a few DNA markers. A recent meta-analysis of QTLs in pepper with special emphasis on disease resistance is available in Lozada et al. (2021c).

10.4 IDENTIFICATION OF QUANTITATIVE TRAIT LOCI (QTLS) FOR DESIRABLE TRAITS USING GENOME WIDE ASSOCIATION STUDIES (GWAS)

In association mapping method, "linkage disequilibrium" (LD) a measure of non-random association between alleles at two or more loci, is used for mapping. The key difference from the conventional QTL methodology is that the starting population for GWAS is a diverse collection of accessions with multiple variations whereas conventional method uses biparental populations having less genetic variability (Figure 10.3). GWAS thus lends itself to identify a greater number of polymorphic SNPs and indeed detect a greater number of significant associations between markers and the phenotypic differences (Table 10.3 versus Table 10.4). As multiple QTLs are detected, GWAS also lends itself to verify previously detected QTLs found using biparental populations and identify interactions between QTLs. For example, in one study interactions between two sets of QTLs influencing *Phytophthora* resistance was reported with additive and epistatic effects (Lu et al. 2012a). In their study 98% of high-level resistance related variation can be explained by three QTLs. Reddy et al. (2014) used gene-specific polymorphisms to find associations to metabolite levels of capsaicin, dihydrocapsaicin, and their precursors.

10.5 DISCOVERY AND USE OF ALLELIC VARIATION FOR PEPPER IMPROVEMENT

Next, we discuss aspects of methodologies and approaches used in utilizing alleleic variations for pepper crop improvement focusing on the current and future needs.

10.5.1 DNA Extractions from Large Populations of Plants

Most plant breeding projects involve genotyping large numbers of plants. For this purpose, genomic DNA samples of high quality need to be prepared. Several commercial kits are available. However, most studies have used modified CTAB protocol to prepare DNA from leaves (Schenk et al. 2023). Improvements in making this step simpler, cost efficient with high yields of quality DNA are always in need.

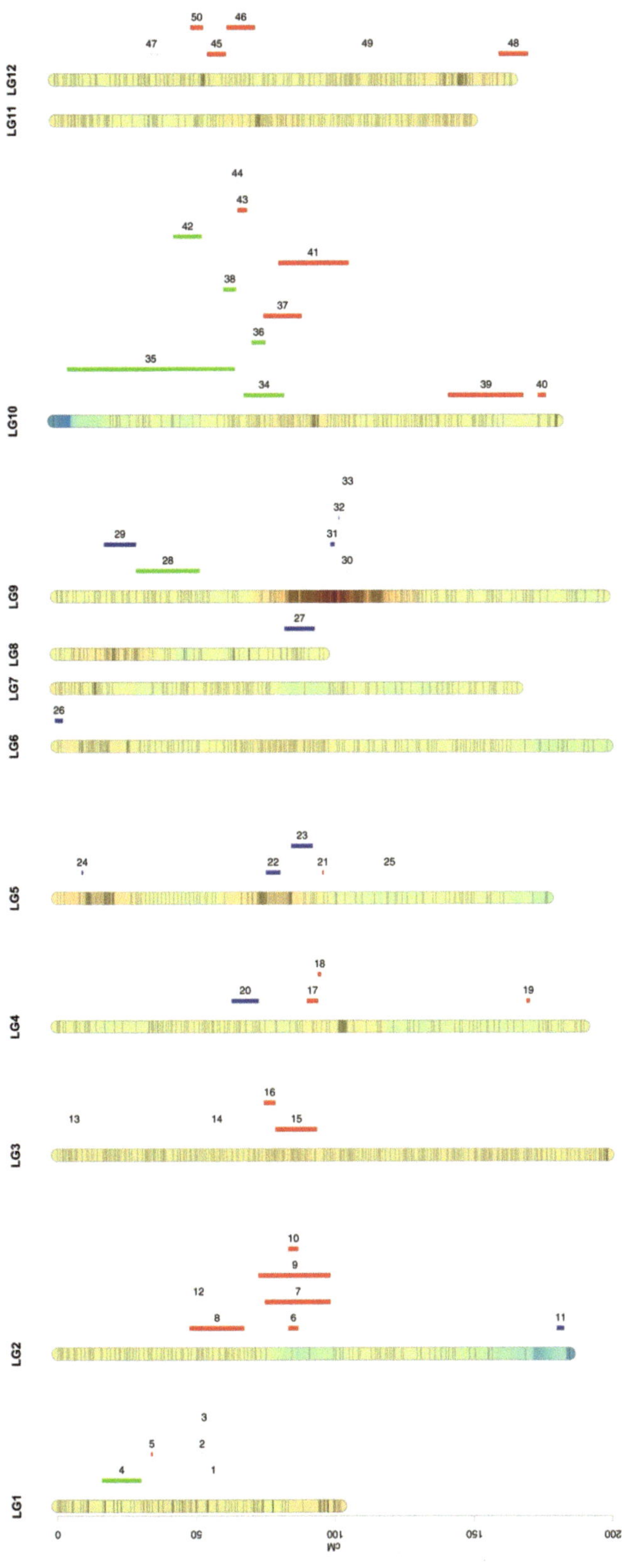

FIGURE 10.4 An illustration of the 12 chromosomes of *Capsicum annuum*. Positions for a selected set of SLAF markers from Zhang et al., (2019) are indicated using lines. Marker density is shown using color intensity from red (low) to blue/black (high). The chromosomal regions where major QTLs reported are indicated using colored bars next to the chromosomes, marking vegetative (green), fruit traits (red) and disease resistance (blue). Fifty major QTLs are numbered: 1-3 - Bacterial wilt disease resistance, 4 - growth habit, 5 - carotenoids, 6-10, 12, 14-19, 21, 34-41 43-48, 50 - fruit quality traits, 11, 24 - Powdery mildew resistance, 20, 22, 23, 27, 29 - Phytophthora resistance, 25, 30 - Ralstonia resistance, 26 - Bacterial Spot resistance, 28, 42 - Stem traits, 31-33 Root-knot nematode resistance, 49 - Yellow leaf curl virus resistance. Refer to Tables 3, 4 and 5 for QTL reports.

TABLE 10.4
Examples for Genome-Wide Association Studies (GWAS). Germplasm Used, Traits, Major QTLs or Significant SNPs and Chromosomal Regions and Genotyping Methods Used and their SNP Outputs. Abbreviations: GBS – Genotyping-by-Sequencing, HRM – High Resolution Melting, SLAF – Specific Locus Amplified Fragment

Germplasm	Traits, QTLs/Significant SNPs and Chromosomes	Genotyping Method & Number of Markers	References
Collection of 94 accessions	Fruit capsaicin and dihydrocapsaicin content, fruit weight. Significant SNPs 30 for capsaicin, 50 for dihydrocapsaicin and 15 for fruit weight.	GBS, 7331 SNPs.	Nimmakayala et al. (2016)
Collection of 94 accessions	*Pun1*was associated with capsaicinoids and their precursors.	GBS, SNP discovery by cloning specific sequences	Reddy et al. (2014)
Collection of 287 accessions	Significant SNPs 1487, candidate genes 2126, significant QTLs for 36 horticultural traits, 13 major association peaks for 11 traits.	SLAF-Seq based GWAS 594,429 SNPs	Wu et al. (2019)
Diverse accessions of 42 from Ethiopia	Five hundred and nine significant SNPs for fruit, leaf, petal, and stem-related traits. 27 candidate genes for fruit weight on chromosomes 3, 7, 8, 10, and 11.	GBS, HRM, 2.8 million SNPs	Solomon et al. (2019)
Core collection of 256 accessions	Resistance to Potato Virus Y	GBS,10,307 SNPs	Tamisier et al. (2020)

10.5.2 Technologies for Genotyping

Depending upon the DNA markers chosen, multiple genotyping platforms are available for plant breeders. Genotyping-by-sequencing (GBS) is the most economical method to generate SNP marker information in a breeding population. In this method complexity of the genome is reduced using restriction enzyme digestion and size selection of DNA. Then, PCR-amplified DNA is made with "bar-coded" adapters to tag the plants. Pooled libraries are then sequenced using the Illumina high-throughput sequencing platform. The results are then analyzed using a bioinformatic pipeline to sort the DNA sequences and identify the SNP polymorphisms. Alternatively, arrays (Cheng et al. 2016; Hulse-Kemp et al. 2016) offer reproducible platforms for genotyping.

Researchers have shown the utility of CAPS, SSRs, SCAR, KASP, and HRM markers (Table 10.2) to identify plants with specific phenotypes. Though many different technologies for genotyping vary in their cost, requirements of equipment, ease of use, and accuracy, progress in the last decade especially in developing markers for specific traits is impressive.

In sequence characterized amplified region (SCAR)-based markers, the technology depends on previous determination of alleles in a locus. Allele-specific amplicons are tested using gel electrophoresis. Output from many QTL mapping studies, GWAS research and the genome sequencing projects together make it easier for researchers to develop easily usable markers for breeding.

KASP (Kompetitive Allele specific PCR) methodology involves one reverse primer and two forward primers that are allele specific designed to distinguish the varying SNP or deletion in the amplicon measured using fluorescence resonance energy transfer (FRET)-based methods (Kaur et al. 2020). Many studies have focused on developing user-friendly markers for disease resistance. For example, Holdsworth and Mazourek (2015) developed markers for potyvirus resistance *pvr1* and *Bs3* a gene for resistance to bacterial spot disease. Loci for Phytophthora resistance are known in all 12 chromosomes of *Capsicum.* Barchenger et al. (2017) developed an allele-specific CAPS marker for *Ve1,* a gene homologous to tomato *Ve1* for *Verticilliam dahlia* resistance and the marker can predict the resistance with 48% accuracy. In a recent GWAS study, Zhang et al. (2023) reported identifying 1683 kb region on chromosome 5 for Phytophthora resistance, in support of previous identification of resistance QTLs on chromosome 5 by others and they developed four KASP markers to distinguish resistant and sensitive plants. Lee et al. (2008) reported a CAPS marker associated with the partial restoration of cytoplasmic male sterility controlled by the *pr* locus. As opposed to the cytoplasmic male sterility, there are multiple sources of genic male sterility known in Capsicums. DNA-based markers, which are applicable to specific populations are available (Lee et al. 2012; Bartoszewski et al. 2012).

10.5.3 Smart Phenotyping Pipelines

Studies on pepper that examined the inheritance of yield, fruit traits, stem traits, leaf traits, etc., have used manual collection of phenotypic data via quantitative methods. In several studies, the analyses of images of pepper fruit using Tomato Analyzer software have been used to infer genetic variation for fruit shape (Nankar et al. 2020; Khokhar et al. 2022). However, such data do require prior manual work of selecting, harvesting, and scanning the fruit and hence only partially high throughput. Such image data were not reported to be used for mapping or selections in pepper breeding so far. However, phenomics researchers have used pepper as a model for constructing 3D models of pepper leaf (Moon et al. 2022), collecting large-scale phenotype data from the field (Tripodi et al. 2021) and evaluating fruit characteristics (Pereira-Dias et al. 2020; Tripodi and Greco, 2018). Under controlled conditions monitoring of plant movements had been shown to be useful to track drought stress (Dalal et al. 2019). Phenomic research when combined with genotyping will help us pinpoint markers for subtle but useful phenotypic variations. Considering the rapid progress in genotyping plants, we urgently need a new paradigm to phenotype plants more efficiently. It will be helpful to revise the Capsicum descriptor (International Board for Plant Genetic Resources et al. 1995) to add guidelines and best practices about phenotyping plants by adding normalized difference vegetation index (NDVI) to estimate growth or yield (Chang et al. 2022), and the use of Tomato Analyzer (Khokhar et al. 2022) and other automated or semi-automated data collection tools to describe phenotypes.

10.5.4 Expression QTLs Using Metabolomic, Transcriptomics, and Proteomic Tools

Quantitative trait loci could also be recognized based on differential expression levels of specific genes controlling the phenotype. Then they should also be identifiable by analyzing the association of DNA markers to expression phenotypes such as metabolite levels (mQTLs), protein levels determined from proteomic experiments (pQTL), or transcript levels (eQTL). While no studies have shown the feasibility of mQTLs and pQTLs in *Capsicums*, some progress has been made in the development of methods to identify eQTLs from RNA-Seq data (Escoto-Sandoval et al. 2021; Martínez et al. 2021). A software tool "Salsa" developed to mine transcriptomic data simplifies the process (Martínez et al. 2021)

To estimate effective allele diversity in *C. annuum* L., Escoto-Sandoval et al (2021) used genome-wide standardized expression profiles (SEPs) from different accessions of chili pepper.

They measured gene expression at seven times of fruit development, 0, 10, 20, 30, 40, 50, and 60 days after anthesis (DAA) in 12 accessions: six domesticated (D), four wild (W), and two F_1 crosses between a D and a W accession. Data and functions to mine such data are publicly available in the R package "*Salsa*" (Escoto-Sandoval et al. 2021). The authors found five ternary models (TM) that were most abundant, the first one implies that gene expression decreased (D) in the first interval, from 0 to 10 DAA, and then stayed steady (S) in all the subsequent five-time intervals. Genes with this expression pattern can be labeled as preferentially expressed in the open flower (at 0 DAA). The second more frequent TM stayed steady (S) at all the stages. The third model implied that the genes were not expressed, or expressed at very low frequency, in the flower at 0 DAA, and then increased at 10 DAA and stayed steady over all subsequent stages of fruit development, from 10 up to 60 DAA. Genes with the fourth model, had maximum expression at 10 DAA where the fruit is growing at its higher rate, and then stayed steady state from 20 up to 60 DAA. Finally, genes fitting the fifth model, represent genes that are highly expressed only in the last stage of development at 60 DAA, where the fruit is entering senescence.

Martínez et al. (2021) used a set of 12 accessions containing domesticated (D: 6), wild (W: 4), and crosses between D and W (2). They detected 28 genes preferentially expressing in the mature flower and remaining in a steady state during the whole of the fruit development process. Out of these, the only transcription factor (TF) was "*VOZ1* ", that had been reported to mediate heat stress responses in *Arabidopsis* (Song et al. 2018) and to regulate drought-mediated flowering in tomato (Chong et al., 2022). The expression pattern of this TF, remaining similar in domesticated and wild accessions, is consistent with TF playing a key role and not being modified by the domestication process. In traditional plant improving strategies, breeders use non-molecular phenotypes to design crosses for selection of targeted horticultural traits. Since there is high diversity in gene expression profiles, they can be exploited for plant improvement for specific traits. E-QTLs can reveal the architecture of gene regulation (Gilad et al. 2008). For example, eQTL analyses had revealed candidate genes associated with fruit flesh softening rate in peach (Carrasco-Valenzuela et al. 2019). We hope that in the near future they will be important components of marker-assisted selection in plant breeding (Xu and Crouch, 2008).

10.5.5 Marker-Assisted Selection

Although specific marker–phenotype associations have been reported for multiple traits, few studies used the markers in the context of breeding to develop new cultivars. It is likely that many plant breeding programs utilize marker-assisted selection but not report cultivar development details in the literature. Sran et al. (2023) reported the introgression of Leaf Curl Virus disease (LCVD) resistant locus from a pungent pepper variety into a heat tolerant bell pepper line where a SCAR marker for pungency was used to eliminate the pungent lines. Markers can also be used to identify useful germplasm for breeding as shown in the study by Channabasava et al. (2023) who determined pollen-fertility restorer alleles by screening germplasm using DNA markers.

10.5.6 Plant Breeding Paradigms Without Reducing Genetic Diversity

Progress made in identifying loci involved and candidate genes for genes involved in genic male sterility and restorer genes should accelerate their utility for F1 hybrid production. Identification of allelic markers and their association to phenotypes in a few lines of peppers would likely have a skew toward pepper breeders using those lines as sources of specific traits. However, efforts should be made to search for new alleles for genes of interest to broaden the genetic pool. Eggink et al. (2014) showed that by combining biochemical characterization of a breeding population to marker data, flavor profiles, terpenoid levels, and total soluble solids content traits could be introgressed into *C. annuum* from *C. baccatum*. For traits such as disease and pest resistance, it will be helpful to pyramid genes that provide resistance via different mechanisms to improve the durability of resistance. Such pyramiding experiments should be possible via use of DNA-markers for candidate genes. QTL analyses suggest that such pyramiding for resistance QTLs has already happened for resistance or tolerance to Potato Virus Y (Tamisier et al. 2020). Though multiple studies are available on plant, fruit traits, and disease resistance, studies linking markers to insect pest resistance traits, abiotic stress resistance, and postharvest quality traits are not available in Capsicum and will be needed for breeding peppers adapted to varying climatic zones.

10.5.7 Validation of Gene Function

In QTL and GWAS studies, the chromosomal region identified to be associated with a phenotype often contains multiple candidate genes. Hence, further investigation and validation on function of identified candidate genes within a locus or loci are required. In addition, although conventional plant breeding significantly contributed to our society over the past hundred years, biotechnological applications are important to meet the increasing demands for crop quality and quantity, tolerance to abiotic and biotic stresses, improved food safety, and ecological conservation. Genetic transformation provides an excellent platform to break through the bottleneck in gene function studies and for the improvement of crops. Genome editing provides a possibility to generate precise mutation(s) within the target gene (Araki and Ishii, 2015; Altpeter et al. 2016). However, the hinderance for genome editing is the slower development of genetic transformation (Altpeter et al. 2016). *Capsicum* is difficult to be manipulated *in vitro* because of its recalcitrant morphogenic nature, difficulty in shoot regeneration, and genotype dependence for tissue culture response, although there are several reports of successful transformation (Kothari et al. 2010). Since the first genetic transformation in pepper was performed in early 90s (Liu et al. 1990), there are few studies reported to improve transformation efficiency in pepper (Hasnat et al. 2008; Jung et al. 2011). Recently, ethylene response factor 28 was modified through genome editing in chile pepper to improve its resistance to anthracnose (Mishra et al. 2021), a promising application of gene editing.

Considering the limitations in the use of genetic transformation in pepper, virus-induced gene silencing (VIGS) is a fast, and powerful alternative tool to study gene function (Courdavault et al. 2020; Rössner et al. 2022). VIGS was developed from plant defense mechanisms against exotic nucleic acid, in which viral replicative form in the host plant is recognized by dicer-like ribonucleases and dsRNA-binding protein followed by being cleaved into 21–24 nucleotide fragments (siRNA). Subsequently, plant mRNA is targeted by siRNA based on sequence homology, and then loaded into RNA-induced silencing complex (RISC) for degradation (Senthil-Kumar and Mysore, 2014; Courdavault et al., 2020).

Tobacco rattle virus (TRV) is a bipartite positive sense RNA virus (Macfarlane, 1999), and was developed into an excellent VIGS vector for its milder symptoms, better infectivity, and silencing effect, compared to other viruses, including Tobacco Mosaic Virus (TMV) (Kumagai et al. 1995), Potato Virus X (PVX) (Ruiz et al. 1998) and Tomato Golden Mosaic Virus (TGMV) (Kjemtrup et al. 1998). TRV-based VIGS is already in use in many plant species including pepper since its first application in tobacco in 2001 (Liu et al. 2002; Ratcliff et al. 2001; Kim et al. 2017) The database, Sol Genomics Network (https://vigs.solgenomics.net), provides a useful platform for researchers to predict the potential silencing efficiency and possible off-target of the designed inserts (Fernandez-Pozo et al. 2015).

10.5.8 Synthetic Biology Applications

Knowledge of specific genes underlying phenotypes could help scientists to put together "novel" combinations of genes to create new phenotypes. Lack of easily usable gene transfer methods is an impediment for making progress using transgenic technology. Tissue-culture independent

TABLE 10.5
Candidate Genes for Specific Traits in *Capsicum* for which Multiple Lines of Evidence Support the Connection Between Gene Expression and the Phenotype. RNA-Seq – RNA sequencing; VIGS – virus-induced gene silencing

Trait	Candidate Gene	Lines of Evidence	References
Resistance to Bacterial Spot Disease	*Bs5*	Association between locus and phenotype; gene transfer to reverse complement *bs5*.	Szabó et al. (2023)
Pendent fruit orientation	*Up*	Association between locus and phenotype; gene silencing	Cao et al. (2022)
Root knot nematode resistance	*Me3*	Bulk segregant analysis–seq, Transgenic Arabidopsis expressing candidate gene correlating to expected phenotype	Liu et al. (2023)
Purple stripe on unripe fruit	CaP_5	Fine mapping, VIGs, transgenic expression	Li et al. (2023)
Nuclear restorer of fertility	*Rf*	Fine mapping, expression analysis	Wu et al. (2019)
Fruit elongation, ovate family protein	*fs10.1*	Fine mapping, VIGS, expression analysis	Borovsky et al. (2022); Borovsky and Paran (2011); Tsaballa et al. (2011)
Fruit locule number	*CaBRX*	Mapping, expression analysis and VIGS	Ma et al. (2022)
Pollen fertility and flavonoid synthesis *Rpf1* gene	*CabHLH1*	Bulk segregant analysis, KASP genotyping, gene expression analysis	Zhang et al. (2022)
Purple coloration on mature unripe fruit	*Ca3GT*	Fine mapping to 110.5 kb region with three open reading frames, co-segregation, and identification of mutations	Liu et al. (2020)
Dark purple coloration of immature fruit	*CaMYBA*	Gene expression, co-segregation with phenotype, cloning and sequence analyses	Ohno et al. (2020)
Dark purple coloration of immature fruit	*CaAN3*	BSA, RNA-Seq	Byun et al. (2022)
Chlorophyll and Carotenoid accumulation in fruit, Pseudo response regulator 2	*CaPRR2*	BSA, RNA-Seq	Lee et al. (2020)
Chlorophyll content of immature fruit	*CaGLK2*	Segregation, sequencing, and expression profiles.	Brand et al. (2014)
Fruit cuticle biosynthesis	*CaFcd1*	Mutant analysis, fFine mapping, eExpression profiling	Wang et al. (2023)

methods of gene transfer would be advantageous. While the use of marker-assisted selection will be important for pyramiding key traits in breeding lines, generation of allelic variation for synthetic biology applications can result in useful genome-edited improved varieties. Table 10.5 lists examples for *Capsicum* genes that have been well characterized for their functional impacts in explaining phenotypic diversity in the germplasm. These genes most likely will remain the targets for genome editing in near future.

ACKNOWLEDGEMENTS

We acknowledge funding support for Capsicum breeding research to B.R. from Florida Department of Agriculture & Consumer Services via the USDA Specialty Crops Block Grant program. We thank Mr. Tyler Jones (IFAS Photography, University of Florida) and Ms. Sonya Stahl for their help with the photo shown in Figure 10.1.

REFERENCES

Ahn, Y.K., Manivannan, A., Karna, S., Jun, T.H., Yang, E.Y., Choi, S., Kim, J.H., Kim, D.S., Lee, E.S., 2018. Whole genome resequencing of *Capsicum baccatum* and *Capsicum annuum* to discover single nucleotide polymorphism related to powdery mildew resistance. *Sci Rep* 8, 1–11. https://doi.org/10.1038/s41598-018-23279-5

Ahn, Y.K., Tripathi, S., Kim, J.H., Cho, Y. Il, Lee, H.E., Kim, D.S., Woo, J.G., Yoon, M.K., 2014. High-throughput microsatellite marker information from *Capsicum annuum* varieties mandarin and blackcluster next-generation sequence data. *Sci Hortic* 170, 123–130. https://doi.org/10.1016/j.scienta.2014.03.007

Alimi, N.A., Bink, M.C., Dieleman, J.A., Magán, J.J., Wubs, A.M., Palloix, A., van Eeuwijk, F.A., 2013a. Multi-trait and multi-environment QTL analyses of yield and a set of physiological traits in pepper. *Theor Appl Genet* 126, 2597–2625. https://doi.org/10.1007/s00122-013-2160-3

Alimi, N.A., Bink, M.C.A.M., Dieleman, J.A., Nicolaï, M., Wubs, M., Heuvelink, E., Magan, J., Voorrips, R.E., Jansen, J., Rodrigues, P.C., van der Heijden, G.W.A.M., Vercauteren, A., Vuylsteke, M., Song, Y., Glasbey, C., Barocsi, A., Lefebvre, V., Palloix, A., van Eeuwijk, F.A., 2013b. Genetic and QTL analyses of yield and a set of physiological traits in pepper. *Euphytica* 190, 181–201. https://doi.org/10.1007/s10681-012-0767-0

Altpeter, F., Springer, N.M., Bartley, L.E., Blechl, A.E., Brutnell, T.P., Citovsky, V., Conrad, L.J., Gelvin, S.B., Jackson, D.P., Kausch, A.P., Lemaux, P.G., Medford, J.I., Orozco-Cárdenas, M.L., Tricoli, D.M., Van Eck, J., Voytas, D.F., Walbot, V., Wang, K., Zhang, Z.J., Neal Stewart, C., 2016. Advancing crop transformation in the era of genome editing. *Plant Cell* 28, 1510–1520. https://doi.org/10.1105/tpc.16.00196

Araki, M., Ishii, T., 2015. Towards social acceptance of plant breeding by genome editing. *Trends Plant Sci.* https://doi.org/10.1016/j.tplants.2015.01.010

Arjun, K., Dhaliwal, M.S., Jindal, S.K., Fakrudin, B., 2018. Mapping of fruit length related qtls in interspecific cross (*Capsicum annuum* l. × Capsicum galapagoense hunz.) of chilli. *Breed Sci* 68, 219–226. https://doi.org/10.1270/jsbbs.17073

Arumuganathan, K., Earle, E.D., 1991. Genetic Resources Nuclear DNA Content of Some Important Plant Species. *Plant Mol Biol Rep* 9, 208–218.

Bai, J., Chen, B., Qi, M., Du, H., Wang, Q., Geng, S., Zhang, X., 2023. QTL mapping and candidate gene analysis of microspore embryogenesis in Capsicum. *Euphytica* 219, 1–10. https://doi.org/10.1007/s10681-022-03100-1

Barchenger, D.W., Rodriguez, K., Jiang, L., Hanson, S.F., Bosland, P.W., 2017. Allele-specific CAPS marker in a Ve1 homolog of *Capsicum annuum* for improved selection of *Verticillium dahliae* resistance. *Mol Breeding* 37: 134. https://doi.org/10.1007/s11032-017-0735-4

Barchi, L., Bonnet, J., Boudet, C., Signoret, P., Nagy, I., Lanteri, S., Palloix, A., Lefebvre, V., 2007. A high-resolution, intraspecific linkage map of pepper (*Capsicum annuum* L.) and selection of reduced recombinant inbred line subsets for fast mapping. *Genome* 50, 51–60. https://doi.org/10.1139/G06-140

Barchi, L., Lefebvre, V., Sage-Palloix, A.M., Lanteri, S., Palloix, A. 2009. QTL analysis of plant development and fruit traits in pepper and performance of selective phenotyping. *Theor Appl Genet.*118,1157–71. doi: 10.1007/s00122-009-0970-0.

Bartoszewski, G., Waszczak, C., Gawroński, P., Stępień, I., Bolibok-Brągoszewska, H., Palloix, A., Lefebvre, V., Korzeniewska, A., Niemirowicz-Szczytt, K., 2012. Mapping of the ms8 male sterility gene in sweet pepper (*Capsicum annuum L.*) on the chromosome P4 using PCR-based markers useful for breeding programmes. *Euphytica* 186, 453–461. https://doi.org/10.1007/s10681-012-0637-9

Borovsky, Y., Paran, I., 2011. Characterization of fs10.1, a major QTL controlling fruit elongation in Capsicum. *Theor Appl Genet* 123, 657–665. https://doi.org/10.1007/s00122-011-1615-7

Borovsky, Y., Raz, A., Doron-Faigenboim, A., Zemach, H., Karavani, E., Paran, I., 2022. Pepper fruit elongation is controlled by *Capsicum annuum* ovate family protein 20. *Front Plant Sci* 12: 815589. https://doi.org/10.3389/fpls.2021.815589

Brand, A., Borovsky, Y., Hill, T., Rahman, K.A.A., Bellalou, A., Van Deynze, A., Paran, I., 2014. CaGLK2 regulates natural variation of chlorophyll content and fruit color in pepper fruit. *Theor Appl Genet* 127, 2139–2148. https://doi.org/10.1007/s00122-014-2367-y

Byun, J., Kim, T.G., Lee, J.H., Li, N., Jung, S., Kang, B.C., 2022. Identification of CaAN3 as a fruit-specific regulator of anthocyanin biosynthesis in pepper (*Capsicum annuum*). *Theor Appl Genet* 135, 2197–2211. https://doi.org/10.1007/s00122-022-04106-y

Cao, Y., Zhang, K., Yu, H., Chen, S., Xu, D., Zhao, H., Zhang, Z., Yang, Y., Gu, X., Liu, X., Wang, H., Jing, Y., Mei, Y., Wang, X., Lefebvre, V., Zhang, W., Jin, Y., An, D., Wang, R., Bosland, P., Li, X., Paran, I., Zhang, B., Giuliano, G., Wang, L., Cheng, F., 2022. Pepper variome reveals the history and key loci associated with fruit domestication and diversification. *Mol Plant* 15, 1744–1758. https://doi.org/10.1016/j.molp.2022.09.021

Carrasco-Valenzuela, T., Muñoz-Espinoza, C., Riveros, A., Pedreschi, R., Arús, P., Campos-Vargas, R., Meneses, C. 2019. Expression QTL (eQTLs) Analyses Reveal Candidate Genes Associated With Fruit Flesh Softening Rate in Peach [Prunus persica (L.) Batsch]. *Front Plant Sci.* 10:1581. doi: 10.3389/fpls.2019.01581

Celik, I., Sogut, M.A., Ozkaynak, E., Doganlar, S., Frary, A., 2016. Physical mapping of NBS-coding resistance genes to the Me-gene cluster on chromosome P9 reveals markers tightly linked to the N gene for root-knot nematode resistance in pepper. *Mol Breed* 36, 1–7. https://doi.org/10.1007/s11032-016-0560-1

Chaim, A. Ben, Paran, I., Grube, R.C., Jahn, M., van Wijk, R., Peleman, J., 2001. QTL mapping of fruit-related traits in pepper (*Capsicum annuum*). *Theor Appl Gene* 102, 1016–1028. https://doi.org/10.1007/s00122 0000461

Chang, S., Lee, U., Kim, J.B., Jo, Y.D., 2022. Application of 3D-volumetric analysis and hyperspectral imaging systems for investigation of heterosis and cytoplasmic effects in pepper. *Sci Hortic* 302. https://doi.org/10.1016/j.scienta.2022.111150

Channabasava, Ganesh, P., Siddu, C.B., Susmitha, B., Rao, A.M., Ramesh, S., 2023. DNA marker-assisted identification of promising exotic advanced breeding lines as pollen fertility restorers and their validation based on field and laboratory assayable traits in chilli (*Capsicum annuum* L.). *Genet Resour Crop Evol* 70, 1957–1969. https://doi.org/10.1007/s10722-023-01547-2

Cheng, J., Qin, C., Tang, X., Zhou, H., Hu, Y., Zhao, Z., Cui, J., Li, B., Wu, Z., Yu, J., Hu, K., 2016. Development of a SNP array and its application to genetic mapping and diversity assessment in pepper (*Capsicum* spp.). *Sci Rep* 6: 1–11.

Chong L, Xu R, Huang P, Guo P, Zhu M, Du H, Sun X, Ku L, Zhu JK, Zhu Y. 2022. The tomato OST1-VOZ1 module regulates drought-mediated flowering. *Plant Cell.* 34, :2001–2018. doi: 10.1093/plcell/koac026

Courdavault, V., Besseau, S., Oudin, A., Papon, N., O'Connor, S.E., 2020. Virus-induced gene silencing: Hush genes to make them talk. *Trends Plant Sci* 25:714–715. https://doi.org/10.1016/j.tplants.2020.02.013

Dalal, A., Bourstein, R., Haish, N., Shenhar, I., Wallach, R., Moshelion, M., 2019. Dynamic physiological phenotyping of drought-stressed pepper plants treated with "productivity-enhancing" and "survivability-enhancing" biostimulants. *Front Plant Sci* 10: 905. https://doi.org/10.3389/fpls.2019.00905

Eggink, P.M., Tikunov, Y., Maliepaard, C., Haanstra, J.P., De Rooij, H., Vogelaar, A., Gutteling, E.W., Freymark, G., Bovy, A.G., Visser, R.G., 2014. Capturing flavors from *Capsicum baccatum* by introgression in sweet pepper. *Theor Appl Gene* 127, 373–390. https://doi.org/10.1007/s00122-013-2225-3

Escoto-Sandoval, C., Ochoa-Alejo, N., Martínez, O., 2021. Inheritance of gene expression throughout fruit development in chili pepper. *Sci Rep* 11: 22647. https://doi.org/10.1038/s41598-021-02151-z

Fernandez-Pozo, N., Rosli, H.G., Martin, G.B., Mueller, L.A., 2015. The SGN VIGS tool: User-friendly software to design virus-induced gene silencing (VIGS) constructs for functional genomics. *Mol Plant* 8, 486–488. https://doi.org/10.1016/j.molp.2014.11.024

Gilad, Y., Rifkin, S.A., Pritchard, J.K. 2008. Revealing the architecture of gene regulation: the promise of eQTL studies. *Trends Genet.* 2008 24:408–415. doi: 10.1016/j.tig.2008.06.001

González-Pérez, S., Garcés-Claver, A., Mallor, C., Sáenz De Miera, L.E., Fayos, O., Pomar, F., Merino, F., Silvar, C., 2014. New insights into Capsicum spp relatedness and the diversification process of *Capsicum annuum* in Spain. *PLoS One* 9: e116276. https://doi.org/10.1371/journal.pone.0116276

Han, K., Jeong, H.J., Yang, H.B., Kang, S.M., Kwon, J.K., Kim, S., Choi, D., Kang, B.C., 2016. An ultra-high-density bin map facilitates high-throughput QTL mapping of horticultural traits in pepper (*Capsicum annuum*). *DNA Res* 23, 81–91. https://doi.org/10.1093/dnares/dsv038

Hasnat, R., Abbasi, N.A., Hafiz, I.A., Ahmad, T., Chudhary, Z., 2008. Effect of different bacterial dilutions on transformation efficiency of hot chilli (*Capsicum frutescens L.*) varieties. *Pak J Bot* 40, 2655–2662.

Hill, T., Ashrafi, H., Chin-Wo, S.R., Stoffel, K., Truco, M.J., Kozik, A., Michelmore, R., Van Deynze, A., 2015. Ultra-high density, transcript-based genetic maps of pepper define recombination in the genome and synteny among related species. *G3: Genes Genomes Genet* 5, 2341–2355. https://doi.org/10.1534/g3.115.020040

Holdsworth, W.L., Mazourek, M., 2015. Development of user-friendly markers for the pvr1 and Bs3 disease resistance genes in pepper. *Mol Breed* 35: 28. https://doi.org/10.1007/s11032-015-0260-2

Hulse-Kemp, A.M., Ashrafi, H., Plieske, J., Lemm, J., Stoffel, K., Hill, T., Luerssen, H., Pethiyagoda, C.L., Lawley, C.T., Ganal, M.W., Van Deynze, A., 2016. A HapMap leads to a Capsicum annuum SNP infinium array: A new tool for pepper breeding. *Hortic Res* 3,16036. https://doi.org/10.1038/hortres.2016.36

International Board for Plant Genetic Resources., Asian Vegetable Research and Development Center., Centro Agronómico Tropical de Investigación y Enseñanza., 1995. Descriptors for Capsicum (Capsicum spp.) = Descriptores para Capsicum (Capsicum spp.). International Plant Genetic Resources Institute.

Jarret, R.L., Barboza, G.E., Batista, F.R. da C., Berke, T., Chou, Y.Y., Hulse-Kemp, A., Ochoa-Alejo, N., Tripodi, P., Veres, A., Garcia, C.C., Csillery, G., Huang, Y.K., Kiss, E., Kovacs, Z., Kondrak, M., Arce-Rodriguez, M.L., Scaldaferro, M.A., Szoke, A., 2019. Capsicum-an abbreviated compendium. *J Am Soc Hortic Sci* 144, 3–22. https://doi.org/10.21273/JASHS04446-18

Jung, J. kee, Park, S.W., Liu, W.Y., Kang, B.C., 2010. Discovery of single nucleotide polymorphism in Capsicum and SNP markers for cultivar identification. *Euphytica* 175, 91–107. https://doi.org/10.1007/s10681-010-0191-2

Jung, M., Shin, S.H., Park, J.M., Lee, S.N., Lee, M.Y., Ryu, K.H., Paek, K.Y., Harn, C.H., 2011. Detection of transgene in early developmental stage by GFP monitoring enhances the efficiency of genetic transformation of pepper. *Plant Biotechnol Rep* 5, 157–167. https://doi.org/10.1007/s11816-011-0168-1

Kaur, B., Mavi, G.S., Gill, M.S., Saini, D.K., 2020. Utilization of KASP technology for wheat improvement. *Cereal Res Commun* 48: 409–421. https://doi.org/10.1007/s42976-020-00057-6

Khokhar, E.S., Lozada, D.N., Nankar, A.N., Hernandez, S., Coon, D., Kaur, N., Nourkbakshs, S.S., 2022. High-throughput characterization of fruit phenotypic diversity among New Mexican Chile Pepper (*Capsicum spp.*) using the Tomato Analyzer Software. *HortScience* 57, 1507–1517. https://doi.org/10.21273/HORTSCI16815-22

Kim, J., Park, M., Jeong, E.S., Lee, J.M., Choi, D., 2017. Harnessing anthocyanin-rich fruit: A visible reporter for tracing virus-induced gene silencing in pepper fruit. *Plant Methods* 13: 3. https://doi.org/10.1186/s13007-016-0151-5

Kim, S., Park, M., Yeom, S.I., Kim, Y.M., Lee, J.M., Lee, H.A., Seo, E., Choi, J., Cheong, K., Kim, K.T., Jung, K., Lee, G.W., Oh, S.K., Bae, C., Kim, S.B., Lee, H.Y., et al., 2014. Genome sequence of the hot pepper provides insights into the evolution of pungency in Capsicum species. *Nat Genet* 46, 270–278. https://doi.org/10.1038/ng.2877

Kjemtrup, S., Sampson, K.S., Peele, C.G., Nguyen, L. V., Conkling, M.A., Thompson, W.F., Robertson, D., 1998. Gene silencing from plant DNA carried by a Geminivirus. *Plant J* 14, 91–100. https://doi.org/10.1046/j.1365-313X.1998.00101.x

Kothari, S.L., Joshi, A., Kachhwaha, S., Ochoa-Alejo, N., 2010. Chilli peppers–A review on tissue culture and transgenesis. *Biotechnol Adv*. 28, 35–48. https://doi.org/10.1016/j.biotechadv.2009.08.005

Kumagai, M.H., Donson, J., Della-Cioppa, G., Harvey, D., Hanley, K., Grill, L.K., 1995. Cytoplasmic inhibition of carotenoid biosynthesis with virus-derived RNA. *Proc Natl Acad Sci USA* 92, 1679–1683. https://doi.org/10.1073/pnas.92.5.1679

Kwon, Y.S., Hong, J.H., Choi, K.J., 2013. Construction of a microsatellite marker database of commercial pepper cultivars. *Kor J Hortic Sci Technol* 31, 580–589. https://doi.org/10.7235/hort.2013.13026

Lee, H.R., An, H.J., Yang, D.C., Choi, S.H., Kim, H.J., Rhee, H.G., Harn, C.H., 2012. Development of a high resolution melting (HRM) marker linked to genic male sterility in *Capsicum annuum L. Plant Breed* 131, 444–448. https://doi.org/10.1111/j.1439-0523.2012.01956.x

Lee, H.R., Bae, I.H., Park, S.W., Kim, H.J., Min, W.K., Han, J.H., Kim, K.T., Kim, B.D., 2009. Construction of an integrated pepper map using RFLP, SSR, CAPS, AFLP, WRKY, rRAMP, and BAC end sequences. *Mol Cells* 27, 21–37. https://doi.org/10.1007/s10059-009-0002-6

Lee, H.R., Kim, K.T., Kim, Hyoun Joung, Han, J.H., Kim, J.H., Yeom, S.I., Kim, Hyun Jung, Kang, W.H., Jinxia-Shi, Park, S.W., Bae, I.H., Lee, S., Cho, J., Oh, D., Kim, B.D., 2011. QTL analysis of fruit length using rRAMP, WRKY, and AFLP markers in chili pepper. *Hortic Environ Biotechnol* 52, 602–613. https://doi.org/10.1007/s13580-011-0002-2

Lee, J., Yoon, J.B., Park, H.G., 2008. A CAPS marker associated with the partial restoration of cytoplasmic male sterility in chili pepper (*Capsicum annuum L.*). *Mol Breed* 21, 95–104. https://doi.org/10.1007/s11032-007-9111-0

Lee, S. Bin, Kim, J.E., Kim, H.T., Lee, G.M., Kim, B.S., Lee, J.M., 2020. Genetic mapping of the c1 locus by GBS-based BSA-seq revealed Pseudo-Response Regulator 2 as a candidate gene controlling pepper fruit color. *Theor Appl Genet* 133, 1897–1910. https://doi.org/10.1007/s00122-020-03565-5

Li, N., Liu, Y., Yin, Y., Gao, S., Wu, F., Yu, C., Wang, F., Kang, B., Xu, K., Jiao, C., Yao, M., 2023. Identification of CaPs locus involving in purple stripe formation on unripe fruit, reveals allelic variation and alternative splicing of R2R3-MYB transcription factor in pepper (*Capsicum annuum L.*). *Front Plant Sci* 14, 1–18. https://doi.org/10.3389/fpls.2023.1140851

Liu, J., Ai, X., Wang, Y., Lu, Q., Li, T., Wu, L., Sun, L., Shen, H., 2020. Fine mapping of the Ca3GT gene controlling anthocyanin biosynthesis in mature unripe fruit of *Capsicum annuum* L. *Theor Appl Genet* 133: 2729–2742. https://doi.org/10.1007/s00122-020-03628-7

Liu, W., Parrott, W.A., Hildebrand, D.E., Collins, G.B., Williams, E.G., Phillips, G.C., 1990. *Agrobacterium* induced gall formation in bell pepper (*Capsicum annuum* L.) and formation of shoot-like structures expressing introduced genes. *Plant Cell Reps* 9, 360–364.

Liu, Y., Cao, H., Ling, J., Yang, Y., Li, Y., Xie, B., Zhao, J., Mao, Z., 2023. Molecular cloning and functional analysis of the pepper resistance gene Me3 to root-knot nematode. *Hortic Plant J* 9, 133–144. https://doi.org/10.1016/j.hpj.2022.12.003

Liu, Y., Schiff, M., Marathe, R., Dinesh-Kumar, S.P., 2002. Tobacco Rar1, EDS1 and NPR1/NIM1 like genes are required for N-mediated resistance to tobacco mosaic virus. *Plant J* 30, 415–429. https://doi.org/10.1046/j.1365-313X.2002.01297.x

Livingstone, K.D., Lackney, V.K., Blauth, J.R., Van Wijk, R., Jahn, M.K., 1999. Genome mapping in capsicum and the evolution of genome structure in the Solanaceae. *Genetics* 152, 1183–1202. https://doi.org/10.1093/genetics/152.3.1183

Lozada, D.N., Bhatta, M., Coon, D., Bosland, P.W., 2021a. Single nucleotide polymorphisms reveal genetic diversity in New Mexican chile peppers (*Capsicum spp.*). *BMC Genomics* 22: 356. https://doi.org/10.1186/s12864-021-07662-7

Lozada, D.N., Nunez, G., Lujan, P., Dura, S., Coon, D., Barchenger, D.W., Sanogo, S., Bosland, P.W., 2021b. Genomic regions and candidate genes linked with *Phytophthora capsici* root rot resistance in chile pepper (*Capsicum annuum L.*). *BMC Plant Biol* 21, 1–14. https://doi.org/10.1186/s12870-021-03387-7

Lozada, D.N., Whelpley, M., Acuña-Galindo, A., 2021c. Genetic architecture of chile pepper (*Capsicum spp.*) qtlome revealed using meta-qtl analysis. *Horticulturae* 7: 227. https://doi.org/10.3390/horticulturae7080227

Lu, F.H., Kwon, S.W., Yoon, M.Y., Kim, K.T., Cho, M.C., Yoon, M.K., Park, Y.J., 2012a. SNP marker integration and QTL analysis of 12 agronomic and morphological traits in F8 RILs of pepper (*Capsicum annuum L.*). *Mol Cells* 34, 25–34. https://doi.org/10.1007/s10059-012-0018-1

Ma, X., Qiao, Y.M., Li, Y., Yu, Y.N., Gong, Z.H., 2022. Identification of Fruit Traits Related QTLs and a Candidate Gene, CaBRX, Controlling Locule Number in Pepper (*Capsicum annuum L.*). *Horticulturae* 8: 146. https://doi.org/10.3390/horticulturae8020146

Macfarlane, S.A., 1999. Molecular biology of the tobraviruses. *J Gen Virol* 80, 2799–2807.

Martínez, O., Arce-Rodríguez, M.L., Hernández-Godínez, F., Escoto-Sandoval, C., Cervantes-Hernández, F., Hayano-Kanashiro, C., Ordaz-Ortiz, J.J., Humberto Reyes-Valdés, M., Razo-Mendivil, F.G., Garcés-Claver, A., Ochoa-Alejo, N., 2021. Transcriptome analyses throughout chili pepper fruit development reveal novel insights into the domestication process. *Plants* 10, 1–20. https://doi.org/10.3390/plants10030585

Mimura, Y., Inoue, T., Minamiyama, Y., Kubo, N., 2012. An SSR-based genetic map of pepper (Capsicum annuum L.) serves as an anchor for the alignment of major pepper maps. *Breed Sci* 62, 93–98. https://doi.org/10.1270/jsbbs.62.93

Mimura, Y., Minamiyama, Y., Sano, H., Hirai, M., 2010. Mapping for axillary shooting, flowering date, primary axis length, and number of leaves in Pepper (*Capsicum annuum*). *J Jpn Soc Hortic Sci* 79, 56–63. https://doi.org/10.2503/jjshs1.79.56

Mishra, R., Mohanty, J.N., Mahanty, B., Joshi, R.K., 2021. A single transcript CRISPR/Cas9 mediated mutagenesis of CaERF28 confers anthracnose resistance in chilli pepper (*Capsicum annuum L.*). *Planta* 254: 5. https://doi.org/10.1007/s00425-021-03660-x

Moon, T., Choi, H., Kim, D., Hwang, I., Kim, J., Shin, J., Son, J.E., 2022. Autonomous construction of parameterizable 3D leaf models from scanned sweet pepper leaves with deep generative networks. *Silico Plants* 4: 1–14. https://doi.org/10.1093/insilicoplants/diac015

Moscone, E.A., Baranyi, M., Ebert, I., Greilhuber, J., Ehrendorfer, F., Hunziker, A.T., 2003. Analysis of nuclear DNA content in Capsicum (Solanaceae) by flow cytometry and feulgen densitometry. *Ann Bot* 92, 21–29. https://doi.org/10.1093/aob/mcg105

Nankar, A.N., Tringovska, I., Grozeva, S., Todorova, V., Kostova, D., 2020. Application of high-throughput phenotyping tool Tomato Analyzer to characterize Balkan Capsicum fruit diversity. *Sci Hortic* 260: 108862. https://doi.org/10.1016/j.scienta.2019.108862

Nimmakayala, P., Abburi, V.L., Saminathan, T., Alaparthi, S.B., Almeida, A., Davenport, B., Nadimi, M., Davidson, J., Tonapi, K., Yadav, L., Malkaram, S., Vajja, G., Hankins, G., Harris, R., Park, M., Choi, D., Stommel, J., Reddy, U.K., 2016. Genome-wide diversity and association mapping for capsaicinoids and fruit weight in *Capsicum annuum L. Sci Rep* 6: 38081. https://doi.org/10.1038/srep38081

Ohno, S., Ueno, M., Doi, M., 2020. Differences in the CaMYBA genome between anthocyanin-pigmented cultivars and non-pigmented cultivars in pepper (*Capsicum annuum*). *Hortic J* 89: 30–36. https://doi.org/10.2503/hortj.UTD-097

Palloix, A., Barchi, L., Lefebvre, V., Sage-Palloix, A.M., Lanteri, S., 2009. QTL analysis of plant development and fruit traits in pepper and performance of selective phenotyping. *Theor Appl Genet* 118, 1157–1171. https://doi.org/10.1007/s00122-009-0970-0

Pereira-Dias, L., Fita, A., Vilanova, S., Sánchez-López, E., Rodríguez-Burruezo, A., 2020. Phenomics of elite heirlooms of peppers (*Capsicum annuum L.*) from the Spanish centre of diversity: Conventional and high-throughput digital tools towards varietal typification. *Sci Hortic* 265: 109245. https://doi.org/10.1016/j.scienta.2020.109245

Qin, C., Yu, C., Shen, Y., Fang, X., Chen, L., Min, J., Cheng, J., Zhao, S., Xu, M., Luo, Y., Yang, Y., Wu, Z., Mao, L., Wu, H., Changying, L.H., Zhou, H., et al., 2014. Whole-genome sequencing of cultivated and wild peppers provides insights into Capsicum domestication and specialization. *Proc Natl Acad Sci U S A* 111, 5135–5140. https://doi.org/10.1073/pnas.1400975111

Ratcliff, F., Martin-Hernandez, A.M., Baulcombe, D.C., 2001. Tobacco rattle virus as a vector for analysis of gene function by silencing. *Plant J* 25, 237–245. https://doi.org/10.1046/j.0960-7412.2000.00942.x

Rathinasabapathi, B (2020) Improving vegetable Capsicums for fruit yield, quality, and tolerance to biotic and abiotic stresses. In: *Genomic Designing of Climate-Smart Vegetable Crops*, Edited by Kole, C., Springer, pp. 277–308.

Reddy, U.K., Almeida, A., Abburi, V.L., Alaparthi, S.B., Unselt, D., Hankins, G., Park, M., Choi, D., Nimmakayala, P., 2014. Identification of gene-specific polymorphisms and association with capsaicin pathway metabolites in *Capsicum annuum L.* collections. *PLoS One* 9: e86393. https://doi.org/10.1371/journal.pone.0086393

Rössner, C., Lotz, D., Becker, A., 2022. VIGS goes viral: How VIGS transforms our understanding of Plant Science. *Annu Rev Plant Biol* 73, 703–728. https://doi.org/10.1146/annurev-arplant-102820

Ruiz, M.T., Voinnet, O., Baulcombe, D.C., 1998. Initiation and maintenance of Virus-Induced gene silencing. *Plant Cell* 10, 937–946.

Sato, S., Tabata, S., Hirakawa, H., Asamizu, E., Shirasawa, K., Isobe, S., Kaneko, T., Nakamura, Y., Shibata, D., Aoki, K., Egholm, M., Knight, J., Bogden, R., Li, Changbao, Shuang, Y., Xu, X., et al., 2012. The tomato genome sequence provides insights into fleshy fruit evolution. *Nature* 485, 635–641. https://doi.org/10.1038/nature11119

Schenk, J.J., Becklund, L.E., Carey, S.J., Fabre, P.P., 2023. What is the "modified" CTAB protocol? Characterizing modifications to the CTAB DNA extraction protocol. *Appl Plant Sci* 11: e11517. https://doi.org/10.1002/aps3.11517

Senthil-Kumar, M., Mysore, K.S., 2014. Tobacco rattle virus-based virus-induced gene silencing in *Nicotiana benthamiana. Nat Protoc* 9, 1549–1562. https://doi.org/10.1038/nprot.2014.092

Sharma, A., Li, J., Wente, R., Minsavage, G. V., Gill, U.S., Ortega, A., Vallejos, C.E., Hart, J.P., Staskawicz, B.J., Mazourek, M.R., Stall, R.E., Jones, J.B., Hutton, S.F., 2023. Mapping of the *bs5* and *bs6* non-race-specific recessive resistances against bacterial spot of pepper. *Front Plant Sci* 14: 1061803. https://doi.org/10.3389/fpls.2023.1061803

Siddique, M.I., Lee, J.-H., Ahn, J.-H., Kusumawardhani, M.K., Safitri, R., Harpenas, A., Kwon, J.-K., Kang, B.-C., 2022. Genotyping-by-sequencing-based QTL mapping reveals novel loci for Pepper yellow leaf curl virus (PepYLCV) resistance in *Capsicum annuum. PLoS One* 17: e0264026. https://doi.org/10.1371/journal.pone.0264026

Solomon, A.M., Han, K., Lee, J.H., Lee, H.Y., Jang, S., Kang, B.C., 2019. Genetic diversity and population structure of Ethiopian Capsicum germplasms. *PLoS One* 14, 1–20.

Song, C., Lee, J., Kim, T. et al. 2018. VOZ1, a transcriptional repressor of DREB2C, mediates heat stress responses in Arabidopsis. *Planta* 247, 1439–1448. https://doi.org/10.1007/s00425-018-2879-9

Sran, T.S., Jindal, S.K., Sharma, A., Bhatia, D., 2023. LCVD resistance breeding in heat tolerant bell pepper: Combined phenotypic and marker-assisted backcrossing for introgression of LCVD resistance from hot pepper. *Euphytica* 219, 50. https://doi.org/10.1007/s10681-023-03182-5

Szabó, Z., Balogh, M., Domonkos, Á., Csányi, M., Kaló, P., Kiss, G.B., 2023. The *bs5* allele of the susceptibility gene Bs5 of pepper (Capsicum annuum L.) encoding a natural deletion variant of a CYSTM protein conditions resistance to bacterial spot disease caused by *Xanthomonas* species. *Theor Appl Genet* 136, 64. https://doi.org/10.1007/s00122-023-04340-y

Tamisier, L., Szadkowski, M., Nemouchi, G., Lefebvre, V., Szadkowski, E., Duboscq, R., Santoni, S., Sarah, G., Sauvage, C., Palloix, A., Moury, B., 2020. Genome-wide association mapping of QTLs implied in potato virus Y population sizes in pepper: evidence for widespread resistance QTL pyramiding. *Mol Plant Pathol* 21, 3–16. https://doi.org/10.1111/mpp.12874

Tripodi, P., Greco, B., 2018. Large scale phenotyping provides insight into the diversity of vegetative and reproductive organs in a wide collection of wild and domesticated peppers (*Capsicum spp.*). *Plants* 7: 103. https://doi.org/10.3390/plants7040103

Tripodi, P., Schiavi, M., Lo Scalzo, R., 2021. Multi-scale evaluation on two locations and digital fruit imaging highlight morpho-agronomic performances and antioxidant properties in chilli pepper hybrids. *Agronomy* 11, 805. https://doi.org/10.3390/AGRONOMY11040805

Tsaballa, A., Pasentsis, K., Darzentas, N., Tsaftaris, A.S., 2011. Multiple evidence for the role of an Ovate-like gene in determining fruit shape in pepper. *BMC Plant Biol* 11: 46.

Wang, D., Bosland, P.W., 2006. The genes of Capsicum. *HortScience* 41:1169–1187.

Wang, J., Shan, Q., Yi, T., Ma, Y., Zhou, X., Pan, L., Miao, W., Zou, X., Xiong, C., Liu, F., 2023. Fine mapping and candidate gene analysis of CaFCD1 affecting cuticle biosynthesis in *Capsicum annuum* L. *Theor Appl Genet* 136, 46. https://doi.org/10.1007/s00122-023-04330-0

Wang, X., Fazari, A., Cao, Y., Zhang, Z., Palloix, A., Mao, S., Zhang, B., Djian-Caporalino, C., Wang, L., 2018. Fine mapping of the root-knot nematode resistance gene Me1 in pepper (Capsicum annuum L.) and development of markers tightly linked to Me1. *Mol Breed* 38, 39. https://doi.org/10.1007/s11032-018-0793-2

Wu, L., Wang, P., Wang, Y., Cheng, Q., Lu, Q., Liu, J., Li, T., Ai, Y., Yang, W., Sun, L., Shen, H., 2019. Genome-wide correlation of 36 agronomic traits in the 287 pepper (Capsicum) accessions obtained from the SLAF-seq-based GWAS. *Int J Mol Sci* 20: 5675. https://doi.org/10.3390/ijms20225675

Xu, Y., Crouch, J.H. 2008. Marker-assisted selection in Plant Breeding: From Publications to Practice. *Crop Sci.* 48:391–407. Doi:10.2135/cropsci2007.04.0191.

Zhang, X.F., Wang, G.Y., Dong, T.T., Chen, B., Du, H.S., Li, C.B., Zhang, F.L., Zhang, H.Y., Xu, Y., Wang, Q., Geng, S.S., 2019. High-density genetic map construction and QTL mapping of first flower node in pepper (*Capsicum annuum L.*). *BMC Plant Biol* 19, 1–13. https://doi.org/10.1186/s12870-019-1753-7

Zhang, Z., Cao, Y., Wang, Y., Yu, H., Wu, H., Liu, J., An, D., Zhu, Y., Feng, X., Zhang, B., Wang, L., 2023. Development and validation of KASP markers for resistance to *Phytophthora capsici* in *Capsicum annuum L. Mol Breed* 43, 1–11. https://doi.org/10.1007/s11032-023-01367-3

Zhang, Z., Liu, Y., Yuan, Q., Xiong, C., Xu, H., Hu, B., Suo, H., Yang, S., Hou, X., Yuan, F., Pei, Z., Dai, X., Zou, X., Liu, F., 2022. The bHLH1-DTX35/DFR module regulates pollen fertility by promoting flavonoid biosynthesis in Capsicum annuum L. *Hortic Res* 9, 172. https://doi.org/10.1093/hr/uhac172

11 Allele Mining in *Cucurbita*
A Mini-Review

Swati Shrestha,[1,2] *Prerna Sabharwal,*[1] *Shailesh Acharya,*[1,3] *and Geoffrey Meru*[1]
[1]Tropical Research and Education Center, University of Florida, Homestead, FL, USA
[2]Current affiliation: Department of Plant and Soil Sciences, Oklahoma State University, Stillwater, OK, USA
[3]Current affiliation: Department of Plant Sciences, North Dakota State University, Fargo, ND, USA

11.1 INTRODUCTION

Cucurbita is a diverse dicotyledonous genus consisting of 27 species including five cultivated species, namely, *C. moschata, C. pepo, C. maxima*, C. *argyrosperma,* and *C. ficifolia* (Whitaker 1974). Of the five cultivated species, *C. moschata, C. pepo,* and *C. maxima* are the most popular worldwide in terms of production and consumption (Whitaker and Bemis 1975). *Cucurbita* crops are diploid with 20 chromosomes ($2n = 2x = 40$). *Cucurbita* species, which include squash, pumpkin, and gourd, are indigenous to Americas, and are currently cultivated throughout the temperate, subtropical, and tropical regions of the world. Squash and pumpkins come in a variety of shapes, sizes, and colors, and the term "squash" can be collectively used for all the *Cucurbita* species. Cucurbitaceous fruits are valued for long shelf-life, yield, and nutrients. The flesh of *Cucurbita* fruit is low in saturated fats, cholesterol, and sodium, but has high levels of vitamins A, C, E, and minerals like potassium, phosphorous, and magnesium thus making them popular among consumers (Pratt and Matthews 2006; Blessing et al. 2011). On the other hand, *Cucurbita* seeds are a rich source of oil (50% w/w), protein (35%), and unsaturated fatty acids (>78%) (Meru et al. 2018). Consequently, *Cucurbita* crops are important for human nutrition, particularly in countries where their production and consumption are high.

There are various culinary ways of utilizing squashes, including use in pumpkin pie, desserts, soups, and puddings. Flowers, young leaves, and shoots of squash are edible and used as vegetables in Asian countries (Holmer et al. 2013). Marinated and roasted pumpkin seeds are popular snacks in many countries, such as the United States and Mexico (Syed et al. 2019). A striking feature of *Cucurbita* crops is their high adaptation to a wide range of agricultural conditions. Their production ranges from extensive monoculture in commercial farms to growing in restricted space/pots with low inputs in urban agriculture (Ferriol and Pico 2008).

Cucurbita are one of the most variable genera in the plant kingdom and have been studied extensively for both quantitative and qualitative traits. Because of high economic and nutritional value of squash, many breeding programs focusing on squash improvement have been established worldwide. Advances in next-generation sequencing has ushered a genomics era for cucurbit crops including the *Cucurbita* genus. The development of genomic tools such as reference genomes, transcriptome assemblies, genetic linkage maps, QTL mapping, marker-assisted selection, genome-wide association studies, gene expression, genomic selection, and gene editing have facilitated the discovery and characterization of genes underlying key biological processes as well as those

DOI: 10.1201/9781003376439-11

governing economically important traits in *Cucurbita* crops. Herein, a review of historical and current advances in allele mining within *Cucurbita* is provided, with an emphasis on disease resistance traits, morphological traits and fruit quality traits. Collectively, these discoveries provide a platform for accelerated breeding processes as well as important insights into the functional mechanisms underlying economically important traits in *Cucurbita*.

11.2 LINKAGE MAPPING IN *CUCURBITA*: A HISTORICAL PERSPECTIVE

Despite their economic importance, genomic studies in *Cucurbita* spp. have lagged behind those of other Cucurbits, such as cucumber, melon and watermelon (Gonzalo and Monforte 2016). The first genetic map of *Cucurbita* was published by Weeden and Robinson (1986). It was based on isozyme markers within an interspecific population between *C. maxima* and *C. ecuadorensis*. The study identified 20 segregating loci that were assembled into 14 independently assorting groups and five linkage groups. Hi et al. (1995) developed a molecular map for an F_2 population derived by crossing *C. pepo* X *C. moschata* using random amplified polymorphic DNA (RAPD); however, the map had low coverage with only 28 markers across five linkage groups. Following these primary studies, Brown and Myers (2002) developed a more detailed genetic map based on molecular and morphological markers in *Cucurbita*. In this study, they employed a BC_1 population developed by crossing *C. pepo* (yellow Straightneck) X *C. moschata* (Nigerian Local), with the former as the recurrent parent. The map consisted of 148 RAPD markers across 28 linkage groups with an estimated genome coverage of 75%. Morphological traits mapped included *B* gene for flesh color, *M* gene for silver leaf mottling, locus controlling intensity of fruit rind color, fruit shape, and depth of indentations between primary leaf veins. This map was a key milestone towards the development of high-density genetic maps in *Cucurbita*.

Genetic maps using a combination of RAPD, amplified fragment length polymorphism (AFLP) and simple sequence repeat (SSR) markers were first constructed in *C. pepo* by Zraidi et al. (2007). In this study, mapping was conducted in two populations, one developed by crossing Oil Seed pumpkin by Zucchini, and the other by crossing Oil Seed pumpkin by Crookneck. A total of 333 markers (247 RAPD, 82 AFLP, 3 SSR and the *n locus–seed coat*) were integrated in the Pumpkin-Zucchini map, which was distributed across 21 major and three minor linkage groups. On the other hand, the Pumpkin-Crookneck map consisted of 322 markers (196 RAPD, 125 AFLP, and the *n locus*) and covered 21 major and three minor linkage groups. The first SSR-based genetic linkage map for *C. moschata* was created by Gong et al. (2008a) using two populations developed from crosses between Waltham Butternut and Nigerian Local, and ZHOU (hull-less) and Waltham Butternut. The integrated map consisted of 205 SSR markers and two morphological traits [hull-less (*n*) and rind color (*Gr*)] distributed across 27 linkage groups with a marker density of 7 cM. Later, Gong et al. (2008b) developed a detailed microsatellite-based map for *Cucurbita* spp. In this study, SSR-based genomic libraries were created and 2,400 clones were sequenced to generate 1,058 SSR markers with at least four repeats long. Of these, 405 SSR primers amplified polymorphic fragments in three *C. moschata* (Nigerian Local, Waltham Butternut and ZHOU), one *C. ecuadorensis* (Grif 9446 USDA) and eight *C. pepo* (True French, Striato di Italia, General Patton, Bianco Friulano, Lady Godiva, Early White Bush, Tay Belle, and Alba) genotypes. Gong et al. (2008b) also updated the *C. pepo* genetic map developed by Zraidi et al. (2007) in a population derived from a cross between Lady Godiva (*C. pepo* sub sp. *pepo)* and Italian Crookneck (*C. pepo* subsp. *ovifera*). The new map consisted of 657 markers (178 SSR, 244 AFLP, 230 RAPD and five sequence characterized amplified region–SCAR markers) distributed across 20 linkage groups with an average distance of 2.9 cM between markers.

Although linkage mapping in *Cucurbita* spp. was recorded as early as 1986, it was only recently that the first intraspecific linkage map for *C. maxima* was developed (Ge et al. 2015). In the study,

57 SSR, 21 AFLP, three RAPD, and one morphological (rind color) marker were used to construct a genetic map consisting of 20 linkage groups with an average distance of 12.1 cM between adjacent markers. A summary of genetic maps developed in *Cucurbita* using AFLP, RAPD and SSR markers has been provided in Table 11.1.

In the last decade, advances in next-generation sequencing technologies, Roche/454, Illumina/Solexa, Life/APG, and HelicosBioSciences, allowed for massive large-scale discovery of genetic variants, even in non-model genera like *Cucurbita* (Esteras et al. 2012). Single nucleotide polymorphisms (SNPs) are the most abundant variations in genomes and a powerful tool for genetic mapping and marker-assisted selection. In *Cucurbita* species, SNP markers in combination with SSR markers have predominantly been used in genetic analysis since 2011 (Esteras et al. 2012). Initial work exploring allelic variation using whole genome sequencing technology was performed by Blanca et al. (2011). In the study, a total of 512,751 *C. pepo* expressed sequence tag (EST) sequences were generated using Roche/454 GS FLX Titanium massive parallel pyrosequencing technology. ESTs were obtained from cDNA libraries of roots, leaves, and flower tissue prepared from two *C. pepo* varieties; MU16 (*Cucurbita pepo* ssp. *pepo* var. Zucchini) and UPV196 (*C. pepo* L. *ovifera* var. Scallop). Unigenes were assembled from the ESTs with an average of 9.3 ESTs per unigene. A scan of unigenes for microsatellites yielded 1,936 SSRs. ESTs were also explored for SNPs, and a total of 3,538 SNPs and 1,174 indels were identified in the study (Blanca et al. 2011). Genomic analysis by Blanca et al. (2011) provided invaluable tools for future molecular breeding

TABLE 11.1
Summary of Genetic Maps Developed in *Cucurbita* Using AFLP, RAPD, and SSR Markers

Cross	Population Type	Number and Type of Markers	Number of Linkage Groups	Morphological Traits Mapped	Reference
C. maxima X *C. ecuadorensis*	F_2	Isozyme (12)	5 linkage groups	None	Weeden and Robinson 1986
C. pepo X *C. moschata*	F_2	RAPD (28)	5 linkage groups	None	Hi et al. 1995
C. pepo X *C. moschata*	BC_1	RAPD (148)	28 linkage groups	Fruit flesh color, rind color, fruit shape, leaf indentations, silver leaf mottling	Brown and Myers 2002
Oil Seed pumpkin X Zucchini	F_2	RAPD (247) AFLP (82) SSR (3)	21 major and 3 minor linkage groups	Hull less seed trait (*n*)	Zraidi et al. 2007
Oil Seed pumpkin X Crookneck (*C. pepo*)	F_2	RAPD (196) AFLP (125)	21 major and 3 minor linkage groups	Hull less seed trait (*n*) Growth habit (*Bu*)	Zraidi et al. 2007
Waltham Butternut (WB) X Nigerian Local and ZHOU X WB (*C. moschata*)	F_2	SSR (205)	27 linkage groups	Hull less seed trait (*n*) Rind color of mature fruit (*Gr*)	Gong et al. 2008a
Lady Godiva (*C. pepo* subsp. *pepo*) X Italian Crookneck (*C. pepo* sub sp. *ovifera*)	F_2	SSR (178) AFLP (244) RAPD (230) SCAR (5)	20 linkage groups	Hull less seed trait (*n*) Bush growth habit (*B*)	Gong et al. 2008b
'98- 2-351' X'06820-1' (*C. maxima*)	F_2	SSR (57) AFLP (21) RAPD (3)	20 linkage groups	Rind color (*Rc*)	Ge et al. 2015

and marker-assisted selection in *Cucurbita* spp. Subsequently, a SNP-based genetic map of *C. pepo* was developed by Esteras et al. (2012) from a population derived from a cross between MU-CU-16 and UPV-196. The population was genotyped using 384 SNPs identified by Blanca et al. (2011) and 25 SSR identified by Gong et al. (2008a). The genetic map covered 1,740.8 cM and was composed of 22 major linkage groups and a minor group (LG23) with only two markers. The map was also used to successfully map and validate QTLs associated with plant flowering and fruit traits that will be discussed later in the chapter. Later, Montero-Pau et al. (2017) published a more detailed linkage map for *C. pepo* from a population generated from a Zucchini X Scallop (ssp. *pepo* × ssp. *ovifera*) cross. This map consisted of 7,718 SNPs (average of 386 markers/linkage group) distributed across 21 linkage groups covering 2,817.6 cM.

For *C. maxima*, Zhang et al. (2015) developed a high-density genetic map using next-generation genotyping by sequencing approach. In this study, an F_2 mapping population was developed by crossing inbred lines Rimu and SQ026, and a high-density genetic map containing 458 bin-markers was constructed with 20 linkage groups spanning a total genetic distance of 2,566.8 cM and a mean marker density of 5.60 cM. This map was used for QTL analysis for several morphological traits in *C. maxima*. In *C. moschata*, Zhong et al. (2017) developed a detailed high-density genetic map using SNP markers developed using the ddRAD sequencing technology. The map consisted of 20 linkage groups and 3,470 SNPs in an F_2 population developed by crossing *C. moschata* inbred lines, CMO-1 and CMO-97. Xu et al. (2017) used 95 SSR markers to develop a linkage map and identify three QTLs associated with cold tolerance in *C. moschata* in a chilling-stress-tolerant inbred line ('8-3-7'). This linkage map consisted of 15 linkage groups with a total length of 830.7 cM.

Overall, the advent of next-generation sequencing has allowed for the rapid development of high-density genetic maps in *Cucurbita*. These maps have facilitated QTL analysis, genetic dissection of complex traits and identification of candidate genes governing traits of interest. In the following sections, allele mining for economically important traits using QTL and genetic analysis in *Cucurbita* is described.

11.3 BREEDING OBJECTIVES AND GENETIC DIVERSITY IN *CUCURBITA*

The goal of squash breeding programs is the development of germplasm with desirable horticultural traits, such as increased fruit or seed yield, earliness, climatic resilience, favorable growth habit, fruit color, shape, size, flavor, enhanced nutritional quality, and resistance to biotic and abiotic stresses (Loy et al. 2012). The availability of adequate genetic diversity within a given crop is one of the most important factors determining plant breeding success. Most of the alleles conferring tolerance to biotic and abiotic stresses in *Cucurbita* are present in wild, unadopted germplasm and need to be introgressed into elite cultivars. The *Cucurbita* genus is rich in both genetic and phenotypic diversity as evidenced by many diversity studies (examples in Table 11.2).

11.4 ALLELE DISCOVERY THROUGH QTL MAPPING

Conventional breeding in *Cucurbita* crops is resource intensive, particularly due to the vast space and labor requirements (Hernandez et al. 2020). Marker-assisted selection is an important tool in *Cucurbita* breeding as it improves the rate of genetic gain while reducing the cost associated with phenotyping, especially for complex traits. However, QTL discovery is a pre-requisite to marker assisted selection (Figure 11.1).

11.4.1 Plant Morphology and Flowering Traits

Breeding for dwarf plant architecture is desirable in *Cucurbita* for the improvement of plant density and yield per acre. In contrast, sprawling plants require extensive space and labor resources,

TABLE 11.2
Examples of Genetic Diversity Studies Conducted in *Cucurbita*

Species	Number of Cultivars	Type and Number of Markers Used	References
C. maxima	8	11 RAPD	Ferriol et al. 2001
Cucurbita pepo and C. argyrosperma	*Cucurbita pepo*–53 *C. argyrosperma*–1	26 RAPD	Decker-Walters et al. 2002
C. pepo	69	11 SRAP 7 AFLP	Ferriol et al. 2003a
C. maxima *C. pepo* *C. ficifolia* *C. moschata* *Lagenariasiceraria* (gourd)	*C. maxima*–19 *C. pepo*–2 *C. ficifolia*–2 *C. moschata*–2 *Lagenariasiceraria* (gourd)–2	21 RAPD 24 SBAP	Ferriol et al. 2003b
C. moschata	47	11 SRAP 6 RAPD	Ferriol et al. 2004a
C. maxima	120	9 SRAP 7 AFLP	Ferriol et al. 2004b
C. pepo	45	448 AFLP 147 ISSR 20 SSR	Paris et al. 2003
C. pepo *C. moschata* *C. maxima*	*C. pepo*–9 *C. moschata*–1 *C. maxima*–4	3 RAPD 7 ISSR	Heikal et al. 2008
C. pepo *C. moschata* *C. maxima*	*C. pepo*–4 *C. moschata*–10 *C. maxima*–2	10 RAPD	Tsivelikas et al. 2009
C. moschata	89	9 AFLP	Wu et al. 2011
C. pepo	104	134 SSR 4 SCAR	Gong et al. 2012
C. pepo *C. moschata* *C. maxima*	*C. pepo*–16 *C. moschata*–4 *C. maxima*- 4	8 ISSR 8 SRAP	Inan et al. 2012
C. pepo *C. maxima*	*C. pepo*–10 *C. maxima*–66	35 SSR	Liu et al. 2013
C. moschata	20	21 RAPD	Muralidhara and Narasegowda 2014
C. pepo	7	20 RAPD	Radwan S.A.A 2014
C. pepo	51	21 SSR	Murovec, J. 2015
C. pepo	7	36 RAPD 55 SSR	Ntuli et al. 2015
C. pepo *C. maxima* *C. moshata*X *C. maxima*	*C. pepo*–7 *C. maxima*–11 *C. moshata*X *C. maxima*- 4	300 SSR	Sim et al. 2015
C. maxima	85	23 SSR	Kaźmińska et al. 2017
C. pepo *C. moschata*	*C. pepo*–21 *C. moschata*–11	Fruit and seed traits	Darrudi et al. 2018
C. pepo	30	8 SSR	Verdone et al. 2018
C. pepo	64	45 morphological traits 300 SSR	Yunli et al. 2020
C. pepo	29	7 SSR	Öztürk et al. 2022
C. pepo	140	18 ISSR	Morilipinar et al. 2021

TABLE 11.2 (Continued)
Examples of Genetic Diversity Studies Conducted in *Cucurbita*

Species	Number of Cultivars	Type and Number of Markers Used	References
C. moschata	9	10 ISSR	Mady et al. 2022
C. argyrosperma	440	9 SSR	Sánchez-de la Vega et al. 2018
C. maxima	14	7 SSR	Öztürk et al. 2021
C. pepo	46	14 RAPD	Méndez-López et al. 2018
*Cucurbita*spp	80	5 SSR	Nyabera et al. 2021
C. moschata	*C. moschata–100*	21 SSR	Arbizu et al. 2022
C. maxima and C. pepo	*C. maxima* and *C. pepo*–10		
C. pepo	*C. pepo*–47	85 SSR	Wang et al. 2021
C. foetidissima	*C. foetidissima*–1		
C. moschata	610	2071 SNPs	Lee et al. 2020

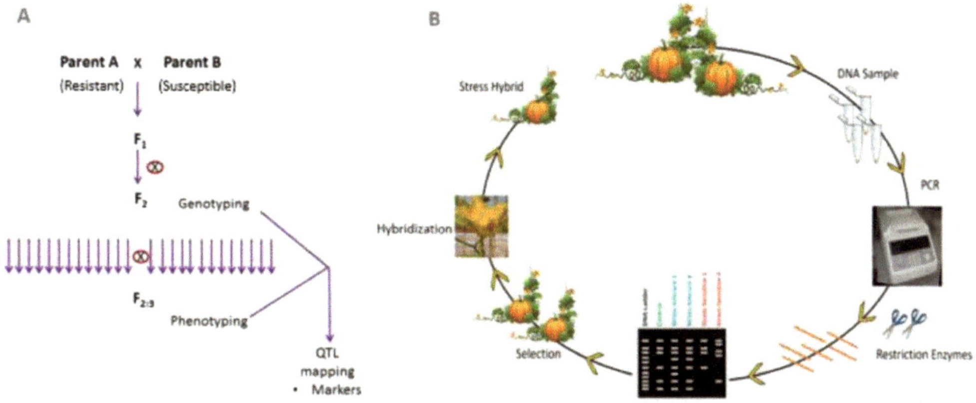

FIGURE 11.1 Schematic for (A) QTL discovery and (B) marker-assisted selection in *Cucurbita.*

especially in commercial production. In *C. maxima*, three QTLs associated with the dwarf phenotype were identified by Zhang et al. (2015) using genotyping by sequencing in an F_2 population ($n = 182$). The major QTL (*qCmB2*) accounted for 21.4% of the phenotypic variation in a region that harbored a gibberellin synthesis gene (*GA20ox*). In a similar study, Xiang et al. (2018) identified three QTLs associated with the dwarf phenotype in *C. pepo* in a cross between Jinganlu (*C. pepo* subsp. *pepo* var. Pumpkin) and HM-S2 (*C. pepo* subsp. *pepo* var. Zucchini). The major QTLs spanned 1.39 Mb and harbored a gene predicted to encode gibberellin 2-oxidase (Xiang et al. 2018). In *C. moschata*, the bush habit was mapped as a morphological trait on linkage group (LG) p12 (Gong et al. 2008a). In a separate study, a RAPD marker, S1225-548, was found to be linked to the trait in an interspecific population between *C. moschata* and *C. maxima* (Li et al. 2007). Later, the *CmV1* gene was found to be responsible for vine elongation in tropical *C. moschata* pumpkin (Wu and Cao 2010). More recently, Wang et al. (2022) reported that the dominant *Bush* locus in pumpkin (*C. moschata*) was governed by cis-regulatory element of a transcription factor gene, *YABBY1*.

Early flowering is a highly valuable trait in *Cucurbita* crops because it directly influences early fruit maturity (Abbas et al. 2022a; Montero-Pau et al. 2017). Flowering time in plants is an oligogenic or quantitative trait governed by multiple genetic factors (Maheswaran et al. 2000). Using a *C. pepo* F_2 population from a cross between Zucchini and Scallop, Esteras et al. (2012) found two major QTLs on LG3 associated with "Days to First Female Flower" and "Days to First Male Flower" that accounted for 59% and 38% of the phenotypic variation, respectively. Other flowering-related traits, including "First Node with a Female Flower", "First Node with a Male Flower", and the ratio of "Male to Female Flowers" were colocalized in the region on LG3, indicating potential pleiotropy and or linkage. Montero-Pau et al. (2017) found two genomic regions controlling female flower earliness on LG12 (R^2=7.9%) and LG 9 (R^2 = 2.6%) in a recombinant inbred line (RIL) population derived from Zucchini X Scallop. More recently, Abbas et al. (2022a), identified QTLs for early male and female flowering traits in *C. pepo* on chromosomes 10 and 11, respectively. This study further identified two Flowering Locus T-like protein (FLT1 and FLT2) genes as the most likely candidates for early flowering in *C. pepo*. In *C. moschata*, Hashimoto et al. (2020) and Abbas et al. (2002) reported that a major QTL was responsible for early female flowering on LG 7 and LG 6 of the genome, respectively.

11.4.2 Fruit and Seed Traits

The *Cucurbita* genus has extensive diversity in fruit characteristics, including size, shape, flesh color, rind pattern and color, flavor and nutritional quality (Ferriol et al. 2003a, 2003b). Many studies have been conducted to elucidate the genetic bases of variation in fruit traits within *Cucurbita* crops. For example, a genetic analysis of *C. pepo* populations (Zucchini X Scallop) by Esteras et al. (2012) and Montero-Pau et al. (2017) revealed several QTLs associated with fruit shape, fruit length and width, intensity of fruit ribbing, cavity thickness, rind color, flesh firmness, and flesh color. In *C. moschata*, diverse populations have been employed to identify many fruit-related QTLs, including fruit length, diameter and weight, fruit shape, rind color, flesh thickness, seed cavity size, and total carotenoids (Echevarria et al. 2020; Han et al. 2022, Hashimoto et al. 2020; Lee et al. 2022; Zhong et al. 2017). Similarly, in *C. maxima*, QTLs for fruit-related traits such as carotenoid content, flesh color and thickness, fruit length and diameter, shape index, sucrose content and dry matter content have been recently identified (Kazminska et al. 2018; Kazminska et al. 2020; Pan et al. 2022; Reddy et al. 2022).

Seeds of *Cucurbita* crops provide a significant source of income and nutrition globally (Baxter et al. 2012; Fruhwirth and Hermetter 2007; Nakić et al. 2006). The seeds are rich in oil (40–50%), protein (35%) and unsaturated fatty acids (Applequist et al. 2006; Bavec et al. 2007; Datta and Lal 1977; Lazos 1986; Meru et al. 2018; Stevenson et al. 2007). Breeding effort in *Cucurbita* crops is primarily focused on the improvement of seed yield, seed nutrition, and seed size (Guo et al. 2020; Jan and Seiler 2007; Lelley et al. 2009; Meru et al. 2018). However, the genetic basis of these traits

FIGURE 11.2 The (A) hull-less seed phenotype in *Cucurbita pepo* and (B) the hulled seed phenotype in *Cucurbita maxima*.

remains largely unknown (McGregor and Meru 2021). In *C. maxima*, Tan et al. (2013) and Wang et al. (2020) identified several genomic regions associated with seed size traits including seed length, seed width, and seed weight. Further, Shi et al. (2021) reported that an MYB transcription factor (*CmaCh15G005270*) was responsible for the white seed coat color in *C. maxima*. In *C. pepo*, a seed-transcriptome study by Wyatt et al. (2015) provided insights into the candidate genes involved in seed and embryo development. Seeds of pumpkin (*C. pepo*) that lack a seedcoat (hull-less) are preferred for snacking as they eliminate the need for manual de-hulling prior to use (Figure 11.2) (Lelley et al. 2009; Loy 2004; Meru et al. 2018).

The hull-less trait in *C. pepo* is controlled by a single recessive gene (designated *n* or *h*) with potential modifiers (Whitaker and Robinson 1986). A single major locus associated with the hull-less trait in *C. pepo* was initially identified by Zraidi et al. (2007) and Gong et al. (2008a) on chromosome 12 of the genome. Subsequent studies validated this locus and identified a NAC domain-containing protein (*Cp4.1LG12g04350*) as the likely candidate gene for the hull-less seed phenotype in *C. pepo* (Fabrizio et al. 2022; Kaur et al. 2022; Meru et al. 2022).

11.4.3 Disease and Insect Resistance

Breeding for resistance or tolerance to biotic stresses (disease and pests) is one of the major objectives of *Cucurbita* breeding programs. *Cucurbita* crops are economically impacted by many fungal, bacterial, and viral diseases, as well as insect pests (Kesh and Yadav 2022).

Powdery mildew is among the most devastating fungal diseases in *Cucurbita* commonly caused by obligate biotrophic pathogen *Podosphaera xanthii* (Bertrand and Pitrat 1989). Various levels of resistance to powdery mildew have been reported in *C. moschata* but not in *C. pepo* or *C. maxima* (Wessel-Beaver 1993; Rhodes 1964). Powdery mildew resistance is believed to be primarily controlled by one major gene with certain modifiers (Jahn et al. 2002). Resistance to powdery mildew in commercial *C. moschata* and *C. pepo* cultivars has been achieved through the introgression of a major resistance gene (*Pm-0*) from wild *C. okeechobeensis* subsp. *Martinezii* (Contin 1978). At present, *Pm-0* is the only known gene conferring resistance in nearly all powdery mildew resistant (PMR) commercial cultivars of *C. moschata* and *C. pepo*. The *Pm-0* gene in *C. pepo* was recently mapped on linkage group 10 using genotyping by sequencing approach (Holdsworth et al. 2016). Recently, Park et al. (2020) reported that the *C. moschata* heirloom, TG10, carried *Pm-0* resistance gene from *C. okeechobeensis* subsp. *Martinezii* and mapped the locus on chromosome 3 of the genome.

Crown rot, fruit rot, root rot, and seedling damping caused by *Phytophthora capsici* are serious constraints to squash production worldwide. Padley et al. (2009) introgressed resistance to crown rot from two wild *Cucurbita* species (*C. lundelliana* and *C. okeechobeensis*) into *C. moschata* background (breeding line #394-1-27-12) and reported that the resistance was controlled by three dominant genes. Subsequently, Ramos et al. (2020) identified three QTLs linked to crown rot resistance in #394-1-27-12 on chromosome 4 (*QtlPC-C04*), 11 (*QtlPC-C11*) and 14 (*QtlPC-C14*). In *C. pepo*, Padley et al. (2009) identified several accessions tolerant to crown rot, including PI 181761. Michael et al. (2019) reported that the resistance in PI 181761 was controlled by three dominant genes, and later mapped a major locus on chromosome 13 (*QtlPC-C13*) associated with crown rot resistance in *C. pepo* (Michael et al. 2021). Vogel et al. (2021) further identified several novel QTLs associated with root and crown rot resistance in *C. pepo* (PI 615089) using biparental mapping populations.

Zucchini Yellow Mosaic Virus (ZYMV) is an aphid transmitted single-stranded RNA virus that was first reported in melon in 1979 but currently infects all the major *Cucurbita* worldwide (Lecoq et al. 1981). Natural source of resistance to ZYMV has been reported in wild unadopted genotypes of *C. moschata*, Nigerian Local, Menina, Soler and Bolina (Provvidenti 1997; Wessel-Beaver 2000; Brown et al. 2003). No natural source of resistance to the virus has been reported in *C. pepo*. Resistance in Nigerian Local is conferred by at least two genes, *Zym-0* and *Zym-4* (Brown et al. 2003; Pachner et al. 2011). Using an F_2 population between Nigerian Local (resistant) and Waltham Butternut (susceptible), Pachner et al. (2015) identified a single SSR marker (TGM-Zym-0) linked to ZYMV resistance in *C. moschata*. Subsequently, Shrestha et al. (2021) identified four QTLs associated with ZYMV resistance in Nigerian Local on chromosome 2 (*QtlZYMV-C02*), 4 (*QtlZYMV-C04*), 8 (*QtlZYMV-C08*), and 20 (*QtlZYMV-C20*) using an F_2 population between Nigerian Local and Butterbush. Resistance to ZYMV in Menina is oligogenic (Paris and Cohen 2000; Paris et al. 1988). Pachner et al. (2015) identified two markers TGM-Zym-1 and TGS-Zym-2 significantly associated with ZYMV resistance in *C. pepo* Accession 381e (carrying ZYMV resistance from Menina).

The term papaya ringspot virus (PRSV) was first coined by Jensen in 1949 to describe the viral disease of papaya in Hawaii (Gonsalves and Ishii 1980), and later, the PRSV-Watermelon strain (PRSV-W) was found to affect cucurbits globally. Resistance to PRSV-W exists in *C. moschata* accessions Nigerian Local and Menina and is believed to be controlled by a major gene with potential modifiers (Provvidenti 2000; Miranda-Vélez et al. 2019; McPhail et al. 2012). Recently, Shrestha et al. (2022) identified a single QTL (*QtlPRSV-C09*) associated with PRSV resistance in Nigerian Local on chromosome 9 using whole genome re-sequencing-based bulked segregant analysis.

Tomato leaf curl New Delhi virus (ToLCNDV) is an economically important whitefly transmitted virus with two circular single-stranded DNA genome components designated as DNA-A and DNA-B (Jyothsna et al. 2013). It is highly damaging to crops of the Cucurbitaceae family, including luffa (*Luffa cylindrica*), cucumber (*Cucumis sativus* L.), watermelon (*Citrullus lanatus* L), melon (*C. melo* L.), and different types of squash (*Cucurbita* spp.) (Sohrab et al. 2003; Singh et al.2009; Chang et al. 2010; Roy et al. 2013). Natural sources of resistance to ToLCNDV has been reported in *C. moschata* accessions PI 604506 and PI 381814 (Saez et al. 2016), and a major QTL for the resistance identified on chromosome 8 (Saez et al. 2020). This QTL was found to be syntenic to the QTL associated with ToLCNDV resistance in melon on chromosome 11.

Breeding for insect resistance in *Cucurbita* is limited despite the high cost of production incurred by growers due to insect-pest problems (Seal et al. 2022). Major insects of *Cucurbita* include silver leaf whitefly (*Bemesia argentifolii*), cucumber beetles (*Acalymma vittatum*), fruit fly (*Dacus cucurbitae*), melon worms (*Diaphania hyalinata*), pickle worms *(Diaphania nitidalis)*, melon thrips (*Thrips palmi*), leaf miners (*Liriomyza trifolii*), aphids (*Myzus persicae*), and silver leaf whitefly (*Bemisia argentifolii*) (Seal et al. 2022; Sharma et al. 2016). Resistance to insects has been reported in some *Cucurbita* species. Resistance to whitefly has been reported in *C. moschata* *"Paraguayan landrace"*, resistance to fruit fly (*Dacus cucurbitae*) in *C. maxima*, and squash vine borer in *C. moschata* (Gonzalez-Roman and Wessel-Beaver 2002; Dhatt et al. 2020). Kabelka and Young (2010) identified four markers associated with resistance to silver-leaf disorder (caused by whitefly feeding) in summer squash. Recently, Del Valle Echevarria et al. (2020) identified three QTLs associated with pickle worm tolerance in *C. moschata* on LG 1, 9, and 15.

11.5 TRANSCRIPTOMICS FOR ELUCIDATING GENETIC BASIS OF TRAITS IN *CUCURBITA*

Transcriptomes are the sum total of all RNA molecules in a cell or the population of cells (Wang et al. 2009). Understanding transcriptomes of an organism is essential for (i) interpreting functional genes related to a particular life stage; (ii) quantifying changing expression level of transcripts under the influence of external factors; and (iii) determining the transcriptional structure of a gene in terms of splicing and post-transcriptional modifications (Wang et al. 2009). Over the past two decades, the field of gene expression profiling has undergone dynamic advances. Several transcriptome studies have been conducted in squash to assess gene function and understand the molecular basis of various traits.

TABLE 11.3
A List of Published Transcriptomics Studies on *Cucurbita*

Species	Sequencing Platform	Total Reads Produced	Trait Under Study	Reference
C. pepo	Illumina HiSeq 2500	240 million	Nectar synthesis and secretion	Solhaug et al. 2019
C. pepo	BGISEQ-500 sequencing platform	82,900 million	Fruit morphology and ripening	Xanthopoulou et al. 2021
C. pepo	Illumina HiSeq 2000	600 million	Carotenoid, starch, and sucrose synthesis	Wyatt et al. 2016
C. maxima	Illumina HiSeq 4000	132 million	Regulation of fruit development	Huang et al. 2019
C. pepo	Illumina HiSeq4000	170.47 million	Peel coloration, carotenoid	Xu et al. 2021
C. pepo	Illumina HiSeqTM2000	12,842 million	Fruit characteristic	Xanthopoulou et al. 2016
C. pepo	Illumina NextSeq 500	16.6 million	Fruit size	Xanthopoulou et al. 2017
C. maxima	Illumina HiSeq 2000	525.51 million	Sex expression	Wang et al. 2019
C. pepo	Illumina HiSeq1500	150.9 million	ZYMV tolerance	Amoroso et al. 2022
C. moschata	Illumina HiSeq 2000	423.2 million	Powdery mildew	Guo et al. 2018
C. pepo	Illumina HiSeq 2000.	695 million	Proteome annotation	Andolfo et al. 2017
C. pepo	Illumina HiSeq 4000 platform	439.2 million	Fruit set	Luo et al. 2021
C. pepo	Illumina HiSeq 2500	34 million	*Aphis gossypii* infestation	Vitiello et al. 2018
C. pepo	Illumina HiSeq 2500	403.7 million	Parthenocarpy	Pomares-Viciana et al. 2019
C. pepo	Illumina HiSeq 2000	156.5 million	Fruit and seed development	Wyatt et al. 2015

The first transcriptomic study in squash was conducted in *C. pepo*, where Blanca et al. (2011) used GS-FLX Titanium technology to generate 512,751 ESTs assembled into 49,610 unigenes. Among 49,610 unigenes, 34% were reported to have known orthologs in Arabidopsis and melon. On the other hand, the first transcriptomic study in *C. moschata* was performed by Wu et al. 2014. In the study, squash transcriptome was sequenced using RNA-seq Illumina Hiseq 2000. A total of 52,849,316 clean reads, 66,621 contigs and 62,480 unigenes were generated. A total of 7,814 SSRs were identified in the unigenes, and 4,794 primer pairs were designed for molecular applications. Xie et al. (2015) used the Illumina sequencing technology to determine RNA transcripts in root tissue of *C. moschata* and *C. maxima*, 4 hours after salt treatment. A total of 21,548,326 and 19,394,108 reads were generated from the control and salt-treated *C. maxima* while 19,108,240 and 20,546,052 reads were obtained from the control and salt-treated *C. moschata*. Fifty-eight conserved miRNA families and 33 novel miRNAs were identified in the two species. Following these early transcriptomics studies, other researchers have conducted many studies in squash to understand the molecular mechanism governing economically important traits and identify candidate genes controlling these traits. Examples of published transcriptomics studies in *Cucurbita* are summarized in Table 11.3.

11.6 GENOMIC SELECTION AND GENE EDITING

Due to the rapid advances in sequencing technologies, high-quality reference genome sequences of many crops are currently available (Le Nguyen et al. 2019). Genome sequences are available for all three major *Cucurbita* species (*C. maxima* cv. Rimu, *C. moschata* cv. Rifu, and *C. pepo* cv.

MU-CU-16) (Sun et al. 2017; Montero-Pau et al. 2018). In addition to sequencing data, various transcriptome and genetic data have been generated for *Cucurbita* and are provided in a central portal for researchers and breeders in Cucurbit Genomics Database (CuGenDB; http://cucurbitgenomics.org) (Zheng et al. 2019). Additionally, with the availability of technologies that enable high-throughput phenotyping of complex traits, there has been rapid improvement in plant phenotyping. The availability of accurate genomic and phenotyping data coupled with advances in biotechnology has enabled a wide range of new "Plant Breeding Techniques" for rapid breeding, such as genomic selection and clustered regularly interspaced short palindromic repeats-CAS (CRISPR/Cas) gene editing technology.

Although marker-assisted selection has become routine for many crops, including species of *Cucurbita*, it often targets QTLs of major effect while excluding those of minor effect (Shikha et al. 2017). However, using genome-wide marker information, breeders can deploy genomic selection to predict progeny performance through genomic estimated breeding values (Meuwissen et al. 2001). However, to-date, genomic selection in *Cucurbita* is in its infancy. Nonetheless, a few studies have demonstrated the potential of genomic selection as a promising breeding tool for *Cucurbita* improvement. Hernandez et al. (2020) tested the prediction accuracy for qualitative (flesh color and shape) and quantitative (fruit length, width, % dry matter, Brix, and fruit yield) traits in *C. moschata* using the GBLUP genome-wide model. The prediction accuracy for the traits ranged from 0.31 to 0.62 and generally followed the trend for heritability values. In a separate study, Brzozowski and Mazourek (2020) tested the efficacy of six genomic prediction models (BayesA, BayesB, BayesC, Bayesian Lasso BL, Bayesian Ridge Regression BRR, and rrBLUP) in the selection of resistance to striped cucumber beetle (*Acalymma vittatum*) in *C. pepo*. The predictive accuracy was low (0.248 ± 0.010) across the models. However, this result was not surprising given the low heritability ($h^2 = 0.192$) for striped cucumber beetle resistance in *C. pepo*.

Since the discovery of CRISPR technology, targeted genome editing has been widely utilized in plant improvement programs (Jinek et al. 2012). First-generation genome editing technologies, which include meganucleases, ZFNs and TALENs are tedious and have low target specificity. In contrast, CRISPR/Cas9 (second-generation genome editing technology) is highly target-specific and allows easier design and methodology (Jaganathan et al. 2018). CRISPR was first identified as a part of bacterial defense system defending against invading viruses (Mojica et al. 2009). Cas9 (or "CRISPR-associated protein 9") is an enzyme that uses CRISPR sequences as a guide to recognize and cleave-specific DNA strands that are complementary to the CRISPR sequence. CRISPR/Cas9 combination forms the basis of CRISPR technology that can be used to edit targeted genes within an organism (Jinek et al. 2012). Even though CRISPR technique has been utilized in plant breeding widely, the first CRISPR-based genome editing in squash was only recently achieved, particularly in *C. moschata*. Xin et al. (2022) deployed CRISPR/Cas9 to create mutants of *C. moschata* with compact plant architecture by targeting homologs of the *ERECTA* family of receptor kinase genes. Similarly, Wang et al. (2022) used CRISPR/Cas9 to generate novel alleles conferring bush trait in *C. moschata* by targeting the *CmoYABBY* gene.

REFERENCES

Abbas HMK, Huang H, Wu T, Wang R, Du H, Lu S et al. (2022) High-density genetic mapping identified a major locus for environmental sex expression in pumpkin (*Cucurbita moschata* Duch.). *Hortic Plant J* 8:593–601.

Abbas HMK, Zhou YY, Huang HX, Qamar MTU, Wang MM, Lu S et al. (2022a) QTL mapping, whole genome resequencing, and marker-assisted selection provide basics of early flowering in pumpkin. *Plant Breed* 141:266–276.

Amoroso CG, Andolfo G, Capuozzo C, Di Donato A, Martinez C, Tomassoli L, Ercolano, MR (2022) Transcriptomic and genomic analysis provides new insights in molecular and genetic processes involved in zucchini ZYMV tolerance. *BMC Genomics* 23:1–11.

Andolfo G, Di Donato A, Darrudi R, Errico A, Aiese Cigliano R, Ercolano MR (2017) Draft of Zucchini (*Cucurbita pepo* L.) proteome: A resource for genetic and genomic studies. *Front Genet* 8:181.

Applequist WL, Avula B, Schaneberg BT, Wang YH, Khan IA (2006) Comparative fatty acid content of seeds of four *Cucurbita* species grown in a common (shared) garden. *J Food Compos Analys* 19:606–611.

Arbizu CI, Blas RH, Ugás R (2022) Genetic diversity and population structure assessed by SSR in a Peruvian germplasm collection of Loche squash (*Cucurbita moschata*, Cucurbitaceae). 2nd International Electronic Conference on Diversity.

Bavec F, Grobelnik SM, Rozman C, Bavec M (2007) Oil pumpkins: Niche for organic producers. In J Janick & A. Whipkey (eds.), Issues in new crops and new uses. ASHS Press, 185–189

Baxter GG, Murphy K, Paech A (2012) The Potential to Produce Pumpkin Seed for Processing in Northeast Victoria. RIRDC.

Bertrand, F., & Pitrat, M. (1989). Screening of a muskmelon germplasm for susceptibility to 5 pathotypes of powdery mildew. In *Cucurbitaceae* 89. USDA.

Blanca J, Cañizares J, Roig C, Ziarsolo P, Nuez F, Picó B (2011) Transcriptome characterization and high throughput SSRs and SNPs discovery in *Cucurbita pepo* (Cucurbitaceae). *BMC Genomics* 1:21–15.

Blessing AC, Ifeanyi UM, Chijioke OB (2011) Nutritional evaluation of some Nigerian pumpkins (*Cucurbita* spp.). *Fruit Veg Cereal Sci Biotechnol* 5:64–71.

Brown RN, Bolanos-Herrera A, Myers JR, Miller Jahn M (2003) Inheritance of resistance to four cucurbit viruses in *Cucurbita moschata. Euphytica* 129:253–258.

Brown RN, Myers JR (2002) A genetic map of squash (*Cucurbita* sp.) with randomly amplified polymorphic DNA markers and morphological markers. *Am Soc Hortic Sci* 127: 568–575.

Brzozowski LJ, Mazourek M (2020) Evaluation of selection methods for resistance to a specialist insect pest of squash (*Cucurbita pepo*). *Agronomy* 10: 847.

Chang HH, Ku HM, Tsai WS, Chien RC, Jan FJ (2010) Identification and characterization of a mechanical transmissible begomovirus causing leaf curl on oriental melon. *Eur J Plant Pathol* 127:219–228.

Contin M (1978) Interspecific transfer of powdery mildew resistance in the genus *Cucurbita*. PhD Dissertation Cornell University.

Darrudi R, Nazeri V, Soltani F, Shokrpour M, Ercolano MR (2018) Genetic diversity of *Cucurbita pepo* L. and *Cucurbita moschata* Duchesne accessions using fruit and seed quantitative traits. *J Appl Res Med Arom Plants* 8:60–66.

Datta N, Lal BM (1977) Distribution of oil in different anatomical parts of some cucurbit kernels. *J Food Sci Technol* 14:24–25.

Decker-Walters DS, Staub JE, Chung SM, Nakata E, Quemada HD (2002) Diversity in free-living populations of *Cucurbita pepo* (Cucurbitaceae) as assessed by random amplified polymorphic DNA. *Syst Bot* 27:19–28.

Del Valle Echevarria AR, Campbell A, Radovich TJ, Silvasy T, Moore S, Kantar MB (2020) Quantitative trait loci (QTL) analysis of fruit and agronomic traits of tropical pumpkin (*Cucurbita moschata*) in an organic production system. *Horticulturae* 6:14.

Dhatt AS, Sharma M, Kaur B (2020) Advances in Improvement of Pumpkin and Squashes. Accelerated Plant Breeding, 2nd edn. In: S Gosal and S Wani (eds.) Springer, Cham.

Esteras C, Gómez P, Monforte AJ, Blanca J, Vicente-Dólera N, Roig C, Picó, B (2012) High-throughput SNP genotyping in *Cucurbita pepo* for map construction and quantitative trait loci mapping. *BMC Genomics* 13:1–21.

Fabrizio J, LaPlant K, Wyatt L, Inzinna G, Li L, Mazourek M (2022) Fine mapping and identification of candidate genes for the hull-less seed phenotype in *Cucurbita pepo*. *Euphytica* 218:1–12.

Ferriol M, Belén Picó M, Nuez F (2003b) Genetic diversity of some accessions of *Cucurbita maxima* from Spain using RAPD and SBAP markers. *Genet Resour Crop Evol* 50:227–238.

Ferriol M, Picó B (2008) Pumpkin and winter squash. Vegetables 1st Edn. In: Prohens J and Nuez F (eds.) Springer, New York.

Ferriol M, Pico B, De Cordova PF, Nuez F (2004a) Molecular diversity of a germplasm collection of squash (*Cucurbita moschata*) determined by SRAP and AFLP markers. *Crop Sci* 44: 653–664.

Ferriol M, Picó B, Nuez F (2001) Genetic variability in pumpkin (*Cucurbita maxima*) using RAPD markers. *Cucurbit Genet Coop Rep* 24:94–96.

Ferriol M, Picó B, Nuez F (2003a) Genetic diversity of a germplasm collection of *Cucurbita pepo* using SRAP and AFLP markers. *Theor Appl Genet* 107:271–282

Ferriol M, Picó B, Nuez F (2004b) Morphological and molecular diversity of a collection of *Cucurbita maxima* landraces. *J Am Soc Hortic Sci* 129: 60–69.

Fruhwirth GO, Hermetter A (2007) Seeds and oil of the Styrian oil pumpkin: Components and biological activities. *Eur J Lipid Sci Technol* 109: 1128–1140.

Ge Y, Li X, Yang XX, Cui CS, Qu SP (2015) Genetic linkage map of *Cucurbita maxima* with molecular and morphological markers. *Genet Mol Res* 14: 5480–4.

Gong L, Pachner M, Kalai K, Lelley T (2008a) SSR-based genetic linkage map of *Cucurbita moschata* and its synteny with *Cucurbita pepo*. *Genome* 51: 878–887.

Gong L, Paris HS, Nee MH, Stift G, Pachner M, Vollmann J, Lelley T (2012) Genetic relationships and evolution in *Cucurbita pepo* (pumpkin, squash, gourd) as revealed by simple sequence repeat polymorphisms. *Theor Appl Genet* 124:875–891.

Gong L, Stift G, Kofler R, Pachner M, Lelley T (2008b) Microsatellites for the genus Cucurbita and an SSR-based genetic linkage map of *Cucurbita pepo* L. *Theor Appl Genet* 117:37–48.

Gonsalves D, Ishii M (1980) Purification and serology of papaya ringspot virus. *Phytopathology* 70:1028–1032.

González-Román M, Wessel-Beaver L. (2002) Resistance to silver leaf disorder is controlled by a single recessive gene in *Cucurbita moschata* Duchesne. *Rep Cucurbit Genet Coop* 25:49–50.

Gonzalo MJ, Monforte AJ (2016) Genetic mapping of complex traits in cucurbits. In: Grumet, R., Katzir, N., Garcia-Mas, J. (eds.) *Genetics and Genomics of Cucurbitaceae*. 20th Edn. Springer, Cham.

Guo WL, Chen BH, Chen XJ, Guo YY, Yang HL, Li XZ, Wang GY (2018) Transcriptome profiling of pumpkin (*Cucurbita moschata* Duch.) leaves infected with powdery mildew. *PLoS One* 13:e0190175.

Guo Y, Gao M, Liang X, Xu M, Liu X, Zhang Y et al. (2020) Quantitative trait loci for seed size variation in cucurbits–a review. *Front Plant Sci* 11:304.

Hashimoto T, Kubo N, Nishimura K, Nagano AJ, Sasaki A, Nakamura Y, Mimura Y (2020) Quantitative trait locus analysis in squash (*Cucurbita moschata*) based on simple sequence repeat markers and restriction site-associated DNA sequencing analysis. *Horticulturae* 6:71.

Heikal AH, Abdel-Razzak HS, Hafez EE (2008) Assessment of genetic relationships among and within *Cucurbita* species using RAPD and ISSR markers. *J Appl Sci Res* 4:515–525.

Hernandez CO, Wyatt LE, Mazourek MR (2020) Genomic prediction and selection for fruit traits in winter squash. *Genes Genomes Genet* 10:3601–3610.

Hi LY, Jeong JH, Hyon HK, Dong KB (1995) Use of random amplified polymorphic DNAs for linkage group analysis in interspecific hybrid F_2 generation of cucurbita. *Hortic Environ Biotechnol* 36:323–330.

Holdsworth WL, LaPlant KE, Bell DC, Jahn MM, Mazourek M (2016) Cultivar-based introgression mapping reveals wild species-derived *Pm-0*, the major powdery mildew resistance locus in squash. *PLoS One* 11:e0167715.

Holmer R, Linwattana G, Nath P, Keatinge JDH (2013) High value vegetables in Southeast Asia: production, supply and demand. Proceedings of the Regional Symposium on High Value Vegetables in Southeast Asia: Production, Supply and Demand. AVRDC-The World Vegetable Center.

Huang HX, Yu T, Li JX, Qu SP, Wang MM, Wu TQ, Zhong YJ (2019) Characterization of *Cucurbita maxima* fruit Metabolomic profiling and transcriptome to reveal fruit quality and ripening gene expression patterns. *J Plant Biol* 62:203–216.

Inan N, Yildiz M, Sensoy S, Kafkas SALİH, Abak K (2012) Efficacy of ISSR and SRAP techniques for molecular characterization of some Cucurbita genotypes including naked (hull-less) seed pumpkin. *J Anim Plant Sci* 22:126–136.

Jaganathan D, Ramasamy K, Sellamuthu G, Jayabalan S, Venkataraman G (2018) CRISPR for crop improvement: an update review. *Front Plant Sci* 9:985.

Jahn M, Munger HM, McCreight JD (2002) Breeding cucurbit crops for powdery mildew resistance. In: R. R. Belanger, W. R. Bushnell, A. J. Dik, and T. L. W. Carver (eds.). The powdery mildews: a comprehensive treatise. The Amer Phytopathol. Soc., St. Paul, Minn. 239–248.

Jan, C. and G. Seiler. 2007. Sunflower, p. 137. In: R.J. Singh (ed). *Genetic resources, chromosome engineering and crop improvement*. CRC Press. New York, NY.

Jinek M, Chylinski K, Fonfara I, Hauer M, Doudna JA, Charpentier E (2012) A programmable dual-RNA–guided DNA endonuclease in adaptive bacterial immunity. *Science* 337(6096):816–821.

Jyothsna P, Haq QMI, Singh P, Sumiya KV, Praveen S, Rawat R et al. (2013) Infection of tomato leaf curl New Delhi virus (ToLCNDV), a bipartite begomovirus with betasatellites, results in enhanced level of helper virus components and antagonistic interaction between DNA B and betasatellites. *Appl Microbiol Biotechnol* 97:5457–5471.

Kabelka EA, Young K (2010) Identification of molecular markers associated with resistance to squash silverleaf disorder in summer squash (*Cucurbita pepo*). *Euphytica* 173:49–54.

Kaur B, Garcha KS, Bhatia D, Khosa JS, Sharma M, Mittal A et al. (2022) Identification of single major QTL and candidate gene(s) governing hull-less seed trait in pumpkin. *Front Plant Sci* 13:948106.

Kaźmińska K, Hallmann E, Korzeniewska A, Niemirowicz-Szczytt K, Bartoszewski G (2020). Identification of fruit-associated QTLs in Winter Squash (*Cucurbita maxima* Duchesne) using recombinant inbred lines. *Genes* 11:419.

Kaźmińska K, Hallmann E, Rusaczonek A, Korzeniewska A, Sobczak M et al. (2018). Genetic mapping of ovary colour and quantitative trait loci for carotenoid content in the fruit of *Cucurbita maxima* Duchesne. *Mol Breed* 38:1–14.

Kaźmińska K, Sobieszek K, Targońska-Karasek M, Korzeniewska A, Niemirowicz-Szczytt K, Bartoszewski G (2017) Genetic diversity assessment of a winter squash and pumpkin (*Cucurbita maxima* Duchesne) germplasm collection based on genomic Cucurbita-conserved SSR markers. *Sci Hortic* 219:37–44.

Kesh H, Yadav S (2022) Recent advances in genetics and breeding of pumpkin (*Cucurbita moschata* Duch.). *J Hortic Sci Biotechnol* 98: 1–18. DOI: 10.1080/14620316.2022.2097961

Lazos ES (1986) Nutritional, fatty acid, and oil characteristics of pumpkin and melon seeds. J Food Sci 51:1382–1383.

Le Nguyen K, Grondin A, Courtois B, Gantet P (2019) Next-generation sequencing accelerates crop gene discovery. *Trends Plant Sci* 24:263–274.

Lee ES, Lee YR, Lee O, Yang HB, Lee H. E., Han, K., & Kim, D. S. (2022). Genetic Diversity and Genome-Wide Association Study of Pumpkins (Cucurbita moschata) Originating from East Asia. https://ssrn.com/abstract=4160303

Lee HY, Jang S, Yu CR, Kang BC, Chin JH, Song K (2020) Population structure and genetic diversity of *Cucurbita moschata* based on genome-wide high-quality SNPs. *Plants* 10:56.

Lelley T, Loy B, Murkovic M (2009) Hull-less oil seed pumpkin. In: Vollmann J, Rajcan I (eds.). *Oil Crops*, 4th Edn. Springer, New York, NY.

Li, Y. L., Li, H. Z., Cui, C. S., Zhang, H. Y., & Gong, G. Y. (2007). Molecular markers linked to the dwarf gene in squash. *J Agr Biotech*, 15(2), 279–82.

Liu C, Ge Y, Wang DJ, Li X, Yang XX, Cui CS, Qu SP (2013) Morphological and molecular diversity in a germplasm collection of seed pumpkin. *Sci Hortic* 154:8–16.

Loy JB (2004) Morpho-physiological aspects of productivity and quality in squash and pumpkins (*Cucurbita* spp.). *Cri Rev Plant Sci* 23:337–363.

Loy JB (2012) Breeding squash and pumpkins. In: Wang, Y.-H., Behera, T.K., & Kole, C. (eds.) *Genetics, Genomics, and Breeding of cucurbits*, 1st Edn. CRC Press, Boca Raton, FL.

Luo W, Li Y, Sun Y, Lu L, Zhao Z, Zhou J, Li X (2021) Comparative RNA-seq analysis reveals candidate genes associated with fruit set in pumpkin. *Sci Hortic* 288:110255.

Mady E, Ibrahim SD, Randhir R, El-Hakim A, Ahmed F, Randhir TO (2022) Genetic variation among pumpkin landraces based on seed qualities and molecular markers. *Mol Biol Rep* 1:11.

Maheswaran M, Huang N, Sreerangasamy SR, McCouch SR (2000) Mapping quantitative trait loci associated with days to flowering and photoperiod sensitivity in rice (*Oryza sativa* L.). *Mol Breed* 6:145–155.

McGregor C, Meru G (2021) Translational genomics of cucurbit oil seeds. In: Tombuloglu, H., Unver, T., Tombuloglu, G., Hakeem, K.R. (eds) *Oil Crop Genomics*. Springer, Cham, pp 89–111.

McPhail-Medina R, Wessel-Beaver L, Rodrigues JCV (2012) Inheritance of resistance to Papaya ringspot virus in tropical pumpkin is controlled by at least two genes. In Cucurbitaceae 2012. Proceedings of the Xth EUCARPIA Meeting on Genetics and Breeding of Cucurbitaceae, Antalya, Turkey, 15–18 October, 2012 . University of Cukurova, Ziraat Fakultesi, pp 697–701.

Méndez-López A, Sánchez-Vega M, Villanueva-Verduzco C, Rojas-Martínez RI, Rodríguez-Pagaza Y (2018) Genetic diversity in parthenocarpic varieties of pumpkin (*Cucurbita pepo* L.) BY RAPD. *Trop and Subtrop Agroecosystems* 21(3):447–455.

Meru G, Fu Y, Leyva D, Sarnoski P, Yagiz Y (2018) Phenotypic relationships among oil, protein, fatty acid composition and seed size traits in *Cucurbita pepo*. *Sci Hortic* 233:47–53.

Meru G, Fu Y, Shrestha S, Michael VNE, Dorval M, Mainviel R (2022) Genomic position and markers associated with the hull-less seed trait in pumpkin. *Plants* 11:1238.

Meuwissen TH, Hayes BJ, Goddard M (2001) Prediction of total genetic value using genome-wide dense marker maps. *Genetics* 157:1819–1829.

Michael VNE, Fu Y, & Meru G (2019). Inheritance of resistance to phytophthora crown rot in *Cucurbita pepo. HortScience* 54(7):1156–1158.

Michael VNE, Fu Y, Shrestha S, Meru G (2021) A novel QTL for resistance to phytophthora crown rot in squash. *Plants* 10:2115.

Miranda-Vélez MA, Wessel-Beaver L, Rodrigues JCV (2019) Non-transmission of ZYMV and PRSV through resistant *Cucurbita moschata* genotypes "Nigerian Local" and "Menina". *Cucurbit Genet Coop Rep* 42:37.

Mojica FJ, Díez-Villaseñor C, García-Martínez J, Almendros C (2009) Short motif sequences determine the targets of the prokaryotic CRISPR defense system. *Microbiology* 155:733–740.

Montero-Pau J, Blanca J, Bombarely A, Ziarsolo P, Esteras C, Martí-Gómez C et al. (2018). De novo assembly of the zucchini genome reveals a whole-genome duplication associated with the origin of the *Cucurbita* genus. *Plant Biotechnol J* 16:1161–1171.

Montero-Pau J, Blanca, Esteras C, Martínez-Pérez EM, Gómez P, Monforte AJ et al. (2017) An SNP-based saturated genetic map and QTL analysis of fruit-related traits in Zucchini using Genotyping-by-sequencing. *BMC Genomics* 18:1–21.

Morilipinar EO, Şekerci AD, Coşkun ÖF, Gülşen O (2021) Genetic analysis of local pumpkin populations. *Int J Agric Nat Sci* 14:264–272.

Muralidhara MS, Narasegowda NC (2014) Genetic diversity analysis of pumpkin genotypes (*Cucurbita moschata* Duch ex. Poir) using morphological and RAPD markers. *Asian J Biol Sci* 9:188–194.

Murovec J (2015) Phenotypic and genetic diversity in pumpkin accessions with mutated seed coats. *HortScience* 50:211–217.

Nakic SN, Rade D, Skevin D, Strucelj D, Mokrovcak Z, Bartolic M (2006) Chemical characteristics of oils from naked and husk seeds of *Cucurbita pepo* L. *Eur J Lipid Sci Technol* 108:963–943.

Ntuli NR, Tongoona PB, Zobolo AM (2015) Genetic diversity in *Cucurbita pepo* landraces revealed by RAPD and SSR markers. *Sci Hortic* 189:192–200.

Nyabera LA, Nzuki IW, Runo SM, Amwayi PW (2021) Assessment of genetic diversity of pumpkins (*Cucurbita* spp.) from western Kenya using SSR molecular markers. *Mol Biol Rep* 48:2253–2260.

Öztürk Hİ, Dönderalp V, Bulut H, Korkut R (2022) Morphological and molecular characterization of some pumpkin (*Cucurbita pepo L.*) genotypes collected from Erzincan province of Turkey. *Sci Rep* 12:1–11.

Öztürk HI, Dönderalp V, Bulut H, Korkut R, Hosseinpour A, Dursun A (2021) Exploring the genetic diversity of some squash (*Cucurbita maxima L.*) germplasm using morphological and molecular markers in Erzincan. *Res Square* 848708: 1–19.

Pachner M, Paris HS, Lelley T (2011) Genes for resistance to zucchini yellow mosaic in tropical pumpkin. *J Hered* 102:330–335.

Pachner M, Paris HS, Winkler J, Lelley T (2015) Phenotypic and marker-assisted pyramiding of genes for resistance to zucchini yellow mosaic virus in oilseed pumpkin (*Cucurbita pepo*). *Plant Breed* 134:121–128.

Padley LD, Kabelka EA, Roberts PD (2009) Inheritance of resistance to crown rot caused by *Phytophthora capsici* in Cucurbita. *HortScience* 44:211–213.

Pan L, Wang M, Yang Y, Chen C, Dai H, Zhang Z et al. (2022). Whole-genome resequencing identified QTLs, candidate genes and kompetitive allele-specific PCR markers associated with the large fruit of Atlantic Giant (*Cucurbita maxima*). *Front Plant Sci* 13: 942004.

Paris HS, Cohen S (2000). Oligogenic inheritance for resistance to zucchini yellow mosaic virus in *Cucurbita pepo*. *Ann Appl Biol* 136:209–214.

Paris HS, Cohen S, Burger Y, Yoseph R (1988) Single-gene resistance to zucchini yellow mosaic virus in *Cucurbita moschata*. *Euphytica* 37:27–29.

Paris HS, Yonash N, Portnoy V, Mozes-Daube N, Tzuri G, Katzir N (2003) Assessment of genetic relationships in *Cucurbita pepo* (Cucurbitaceae) using DNA markers. *Theor Appl Genet* 106:971–978.

Park BS, Jang S, Yu Y, Choi GJ, Kang BC, Seo S (2020) QTL Mapping and molecular markers of Powdery Mildew Resistance in Pumpkin (*Cucurbita moschata*). *Hort Sci Technol* 38(5):717–729.

Pomares-Viciana T, Río-Celestino D, Román B, Die J, Pico B, Gómez P (2019) First RNA-seq approach to study fruit set and parthenocarpy in zucchini (*Cucurbita pepo* L.). *BMC Plant Biol* 19:1–20.

Pratt S, Matthews K (2006) Superfoods: Fourteen foods that will change your life. Random House.

Provvidenti R (1997) New American summer squash cultivars possessing a high level of resistance to a strata of zucchini yellow mosaic virus from China. *Report-Cucurbit Genet Coop* 20:57–57.

Provvidenti R (2000) A strain of watermelon mosaic virus from Massachusetts causes prominent symptoms on squashes and systemically infects *Cucurbita equadorensis and C.* maxima PI 419801-1. Report–*Cucurbit Genet Coop* 23:68–68.

Radwan SAA (2014). Molecular discrimination and genetic relationships between some cultivars of *Cucurbita pepo* ssp. pepo using random amplification of polymorphic DNA (RAPD) analysis. *Afr J Biotechnol* 13(11).

Ramos A, Fu Y, Michael V, Meru G (2020) QTL-seq for identification of loci associated with resistance to *Phytophthora* crown rot in squash. *Sci Rep* 10:1–8.

Reddy UK, Natarajan P, Abburi VL, Tomason Y, Levi A, Nimmakayala P (2022) What makes a giant fruit? Assembling a genomic toolkit underlying various fruit traits of the mammoth group of *cucurbita maxima. Front Genet* 13:1005158–1005158.

Rhodes AM (1964) Inheritance of Powdery mildew resistance. *Plant Dis Reporter*, 54: 54–60.

Roy A, Spoorthi P, Panwar G, Bag MK, Prasad TV, Kumar G et al. (2013) Molecular evidence for occurrence of tomato leaf curl New Delhi virus in ash gourd (*Benincasa hispida*) germplasm showing a severe yellow stunt disease in India. *Indian J Virol* 24:74–77.

Saez C, Martínez C, Ferriol M, Manzano S, Velasco L, Jamilena M., et al. (2016). Resistance to tomato leaf curl New Delhi virus in *Cucurbita* spp. *Ann Appl Biol* 169:91–105.

Saez C, Martínez C, Montero-Pau J, Esteras C, Sifres A, Blanca J, Picó B (2020) A major QTL located in chromosome 8 of *Cucurbita moschata* is responsible for resistance to Tomato leaf curl New Delhi virus. *Front Plant Sci* 11:207.

Sánchez-de la Vega G, Castellanos-Morales G, Gámez N, Hernández-Rosales HS, Vázquez-Lobo A, Aguirre-Planter E, Eguiarte LE (2018) Genetic resources in the “Calabaza Pipiana” squash (*Cucurbita argyrosperma*) in Mexico: genetic diversity, genetic differentiation and distribution models. *Front Plant Sci* 9:400.

Seal D, Zhang S, Dittmar P, Li Y, Klassen W, Wang Q, Olczyk T (2022) Summer Squash Production in Miami-Dade County, Florida: HS-861/TR012, rev. 6/2022. EDIS: 16–16.

Sharma A, Rana C, Shiwani K (2016) Important Insect Pests of Cucurbits and Their Management. In: *Handbook of Cucurbits: Growth, Cultural Practices, and Physiology*. CRC Press. 327–360

Shi Y, Zhang M, Shu Q, Ma W, Sun T, Xiang C et al. (2021). Genetic mapping and identification of the candidate gene for white seed coat in *Cucurbita maxima. Int J Mol Sci* 22:2972.

Shikha M, Kanika A, Rao AR, Mallikarjuna MG, Gupta HS, Nepolean T (2017) Genomic selection for drought tolerance using genome-wide SNPs in maize. *Front Plant Sci* 8:550.

Shrestha S, Fu, Y, Michael VNE, Meru G (2022) Whole genome re-sequencing and bulk segregant analysis reveals chromosomal location for papaya ringspot virus W resistance in squash. *Front Plant Sci* 13. DOI: 10.3389/fpls.2022.848631

Shrestha S, Michael VNE, Fu Y, Meru G (2021) Genetic loci associated with resistance to zucchini yellow mosaic virus in squash. *Plants* 10:1935.

Sim SC, Hong JH, Kwon YS (2015) DNA profiling of commercial pumpkin cultivars using simple sequence repeat polymorphisms. *Hortic Environ Biotechnol* 56:811–820.

Singh AK, Mishra KK, Chattopadhyay B, Chakraborty S (2009) Biological and molecular characterization of a begomovirus associated with yellow mosaic vein mosaic disease of pumpkin from Northern India. *Virus Genes* 39:359–370.

Sohrab SS, Mandal B, Pant RP, Varma A (2003) First report of association of Tomato leaf curl virus-New Delhi with yellow mosaic disease of *Luffa cylindrica* in India. *Plant Dis* 87:1148–1148.

Solhaug EM, Roy R, Chatt EC, Klinkenberg PM, Mohd-Fadzil NA, Hampton M, Carter CJ (2019) An integrated transcriptomics and metabolomics analysis of the *Cucurbita pepo* nectary implicates key modules of primary metabolism involved in nectar synthesis and secretion. *Plant Direct* 3:e00120.

Stevenson DG, Eller FJ, Wang L, Jane JL, Wang T, Inglett GE (2007) Oil and tocopherol content and composition of pumpkin seed oil in 12 cultivars. *J Agric Food Chem* 55:4005–4013.

Sun H, Wu S, Zhang G, Jiao C, Guo S, Ren Y, Xu Y (2017) Karyotype stability and unbiased fractionation in the paleo-allotetraploid *Cucurbita* genomes. *Mol Plant* 10:1293–1306.

Syed QA, Akram M, Shukat R (2019) Nutritional and therapeutic importance of the pumpkin seeds. *Seed* 21:5798–15803.

Tan XZ, Ge Y, Xu WL, Cui CS, Qu SP (2013) Construction of genetic linkage map and QTL analysis for seed width in pumpkin (*Cucurbita maxima*). *Acta Bot Borealioccidentalia Sin* 33:697–702.

Tsivelikas AL, Koutita O, Anastasiadou A, Skaracis GN, Traka-Mavrona E, Koutsika-Sotiriou M (2009) Description and analysis of genetic diversity among squash accessions. *Braz Arch Biol Technol* 52: 271–283.

Verdone M, Rao R, Coppola M, Corrado G (2018) Identification of zucchini varieties in commercial food products by DNA typing. *Food Control* 84: 197–204.

Vitiello A, Rao R, Corrado G, Chiaiese P, Digilio MC, Cigliano RA, D'Agostino N (2018) De novo transcriptome assembly of *Cucurbita pepo* L. leaf tissue infested by *Aphis Gossypii*. *Data* 3:36.

Vogel G, Gore MA, Smart CD (2021) Genome-wide association study in New York *Phytophthora capsici* isolates reveals loci involved in mating type and mefenoxam sensitivity. *Phytopathology* 111:204–216.

Wang S, Wang K, Li Z, Li Y, He J, Li H et al. (2022) Architecture design of cucurbit crops for enhanced productivity by a natural allele. *Nat Plants* 8(12):1–14.

Wang W, Shi Y, Liu Y, Xiang C, Sun T, Zhang M et al. (2021). Genetic relationships among *Cucurbita pepo* ornamental gourds based on EST-SSR markers. *Czech J Genet Plant Breed* 57:125–139.

Wang Y, Wang C, Han H, Luo Y, Wang Z, Yan C et al. (2020) Construction of a high-density genetic map and analysis of seed-related traits using specific length amplified fragment sequencing for *Cucurbita maxima*. *Front Plant Sci* 10:1782.

Wang Y, Yan C, Zou B, Wang C, Xu W, Cui C, Qu S (2019) Morphological, transcriptomic and hormonal characterization of trimonoecious and subandroecious pumpkin (*Cucurbita maxima*) suggests important roles of ethylene in sex expression. *Int J Mol Sci* 20:3185.

Wang Z, Gerstein M, Snyder M (2009) RNA-Seq: a revolutionary tool for transcriptomics. *Nat Rev Genet* 10:57–63.

Weeden NF, Robinson RW (1986) Allozyme segregation ratios in the interspecific cross *Cucurbita maxima x C. ecuadorensis* suggest that hybrid breakdown is not caused by minor alterations in chromosome structure. *Genetics* 114:593–609.

Wessel-Beaver L (2000) Evidence for the center of diversity of *Cucurbita moschata* in Colombia. *Cucurbit Genet Coop Rep* 23: 54–55.

Wessel-Beaver, L. (1993). Powdery and downy mildew resistance in *Cucurbita moschata* accessions. *Cucurbit Genet Coop Rep* 16:73–74.

Whitaker TW (1974) *Cucurbita*. In: King RC (ed) *Handbook of Genetics*. Springer, New York, NY.

Whitaker TW, Bemis WP (1975) Origin and evolution of the cultivated Cucurbita. *Bull Torrey Bot Club* 102(6):362–368.

Whitaker TW, Robinson RW (1986) Squash breeding. In: Basset MJ (ed) *Breeding Vegetable Crops*. Avi Publishing Company, Westport, CT 209–238.

Wu J, Chang Z, Wu Q, Zhan H, Xie S (2011). Molecular diversity of Chinese *Cucurbita moschata* germplasm collections detected by AFLP markers. *Sci Hortic* 128:7–13.

Wu T, Cao J (2010) Molecular cloning and expression of a bush related CmV1 gene in tropical pumpkin. *Mol Biol Rep* 37:649–652.

Wu T, Luo S, Wang R, Zhong Y, Xu X, Lin YE, Huang H (2014) The first Illumina-based de novo transcriptome sequencing and analysis of pumpkin (*Cucurbita moschata Duch.*) and SSR marker development. *Mol Breed* 34:1437–1447.

Wyatt LE, Strickler SR, Mueller LA, Mazourek M (2015) An acorn squash (*Cucurbita pepo* ssp. *ovifera*) fruit and seed transcriptome as a resource for the study of fruit traits in Cucurbita. *Hortic Res* 2:14070.

Wyatt LE, Strickler SR, Mueller LA, Mazourek M (2016) Comparative analysis of *Cucurbita pepo* metabolism throughout fruit development in acorn squash and oilseed pumpkin. *Hortic Res* 3:16045.

Xanthopoulou A, Ganopoulos I, Psomopoulos F, Manioudaki M, Moysiadis T, Kapazoglou A et al. (2017) De novo comparative transcriptome analysis of genes involved in fruit morphology of pumpkin cultivars with extreme size difference and development of EST-SSR markers. *Gene* 622:50–66.

Xanthopoulou A, Montero-Pau J, Picó B, Boumpas P, Tsaliki E, Paris HS et al. (2021). A comprehensive RNA-Seq-based gene expression atlas of the summer squash (*Cucurbita pepo*) provides insights into fruit morphology and ripening mechanisms. *BMC Genomics* 22:1–17.

Xanthopoulou A, Psomopoulos F, Ganopoulos I, Manioudaki M, Tsaftaris A, Nianiou-Obeidat I, Madesis P (2016) De novo transcriptome assembly of two contrasting pumpkin cultivars. *Genom Data* 7:200–201.

Xiang C, Duan, Y, Li H, Ma W, Huang S, Sui X et al. (2018). A high-density EST-SSR-based genetic map and QTL analysis of dwarf trait in *Cucurbita pepo* L. *Int J Mol Sci* 19:3140.

Xie J, Lei B, Niu M, Huang Y, Kong Q, Bie Z (2015) High throughput sequencing of small RNAs in the two Cucurbita germplasm with different sodium accumulation patterns identifies novel microRNAs involved in salt stress response. *PLoS One* 10:e0127412.

Xin, T., Tian, H., Ma, Y., Wang, S., Yang, L., Li, X., ... & Yang, X. (2022). Targeted creation of new mutants with compact plant architecture using CRISPR/Cas9 genome editing by an optimized genetic transformation procedure in cucurbit plants. *Hortic Res*, 9, uhab086.

Xu X, Lu X, Tang Z, Zhang X, Lei F, Hou L, Li M (2021) Combined analysis of carotenoid metabolites and the transcriptome to reveal the molecular mechanism underlying fruit colouration in zucchini (*Cucurbita pepo* L.). *Food Chem: Mol Sci* 2:100021.

Xu Y, Guo SR, Shu S, Ren Y, Sun J (2017) Construction of a genetic linkage map of rootstock-used pumpkin using SSR markers and QTL analysis for cold tolerance. *Sci Hortic* 220:107–113.

Yunli W, Yangyang WANG, Wenlong X, Chaojie W, Chongshi C, Shuping Q (2020) Genetic diversity of pumpkin based on morphological and SSR markers. *Pak J Bot* 52:477–487.

Zhang G, Ren Y, Sun H, Guo S, Zhang F, Zhang J et al. (2015). A high-density genetic map for anchoring genome sequences and identifying QTLs associated with dwarf vine in pumpkin (*Cucurbita maxima* Duch.). *BMC Genomics* 16:1–13.

Zheng Y, Wu S, Bai Y, Sun H, Jiao C, Guo S et al. (2019) Cucurbit Genomics Database (CuGenDB): a central portal for comparative and functional genomics of cucurbit crops. *Nucleic Acids Res* 47(D1):D1128–D1136.

Zhong YJ, Zhou YY, Li J X, Yu T, Wu TQ et al. (2017) A high-density linkage map and QTL mapping of fruit-related traits in pumpkin (*Cucurbita moschata* Duch.). *Sci Rep* 7:1–12.

Zraidi A, Stift G, Pachner M, Shojaeiyan A, Gong L, Lelley T (2007) A consensus map for *Cucurbita pepo*. *Mol Breed* 20:375–388.

12 Allele Mining for Genome Designing in Radish (*Raphanus sativus* L.)

Raman Selvakumar[1] *and Gangadhara Rao P.*[2]
[1]Center for Protected Cultivation Technology, ICAR-IARI, New Delhi 110012, India
[2]College of Horticulture, Venkataramannagudem, Dr. YSR Horticultural University, Andhra Pradesh 534101, India
Corresponding author: selvakumrsingai@gmail.com

12.1 INTRODUCTION

New cultivars have steadily supplanted conventional agricultural practises, plant landraces, and vegetable variations during the previous century. This hastens the loss of genetic variety and financial resources in the vegetable business. However, certain vegetable crop-focused gene banks have been able to recover a significant percentage of the critical genetic capital lost as a result of the reduction of elite content (Till et al. 2003a). Vegetable crops are genetically improved through breeding for specific features, and this improvement is based on conserved accessions from gene banks. Furthermore, they give a wealth of unusual allelic diversity that may be employed in food crop production. This is a significant benefit. Genomic advances over the past quarter-century have made genetic manipulation of vegetable crops more feasible. This will help to solve challenges about the long-term security of the world's food supply and the health of its growing population. The development of novel molecular markers for commercially valuable vegetable crops, as well as transgenic gene expression changes, need a detailed understanding of a vegetable crop's genomic sequence. TILLING and Eco-TILLING are only two of the allele mining technologies that have been developed in recent years to take use of widely accessible DNA sequence data to test specific gene functions. These strategies are intended to capitalize on the availability of such information. The expansion of the gene sequence collection, which now includes data on vegetable crop genes, has opened up new possibilities for studying gene function. Forward genetic analysis of a phenotype is a tried and precise method for discovering the genetic basis of a character. This approach has a lengthy history of being used. In inverse genetics, researchers change the gene or its expression level and then evaluate the phenotype that results (Portis et al. 2016). The forward genetic method is promoted because it enables one to determine how each gene contributes to the overall function of the organism. The use of reverse genetic approaches has tremendously assisted the investigation of fundamental gene function. As a consequence, using strategies based on forward genetics is the wisest choice. New complementary DNA (cDNA) and expressed sequence tag (EST) sequences are being developed, and additional vegetable crop genome sequencing experiments are being proposed (www.ncbi.nlm.nih.gov/sites/entrez?db=genomeplj). The terms (EST) and cDNA refer to the same entity. As a result, comparative genomic investigations may continue, perhaps leading to proof of genomics (Prochnik et al. 2012). Targeted induced local lesions in genomes (TILLING) was developed to complete tasks that other, more well-known technologies, such as

DOI: 10.1201/9781003376439-12

gene attribute projections and genome sequencing, could not. This shows that it may be beneficial for vegetable crops that have not yet been well investigated. Because it contains no genetically modified organisms (GMOs), its use in research aimed at improving vegetable crops is permissible and has no legal ramifications (McCallum et al. 2000a). This permits it to be used without fear of violating any rules. It is feasible that functional genomics results relating to critical features of one crop will have an impact on another crop. TILLING is the finest technology for enhancing vegetable crops that does not need the use of transgenic organisms since it allows access to practically every gene for functional analysis. Increase the genetic variety of domesticated wild vegetable crop relatives to accomplish this. TILLING is an alternative way of growing vegetables that does not need the use of transgenics. In this chapter we discuss about different TILLING approaches and applications in radish.

12.2 EVOLUTION OF ALLELE MINING

An "allele" is a variation form of a gene that may be found on a certain chromosome. Mining is the process of searching for new alleles in wild germplasm. Kumari et al. (2018) define "allele mining" as the act of looking for potentially advantageous alleles of genes across numerous cultivars, related species, and even unrelated species. The goal of this research is to find allelic variation in fundamental aspects of genetic resource sets. Early investigations of allele mining focused only on identifying single nucleotide polymorphisms (SNPs) and insertion/deletion (InDels) polymorphisms in gene coding portions or exons, since these changes were thought to affect the encoded protein structure and/or function. The effect of such genic sequence changes on phenotypes may be seen in a variety of contexts. Nucleotide changes in noncoding regions (5′ UTR including promoter, introns, and 3′ UTR) have been shown to have a significant influence on transcript production and accumulation, and hence trait expression.

Mining for sequence variation has grown increasingly significant in the context of gene expression in recent years, especially in regulatory regions of genes. Changes to promoter element sequences, which are well-known to play a key role in gene regulation, will have a significant impact on gene expression and hence trait expression. Variation in the binding locations of upstream cisacting and transacting regulatory elements is often at the root of such gene expression variations. It has been proposed that "real" allele mining go beyond only looking at sequence variations in candidate gene coding areas to capture the majority of functional variants of relevance in the genes (Latha et al. 2004). Promoters of genes have been identified as a possible target for allele mining, sometimes known as "promoter mining," because of the relevance of "mining" for variation in regulatory domains. The goal of promoter mining and *in silico* analysis was to identify the number, frequency, and location of regulatory element binding sites in promoter regions, as well as the transcription factors associated with specific expression patterns in response to constitutive, developmental, tissue-specific, hormonal, and environmental regulation (Veerla and Hoglund 2006). This sort of promoter sequence analysis may also reveal the conservation of regulatory patterns unique to differentiated cell/tissue types. Extraction of innovative and effective promoters for use in genetic engineering applications; insights into gene expression molecular principles. The incorporation of methods such as *in vitro* and *in silico* promoter mining into regular allele mining programs to identify and categorize cisregulatory elements and understand their role in gene regulation is becoming increasingly important as the number of global gene expression experiments to identify sets of genes with similar expression patterns grows.

12.3 APPROACHES OF ALLELE MINING

Allele mining may be done in a variety of methods, ranging from the quick and inexpensive polymerase chain reaction (PCR) to the more time- and cost-intensive next-generation sequencing (NGS)

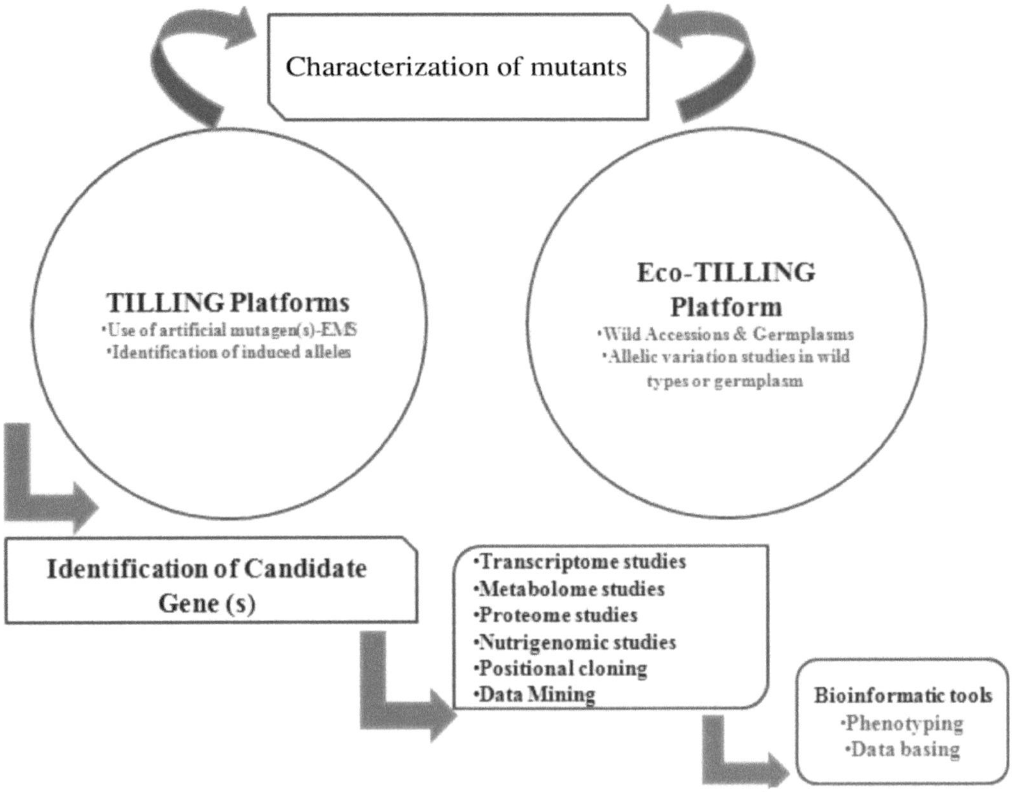

FIGURE 12.1 Allele mining approaches in radish.

Source: Selvakumar et al. 2023.

and everything in between. There are several applications for allele molecular polymorphism and incomplete sequence information. Additional uses need complete cloning of the allele. All DNA-based technologies rely on precise targeting of target genes. Following the validation of the target gene model, a complete set of selective primers may be built to search for new alleles in the whole or core germplasm collection. PCR amplification of a given allele or collection of alleles is included in all molecular marker techniques. Polymorphisms between amplified alleles might be detected using single-strand specific nucleases. This method is often employed in TILLING processes because it allows for the simple identification of nicking of PCR product heteroduplexes. The results might appear as a certain size DNA band on an agarose gel. However, when using gene-specific markers in diverse germplasm material, specific annealing of the primers may occur, necessitating the sequencing of the produced PCR fragments to validate their legitimacy and uniformity. Figure 12.1 from Selvakumar et al. (2023) shows the advantages of conventional and eco-tilling on vegetable crops.

12.3.1 Targeted Induced Local Legions in Genome

TILLING is often used to detect mutations as a kind of reverse genetics. This technique combines recurrent mutagenesis by nuclease-deficient DNA cleavage with high-throughput mutation detection methodologies (Till et al. 2003b; Till et al. 2007a, b; Tadele 2016). The Howard Hughes Medical Institute and the Fred Hutchinson Cancer Research Centre were the first to use it on Arabidopsis thaliana (McCallum et al. 2000a, b; Tadele 2016). TILLING includes a diverse collection of germplasm, including seeds and genomic DNA derived from mutagenic agents. Long-term storage of DNA and other biological resources, such as seeds, may be possible. The storage contents of the library may be simply transmitted across the various breeders working on vegetable genetic improvement, making screening for many traits simpler (Till et al. 2003b). This is a significant advancement above standard forward genetics or mutant breeding procedures. These approaches only maintain a subset of the mutant alleles and discard the rest of the manufactured modifications. Till et al. 2003b; Greene et al. 2003) expect that a screening of 6,000 chemically mutagenic Arabidopsis plants from a diploid generation would reveal around 30 potentially damaging mutations. TILLING may be used to amplify putatively favorable alleles that do not generate phenotypic changes on their own but do when coupled with other alleles. This strategy requires substantial plant phenotypic research; nonetheless, concentrated mutant breeding combined with genomics may boost agricultural production. For decades, plant species have been mined for mutants for reverse-genesis methods, and chemical mutagenesis on *Drosophila* has been employed since the 1940s (Natarajan 2005). Mutants have resulted in the development of new plant species such as tomatoes, chilli peppers, okra, sweet potatoes, garden peas, dolichos beans, and French beans (Ahloowalia et al. 2004). These synthetic mutations have the potential to combine a desired phenotype and genotype. Ionizing radiation, ethylmethane sulfonate (EMS), N-ethyl-N-nitrosoure (ENU), and N-nitroso-N-methylurea (NMU) are some of the chemical agents utilized in the generation of mutants. Mutagenic chemicals may engage in base-pairing interactions and integrate their modifications into the replicating DNA sequence as a consequence of EMS's capacity to form nuclear adducts (Haughn and Somerville 1987; Ashburner et al. 2005). Because of the alkylation, guanine bases are wrongly paired with thymine rather than cytosine. As a consequence, G/C to A/T mutations will follow (Henikoff and Comai 2003; Nagy et al. 2003). TILLING can identify the allelic sequence of missense and truncation mutations using chemical mutagens that primarily generate random point mutations at high density (Greene et al. 2003). TILLING is a technique for categorizing single nucleotide polymorphisms (SNPs) and insertions/deletions (InDels) in genes of interest in a mutagenesis population. To initiate TILLING, a mutagenic species must first be created and then exposed to a chemical mutagen such as ethylene dioxychloride (EMS). TILLING's capacity to generate single nucleotide modifications or small insertions/deletions at a high frequency (30 nucleotides) is what makes it such a potent mutagen (Till et al. 2007a). TILLING uses a chemical muagenesis-based reverse genetics approach to uncover phenotype-altering point mutations via heterogeneity and adaptive molecular screening.

To find mediated polymorphism or mutations, approaches that rely on the structural features of single-stranded DNA are often used (Myers et al. 1987; Nataraj et al. 1999; Rozycka et al. 2000). Because the detection sensitivity is limited to 450 bp dHPLC, which finds mismatches in heteroduplexes created by melting and annealing heteroallele DNA, target genes with small exons separated by large introns are best suited for this method. Among the faults uncovered (Jeong et al. 2012), major changes were observed to induce dishonesty and irrational behavior. These modifications may enhance phenotypic analysis. This mutation has been found in all TILLING tests, which refer to basic nuclease cleavage (Figure 12.2).

12.3.1.1 Conventional TILLING

Because most plant species are capable of self-fertilization, this method is appropriate for their cultivation. These plants' seeds may be stored in a cold, dry environment for several years (McCallum et al. 2000a, b; Colbert et al. 2001; Borevitz et al. 2003). TILLING's most recent version combines traditional mutagenesis processes with cutting-edge mutation detection algorithms. TILLING's most significant components are mutagenesis, non-chemical culture, germplasm collection, DNA isolation, sample pooling, mutation detection (by sampling a population for induced mutations), and mutant assessment. TILLING is a popular approach since its operations may be used to a variety of vegetable crop kinds. TILLING-revealed mutants might be utilized to investigate the structure and function of genes that could be incorporated in crop breeding systems via genetic enhancement.

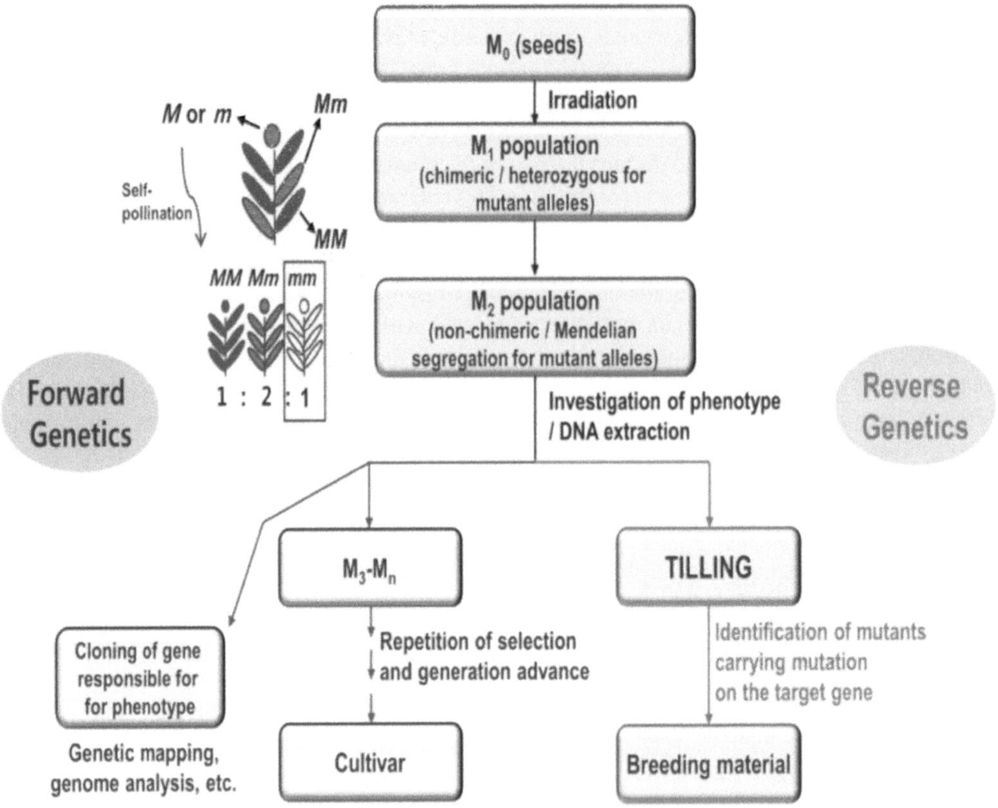

FIGURE 12.2 Steps involved in the TILLING approaches.

Plants that create a high level of chemical mutagens during TILLING may transfer unique single basis pair mutations into a culture via their seeds and pollen. At the same time, these cultures are nearing a point in time where mutations are passed on to succeeding generations quite efficiently. To promote the expansion of the M_2 population, the population's seeds are fed mutagens, fostered into M_1 seeds, and then distributed to the general population. In the mutation screening technique, DNA isolated from M_2 plants was employed (Colbert et al. 2001). The seeds' widespread distribution will aid in the establishment of a sustainable food supply. To avoid reintroducing a deleted mutation, just one M_2 plant's DNA is retrieved for each M_1 plant (Till et al. 2007a, b). M_3 seeds, which are the offspring of M_2 plants, have a lengthy storage life. Mutations in diploid animals may be identified using a DNA pool made up of eight human samples (Henikoff and Comai 2003). The SNP's ploidy level, heterozygosity, and beneficial effects should be utilized to detect a significant pooling of the problematic plant species in the wild. These elements must be considered. This DNA mixture may be cultivated on microtitre plates for further investigation. Primers for polymerase chain reactions (PCRs) are particularly designed for amplification of target genes prior to TILLING investigations. Exploring possible targets in unknown plant species may potentially benefit from integrating relevant genomes data from a wide range of organisms using bioinformatics techniques. The use of dye-labeled primers may enhance the outcomes of a population-scale DNA PCR experiment. The desired genes must be amplified using primers designed to detect mutations. The use of IR dye-labeled primers allows for automated capillary electrophoresis amplification, and PCR products may be identified using equipment such as LI-COR. In both cases, primers must be carefully indicated. Heteroduplexes, which may be created by denaturing and reannealing PCR products, are the cornerstone of mediated natural PCRs, which are used to discover disparities. TILLING technique has made use of a variety of endonucleases, including celery-isolated CEL-1 endonuclease, S1 nucleases, and T4 endonuclease. When pools are driven to mutate, homoduplexes and heteroduplexes develop. Till et al. (2007a) employed cleaved heteroduplexes to calculate the full-length product by summing the diameters of the two separate products. IRDye 700 and IRDye 800 were given to the two substances with lower molecular weights. Based on the size of the pieces breaking off from the whole in relation to the size norm, the series will be able to roughly determine the location of the mutation. Users of the PARSESNP (Project Compatible Linked Sequences and SNPs Evaluation) software, which can be found at www.proweb.org/parsesnp, will be directed through the process of inputting a certain number of nucleotide gene variants. The program monitors how each variation affects the expression of a gene variable by comparing it to a reference DNA sequence and an exon/intron position model. PARSESNP (Taylor et al. 2003) is a software that classifies polymorphisms within genes and shows their roles. CODDLE (Codons Programmed to Identify Deleterious Lesions) (www.proweb.org/input) and SIFT (Sorting Intolerant from Tolerant) were developed by Arabidopsis TILLING project (ATP) researchers to assess the impact of mediated or normal polymorphisms on gene expression (Gilchrist et al. 2005; Ng and Henikoff et al. 2003). When the DNA of four to eight distinct species is swapped, the throughput is maximized. The genetic sequences examined varied in size from 0.3 to 1.6 kilobytes. By increasing pooling by 8%, cutting the fragments into 1.4 kb units (excluding the ends of the fragments because those are the only places where SNP detection is noiseless), and running the experiment with 96 lanes, it is possible to scan up to one million base pairs of genomic DNA in a single experiment. The majority of existing TILLING systems are divided into two phases: (1) pool scanning and (2) individual marking.

12.3.1.2 TILLING by Sequencing

This method may rule out errors in the sequence due by anything other than random chance (Bansal 2010; Druley et al. 2009; Koboldt et al. 2009; Bansal et al. 2010). The ability of high-throughput sequencing to give wide parallel sequencing allows for the development of mutation detection approaches (Gilchrist and Haughn 2005). Scientists recently resequenced Arabidopsis' whole genome to better understand the extent and frequency of spontaneous mutations (Ossowski et al. 2010). One-nucleotide polymorphisms, on the other hand, can only be definitively discovered with extensive sequencing coverage. The cost of resequencing a population of such magnitude is too high. By concentrating on fewer genes, the cost of resequencing might be decreased. This may be accomplished by sequencing PCR products (Porreca et al. 2007, Rigola et al. 2009) or by releasing an oligonucleotide-induced sequence that can then be sequenced (Nijman et al. 2010). The creation of standards has also addressed the difficulty of finding relatively rare polymorphisms in live organisms. Rigola et al. (2009) demonstrated the use of 454 sequences from overlapping pools in conjunction with a computational technique for mutation study using tomatoes (*Solanum lycopersicum*). Next-generation methods for SNP identification that do not overlap include CRISP (Bansal 2002; Bansal et al. 2010), SNP Seeker (Druley et al. 2009), SAM tools, and MAQ (Li et al. 2009). These methodologies and computational procedures for a concurrent multi-dimensional Illumina-GA-based system, however, have yet to be established. TILLING by sequencing is a versatile strategy that offers a thorough search for mutations in specific areas by using next-generation sequencing and a mix of molecular and bioinformatics tools. The approach is now ready for usage by effective TILLING populations across a variety of crops. Even in a muddled sample, signs of uncommon SNPs and mutations may be discovered. In this strategy, the combination of multidimensional pooling and a random threshold allows for scalable design. It identifies a mutation with a matching base change. It does not depend on fluorescence or endonuclease digestion, which are both potentially unreliable. It is versatile, enabling you to mix and match various swimming pool techniques and animals.

The TILLING method (Colbert et al. 2001; Henikoff et al. 2004) may be used to locate the induced mutation without sequencing the whole fragment. The resultant mutation may be selectively sequenced. The approach is flexible enough to identify SNPs that arise naturally as well as those that are artificially inserted (Dahm and Geisler 2006), as well as the ability to distinguish heterozygotes. Instead of resequencing an entire population, mining for uncommon SNPs is more practicable and cost-effective. This method may be adapted to categorise homozygous and heterozygous mutations in a pool of eight copies of any of the diploid species' 16 genomes (Colbert et al. 2001; Slade and Knauf 2005). One of the primary goals of TIILLING is to concentrate the desired genotype sequence using chemically produced variant alleles. With this TILLING process, no intricate or expensive changes are required. It speeds up the screening of mutant pools to find gene activity and eliminates the need for the time-consuming and puzzling methods of gene isolation and tissue culture. Furthermore, it has the ability to categorize potentially significant genotypes based on biotic and abiotic gene voltage. Backcross breeding and mutation selection utilizing SNP-based markers for each generation may readily erase a mutant's past if it is to be employed in crop breeding (Allen et al. 2011). This TILLING approach may be used on any plant species at extremely low densities without the need for a plant growing facility. Because most positive pools lack target gene knockouts, a key disadvantage of TILLING is that the vast majority of mutations are quiet. Costly gear is required to detect mutations.

12.3.1.3 TILLING by Next-Generation Sequencing

For most genome and transcriptome sequencing research, Sanger sequencing methods are being phased out in favor of NGS technology. Until recently, Sanger sequencing was the gold standard for DNA sequencing. These next-generation sequencing approaches may produce data at a pace many orders of magnitude quicker than earlier technologies, and at a far lower cost. According to Voelkerding et al. (2009), various next-generation sequencers are commercially available at this moment. Although all of these methods sequence populations of PCR-amplified DNA molecules, their underlying biology is somewhat different. The Oxford Nanopore and Pacific Biosciences systems, both of which can sequence individual molecules, stand out. Sequencing-based allele mining might result in faster capture of allelic data at a cheaper cost due to the future availability of low-cost, longer-read-length, and high-throughput sequencing technologies. Furthermore, the mining of particular alleles for specific genes in recognized accessions based on sequencing, as well as the correlation of these alleles with phenotypic changes, will considerably help precision breeding efforts in agricultural plants. Sequencing-based genotyping is gaining popularity as a tool for characterizing genomic variation and identifying SNPs.

12.3.2 Eco-TILLING

Similar to the TILLING process, this molecular strategy has been used to classify natural polymorphisms in target alleles in the primary and secondary gene pools of melons, beans, and peppers (Comai et al. 2004; Nieto et al. 2007; Comai and Henikoff 2006; Gilchrist et al. 2006a, b). Arabidopsis thaliana was the first plant species to profit from this technique for extracting natural polymorphisms (Comai et al. 2004a). Rather than depending on GMOs, this strategy leverages the DNA of currently existing vegetable species. TILLING compares DNA samples to a reference DNA collection to find polymorphisms. The method was used to merge ecotypes in a 1:1 ratio (Comai et al. 2004). When there are fewer haplotypes detected in the population than individuals, Eco-TILLING outperforms resequencing. Crop variety trials (Gilchrist et al. 2006a), potato polyphenol oxidase (Jin et al. 2010), and mining of several brassica vegetable varieties (Wang and al. 2010) were all successfully evaluated for their effects on viral resistance in melons and capsicums. It is useful for screening genetic variation in species due to the problems inherent in standard genetic analysis (Gilchrist et al. 2006b). This strategy might be utilized to learn more about these variations and the possible functions they play in the genome in wild crop species that demonstrate resistance to induced mutations. The SNP/haplotyping approaches may be used to identify naturally occurring SNPs and small InDels at a low cost. Using this strategy, the presence of a target gene may be evaluated in a large number of samples (Comai et al. 2004b). It is now feasible to identify distinctive traits quickly. The approach is also cost-effective since it only requires one person's sequencing to identify a haplotype. Outcrossed vegetable species, as well as heterozygous and polyploid individuals, benefit considerably from this strategy. The identification of DNA polymorphisms, such as variations in the number of satellite repeats, may be used to quantify population genetic diversity (Gilchrist et al. 2006b; Zerr and Henikoff 2005). The focus of Eco-TILLING on identifying natural alleles in vegetable crop genomic resources makes its premise to improve economic properties, such as fruit size, fruit structure, fruit color, fruit consistency, nutritional, and nutraceutical capacity, shelf life, pest and disease resistance, drought tolerance, and so on clear. Furthermore, this approach is applicable to both heterozygous and polyploid species. The key difference between TILLING and Eco-TILLING is the amount of information collected. The discovery of four induced mutations in a 1.5 kb region by 768 people is consistent with a sufficiently mutagenized diploid population. We screened 96 samples using Eco-TILLING to uncover a 1.5 kb target gene and discovered 100 polymorphisms between Arabidopsis and humans. GelBuddy was developed to assist researchers in examining Li-Cor gel images obtained from Eco-TILLING studies (Zerr and Henikoff 2005). Gelbuddy.org is a free program that can do automated lane and molecular weight identification as well as manual and automatic polymorphism finding. Till and colleagues (2006) used Eco-TILLING technology to create multiple alleles for two resistance genes. These genes might be exploited to boost the resistance of certain vegetable crops to biotic and abiotic stressors. Eco-TILLING was utilized to explain the natural allelic variation in resistance to Melon Necrotic Spot Virus (MNSV) in Cucumis species. Eco-TILLING has been proven to accelerate the finding of beneficial alleles in superior vegetable and hybrid varieties (Nieto et al. 2007). SNP mining in diverse vegetable germplasms, measuring heterozygozity, exposing changes in disease resistance, and discovering gene importance by unearthing natural variations are all possible uses of this technique, although they are not presently being employed. Morphological data may be characterized effectively.

12.4 APPLICATION OF NEXT-GENERATION SEQUENCING (NGS) IN TILLING EXPERIMENTS

Traditional breeding techniques as well as recently discovered transgenics technologies have both contributed greatly to the growth in agricultural productivity. However, the broad commercialization of genetically modified (GM) crops has been delayed by regulatory restrictions in a number of countries. As a result, a strategy that does not include genetic alteration of agricultural plants is necessary. Next-generation sequencing, or NGS, is rapidly replacing classical TILLING as the technique of choice for finding mutations. This is because NGS allows for the examination of a large number of amplicons in a very short period of time. When employing NGS methods, the sequencing of specific amplified sections enables for the accurate and rapid detection of rare mutations in target genes (Guo et al. 2015; Gupta et al. 2017). This is one of the advantages of using NGS. Because of the advancement of NGS technology, the cost of DNA sequencing has considerably lowered; as a consequence, there has been a fast rise in the adoption of these approaches into TILLING research programs. Only 5 years ago, the cost of sequencing was about $1,000 for one million base pairs. The identical sequence may now be sequenced for only ten cents (Irshad et al. 2020). A single genotype of a three gigabase genome sequenced at a coverage of 30× is expected to cost roughly $5 for whole genome sequencing (WGS). When compared to WGS, the cost of conducting restriction site associated DNA sequencing, often known as RAD-seq, is reduced by a factor of 35 and has the ability to model 200,000 SNPs in those individuals while offering equivalent coverage. According to Rothberg et al. (2011), depending on whether or not the complete genome sequence of the species

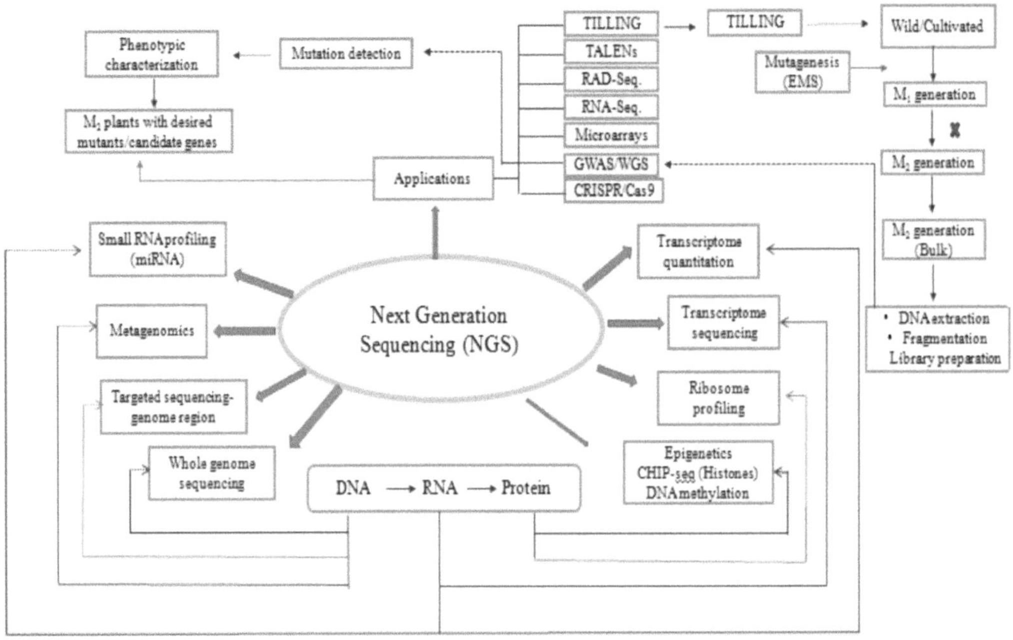

FIGURE 12.3 Flow diagram of next-generation sequencings methods for TILLING experiments.

Source: Adapted from Selvakumar et al. 2023.

being targeted is already available, the cost of genotyping by sequencing (GBS) or multiplexed shotgun genotyping (MSG) may be lowered by 10–14 times. Targeted sequencing will be the best option for discovering large scale markers in the case of a genome that has not yet been decoded. Because the cost of next-generation sequencing (NGS) and whole-genome sequencing (WGS) is predicted to drop many times in the near future, NGS is likely to replace partial genome sequencing as the technique of choice. As a consequence, the evaluation of numerous high-throughput sequencing technologies, including as PacBio and Ion Torrent, is supporting plant researchers in lowering the cost of sequencing, which will coexist with TILLING research (Figure 12.3)

12.5 APPLICATION OF GENOME-EDITING APPROACHES TOWARDS TILLING PLATFORMS

Genome editing (GE) approaches are rapidly being employed in crop improvement efforts because to their many advantages, including high efficiency, simplicity, and durability. CRISPR/Cas9 and clustered regularly interspaced short palindromic repeats (*CRISPR/Cas9*) are two examples of precise genome editing techniques that have lately acquired popularity because of their ease of use, scalability, and low cost (Jaganathan et al. 2018). Other GE techniques that may be used to change gene sequences include ZFNs (zinc finger nucleases), MNs (maganucleases), and TALENs (transcription activator-like effector nucleases); however, their applicability is restricted due to the demand for technical competence. Before new genome editing technologies such as *CRISPR/Cas9* and others can be commercialized, the regulatory framework must support them. Plant breeders may be able to employ TILLING and CRISPR together to boost crop variability. A single guide RNA (sgRNA) directs the *Cas9* nuclease to its target spot in the genome, inducing double-strand breaks (DSBs) and subsequent repair to yield a deletion (InDel) or insertion mutation (Palan et al. 2021). These mutations result in alleles with a loss of function. Mutant screening employing

TILLING DNA libraries and CRISPR/Cas9 enables for the investigation of developmental effects in relation to gene sequencing. TILLING by sequencing may be utilized to speed up the mutation screening process. With the advancement of next-generation sequencing (NGS) technology, improved software, and artificial intelligence (AI), the identification of erroneous mutation rates has been considerably lowered. Mutations occur exclusively in the genes of interest in genetic engineering (GE), while mutations may arise everywhere in the genome in TILLING. TILLING requires researchers to know the gene sequence and copy number for a desired trait in order for it to operate. TILLING the population is required to look for mutations in the gene of interest, and the mutant population is then examined for phenotypic changes that may be related with the mutation. TILLING has a higher rate of off-target mutations than genome editing. As a result, deletions or insertions may be made at exact sites in diploid or polyploid genomes utilizing GE technologies like as *CRISPR/Cas9* (Palan et al. 2021). Rice and wheat have both benefitted from CRISPR/Cas9-induced changes to boost blast resistance (Shan et al. 2014), as has maize protoplast (Xu et al. 2017). Point mutation analysis for the aim of detecting single locus heterosis has recently been employed successfully with both GE and TILLING technologies (Kashtwari et al. 2019). *CRISPR/Cas9* technology has greatly benefited plant research and breeding. In the future, we may use this method to simultaneously modify several members of a gene family on a genome-wide scale by using conserved 20-bp sgRNAs among prolific genes. Given this, it is feasible that combining both technologies may boost the efficiency and accuracy of identifying particular mutants for use in improving crop yields.

12.6 BIOINFORMATIC TOOLS FOR ALLELE MINING

Gene mining relies on comparing strings from individuals with various genotypes to identify variance. In addition to PLACE, plantCARE, TRANSFAC, JASPAR, MEME, Plantprom DB, DCPD, SCPD, BioEdit, ClustalW, etc., there are many more software tools available. These methods may be used to process large amounts of nucleotide data, predict the probable function or structural components of complex macromolecules, identify sequence polymorphisms, predict the effects of amino acid alterations on the structure and/or function of encoded proteins, predict transcription factor binding motif binding sites, and predict transcription factor binding motifs. Using these methods, analyzing DNA sequences is a breeze. They also aid in the detection of sequence polymorphism and the verification that differences in TFBM are consistent with such differences. While defining the primary promoter area, these methods are also useful for locating upstream regions with overrepresented motifs. This helps to streamline the process of discovering novel variants in regulatory components. These computational approaches often provide a set of patterns with the highest likelihood of matching the actual binding sites, combined with a huge number of variations due to random chance. The algorithms used by computers to make educated estimates have already been developed. Most of the time, these algorithms are developed by mining datasets for preexisting patterns. Predictions of this kind should thus always be evaluated in a methodical manner in a wet lab.

12.7 CONSIDERATIONS FOR ALLELE MINING

The genetic background of the crop species in issue is one of the most crucial elements in determining whether or not an allele mining operation will be successful. Access to efficient and reliable phenotyping techniques, genomic resources, high-throughput techniques for the rapid and simple generation of allelic data points, cost-effective sequencing platforms, efficient bioinformatic tools for identifying nucleotide variation, and molecular marker construction technologies are all important factors for marker-assisted selection (MAS).

12.8 UTILIZATION OF RADISH GENETIC RESOURCES IN MINING APPROACHES

The four native *Raphanus* species (*R. raphanistrum, R. maritimus, R. landra,* and *R. rostatus*), are all highly compatible with one another. Furthermore, *R. sativus* crosses easily with a variety of other Brassica species as well as wild mustard, Sinapis arvensis. Radishes are classified according to their root size (small-rooted and large-rooted), root shape (tapered, cylindrical, elliptical, and globose), origin (European, Chinese, Indian, and Japanese), root color (white, red, purple, yellow, and black), growing season (winter, spring, summer, autumn, and all-season) and vernalization requirement (temperate and tropical).

The conservation of genetic resources is required to protect the genetic riches of radish and to promote plant breeding efforts. More than 10,000 radish accessions are conserved in gene banks around the world, including the Centre for Genetic Resources (CGN), Wageningen; the Institute of Plant Genetics and Crop Plant Research (IPK), Gatersleben; the World Vegetable Centre (WVC), Taiwan; the Institute of Crop Germplasm Resources (ICGR), Beijing, China; the Plant Genetic Resources Unit, United States Department of Agriculture (USDA), USA; the Crucifer Genetics Cooperatives, University of Wisconsin, Madison, USA China (>2000 accessions) and Japan (>1000 accessions) have the biggest collections, accounting for over 30% of the total genetic resources preserved globally. Furthermore, Germany, Russia, the United Kingdom, and the United States all have collections with over 500 accessions. The majority of the accessions are varieties and landraces, and the wild relatives of radish are mostly preserved in gene banks in Australia, Germany, and Spain (Yang 2019). In general, gene banks keep their base collections at temperatures below zero, with most of them adhering to the 18°C FAO-IPGRI standard (Food and Agriculture Organization/International Plant Genetic Resources Institute); however, storage temperatures at SASA and WARGRU are 22°C ±3 °C (Thormann et al. 2013).

12.9 CONVENTIONAL TILLING APPROACHES IN RADISH

The term "mutation breeding" is used to describe the process of creating new varieties of plants by inducing mutations in existing ones. It is a powerful tool for enhancing complex features, particularly in vegetatively propagated, self-pollinated, and crops with a limited genetic background. Seven hundred and seventy mutant cultivars have been produced for different nutritional quality features, including minerals, and they have been distributed in more than 60 countries (Jain and Suprasanna 2011). Radish mutants had a 30% increase in net photosynthetic rate and a 36% increase in total chlorophyll content compared to wild-type plants (Wang et al. 2015).

There are around 20 S-alleles found in cultivated radishes, as documented by researchers (Bateman 1955; Okazaki and Hinata 1984). In addition, between 25 and 34 S-alleles were found in *R. raphanistrum* from the wild, while 22 were found in *R. sativus* from the wild (Sampson 1967; Karron et al. 1990). Niikura and Matsuura (1999) and Niikura (2000) recently compared 71 inbred lines and 223 landraces from 18 domestic and five foreign varieties maintained in sib-cross among three to five plants to study the genetic variety of S-alleles. Thus, 37 alleles (labeled S201 to S237) were discovered for the genotypes, with S208, S218, and S219 being the most prevalent among the Japanese radish cultivars investigated. They are able to discover the prominent relationships among the S-alleles by evaluating a subset of lines consisting of 16 similar S-alleles (S201–S217, except S207). Haruta (1962) discovered four different genotypes for the S-alleles. The dominating S-alleles were detected in abundance in numerous types, including Awa bannsei, Moriguchi, Ninengo, Risyo, Shiro-agari, South Kyushu Ji-radish, Korean lines, and Nepal lines, but in low abundance in Hamadaikon, Miyashige, Nerima, Aki-tumari, and Tropical lines. Tropical lineages carried the dominating S-alleles less often.

The Japanese radish, which was studied in the Japanese Kagoshima Prefecture in 1968 (Ogura 1968), is the most likely CMS source among Brassiceae. The Ogura CMS type was introduced to the

Brassica family of plants by the French agricultural research institution INRA (Bannerot et al. 1974, 1977; Dolourme et al. 1999). However, when exposed to cold, this variety developed chlorosis, and its seed fertility was low when produced using standard techniques. Cell fusion (Pelletier et al. 1983; Sakai and Imamura 1994; Dolourme et al. 1999) and recombination between radish and oilseed rape homoeologous domains (Paulmann and Robbelen 1988) were proposed to address the aforementioned deficiencies. Namai (1991) reported two more lines, one from an F_1 hybrid of Chinese *cv.* Daibaika and an unknown cultivar (the CMS UK-1 line) and the other from the Japanese radish cv. Kozena (the KS line). Sakai and Imamura (1994) created a new CMS system by irradiating radish cv. Kozena with a CMS-associated gene, *orf125*, that had a sequence homologous to that of Ogura in the CMS-associated gene, *orf138*, except for two amino acid substitutions and a 39 bp deletion in the *orf138* coding region (Iwabuchi et al. 1999).

CMS Kozena and oilseed rape cv. Wester hybrids resulted in oil seed rape (2n = 38) containing Kozena cytoplasm. A prospective strategy for generating F_1 seeds has been created using the Kozena CMS and the *Rf*-line with restored pollen fertility identified in Kozena (Sakai and Imamura 1994, 2003). Sakai and Imamura (2003) were able to generate a plant with consistent seed fertility and no chlorosis by introducing one pair of radish chromosomes into a rape-like plant. Yamagishi and Terachi (1994) reported a new CMS radish *cv.* Utsugi-gensuke with Ogura cytoplasm.

12.10 APPLICATION OF ALLELE MINING IN RADISH

It is difficult to differentiate between induced mutations and changes that are already present in radish cultivars, which is one of the reasons why radish mutant breeding has not been successful. In addition, since the majority of mutant features are governed by recessive alleles, it is impossible to breed hybrids using them because hybrid breeding mixes the dominant alleles of both parents. Mutant traits do not fit this description. However, all landraces that have unique traits are the consequence of spontaneous mutations occurring in the population throughout time (Ishida et al. 2015). Wang et al. (2015) treated radish with EMS to create a compound-leaf mutant, then compared the mutant's photosynthetic, agronomic, microstructural, and qualitative characteristics to those of the wild type. Total chlorophyll content was about 36% higher in the mutant than in the wild-type, and net photosynthetic rate was around 30% higher in the mutant. The mutant, on the other hand, has half the root weight of the native variety. When compared to the wild type, the mutant showed a 39% increase in total soluble solids, a 12% drop in soluble sugars, and a 75% increase in vitamin C. Stomata are more densely packed in compound leaves than in simple leaves. Each guard cell in complex leaves possessed six chloroplasts, but only five in simple leaves. Compound leaves had increased stomatal conductivity. Compound leaves have more distributed and slender vascular bundles and phloem cells, a narrower petiole, and a higher parenchyma cell density than simple leaves. Sequence-related amplified polymorphism study revealed that ethyl methanesulfonate induced DNA changes in numerous genes.

Recently, a cultivar that has a novel glucosinolate composition was generated by selecting a spontaneous mutation in a landrace. This allowed for the development of the cultivar. Techniques for the selection of mutants via the use of reverse genetics have been developed. Although TILLING (McCallum et al. 2000a, b), which has been used successfully for the reverse-genetic selection of mutants in many plant species, is not suitable for use in radish cultivars with many gene variations, mutant selection using a next-generation sequencer, such as KeyPoint ® of Keygene, may be available for use in radish mutation breeding. This is because TILLING has been used successfully for the selection of mutants through reverse genetics in many plant species. There is a need for an additional method that does not require a lot of money.

After obtaining the correct concentration of EMS for mutant breeding in radish, the mutated phenotypes of Huangzhou taproots were seen in the crop. The application of EMS at a concentration of 0.5 percent to the Huangzhou radish seeds provided a good basis for the formation of the radish breeding program. Even though the altered taproots in Huangzhou were focused on root length, which was not compatible with the purpose of our study, the increase in purity and quality of Huangzhou would be continuing in the offspring of those mutant Huangzhou in our future work (Yu et al. 2016). This was predicted by the researchers who conducted the study. It is possible to generate mutations in radishes by treating them with ethyl methane sulfonate (EMS), and this technique works well for getting desirable characteristics in newly developed radishes. The study of the radishes of the Duan-Ye No. 13 (DY13) variety in order to determine the best concentration of EMS for mutagenesis. When DY13 radish seedlings were exposed to EMS at a concentration of 0.5 percent, sterile mutants developed. (Li et al. 2011) Radishes of the *DY13* variety were used in an experiment to effectively produce male sterile lines, indicating a unique and effective strategy for creating male sterile radish lines. Because the root of radish (*Raphanus sativus* L. var. *sativus*) has a big sink organ, this suggests that modifying both the plant's source capacity and its sink capacity might boost photosynthetic activity. Radish is a species of the genus Raphanus. However, due to the fact that radish is an incompatible plant with itself, improved breeding tactics based on mutation are necessary for this crop. According to Kohzuma et al. (2017), a newly developed TILLING technique that included a two-step mutant selection procedure for mutagenized radish plants resulted in the production of six mutant lines for the six *RsRBCS* genes found in the radish. These mutant lines encode proteins with four distinct kinds of amino acid substitutions.

Ogura CMS is induced by the expression of the orf138 gene in Brassicaceae (Duroc et al. 2006; Grelon et al. 1994). Yamagishi and Terachi (2001) discovered one of at least nine nucleotide sequence variants in the *orf138* gene, a 39-nucleotide deletion (Kosena type). Furthermore, plants with radish Ogura-type CMS contain a 278 bp fragment amplified by PCR using orf138 gene specific primers, 5′-gacatctagagaagttaaaaat-3′ and 5′-agcaattgggttcacaaaagcat-3′ (Iwabuchi et al. 1999; Yamagishi and Terachi 1996). Moreover, Nahm et al. (2005) reported a primer pair at the 3′ region of the atp6 gene (5′-cgcttggactatgctatgtatga-3′) and the 5′ region of the nad3 gene (5′-tcatagagaaatccaatcgtcaa-3′) producing a 2 kbp fragment, specific to NWB CMS type of male sterility which was not present in other CMS types of radish. Nie et al. (2016) discovered critical genes involved in radish bolting and blooming using de novo transcriptome analysis.

12.10.1 Radish Mining Through Sequencing Approaches

Since the publishing of the genome sequences of *A. thaliana* (The Arabidopsis Genome Initiative 2000) and *B. rapa* (The *Brassica rapa* Genome Sequencing Project Consortium 2011), three distinct research groups have published *R. sativus* genome sequences. Kitashiba et al. (2014) published the original research. Genome contig sequences were generated by assembling paired-end sequence reads (insert size: 250 bp) from the "Aokubi" inbred line containing the *S-h* haplotype, then closing gaps using GapCloser. Furthermore, SSPACE 2.0 was used to scaffold the contigs using sequence reads from a mate-pair library (insert size: 5 kb) and BAC-end sequences, generating 76,592 sequences (N50 = 46 kb; 300 bp) encompassing 402 Mb of the radish genome, or about 75.6% of the projected genome size of 530 Mb. The Augustus algorithm predicts 61,572 genes based on the sequencing (Stake and Waack 2003). A high-density genetic map consisting of 2553 markers has been developed by integrating two maps reported by Shirasawa et al. (2011) and Li et al. (2021), as well as an additional 989 DNA markers (221 SNP markers and 768 PCR-RFLP markers). *Raphanus* chromosomes have been identified with 1345 scaffolds totaling 116 Mb of sequence. Mitsui and others published the second genome sequencing in 2015. In their research, they employed both types of sequencing technology. Single-end sequencing on the Roche 454 platform and paired-end sequencing on the Illumina platform (insert sizes of 300, 500 bp, 8, 20, and 40 kb). The sequencing data was combined using Newbler 2.7 and then scaffolded using SSPACE 2.0, yielding 383 Mb of sequences (N50 = 138 kb; 500 bp), from which the Augustus program predicted 54,357 genes. SSR and SNP markers were used to create a genetic map of the radish, and 180 Mb of its genome was sequenced and assigned to those markers.

The most recent one was reported by Jeong et al. (2016). A chromosome-scale draft radish genome sequence was created using data from three different NGS technologies: Illumina (500 bp insert size), Roche 454 (SE and PE for 3-, 8-, 20-, and 30-kb insert sizes), and PacBio long reads, as well as the end sequences of clones from two libraries: fosmid (40-kb insert size) and BAC. The Roche 454 reads were assembled using Newbler 2.6, while the PacBio reads were reassembled using Celera Assembler 7.0. Following the addition of the fosmid and BAC clone-end sequences to the contigs, they were subjected to error correction using Illumina reads. The ultimate size was 426 Mb (N50 = 19.6 kb; 100 bp), with 344 Mb linked to the Raphanus chromosomes (N50 = 1.22 Mb) along a high-density genetic map built by genotyping an F_2 mapping population using NGS technology (Mun et al. 2015). The prediction of 46,514 genes was aided by ab initio, sequence similarity, and transcriptome evidence. *R. raphanistrum*, a wild relative of cultivated radishes, has also had its genome sequence published (Moghe et al. 2014). From Illumina sequencing data collected from a fifth generation inbred line, 68,331 contigs (N50 = 10.1 kb) covering 254.6 Mb were assembled. An annotation study utilizing an MAKER pipeline predicted a total of 38,174 genes.

12.10.2 Challenges

The most challenging aspect of the species in question is the need to collect its germplasm in order to "mine" it. If the amount of inventory is excessively large, micro core selection (1% of the accumulation) is carried out (Upadhyaya and Ortiz 2001; Mazzucato et al. 2015). A micro core selection constitutes 1% of the total collection or 10% of the core. It could take up to 2 years to produce a robust population (Slade and Knauf 2005; Caldwell et al. 2004). Henikoff and Comai (2003) found that the optimal dosage induces numerous mutations with minimal embryo mortality. When chemical mutagens are used above a certain concentration, crop mutation rates skyrocket. This is not a significant concern because the outcomes of the various chemical mutagenesis processes will vary depending on the plant or plant type. According to Slade and Knauf (2005), long-lived perennial vegetables that can also be propagated vegetatively may be able to prevent the spread of mutations. Normal polymorphisms hinder the ability of TILLING and Eco-TILLING to detect mutations in highly heterozygous species. If many fragments are generated, it may be more difficult to identify unusual induced mutations (Till et al. 2003a, b). Numerous procedures, such as the production of test genotypes and DNA collections for analytical purposes, add complexity to this method. This is just one instance of the added complexity that may result from employing such methods. Recognition of novel allelic variations of existing agarose gel candidates and identification of their use in candidate gene mapping (Raghavan et al. 2007) is a recently discovered, rapid, and inexpensive method that could significantly reduce the price of Eco-TILLING. This strategy may prove beneficial on occasion.

A functional TILLING community of vegetatively propagated vegetables requires a number of tasks, the most important of which are the selection of mutagens, the ability to mutagenize the tissues, and the achievement of a sufficient density of dominant mutations or the production of two haploid plants. The study of plants that can propagate from spores is in its infancy, and the only method to determine what works is through trial and error. Irshad et al. (2020) and Szurman-Zubrzycki et al. (2018) note that TILLING is used to rapidly create and screen minuscule populations. This allows only plantlets with intriguing mutations to be selected for future study. This is particularly advantageous in regions where in vitro diagnostics are not commonly available.

Due to unanticipated circumstances, it may be necessary to till certain vegetable harvests. In some instances, polyploidy can be problematic. TILLING is utilized to identify extreme alleles in a variety of polyploid homeologists. Cross-fertilization can be used to produce the desired phenotype once these alleles have been identified. To detect mutations, PCR amplification of a single gene of interest (homologist) is followed by an enzymatic cleavage TILLING test. This is an essential phase. Consequently, multiplexing is fundamentally the challenge of identifying multiple gene targets in a single PCR cycle. It has been demonstrated, however, that this method is ineffectual of detecting mutations (Cooper et al. 2008a, 2008b). Target gene amplification (Cooper et al. 2008a; Prasanna and Jain 2017; Chen et al. 2014) or deletion (Slade and Knauf 2005) is possible with the correct prime design. To accomplish this, we conduct PCR with as little genomic DNA digestion as possible. There have been concerns regarding the dearth of genome sequencing data. Using sequence information from cDNA or EST, ineffective TILLING primers may be generated. The quantity of background mutations present in a plant, known as its "mutational load," is regarded as a formidable obstacle for the TILLING strategy employed in vegetal breeding. Genetic engineering is to blame for the decline in production because earlier mutations can result in phenotypes with both desirable and undesirable traits. (Getahun et al. 2003; Vetter 2000) All plants produce substances that are toxic to humans if ingested. Erosion, soil acidity, salinity, rodents, diseases, and vegetation are environmental factors that may impact production and yield. To increase the food supply of under-researched commodities and satisfy the rising food demand in developing nations, an agricultural revolution is required.

12.11 FUTURE PERSPECTIVES

Many vegetable crops have benefitted from genetic improvement attempts using vegetable mutants, and mutant vegetables are currently actively used in improvement initiatives. However, the TILLING and Eco TILLING mutation screening techniques mentioned in this chapter will hasten advantageous gene modifications with proven horticultural use, as well as the deployment of candidate genes in Daucus species. As a result, the industry will progress quickly. The introduction of next-generation sequencing (NGS) technology may improve this by enabling the sequencing of considerably larger datasets, resulting in an explosion of sequence data for Daucus species. In paralogists or homeologists, for example, point mutagenesis may only impact a single copy of an identical gene, while technologies such as RNAi may target several genes at the same time. This applies to all gene types. Crossing may be required to steer the mutagenesis process in the desired direction, although this is not always the case. Recent reverse genetic techniques, including as TILLING and Eco-TILLING, have shown to be valuable tools for doing practical genomic research in vegetables. TILLING, unlike other reverse genetics methods such as RNAi and homologous recombination, may be relevant to a broad range of radish since its species-specificity has not been shown.

12.12 CONCLUSION

As a result of the world's rising population, climate anomalies, the need to diversify food baskets, and a growing emphasis on healthy nutrition, consumer demand will shift away from traditional cuisine and towards food that is more wellness and health conscious. As a result of the COVID-19 epidemic, agricultural practises throughout the world have changed, and people are becoming more aware of the need to eat nutrient-dense foods and adjusting their dietary habits appropriately. In all honesty, this procedure has already begun. It is critical that we expand production of nutrient-dense vegetable crops in order to sustainably feed the rising population and meet their changing dietary needs. According to this viewpoint, vegetable crops will be grown with more calories, carbohydrates, minerals, and natural therapeutic components that promote the immune system. Breeding strategies and programs make it possible to improve radish resilience to biotic and abiotic stress, reduce pollution, and improve industrial attributes such as fiber intensity and glucosinolates production. Mutations, both natural and engineered, are a powerful tool for improving performance. However, the rapid gathering of plant genomic sequence information and genetic feature predictions opens the way to using mutations in reverse genetic techniques to categorise lesions in specific target genes. These gene-driven strategies aim to accelerate the synthesis of novel phenotypes, and they may even enable the creation of phenotypes that would be difficult to achieve using more traditional, future-oriented methods. TILLING was designed as a reverse genetic technique to uncover mediated mutations at a low cost and with a high yield. Eco-TILLING is a novel technique for NNP categorization that has been introduced to the current toolkit. In the 21st century, growing vegetable crop yield is as crucial as expanding production of modern olives. This is true despite widespread opposition to transgenic or genetically modified crops. Furthermore, the plant species that may benefit from agrobacterium-based insertional mutation techniques is limited. As a result, this gap may be filled by using allele-mining techniques like as TILLING (which prioritizes EMS mutagenesis) and Eco-TILLING. TILLING technique alleles may be used in vegetable crops despite the strict regulatory processes that control genetically modified products. This strategy may be quickly used in vegetable breeding initiatives by simply tilling. These methods have the potential to accelerate data processing from large-scale investigations of induced point mutations in relevant populations. Both natural polymorphism and artificial mutations have been shown to offer potential advantages for radish crop development. TILLING and/or Eco-TILLING are viable approaches for finding alleles linked to significant qualitative and quantitative economic value components in vegetable breeding.

REFERENCES

Ahloowalia BS, Maluszynski M, Nichterlein K (2004) Global impact of mutation-derived varieties. *Euphytica* 135:187–204.

Allen AM, Barker GLA, Berry ST, Coghill JA, Gwilliam R, Kirby S, Robinson P, Brenchley RC, D'Amore R, McKenzie N, Waite D, Hall A, Bevan M, Hall N, Edwards KJ (2011) Transcript specific, single nucleotide polymorphism discovery and linkage analysis in hexaploid bread wheat (*Triticum aestivum* L.). *Plant Biotechnol J* 9:1086–1099.

Ashburner M, Hawley RS, Golic KG (2005) Drosophila. A Laboratory Handbook. Cold Spring Harbor Lab.

Bannerot H, Boulidard L, Cauderon Y, Tempe J (1974) Transfer of cytoplasmic male sterility from *Raphanus sativus* to *Brassica oleracea*. Eucarpia Meet, Cruciferae. Dundee, Scotland, pp 52–54.

Bannerot H, Boulidard L, Chupeau Y (1977) Unexpected difficulties met with the radish cytoplasm in *Brassica oleracea*. *Cruciferae Newslett* 2:16.

Bansal V (2010) A statistical method for the detection of variants from next-generation resequencing of DNA pools. *Bioinformatics* 26:i318–i324.

Bansal V, Harismendy O, Tewhey R, Murray SS, Schork NJ, Topol EJ, Frazer KA (2010) Accurate detection and genotyping of SNPs utilizing population sequencing data. *Genome Res* 20:537–545.

Bateman AJ (1955) Self-incompatibility systems in angiosperms. III. Cruciferae. *Heredity* 9:53–68.

Borevitz JO, Liang D, Plouffe D, Chang HS, Zhu T, Weigel D, Berry CC, Winzeler E, Chory J (2003) Large-scale identification of single-feature polymorphisms in complex genomes. *Genome Res* 13:513–523.

Caldwell DG, McCallum N, Shaw P, Muehlbauer GJ, Marshall DF, Waugh R (2004) A structured mutant population for forward and reverse genetics in Barley (*Hordeum vulgare* L.). *Plant J* 40:143–150.

Chen L, Hao L, Parry MAJ, Phillips AL, Hu YG (2014) Progress in TILLING as a tool for functional genomics and improvement of crops. *J Integr Plant Biol* 56:425–443.

Colbert T, Till BJ, Tompa R, Reynolds S, Steine MN, Yeung AT, McCallum CM, Comai L, Henikoff S (2001) High-throughput screening for induced point mutations. *Plant Physiol* 126:480–484.

Comai L, Henikoff S (2006) TILLING: practical single nucleotide mutation discovery. *Plant J* 45:684–694.

Comai L, Young K, Till BJ, Reynolds SH, Greene EA, Codomo CA, Enns LC, Johnson JE, Burtner C, Odden AR, Henikoff S (2004) Efficient discovery of DNA polymorphisms in natural populations by Ecotilling. *Plant J* 37:778–786.

Cooper JL, Greene EA, Till BJ, Codomo CA, Wakimoto BT, Henikoff S (2008a) Retention of induced mutations in a *Drosophila* reverse-genetic resource. *Genetics* 180:661–667.

Cooper JL, Till BJ, Henikoff S (2008b) Fly-TILL: reverse genetics using a living point mutation resource. *Fly (Austin)* 2:300–302.

Dahm R, Geisler R (2006) Learning from small fry: the zebrafish as a genetic model organism for aquaculture fish species. *Mar Biotechnol* 8:329–345.

Dolourme R, Horvais R, Vallee P, Renard M (1999) Double low restored F_1 hybrids can be produced with the Ogu-INRA CMS in rapeseed. In: Proceedings of the 10th International Rapeseed Congress, Canberra, Australia, pp 1999.

Druley TE, Vallania FLM, Wegner DJ, Varley KE, Knowles OL, Bonds JA, Robison SW, Doniger SW, Hamvas A, Cole FS, Fay JC, Mitra RD (2009) Quantification of rare allelic variants from pooled genomic DNA. *Nat Methods* 6:263–265.

Duroc, Y., C. Gaillard, S. Hiard, C. Tinchant, R. Berthom, G. Pelletier and F. Budar (2006) Nuclear expression of a cytoplasmic male sterility gene modifies mitochondrial morphology in yeast and plant cells. *Plant Sci.* 170: 755–767.

Getahun H, Lambein F, Vanhoorne M, Van der Stuyft P (2003) Food-aid cereals to reduce neurolathyrism related to grass-pea preparations during famine. Lancet 362:1808–1810.

Gilchrist EJ, Haughn GW (2005) TILLING without a plough: a new method with applications for reverse genetics. *Curr Opin Plant Biol* 8:211–215.

Gilchrist EJ, Haughn GW, Ying CC, Otto SP, Zhuang J, Cheung D, Hamberger B, Aboutorabi F, Kalynyak T, Johnson L, Bohlmann J, Ellis BE, Douglas CJ, Cronk QC (2006a) Use of Ecotilling as an efficient SNP discovery tool to survey genetic variation in wild populations of *Populus trichocarpa*. *Mol Ecol* 15:1367–1378.

Gilchrist EJ, O'Neil NJ, Rose AM, Zetka MC, Haughn GW (2006b) TILLING is an effective reverse genetics technique for *Caenorhabditis elegans*. *BMC Genomics* 7:1–16.

Greene EA, Codomo CA, Taylor NE, Henikoff JG, Till BJ, Reynolds SH, Enns LC, Burtner C, Johnson JE, Odden AR, Comai L, Henikoff S (2003) Spectrum of chemically induced mutations from a large-scale reverse-genetic screen in *Arabidopsis*. *Genetics* 164:731–740.

Grelon M, Budar F, Bonhomme S, Pelletier G (1994) Ogura cytoplasmic male-sterility (CMS)–associated *orf138* is translated into a mitochondrial membrane polypeptide in male sterile Brassica cybrids. *Mol Gen Genet* 243:540–547.

Guo Y, Abernathy B, Zeng Y, Ozias-Akins P (2015) TILLING by sequencing to identify induced mutations in stress resistance genes of peanut (*Arachis hypogaea*). *BMC Genomics* 16:157.

Gupta P, Reddaiah B, Salava H, Upadhyaya P, Tyagi K, Sarma S, Datta S, Bharti Malhotra, Thomas S, Sunkum A, Devulapalli S, Till BJ, Sreelakshmi Y Sharma R (2017) Next-generation sequencing (NGS)-based identification of induced mutations in a doubly mutagenized tomato (*Solanum lycopersicum*) population. *Plant J* 92(3):495–508 https://doi.org/10.1111/tpj.13654

Haruta T (1962) Studies on the genetics of self and cross incompatibility in cruciferous vegetables. *Rep Agri Nagaoka Farm Takii Co* 2:1–169.

Haughn G, Somerville CR (1987) *Selection for herbicide resistance at the whole-plant level*. ACS Publications.

Henikoff S, Comai L (2003) Single-nucleotide mutations for plant functional genomics. *Annu Rev Plant Biol* 54:375–401.

Henikoff S, Till BJ, Comai L (2004) TILLING. Traditional mutagenesis meets functional genomics. *Plant Physiol* 135:630–636.

Irshad A, Guo H, Zhang S, Liu L (2020) TILLING in cereal crops for allele expansion and mutation detection by using modern sequencing technologies. *Agronomy* 10:405.

Ishida M, Kakizaki T, Morimitsu Y, Ohara T, Hatakeyama K, Yoshiaki H, Kohori J, Nishio, T (2015) Novel glucosinolate composition lacking 4-methylthio-3-butenyl glucosinolate in Japanese white radish (*Raphanus sativus* L.). *Theor Appl Genet* 128(10):2037–2046. https://doi.org/10.1007/S00122-015-2564-3/METRICS

Iwabuchi M, Koizumi N, Fujimoto H, Sakai T, Imamura J (1999) Identification and expression of the kosena radish (*Raphanus sativus* cv. Kosena) homolog of the ogura radish CMS-associated gene, orf138. *Plant Mol Biol* 39:183–188.

Jaganathan D, Ramasamy K, Sellamuthu G, Jayabalan S, Venkataraman G (2018) CRISPR for crop improvement: An update review. *Front Plant Sci* 9:985. doi: 10.3389/fpls.2018.00985

Jain SM, Suprasanna P (2011) Induced mutations for enhancing nutrition and food production. *Gene Conserv* 40:201–215.

Jeong H-J, Kwon J-K, Pandeya D, Hwang J, Hoang NH, Bae JH & Kang BC (2012) A survey of natural and ethyl methane sulfonate-induced variations of *eIF4E* using high-resolution melting analysis in Capsicum. *Mol Breed* 29:349–360.

Jeong YM, Kim N, Ahn BO, Oh M, Chung WH, Chung H, Jeong S, Lim KB, Hwang YJ, Kim GB, Baek S, Choi SB, Hyung DJ, Lee SW, Sohn SH, Kwon SJ, Jin M, Seol YJ, Chae WB, Choi KJ, Park BS, Yu HJ, Mun JH (2016) Elucidating the triplicated ancestral genome structure of radish based on chromosome-level comparison with the Brassica genomes. *Theor Appl Genet* 129:1357–1372.

Jin J, Chen L, Yao M, Wang X, Chun-lei MA (2010) Simplification of EcoTILLING technique for tea plant. *J Tea Sci* 30:19–26.

Karron JD, Marshall DL, Oliveras DM (1990) Numbers of sporophytic self-incompatibility alleles in populations of wild radish. *Theor Appl Genet* 79:457–460.

Kashtwari M, Wani AA, Rather RN (2019) TILLING: An alternative path for crop improvement. *J Crop Improv* 33:83–109.

Kitashiba H, Li F, Hirakawa H, Kawanabe T, Zou Z, Hasegawa Y, Tonosaki K, Shirasawa S, Fukushima A, Yokoi S, Takahata Y, Kakizaki T, Ishida M, Okamoto S, Sakamoto K, Shirasawa K, Tabata S, Nishio T (2014) Draft sequences of the radish (Raphanus sativus L.) genome. *DNA Res* 21:481–490.

Koboldt DC, Chen K, Wylie T, Larson DE, McLellan MD, Mardis ER, Weinstock GM, Wilson RK, Ding L (2009) VarScan: variant detection in massively parallel sequencing of individual and pooled samples. *Bioinformatics* 25:2283–2285.

Kohzuma K, Chiba M, Nagano S, Anai T, Ueda MU, Oguchi R, Shirai K, Hanada K, Hikosaka K, Fujii N (2017) Mutant selection in the self-incompatible plant radish (*Raphanus sativus* L. var. *sativus*) using two-step TILLING. *Breed Sci* 67(3):268–276. https://doi.org/10.1270/JSBBS.16200

Kumari R, Kumar P, Sharma VK, Kumar H (2018) Allele mining for crop improvement. *Int J Pure App Biosci* 6(1): 1456–1465.

Latha R, Rubia L, Bennett J, Swaminathan MS (2004) Allele mining for stress tolerance genes in *Oryza* species and related germplasm. *Mol Biotechnol* 27:101–108.

Li F, Hasegawa Y, Saito M, Shirasawa S, Fukushima A, Ito T, Fujii H, Kishitani S, Kitashiba H, Nishio T (2011) Extensive chromosome homoeology among Brassiceae species were revealed by comparative genetic mapping with high-density EST-based SNP markers in radish (*Raphanus sativus* L.). *DNA Res* 18:401–411.

Li H, Handsaker B, Wysoker A, Fennell T, Ruan J, Homer N, Marth G, Abecasis G, Durbin R (2009) 1000 genome project data processing subgroup. The sequence alignment/map format and SAM tools. *Bioinformatics* 25:2078–2079.

Li T, Deng YJ, Liu JX, Duan AQ, Liu H, Xiong AS (2021) *DcCCD4* catalyzes the degradation of α-carotene and β-carotene to affect carotenoid accumulation and taproot colour in carrot. *Plant J* 108(4):1116–1130. https://doi.org/10.1111/TPJ.15498

Mazzucato A, Cellini F, Bouzayen M, Zouine M, Mila I, Minoia S, Petrozza A, Picarella ME, Ruiu F, Carriero F (2015) A TILLING allele of the tomato *Aux/IAA9* gene offers new insights into fruit set mechanisms and perspectives for breeding seedless tomatoes. *Mol Breed* 35:1–15.

McCallum CM, Comai L, Greene EA, Henikoff S (2000a) Targeted screening for induced mutations. *Nat Biotechnol* 18(4):455–457. https://doi.org/10.1038/74542

McCallum CM, Comai L, Greene EA, Henikoff S (2000b) Targeting induced local lesions in genomes (TILLING) for plant functional genomics. *Plant Physiol* 123:439–442.

Mitsui Y, Shimomura M, Komatsu K, Namiki N, Shibata-Hatta M, Imai M, Katayose Y, Mukai Y, Kanamori H, Kurita K, Kagami T, Wakatsuki A, Ohyanagi H, Ikawa H, Minaka N, Nakagawa K, Shiwa Y, Sasaki T (2015) The radish genome and comprehensive gene expression profile of tuberous root formation and development. *Sci Rep* 5:10835.

Moghe GD, Hufnagel DE, Tang H, Xiao Y, Dworkin I, Town CD, Conner JK, Shiu SH (2014) Consequences of whole-genome triplication as revealed by comparative genomic analyses of the wild radish *Raphanus raphanistrum* and three other Brassicaceae species. *Plant Cell* 26:1925–1937.

Mun JH, Chung H, Chung WH, Oh M, Jeong YM, Kim N, Ahn BO, Park BS, Park S, Lim KB, Hwang YJ, Yu HJ (2015) Construction of a reference genetic map of Raphanus sativus based on genotyping by whole genome resequencing. *Theor Appl Genet* 128:259–272.

Myers RM, Maniatis T, Lerman LS (1987) Detection and localization of single base changes by denaturing gradient gel electrophoresis. *Methods Enzymol* 155:501–27. doi: 10.1016/0076-6879(87)55033-9

Nagy A, Perrimon N, Sandmeyer S, Plasterk R (2003) Tailoring the genome: the power of genetic approaches. *Nat Genet* 33:276–284.

Nahm SH, Lee HJ, Lee SW, Joo GY, Harn CH, Yang SG, Min BW (2005) Development of a molecular marker specific to a novel CMS line in radish (*Raphanus sativus* L.). *Theor Appl Genet* 111:1191–1200.

Namai H (1991) Breeding using cytoplasmic male sterility (CMS) in radish. *Jpn Soc Seed Prod Breed Engine* in 1991, pp 165–181 (in Japanese).

Nataraj AJ, Olivos Glander I, Kusukawa N, Highsmith Jr WE (1999) Single strand conformation polymorphism and heteroduplex analysis for gel based mutation detection. *Electrophor An Int J* 20:1177–1185.

Natarajan AT (2005) Chemical mutagenesis: From plants to human. *Curr Sci* 89(2): 312–317.

Ng PC, Henikoff S (2003) SIFT: Predicting amino acid changes that affect protein function. *Nucleic Acids Res* 31:3812–3814.

Nie S, Li C, Xu L et al (2016) *De novo* transcriptome analysis in radish (*Raphanus sativus* L.) and identification of critical genes involved in bolting and flowering. *BMC Genomics* 17:389–405.

Nieto C, Piron F, Dalmais M, Marco CF, Moriones E, Gómez-Guillamón ML, Truniger V, Gómez P, Garcia-Mas J, Aranda MA, Bendahmane A (2007) EcoTILLING for the identification of allelic variants of melon *eIF4E*, a factor that controls virus susceptibility. *BMC Plant Biol* 7:1–9.

Niikura S (2000) Studies on the self-incompatibility related characteristics of radish (*Raphanus sativus* L.). Dissertation, University of Hokkaido, Japan, pp 1–137 (in Japanese).

Niikura S, Matsuura S (1999) Genetic variation of the S-alleles and level of self-incompatibility in Japanese cultivated radish (*Raphanus sativus* L.). *Breed Res* 1:211–220.

Nijman IJ, Mokry M, Van Boxtel R, Toonen P, de Bruijn E, Cuppen E (2010) Mutation discovery by targeted genomic enrichment of multiplexed barcoded samples. *Nat Methods* 7:913–915.

Ogura H (1968) Studies on the new male sterility in Japanese radish, with special reference to the utilization of this sterility towards the practical raising of hybrid seeds. *Mem Fac Agri Kagoshima Univ* 6:39–78.

Okazaki K, Hinata K (1984) Analysis of S-alleles and S glycoproteins in F_1-hybrid varieties of Japanese radish (*Raphanus sativus* L.). *Jpn J Breed* 34:237–245.

Ossowski S, Schneeberger K, Lucas-Lledó JI, Warthmann N, Clark RM, Shaw RG, Weigel D, Lynch M (2010) The rate and molecular spectrum of spontaneous mutations in *Arabidopsis thaliana*. *Science* 327:92–94.

Palan B, Bhattacharya A, Char B (2021) TILLING in the era of precise genome editing. *Indian J Biotechnol* 20:9–16.

Paulmann W, Robbelen G (1988) Effective transfers of cytoplasmic male sterility from radish (*Raphanus sativus* L.) to rape (*Brassica napus* L.). *Plant Breed* 100:299–309.

Pelletier G, Primard C, Vedel F, Chetrit P, Remy R, Rousselle P, Renard M (1983) Intergeneric cytoplasmic hybridization in Cruciferae by protoplast fusion. *Mol Gen Genet* 191:244–250.

Porreca GJ, Zhang K, Li JB, Xie B, Austin D, Vassallo SL, LeProust EM, Peck BJ, Emig CJ, Dahl F, Gao Y, Church GM, Shendure J (2007) Multiplex amplification of large sets of human exons. *Nat Methods* 4:931–936.

Portis E, Portis F, Valente L, Moglia A, Barchi L, Lanteri S, Acquadro A (2016) A genome-wide survey of the microsatellite content of the globe artichoke genome and the development of a web-based database. *PLoS One* 11:e0162841.

Prasanna S, Jain SM (2017) Mutant resources and mutagenomics in crop plants. *Emirates J Food Agric* 29(9): 651–657.

Prochnik S, Marri PR, Desany B, Rabinowicz PD, Kodira C, Mohiuddin M, Rodriguez F, Fauquet C, Tohme J, Harkins T, Rokhsar DS, Rounsley S (2012) The cassava genome: current progress, future directions. *Trop Plant Biol* 5:88–94.

Raghavan C, Naredo M, Elizabeth B, Wang H, Atienza G, Liu B, Qiu F, McNally KL, Leung H (2007) Rapid method for detecting SNPs on agarose gels and its application in candidate gene mapping. *Mol Breed* 19:87–101.

Rigola D, van Oeveren J, Janssen A, Bonné A, Schneiders H, van der Poel HJ, van Orsouw NJ, Hogers RC, de Both MT, van Eijk MJ (2009) High-throughput detection of induced mutations and natural variation using KeyPointTM technology. *PLoS One* 4:e4761.

Rothberg JM, Hinz W, Rearick TM, Schultz J, Mileski W, Davey M, Leamon JH, Johnson K, Milgrew MJ, Edwards M (2011) An integrated semiconductor device enabling non-optical genome sequencing. *Nature* 475:348.

Rozycka M, Collins N, Stratton MR, Wooster R (2000) Rapid detection of DNA sequence variants by conformation-sensitive capillary electrophoresis. *Genomics* 70:34–40.

Sakai T, Imamura J (1994) Somatic hybridization between radish (*Raphanus sativus*) and rapeseed (*Brassica napus*). In: Bajaj YPS (ed) *Biotechnology in Agriculture and Forestry, vol 27. Somatic Hybridization in Crop Improvement.* Springer, Berlin Heidelberg New York, pp 320–333.

Sakai T, Imamura J (2003) Development and utilization of hybrid-seed production system in rapeseeds (*Brassica napus* L.). *Breed Res* 5:93–102.

Sampson DR (1967) Frequency and distribution of self incompatibility alleles *in Raphanus raphanistrum. Genetics* 56:241–251.

Selvakumar R, Jat GS, Manjunathagowda DC (2023) Allele mining through TILLING and EcoTILLING approaches in vegetable crops. *Planta* 258: 15. https://doi.org/10.1007/s00425-023-04176-2

Shan Q, Wang Y, Li J, Gao C (2014) Genome editing in rice and wheat using the *CRISPR/Cas* system. *Nat Protoc* 9:2395.

Shirasawa K, Oyama M, Hirakawa H, Sato S, Tabata S, Fujioka T, Kimizuka-Takagi C, Sasamoto S, Watanabe A, Kato M, Kishida Y, Kohara M, Takahashi C, Tsuruoka H, Wada T, Sakai T, Isobe S (2011) An EST-SSR linkage map of *Raphanus sativus* and comparative genomics of the Brassicaceae. *DNA Res* 18:221–232.

Slade AJ, Knauf VC (2005) TILLING moves beyond functional genomics into crop improvement. *Transgen Res* 14:109–115.

Stake M, Waack S (2003) Gene prediction with a hidden Markov model and a new intron submodel. *Bioinformatics* 19(Suppl 2):i215–i225.

Szurman-Zubrzycka ME, Zbieszczyk J, Marzec M, Jelonek J, Chmielewska B, Kurowska MM, Krok M, Daszkowska-Golec A, Guzy-Wrobelska J, Gruszka D, Gajecka M, Gajewska P, Stolarek M, Tylec P, Sega P, Lip S, Kudełko M, Lorek M, Gorniak-Walas M, Malolepszy A, Podsiadlo N, Szyrajew KP, Keisa A, Mbambo Z, Todorowska E, Gaj M, Nita Z, Orlowska-Job W, Maluszynski M, Szarejko I (2018) HorTILLUS—a rich and renewable source of induced mutations for forward/reverse genetics and pre-breeding programs in barley (*Hordeum vulgare* L.). *Front Plant Sci* 9:216.

Tadele Z (2016) Mutagenesis and TILLING to dissect gene function in plants. *Curr Genomics* 17:499–508.

Taylor NE, Greene EA (2003) PARSESNP: a tool for the analysis of nucleotide polymorphisms. *Nucleic Acids Res* 31:3808–3811.

Thormann, I., Yang, Q., Allender, C., Bas, N., Campbell, G., Dulloo, M.E., Ebert, A.W., Lohwasser, U., Pandey, C., Robertson, L.D. and Spellman, O. (2013) Development of best practices for ex situ conservation of radish germplasm in the context of the crop genebank knowledge base. *Genet Resour Crop Ev*, 60: 1251–1262.

Till BJ, Colbert T, Codomo C, Enns L, Johnson J, Reynolds SH, Henikoff JG, Greene EA, Steine MN, Comai L, Henikoff S (2006) High-throughput TILLING for *Arabidopsis. Methods Mol Biol* 323:127–35. doi: 10.1385/1-59745-003-0:127

Till BJ, Colbert T, Tompa R, Enns LC, Codomo CA, Johnson JE, Reynolds SH, Henikoff JG, Greene EA, Steine MN, Comai L, Henikoff S (2003a) High-throughput TILLING for functional genomics. *Methods Mol Biol* 236:205–20. doi: 10.1385/1-59259-413-1:205

Till BJ, Comai L, Henikoff S (2007a) Tilling and Ecotilling for Crop Improvement. In: Varshney RK, Tuberosa R (eds) Genomics-Assisted Crop Improvement. Springer, Dordrecht. https://doi.org/10.1007/978-1-4020-6295-7_15

Till BJ, Cooper J, Tai TH, Colowit P, Greene EA, Henikoff S, Coma L (2007b) Discovery of chemically induced mutations in rice by TILLING. BMC Plant Biol 7:1–12.

Till BJ, Reynolds SH, Greene EA, Codomo CA, Enns LC, Johnson JE, Burtner C, Odden AR, Young K, Taylor NE, Henikoff JG, Comai L, Henikoff S (2003b) Large-scale discovery of induced point mutations with high-throughput TILLING. *Genome Res* 13:524–530.

Upadhyaya HD, Ortiz R (2001) A mini core subset for capturing diversity and promoting utilization of chickpea genetic resources in crop improvement. *Theor Appl Genet* 102:1292–1298.

Veerla S, Hoglund M (2006) Analysis of promoter regions of co-expressed genes identified by microarray analysis. *BMC Bioinformatics* 7:384.

Vetter J (2000) Plant cyanogenic glycosides. *Toxicon* 38:11–36.

Voelkerding K, Dames S, Durtschi J (2009) Next-generation sequencing: From basic research to diagnostics. *Clin Chem* 55: 641–58.

Wang F, Li G, Chen S, Jiang Y, Wang S (2015) Characterization of a new leaf-compound radish mutant (*Raphanus sativus* L.). *J Appl Hortic* 17(2):115–120.

Wang N, Shi L, Tian F, Ning H, Wu X, Long Y, Meng J (2010) Assessment of *FAE1* polymorphisms in three *Brassica* species using EcoTILLING and their association with differences in seed erucic acid contents. *BMC Plant Biol* 10:1–11.

Xu Y, Li P, Zou C, Lu Y, Xie C, Zhang X, Prasanna BM, Olsen M (2017) Enhancing genetic gain in the era of molecular breeding. *J Exp Bot* 68:2641–2666.

Yamagishi H, Terachi T (1994) Molecular and biological studies on male-sterile cytoplasm in Cruciferae. II. The origin of *Ogura* male-sterile cytoplasm inferred from the segregation pattern of male sterility in the F_1 progeny of wild and cultivated radishes (*Raphanus sativus* L.). *Euphytica* 80:201–206.

Yamagishi H, Terachi T (1996) Molecular and biological studies on male-sterile cytoplasm in the Cruciferae. III. Distribution of Ogura-type cytoplasm among Japanese wild radishes and Asian radish cultivars. *Theor Appl Genet* 93:325–332.

Yamagishi H, Terachi T (2001) Intra–and inter-specific variations in the mitochondrial gene *orf138* of Ogura-type male sterile cytoplasm from *Raphanus sativus* and *Raphanus raphanistrum. Theor Appl Genet* 103:725–732.

Yang J, Zhang C, Zhao N, Zhang L, Hu Z, Chen S, et al (2019) Chinese Root-type Mustard Provides Phylogenomic Insights into the Evolution of the Multi-use Diversified Allopolyploid Brassica juncea. *Mol. Plant* 11:512–514.

Yu X, Ge C, Xu J, Zhang N, Li S, Yu X, Ge C, Xu J, Zhang N, Li S (2016) Primary study on mutation effect in taproots of radish cultivar "Huangzhou" by Ethylmethylsulfone (EMS). *Open Access Library J* 3(3):1–5. https://doi.org/10.4236/OALIB.1102420

Zerr T, Henikoff S (2005) Automated band mapping in electrophoretic gel images using background information. *Nucleic Acids Res* 33:2806–2812.